Progress in Brain Research
Volume 229

Motivation: Theory, Neurobiology and Applications

Serial Editor

Progress in Brain Research
Volume 229

Motivation: Theory, Neurobiology and Applications

Edited by

Bettina Studer
Mauritius Hospital, Meerbusch;
Institute of Clinical Neuroscience and Medical Psychology
Heinrich-Heine-University, Düsseldorf, Germany

Stefan Knecht
Mauritius Hospital, Meerbusch;
Institute of Clinical Neuroscience and Medical Psychology
Heinrich-Heine-University, Düsseldorf, Germany

ELSEVIER
AMSTERDAM • BOSTON • HEIDELBERG • LONDON • NEW YORK • OXFORD
PARIS • SAN DIEGO • SAN FRANCISCO • SINGAPORE • SYDNEY • TOKYO

Elsevier
Radarweg 29, PO Box 211, 1000 AE Amsterdam, Netherlands
The Boulevard, Langford Lane, Kidlington, Oxford OX5 1GB, United Kingdom
50 Hampshire Street, 5th Floor, Cambridge, MA 02139, United States

First edition 2016

ISBN: 978-0-444-63701-7
ISSN: 0079-6123

Publisher: Zoe Kruze
Acquisition Editor: Kirsten Shankland
Editorial Project Manager: Hannah Colford
Production Project Manager: Magesh Kumar Mahalingam
Cover Designer: Greg Harris

Contributors

J. Bernacer
Mind-Brain Group (Institute for Culture and Society, ICS), University of Navarra, Pamplona, Spain

V. Bonnelle
University of Oxford, Oxford, United Kingdom

A. Bourgeois
Laboratory for Behavioral Neurology and Imaging of Cognition, University of Geneva, Geneva, Switzerland

C. Burrasch
Technische Universität Dresden, Dresden; University of Lübeck, Lübeck, Germany

L. Chelazzi
University of Verona; National Institute of Neuroscience, Verona, Italy

T.T.-J. Chong
Macquarie University; ARC Centre of Excellence in Cognition and its Disorders, Macquarie University, Sydney, NSW; Monash Institute of Cognitive and Clinical Neurosciences, Monash University, Clayton, VIC, Australia

P.J. Currie
Reed College, Portland, OR, United States

C. Eisenegger
Neuropsychopharmacology and Biopsychology Unit, Faculty of Psychology, University of Vienna, Vienna, Austria

B. Eitam
University of Haifa, Haifa, Israel

L. Font
Area de Psicobiología, Universitat Jaume I, Castellón, Spain

J. Gottlieb
Kavli Institute for Brain Science, Columbia University, New York, NY, United States

R. Handermann
Mauritius Hospital, Meerbusch, Germany

U. Hegerl
Research Center of the German Depression Foundation; University of Leipzig, Leipzig, Germany

J. Held
University Hospital of Zurich, Zurich; Cereneo, Center for Neurology and Rehabilitation, Vitznau, Switzerland

L. Hellrung
Technische Universität Dresden, Dresden, Germany

E.T. Higgins
Columbia University, New York, NY, United States

C.B. Holroyd
University of Victoria, Victoria, BC, Canada

M. Husain
University of Oxford; John Radcliffe Hospital, Oxford, United Kingdom

P. Kenning
Heinrich-Heine-University Düsseldorf, Düsseldorf, Germany

S. Knecht
Mauritius Hospital, Meerbusch; Institute of Clinical Neuroscience and Medical Psychology, Medical Faculty, Heinrich-Heine-University Düsseldorf, Düsseldorf, Germany

N.B. Kroemer
Technische Universität Dresden, Dresden, Germany

M. Lopes
Inria and Ensta ParisTech, Paris, France

A.B. Losecaat Vermeer
Neuropsychopharmacology and Biopsychology Unit, Faculty of Psychology, University of Vienna, Vienna, Austria

A. Luft
University Hospital of Zurich, Zurich; Cereneo, Center for Neurology and Rehabilitation, Vitznau, Switzerland

E. Luis
Neuroimaging Laboratory, Center for Applied Medical Research (CIMA), University of Navarra, Pamplona, Spain

K. Lutz
University Hospital of Zurich; Institute of Psychology, University of Zurich, Zurich; Cereneo, Center for Neurology and Rehabilitation, Vitznau, Switzerland

P. Malhotra
Imperial College London, Charing Cross Hospital, London, United Kingdom

I. Martinez-Valbuena
Mind-Brain Group (Institute for Culture and Society, ICS), University of Navarra, Pamplona, Spain

M. Martinez
Neuroimaging Laboratory, Center for Applied Medical Research (CIMA), University of Navarra, Pamplona, Spain

I. Morales
Reed College, Portland, OR, United States

O. Nafcha
University of Haifa, Haifa, Israel

E. Olgiati
Imperial College London, Charing Cross Hospital, London, United Kingdom

P.-Y. Oudeyer
Inria and Ensta ParisTech, Paris, France

S.Q. Park
University of Lübeck, Lübeck, Germany

M.A. Pastor
Mind-Brain Group (Institute for Culture and Society, ICS); Neuroimaging Laboratory, Center for Applied Medical Research (CIMA); Clínica Universidad de Navarra, University of Navarra, Pamplona, Spain

R. Pastor
Reed College, Portland, OR, United States; Area de Psicobiología, Universitat Jaume I, Castellón, Spain

N. Pujol
Clínica Universidad de Navarra, University of Navarra, Pamplona, Spain

D. Ramirez-Castillo
Mind-Brain Group (Institute for Culture and Society, ICS), University of Navarra, Pamplona, Spain

I. Riečanský
Laboratory of Cognitive Neuroscience, Institute of Normal and Pathological Physiology, Slovak Academy of Sciences, Bratislava, Slovakia; Social, Cognitive and Affective Neuroscience Unit, Faculty of Psychology, University of Vienna, Vienna, Austria

C. Russell
Institute of Psychiatry, Psychology and Neuroscience, King's College London, London, United Kingdom

D. Soto
Basque Center on Cognition, Brain and Language, San Sebastian; Ikerbasque, Basque Foundation for Science, Bilbao, Spain

S. Strang
University of Lübeck, Lübeck, Germany

T. Strombach
Heinrich-Heine-University Düsseldorf, Düsseldorf, Germany

B. Studer
Institute of Clinical Neuroscience and Medical Psychology, Medical Faculty, Heinrich-Heine-University Düsseldorf, Düsseldorf; Mauritius Hospital, Meerbusch, Germany

C. Ulke
Research Center of the German Depression Foundation, Leipzig, Germany

A. Umemoto
Institute of Biomedical and Health Sciences, Hiroshima University, Hiroshima, Japan

H. Van Dijk
Institute of Clinical Neuroscience and Medical Psychology, Medical Faculty, Heinrich-Heine-University Düsseldorf, Düsseldorf, Germany

P. Vuilleumier
Laboratory for Behavioral Neurology and Imaging of Cognition, University of Geneva, Geneva, Switzerland

M. Widmer
University Hospital of Zurich; Neural Control of Movement Lab, ETH Zurich, Zurich; Cereneo, Center for Neurology and Rehabilitation, Vitznau, Switzerland

N. Ziegler
Institute of Human Movement Sciences and Sport, ETH Zurich, Zurich, Switzerland

Contents

SECTION 2 ASSESSING MOTIVATION

SECTION 3 THE NEUROBIOLOGY OF MOTIVATION

SECTION 5 CONCLUSIONS

Preface

Motivation, the driving force of our behavior, is relevant to all aspects of human life and the question how motivation can be enhanced is likewise ubiquitous. As a consequence, motivation is a prominent topic in the psychological, educational, neuroscience, and economic literature and has been subject to both extensive theoretical consideration and empirical research. Yet, motivation and its neural mechanisms are not yet fully understood, and the demand for new tools to enhance motivation in education, health, and work settings remains high. This volume provides an up-to-date overview over theoretical and experimental work on motivation, discusses recent findings about the neurobiological mechanisms underlying motivation and goal-directed behavior, and presents novel approaches targeting motivation in clinical and nonclinical application settings. It contains a mix of review articles and new original research studies, and crosses the boundaries of and connects findings from a range of scientific disciplines, including psychology, economics, behavioral and cognitive neurosciences, and education.

The volume is structured into four sections: The first section discusses theories of motivation. Strombach and colleagues (Chapter 1) review extant psychological and economic theories of motivation and converse the similarities and differences in how motivation is conceptualized in these two scientific traditions. Chapters 2 and 3 present two novel, nonexclusive models of motivation. The first model, proposed by Studer and Knecht (Chapter 2), defines motivation for a given activity as a product of the anticipated subjective benefits and anticipated subjective costs of (performance of) the activity. This benefit–cost model incorporates core concepts of previous motivation theories and allows deriving strategies for how motivation might be increased in application settings. Meanwhile, Nafacha et al. (Chapter 3) focus on the motivation underlying habitual behavior and propose that habitual behavior is motivated by the control it provides over ones environment. They discuss the intrinsic worth of control and in which circumstances an activity may attain control-based motivational value.

The second section of this volume covers the assessment of motivation. One tradition in motivation research is to use questionnaire-based qualitative measures. But, this approach has some limitations, including that questionnaires can only be used to measure motivation in humans, and that these measures rely on adequate insight of responders. In Chapter 4, Chong et al. present an alternative approach to the assessment of motivation, namely use of objective measures of motivation derived from effort-based decision-making paradigms. This behavioral assessment approach allows identifying motivation deficits in clinical populations and investigating neurobiological mechanisms of motivation in both human and nonhuman animals (see also Chapters 5–9).

Section 3 of this volume covers current knowledge about the neurobiological underpinnings of motivation. Chapter 5 by Bernacer et al. presents new original work on the valuation of physical activity in sedentary individuals and on the neural correlates of the subjective cost of physical effort. Kroemer and colleagues (Chapter 6) argue that signal fluctuations in a mesocorticolimbic network underlie and give rise to intraindividual fluctuations in motivation and effort production. The authors review extant empirical support for this proposition and discuss how novel functional magnetic resonance imaging techniques will enable further testing of the suggested neurobehavioral model.

Morales and colleagues (Chapter 7) focus on motivation for seeking and consumption of food. Their chapter reviews the current knowledge about the role of opioid signaling in food motivation gained through laboratory experiments in animals and presents new original data on the effects of opioid receptor antagonists upon food motivation and effort-related behavior.

Umemoto and Holroyd (Chapter 8) explore the role of the anterior cingulate cortex in motivated behavior and theorize that this brain structure contributes to the motivation-related personality traits reward sensitivity and persistence. They also present new data from a behavioral experiment in support of this theory.

Vermeer et al. (Chapter 9) review evidence for the involvement of sex hormones testosterone and estradiol in motivation for partaking in competitions and in performance increases during competitions. They describe how competition-induced testosterone can have long-lasting effects upon behavior and discuss how testosterone might enable neuroplasticity in the adult brain.

In the final chapter of Section 3, Hegerl and Ulke (Chapter 10) describe the clinical symptom fatigue and its neurobiological correlates. They discuss clinical, behavioral, and neurobiological support for why distinguishing between "hyperaroused fatigue" (observed in major depression) and "hypoaroused fatigue" (occurring in the context of inflammatory and immunological processes) is important and propose a clinical procedure to achieve this separation.

The fourth section of this volume showcases recent research on enhancing motivation in education, neurorehabilitation, and other application domains. In Chapter 11, Oudeyer et al. argue that curiosity and learning progress act as intrinsic motivators that foster exploration and memory retention, and discuss how this mechanism can be utilized in education technology applications.

Strang et al. (Chapter 12) review recent work on the use of monetary incentives as a motivation enhancement tool in the context of (laboratory) task performance, prosocial behavior, and health-related behavior, and debate the conditions under which this approach is and is not effective. Meanwhile, new research by Widmer et al. (presented in Chapter 13) tested whether augmentation of striatal activation during a motor learning task through strategic employment of performance feedback and of performance-dependent monetary reward can strengthen motor skill acquisition and consolidation.

Chapters 14 and 15 investigate how motivation influences perception and attention. Bourgeois et al. (Chapter 14) discuss how reward-signaling stimuli attract and bias attention, and which neural mechanisms underlie this impact of motivation upon attention. In Chapter 15, Paresh and colleagues then elaborate on how these effects can be utilized in the treatment of spatial neglect, a disorder of attention common in stroke patients. They cover previous evidence on the effectiveness of motivational stimulation in reducing attention deficits and present a new original study examining the impact of monetary incentives on attentional orienting and task engagement in patients with neglect.

In Chapter 16, we present a proof-of-concept study which shows that competition can be used as a tool to enhance intensity and amount of (self-directed) training in stroke patients undergoing neurorehabilitation.

Chapter 17 by Chong and Husain reviews extant clinical and laboratory evidence for the use of dopaminergic medication in the treatment of apathy, a neuropsychiatric syndrome characterized by diminished motivation. They also discuss how effort-based decision-making paradigms could be used as more objective endpoint measures in future treatment studies.

In Chapter 18, Knecht and Kenning explore how insights gained in neuroeconomic and marketing research into motivation and behavior offer new avenues and models for health facilitation and meeting the challenge of lifestyle-mediated chronic disease.

We hope that this volume will not only provide an up-to-date account on motivation but also help to integrate knowledge gained in the covered disciplines and research fields and to connect basic research on the neurobiological foundations of motivation, clinical work on motivation deficits, and application research. To aid this integration, we reflect on connections between and conclusions derived from the various lines of research presented in the final chapter of this volume (Chapter 19). We also outline open questions for future motivation research.

Bettina Studer
Stefan Knecht

SECTION

1

Theories of Motivation

CHAPTER

1

Common and distinctive approaches to motivation in different disciplines

T. Strombach*[,1], S. Strang[†,1,2], S.Q. Park[†], P. Kenning*

*Heinrich-Heine-University Düsseldorf, Düsseldorf, Germany

[†]University of Lübeck, Lübeck, Germany

[2]Corresponding author: Tel.: +49-451-3101-3611; Fax: +49-451-3101-3604, e-mail address: sabrina.strang@uni-luebeck.de

Abstract

Over the last couple of decades, a body of theories has emerged that explains when and why people are motivated to act. Multiple disciplines have investigated the origins and consequences of motivated behavior, and have done so largely in parallel. Only recently have different disciplines, like psychology and economics, begun to consolidate their knowledge, attempting to integrate findings. The following chapter presents and discusses the most prominent approaches to motivation in the disciplines of biology, psychology, and economics. Particularly, we describe the specific role of incentives, both monetary and alternative, in various motivational theories. Though monetary incentives are pivotal in traditional economic theory, biological and psychological theories ascribe less significance to monetary incentives and suggest alternative drivers for motivation.

Keywords

Incentives, Intrinsic motivation, Extrinsic motivation, Drives, Motives

1 INTRODUCTION

Motivation describes goal-oriented behavior and includes all processes for initiating, maintaining, or changing psychological and physiological activity (Heckhausen and Heckhausen, 2006). The word "motivation" originates from the Latin verb "*movere*," meaning "to move" (Hau and Martini, 2012), which effectively describes what motivation is—the active "movement" of an organism in reaction to a stimulus.

[1]These authors contributed equally to this paper.

Progress in Brain Research, Volume 229, ISSN 0079-6123, http://dx.doi.org/10.1016/bs.pbr.2016.06.007

Assuming that most human behavior is driven by a specific motivation, knowing the underlying motives is crucial to understanding human behavior. While motivation explains desired behaviors, such as striving for a career or finding a partner, it also accounts for maladaptive behaviors, such as drug addiction (eg, Baker et al., 2004; Kalivas and Volkow, 2005; Koob and Le Moal, 2001) or gambling (Clark et al., 2009). During the last ten decades, such disciplines as psychology, economics, biology, and neuroscience have investigated motivation in a variety of contexts, to gain a better understanding of factors that drive human behavior. Because the findings of these studies are inconsistent, however, a general theory of motivation processes remains elusive (Gneezy et al., 2011).

In the following, we present a range of theories of motivation from biological, psychological, and economic perspectives, and discuss both commonalities and differences among the various approaches. The goal of this chapter is (1) to provide a brief and selective overview of current theories on motivation in various disciplines and (2) to discuss important and conflicting aspects of those theories.

1.1 A DEFINITION OF MOTIVATION

Currently, no consensus on a single definition of motivation exists among the disciplines (Gneezy et al., 2011). In general, motivation is defined by a directedness and intensity of behavior and tries to explain how and why goals emerge and how these goals are sustained (Frey and Jegen, 2001; White, 1959). In everyday life, motivation is often used to explain a person's behavior—for example, to explain why people buy a specific product brand, or why students study all night for an upcoming exam. These questions have one thing in common: the goal of the motivated behavior is to fulfill a specific need or desire. Nevid (2013) explains: "The term motivation refers to factors that activate, direct, and sustain goal-directed behavior [...]. Motives are the 'whys' of behavior—the needs or wants that drive behavior and explain what we do. We do not actually observe a motive; rather, we infer that one exists based on the behavior we observe" (p. 288). The forces that drive behavior refer to motives and might have their origin in biological, social, emotional, or cognitive aspects. The observed behavior is understood by inferring the motive behind it. A motive is an isolated factor that drives human behavior (Herkner, 1986). For example, eating a banana is an observed behavior, while hunger might be the inferred motive for the behavior.

The study of motives has revealed a basic distinction between inherent motives and learned motives (Skinner, 1938, 2014). Inherent motives are inborn and central to survival, as can be seen in instincts and drives directed toward fulfilling biological needs (James, 1890). Hunger is a typical inherent motive; it fulfills the biological need to maintain a certain energy level. In contrast, learned motives are formed through experience. The desire to receive money is an illustrative learned motive (Opsahl and Dunnette, 1966). Money cannot directly fulfill any biological need; however, money allows indirect fulfillment of several biological needs (eg, buying food), and social rewards, such as status. Learned motives, therefore, depend strongly on social and cultural influences, as they are formed and framed by experience (White and Lehman, 2005; Zimbardo, 2007).

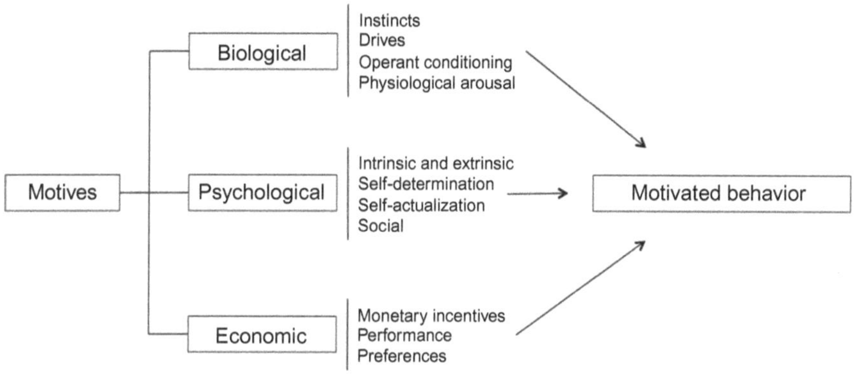

FIG. 1

Overview of the different motives that are used to explain motivated and goal-directed behavior. Motives can be divided into three categories: biological, psychological, and economic motives, covering different aspects of human behavior.

Motives can further be categorized into extrinsic and intrinsic motives (Deci, 1971). A person is said to be intrinsically motivated when performing a behavior simply out of enjoyment of the behavior itself, without receiving reward for the behavior. Alternatively, a person who performs a task only to receive a reward (typically from a second party) is said to be externally motivated (Deci, 1971). This reward can be tangible, such as money, but also nontangible, as in the case of verbal feedback (Deci et al., 1999).

Furthermore, motives are influenced by the context and the situation (Zimbardo, 2007). A situation includes both the objective experience and the subjective interpretation of situational factors. The objective and the subjective component are independent of each other and might be independently consulted in order to explain motivated behavior. A person might not be hungry, but the enticing smell of French fries might provoke a craving for that food, without an actual change in hunger status.

The discussion of theories of motivation begins with biological motives, which were the first theories used to explain goal-directed, motivated behavior. Psychological theories on motivation cover individual differences and aim to explain complex behavior. Finally, management and economic research introduce tangible incentives into motivation theory, equating motivation with performance. Fig. 1 offers an overview of the various approaches to explaining motivated behavior.

2 BIOLOGICAL MOTIVES

The four most prominent biological theories on motivation consider instincts, drives, operant conditioning, and physiological arousal. All biological theories focus on motives that aim to achieve a physical/bodily change. They all build on the premise that physical needs, urges, or deficiencies initiate behavior.

2.1 INSTINCTS AS MOTIVES

Instincts are biologically determined, existing in all species, and are innate drivers of behavior (James, 1890; Kubie, 1948; Sherrington, 1916). Instincts are thus inherent motives; they are fixed, rigid, and predictable patterns of behavior that are not acquired by learning. They are sometimes described as a chain of reflexes initiated by a given stimulus (James, 1890). Accordingly, the observed behavior and the underlying motive are identical and observed behavior is at least clearly attributable to a specific stimulus. For example, newborns exhibit sucking behavior as soon as their lips or tongues are touched. This behavior occurs without any learning (Davis et al., 1948). Instincts as motivation, therefore, suggest that a single stimulus triggers a reflex or chain of reflexes that is genetically preprogrammed (Morgan, 1912). According to instinct theory, humans primarily react to environmental stimuli, precluding explorative and planned behavior (White, 1959). This also implies that instincts cannot readily explain the motivation to learn, as pointed out by Maslow (1954). As early as 1954, Maslow proposed that because humans are able to voluntarily override certain instincts, human behavior is not as rigid and predictable as assumed by instinct theory. In summary, instinct alone cannot sufficiently explain the complexities of human behavior.

2.2 DRIVES AS MOTIVES

In 1943 Clark Hull introduced the drive-reduction theory as explanation for motivated behavior, expanding the idea in 1952. A "drive" is a state of arousal or tension triggered by a person's physiological or biological needs, which might be food, water, or even sex (Hull, 1943). Hull's (1943, 1952) drive-reduction theory states that behavior arises from physiological needs created by a deviation from homeostasis (the tendency to maintain a balance, or an optimal level, within a biological system). This deviation triggers internal drives to push the organism to satisfy the need, and to reduce tension and arousal.

Drive-reduction theory distinguishes between primary or innate drives and secondary or acquired drives. While primary drives are defined by needs of the body such as hunger, thirst, or the desire for sex, secondary drives are not directly linked to bodily states. Instead, they are associated with primary drives via experiences or conditioning procedures (Pavlov, 1941). One example of such secondary drives is a desire to receive money, which helps to pay for the satisfaction of primary drives like food and shelter (Mowrer, 1951; Olds, 1953). Drive-reduction theory thus extends previous approaches by integrating secondary reinforcers into the model. With the introduction of this concept, motives came to be seen as more complex and flexible, in comparison to instinct theory. However, the theory was criticized for lack of ecologic validity and an explanation for the role of secondary reinforcers in regulating tension. Money, as a secondary reinforce, can be used to purchase primary reinforcers such as food and water. However, money in itself cannot reduce an individual's tension. Another shortcoming of this approach is that drive-reduction theory does not provide an explanation for behavior that is not intended to reduce any tension, such as a person eating even if not hungry (Cellura, 1969).

Also based on the idea of drives and biological unconscious needs, Freud's motivation theory is framed on three central elements. First, his idea of psychological determinism suggests that all psychological phenomena, no matter whether only a thought or actual behavior, happen for a reason and the underlying motivation can, therefore, be explained (Freud, 1961). Second, Freud states that the motives of behavior are mainly instinct driven, and drives are dependent on biological processes that are mostly unconscious (Freud, 1952, 1961). Third, behavior does not directly reflect drives, but is a state of conflict that may be internal, or that may directly express a desire contrary to socially accepted behavior (Freud, 1961). Thus, drives are internal energizers and initiate behavior. In Freudian psychoanalysis, the sex drive (the libido) is the most powerful drive. The libido originates in the unconsciousness (Id) and modulates internal and external conditions (Ego and Superego)—thereby also modulating perception and behavior in social settings.

2.3 OPERANT CONDITIONED MOTIVES

Watson (1913) held a view on behavior that opposes the ideas of Hull and Freud, who mainly used introspection, an examination of internal thoughts and feelings, as support for their approaches. Watson, in contrast, voted strongly against the idea of introspection, suggesting a more objective interpretation of human behavior. In his view, contrary to Freud's theory, motives are clearly deducible from the behavior that is observed. The field of research that resulted from Watson's theories can be referred to as behaviorism, highlighting the central and informative aspect of the observable aspect of human behavior (Skinner, 2011; Watson, 1930). Behaviorism was greatly influenced by the research of Skinner, who coined the term "operant conditioning" (Skinner, 1938, 2011). While classical conditioning relies on the presence of a given stimulus that exhibits a natural reaction (Skinner, 1938), operant conditioning refers to the association of a spontaneous behavior with a specific incentive (Flora, 2004).

Skinner differentiated between two kinds of reinforcers—primary and secondary reinforcers (Skinner, 1938; Wike and Barrientos, 1958). Primary reinforcers, or unconditioned reinforcers, are stimuli that do not require pairing to provoke a specific response. Those stimuli, evolved through evolution, play a primary role in human survival. Primary reinforcers include sleep, food, or sex and are quite stable over the human lifetime. Secondary or conditioned reinforcers, in contrast, are stimuli or situations that have acquired their function after pairing with a specific outcome. Therefore, comparable to the primary and secondary reinforcers in drive-reduction theory, the secondary reinforcers are often acquired to fulfill the primary reinforcers, as in the case of gaining money to buy food.

In a similar vein, Hsee and colleagues (2003) describe money and other secondary reinforcers as a medium between effort or performance and a desired mostly primary reinforcer. In his theorizing, people receive a medium as an immediate reward and can then trade this for another desired outcome/primary reinforcer. Money, for example, can be traded for food. Sometimes there are even multiple channels between performance and the outcome/primary reinforce (Hsee et al., 2003). As an example of other mediating elements, money can also be used to buy expensive clothes, with a goal of increasing social status in order to, ultimately, achieve sexual relations.

The reinforcement approach as explanation for motivated behavior was criticized for not sufficiently explaining the link between behavior and reinforcement. The approach basically states that all behavior needs to happen at least once, accidentally or voluntarily, before it can be modulated or altered (Chomsky, 1959; Wiest, 1967). However, in real life that might not always be the case. In a typical reinforcement experiment, a very limited set of choices is offered and one of the choices is rewarded. As an example, a rat is put in a condition where the only choices are to do nothing, or to explore its surroundings, which are empty except for a lever. It is thus very likely that the rat will press the lever at some point, which results in a reward. The action of pressing a lever is thereby strengthened as a behavioral option. In real life, both animals and humans have larger choice sets. Therefore, a more complex explanation for motivated behavior is needed than suggested by Skinner.

2.4 PHYSIOLOGICAL AROUSAL AS MOTIVE

The arousal theory of motivation suggests that people execute a specific behavior in order to maintain an "optimum" level of physiological arousal (Keller, 1981; Mitchell, 1982). That optimal level might vary among people and might also change throughout a lifetime. The theory suggests that whenever the arousal drops below or rises above a specific individual level, people seek stimulation to elevate or reduce it again (Keller, 1981). Thus, commonalities with the drive-reduction theory exist, but instead of tension, arousal theory suggests that humans are motivated to maintain an "ideal" level of arousal and stimulation. No biological balance needs to be maintained.

Consistent with this approach, the Yerkes–Dodson law (Yerkes and Dodson, 1908) states that performance is also related to arousal. In order to maintain an "optimum" arousal level, humans adapt performance in accordance with the current level of arousal. Moderate levels of arousal lead to better performance, compared to performance when arousal levels are too high or too low (Broadhurst, 1959). However, the effect of incentives varies with the difficulty of the task being performed. While easy tasks require a high-to-moderate level of arousal to produce high performance, more difficult tasks require a low-to-moderate level of arousal (Broadhurst, 1959). Thus, arousal theory introduces the concept of performance into motivation theory, proposing direct and measurable outcomes of motivated behavior.

In summary, biological theories on motivation suggest that biologically determined factors such as instinct or drive underlie motivated behavior. While instinct theory regards human behavior as biologically predetermined reactions to stimuli in the environment, drive-reduction theory and arousal theory state that humans behave in a way that attempts to maintain a determined balance. Finally, operant conditioned rewards link behavior to biologically relevant needs. Although biological approaches to motivation can be regarded as simplifications of the actual processes underlying motivated behavior, they inspired many subsequent theories to understanding human behavior. It is worth remembering, however, that despite biological theories lack validity in studies of motivation, biological theories continue to be useful tools in the study of other areas of behavior.

3 PSYCHOLOGICAL MOTIVES

Psychological approaches explaining motivated behavior differ from biological motives, in the sense that they do not focus solely on physiological changes, but go further in their assumption of goal-directed behavior. Psychological theories allow more variables additionally to biological factors in explaining individual behavior. In psychology, theories of motivation propose that behavior can be explained as a response to any stimulus and the individual rewarding properties of that stimulus. However, the difficulty in studying these motives is that humans are often not explicitly aware of the underlying motive. The complexity in psychology is thus based on the assumption that actions of humans cannot be predicted or fully understood without understanding their beliefs and values. Therefore, it is important to understand the association to those beliefs and values, and the associated actions at any given time. It is crucial, as well, to account for individual differences in the motives driving behavior. Furthermore, the investigation of motives sets a challenge because not only is there a single defined motive, but there is often an aggregation of different motives initiating goal-directed behavior. In general, psychological research on motives focuses on systematizing motives in a comprehensive way by accounting for individual and temporary behaviors. The categorization and focus of individualism thereby differs among theories.

3.1 INTRINSIC AND EXTRINSIC MOTIVES

As mentioned previously, one of the most prominent categorizations of psychological motives differentiates between intrinsic and extrinsic motives (Deci and Ryan, 2000). The distinction between the two types of motives is based on the origin of the motive. Intrinsic motives are subjective valuations of a behavior—meaning that the behavior in itself is rewarding. The motivation is thus the inherent value of a specific behavior. In contrast, extrinsic motivation refers to external incentives that are separable from the behavior itself. Here, motivation is thus not inherent, but is induced by the prospect of an external outcome. For example, students showing the same strong academic performance can be motivated either intrinsically or extrinsically. When a specific study topic is interesting to a student, the desire to know about the subject can lead to a good grade. This would be an intrinsic motive and is free of external prompts, pressures, or rewards (Deci and Ryan, 1985; Ryan, 2012; Ryan and Deci, 2000). In other situations, students do face external factors. A student who receives a scholarship or another reward for good grades is extrinsically motivated to perform well and is responding to external cues (Deci and Ryan, 1985; Ryan and Deci, 2000).

Intrinsic motivation has also been acknowledged in animal studies. While biological motives do not account for voluntary behavior executed with no given reward, White (1959) indicates that some animals—cats, dogs, and monkeys, for instance—show curiosity-driven or playful behavior even in the absence of reinforcement. This explorative behavior can be described as "novelty seeking" (Hirschman, 1980). In such cases, intrinsic motivated behavior is performed for the positive experience associated with exercising and extending capabilities, independent of an objective benefit (Deci and Ryan, 2000; Ryan and Deci, 2000). Also humans are active, playful, and curious (Young, 1959) and have an inherent and natural motivation to learn and explore (White, 1959). This natural motivation in humans and several animals is important for cognitive, social, and physical development (White, 1959). As people experience new things and explore their limits, they are learning new skills and extending their knowledge in ways that may be beneficial in the future.

Operant learning, thus the association of a spontaneous behavior with an incentive (as suggested by Skinner), implies that learning and motivated behavior is only initiated by rewards such as food. However, according to intrinsic motivation theory, the behavior in itself is rewarding. Operant learning thus suggests that behavior and consequence (or reward) are separable, while intrinsic motivation implies that behavior and reward are identical. Thus, research on intrinsic motivation focuses on the features that make an activity interesting (Deci et al., 1999). In contrast, learning theory as proposed by Hull (1943) asserts that behavior is always initiated by needs and drives. Intrinsic motivation in this context pursues the goal of satisfying innate psychological needs (Deci and Ryan, 2000).

Although intrinsic motivation is a very important aspect of human behavior, most behavior in our everyday life is not intrinsically motivated (Deci and Ryan, 2000). Extrinsic motives are constructs that pertain whenever an activity is carried out in order to attain a separate outcome. In light of Skinner's use of extrinsic rewards to explain operant conditioning, learning, and goal-directed processes (Skinner, 1938, 2014), extrinsic rewards refer to the instrumental value that is assigned to a specific behavior. However, the experience of an instrumental value is often associated with a perceived restriction of his or her own behavior and their set of choices (Deci and Ryan, 1985).

Comparing both intrinsic and extrinsic motives with biological motives, it becomes evident that most of the earlier theories tended to ignore intrinsic motivation. To a great extent, learning theories, particularly, ignored the influence of innate motives for understanding progress and human development. Theories related to drives and needs integrated psychological aspects into their theories (Hull, 1943). However, the theories are not clearly described and are not sufficient to explain complex human behavior. The concept of intrinsic and extrinsic motives thus extends the previous approaches by explaining more realistic behavior.

3.2 SELF-DETERMINATION MOTIVE

Self-determination as a motive for goal-directed behavior is based on the premise that the organism is an active system with an inherent propensity for growth and for resolution of inconsistencies (Deci and Ryan, 2002). This new approach has many similarities to the assumptions made by drive theories and physiological arousal theory. However, there is one major difference—while biological drive theories assume that the set point is the equilibrium, self-determination theory suggests that the set point is growth oriented, going beyond the initial state. The idea implies an inherent need for development and progress. Deci and Ryan (2002) suggest that motivation is contingent upon the degree to which an individual is self-motivated and self-determined. They identify three innate factors that people try to fulfill in order to develop optimally: (1) competence, (2) relatedness, and (3) autonomy (Deci and Ryan, 2002). Competence refers to the need to feel capable of reliably producing desired outcomes and/or avoiding negative outcomes. Thus, a requirement for competence is an understanding of the relationship between behavior and the resulting consequence, similar to the outcome expectations in Skinner's operant conditioning theory (Chomsky, 1959; Skinner, 1938). An individual strives for successful engagement in the behavior, which is reflected by efficacy expectations. Different from the concept of competence, the concept of relatedness references a social and psychological need to feel close to others, and to be emotionally secure in relationships with others. Individuals seek assurance that other persons care about their well-being. Deci and Ryan's (2002) third factor, autonomy, addresses a person's feeling of acting in accord with his or her own sense of self (Markland, 1999). When acting autonomously, individuals feel that they are causal agents with respect to their actions. Therefore, autonomy implies a sense of determination rather than a feeling of being compelled or controlled by external forces, thus emphasizing the intrinsic aspects of human motivation.

Taken together, self-determination theory comprises three innate needs or motives that must be fulfilled in order to display motivated behavior. Deci and Ryan combine these three different motives into a more general theory (Deci and Ryan, 2000, 2002; Ryan, 2012). However, their theory is not precise, making it difficult to predict behavior based on these categories. Nevertheless, self-determination theory can be used to differentiate between personalities. For example, while autonomy plays a central role for the behavior of some people, other people are motivated more by social aspects and a need for relatedness.

3.3 MOTIVE FOR SELF-ACTUALIZATION

Goldstein coined the term self-actualization (Goldstein, 1939; Modell, 1993), which refers to the idea that people have an inner drive to develop their full potential. The process of development is thus considered to be an important motive for goal-oriented behavior. The implication is not that every person must strive for an objective goal such as a career, but rather that all persons should develop according to their own potential—potential that might be directed toward creativity, spiritual enlightenment, pursuit of knowledge, or the desire to contribute to society (Goldstein, 1939). Self-actualization is related to the concept of self-determination, both built on the assumption that an individual's greatest need is to realize her or his own maximum potential.

One approach systematizing the idea of need for self-actualization was proposed by Maslow (1943). He developed the widely used concept of a hierarchy of needs, a pyramid model aimed toward explaining the order of needs that humans try to satisfy. In Maslow's model, the needs are organized in a sequential manner, such that the lower level of needs—hunger, for example—must be satisfied to enable striving for the next higher motive. His pyramid consists of five levels, with the lowest level addressing basic physiological needs such as water, food, and sleep that are required for human survival. The second level contains the need for security. Only when people feel secure in personal, financial, and health domains they can approach the next level—a level that consists of psychological needs, such as friendship or a feeling of belonging. Humans have a need to belong, to feel connected to friends and family, or to a partner. The fourth level details the need to feel respected, proposing that when people are accepted and valued by others they are capable of attaining the final level, self-actualization. However, while Goldstein understood self-actualization as an inner force that drives people to achieve their maximum performance, Maslow interpreted self-actualization more moderately as a tendency for people to become actualized in what they are capable of becoming (Gleitman et al., 2004).

Although prominent, the pyramid by Maslow is often criticized for not depicting, precisely, how people are motivated in real life. For instance, in some societies people suffer from hunger or are exposed to life-threatening situations on a regular basis. The first two levels of Maslow's pyramid would clearly not be met. However, those same people form strong social bonds, thus fulfilling the need for bonding which is a higher order need. Obviously, the hierarchical nature of Maslow's theory does not account for this behavior (Neher, 1991). Nevertheless, the hierarchy of needs continues to be influential in research in psychology and economics. One reason is that it proposes a model that is applicable for various approaches to motivation, and that systematizes different motives into subgroups—of which some are innate and others can only be satisfied in coordination with other people (Trigg, 2004).

3.4 SOCIAL MOTIVES

With regard to factors driving human behavior, it is not the outcome itself (such as receiving a bonus of $1000 for good job performance) that tends to be most important, but it is, rather, outcome expectancies. Thus, behavior is influenced by expectations. These expectations, moreover, are strongly shaped by social and cultural environments (McClelland, 1987). Theories on social motives maintain a specific focus on social motives to explain motivated behavior.

McClelland (1987), one of the most influential representatives of the social cognitive approach to human motivation, proposes three groups of motives: (1)

achievement, (2) power, and (3) affiliation. Similar to self-determination theory, these groups of motives are used to describe different personalities (Deci and Ryan, 2002). In order to assess these three motives, a picture story test is typically used. For this type of testing, participants receive pictures (for example, the image of a ship's captain explaining something to someone) and are asked to write a story about the pictures. The stories are then rated in accordance with elements included that relate to achievement, power, and affiliation. The first category of motive, achievement, refers to the need for success. People scoring high on this dimension are predominantly motivated to perform well in order to reach high levels of achievement. McClelland (1987) suggests that people with a need for high achievement often also display a need for autonomy—which might present an outcome complication. McClellan's second motive group, power, is not contingent on a person's actual performance. Power refers to the motivation to exert control on other people, thereby reaching a higher level of status or prestige. Consequently, people scoring high on the power dimension have a strong motivation to be influential and controlling. The final motive group, affiliation, refers to a need for membership and strong social relationships with other people (McClelland, 1987). Individuals scoring high on this dimension are motivated to show specific behaviors in order be liked by others.

Although McClelland's theory on social motives reveals a number of similarities with self-determination theory, McClelland's approach assumes that motives are learned and shaped by the environment, while self-determination theory suggests that the need for development and progress is inherent.

4 ECONOMICS AND MOTIVATION

Motivation was, and still is, an important concept in economic research. However, its interpretation varies between different "schools" and "fashions" of economic research. Generally, economic research during the last 150 years can be divided into four such schools: neoclassical economics, information economics, behavioral economics and, very recently, neuroeconomics. The neoclassical school is the oldest and assumes that people behave in a purely selfish, opportunistic, and rational way—meaning that their behavior is determined by utility. Only when benefits outweigh the costs will a given behavior be carried out. According to information economics, people behave rationally whenever possible, meaning that people can only behave rational when they are sufficiently informed about the costs and benefits of their behavior. Both the neoclassical and the information approaches assume that people compare costs and benefits in order to make decisions, though information economics suggests that people do not always have sufficient information in order to make a completely rational decision (Akerlof, 1970). In the context of motivation this means that, according to these two schools, only in the presence of an external reward or in prospect of receiving an incentive (about which people have full information) are people willing to adapt their behavior in order to reach a goal. Accordingly, an individual's performance is understood to be the output variable that depends solely on the size of the incentive. The incentive is thought to influence the degree of motivation to perform well, but this is moderated by information. Although much of the psychological, behavioral, and—most recently—neuroeconomic research in this area empirically demonstrates that behavior cannot be fully explained by a cost–benefit analysis (as indicated by neoclassical and information economics), there are still some, not to say many, proponents of these economic schools. In the early 1970s, however, behavioral economics for the first time broke away from the concept of humans as rational agents and introduced psychological concepts into economic theories. This development moved the focus more toward individual properties and resulting differences in order to explain behavior. As a result, individual differences entered economic motivation theories (Mullainathan and Thaler, 2001). Theories in behavioral economics thus imply that different people might be motivated by different motives, or by more than one motive.

With the introduction of functional neuroimaging methods in the early 1990s, the research field of neuroeconomics developed (Camerer et al., 2005; Kenning and Plassmann, 2005). By investigating the neural basis of economic behavior, the neurological plausibility of theories on human behavior can be determined. Different motives can be ascribed to processes in various brain areas, and the involvement of these brain areas can be tested across contexts and between participants.

The following section comparatively presents different economic approaches to motivation and discusses their ability to explain real-life behavior. A more detailed discussion of neuroeconomic approaches to motivation is developed in chapter "Applied Economics—The Use of Monetary Incentives to Modulate Behavior" by Strang et al.

4.1 MONETARY INCENTIVES AS MOTIVES

For a number of reasons, economists have often proposed that behavior is initiated only when an incentive is available (Camerer and Hogarth, 1999). This idea is supported by a variety of studies, showing that incentives promote effort and performance (Baker, 2000; Baker et al., 1988; Gibbons, 1997; Jenkins et al., 1998). Behavior has thus been shown to be modulated in ways that are desired by employers. However, in addition to the clearly financial properties of monetary incentives, incentives also convey symbolic meaning, such as recognition and status (Benabou and Tirole, 2003). Money allows humans to fulfill multiple needs and, thereby, it serves multiple functions (Hsee et al., 2003; Opsahl and Dunnette, 1966; Steers et al., 1996). For instance, most employees are paid with money and can choose for themselves what to spend the money on. If the financial compensation is high enough, they can, for example, buy a Ferrari or Porsche, which will indicate a high social status. This multifunctionality or—using the terminology of economics—the "utility" makes money a powerful secondary reinforcer.

In addition to the clearly positive effect of monetary incentives on motivation, evidence of negative effects of external rewards also exists (Albrecht et al., 2014; Camerer and Hogarth, 1999; Fehr and Falk, 2002a). For example, receiving very large rewards for a laboratory task (a reward equal to an annual salary) was shown to decrease performance compared to smaller rewards (Ariely et al., 2009). In specific contexts, monetary incentives can thus also have unwanted negative effects on human behavior. (An in-depth discussion of this topic is provided in chapter "Applied Economics—The Use of Monetary Incentives to Modulate Behavior" by Strang et al.)

In summary, many situations exist in which monetary incentives can be powerful and useful for increasing performance in the workplace, as well as other environments. However, the results presented in the previous paragraphs need to be considered with care. The increase in performance cannot invariably be explained by monetary rewards. The incentive may have triggered additional intrinsic or social rewards, such as power or status. The relationship between incentives and intrinsic motivation is not yet completely understood, and the assumption that performance-contingent rewards improve performance may not always hold true (Strombach et al., 2015).

4.2 PERFORMANCE AS MOTIVE

One of the most influential models in economics and management was suggested by Porter and Lawler (Lawler and Porter, 1967; Porter and Lawler, 1982). Their model was supposed to be compatible with work and organizational processes and therefore aimed to explain increases and decreases in performance. Performance, which in this context is synonymous to motivation, depends on the potential reward and on the likelihood of reaching the goal. Motivation is, therefore, also dependent on personal skills and abilities, and on an individual's self-evaluation of the potential to be successful. Contrary to previous theories on motivated behavior, Porter and Lawler are the first to equate motivation with good performance in a given task (Lawler and Porter, 1967). This differentiates the idea of performance as motive from approaches in psychology, because it does not rely on biologically plausible theories. However, while Lawler and Porter's theory clearly predicts that external incentives increase performance in the short run, the theory does not make explicit assumptions about how external incentives modify behavior in the long run, over month and years. Lawler and Porter's theory is based on the classical economic assumption that people are only motivated to perform well when an incentive is available (Kunz and Pfaff, 2002; Schuster et al., 1971). This is one of the central differences between their approach and traditional psychological approaches to motivation that assume that people can be intrinsically motivated in the absence of external rewards.

4.3 PREFERENCES AS MOTIVES

The classical economic approach attempted to solve the motivation problem by applying explicit pay-for-performance incentives. This approach is based on the premise that people are predominantly motivated by self-regarding preferences (eg, receiving money for themselves). An alternative view highlights the influence of additional preferences, called "social preferences," such as fairness, reciprocity, and trust

(Fehr and Falk, 2002a,b). To date, empirical evidence from laboratory and field experiments suggests the importance of these interpersonal or "other-regarding" preferences (Camerer and Hogarth, 1999; Falk et al., 1999; Fehr and Falk, 2002a). Other-regarding preferences are one of the core ideas in behavioral economics by establishing the important implication that self-regarding preferences are not sufficient to explain and motivate behavior of economic man. Additionally, several social preferences were identified that modulate motivation to a significant extent, though not exclusively (Barmettler et al., 2012; Camerer and Fehr, 2006; Fehr and Fischbacher, 2002; Fehr and Gächter, 1998, 2000a,b; Fehr and Schmidt, 1999; Fehr et al., 2014; Fischbacher et al., 2001). Thus, other-regarding preferences are exhibited if a person both selfishly cares about the material resources allocated to him or her, and generously cares about the material resources allocated to another agent. Such a condition implies that humans do not value their own reward in isolation, but they also compare their own set-point with reference to others. Research on the role of social preferences for human behavior has identified three important motives for goal-directed behavior—fairness, reciprocity, and social approval (Baumeister and Leary, 1995; Fehr and Falk, 2002b). When individuals consider their own outcome with regard to the outcome of others, fairness plays an important role (Sanfey, 2007). The other people serve as a reference point for determining whether or not to feel content with the reward. Monetary incentives are less effective when offers are perceived as unfair. Experiments in behavioral economics show that people are willing to punish the opponents for unfair offers, even if the punishment is costly to them—as shown in the Ultimatum Game (Sanfey et al., 2003; Strang et al., 2015). This inequality aversion could motivate specific types of behaviors and feeling (eg, the feeling of envy; Wobker, 2015). On the other hand, according to reciprocity theory, people repay kind as well as unkind behavior. In other words, people are kind to those persons who were previously kind, but are not kind to another unkind person (Falk and Fischbacher, 2006; Falk et al., 2003; Fehr and Gächter, 2000a,b, 2002). Therefore, perceived fairness and reciprocity are tightly connected. If an individual's behavior is perceived to be fair, this behavior is likely to be reciprocated in the future. Reciprocity and fairness are also central in workplace settings. Cooperation is a desired behavior that cannot be evoked by monetary incentives (Fehr and Falk, 2002a). Nevertheless, from the perspective of reciprocity, the higher salary the organization promises, the more is the employee willing to reciprocate by contributing to the organization. Fairness and reciprocation, therefore, are not only important in relationships between individuals, but are also important between company and employee (Fehr and Falk, 2002a,b). Thus, fairness and reciprocity are considered to be powerful motives for cooperation that go beyond monetary incentives (Fehr and Falk, 2002a).

A second type of social preference discussed as a motive for behavior includes social norms and social approval. Social norms are generally defined as unwritten rules that are based on widely shared beliefs about how individual members of a group should behave in specific situations (Elster, 1989). When people behave in accordance with the social norms, they receive social approval from other group members, meaning that they are evaluated positively by other individuals. People use the social information to guide their own behavior. Empirically, Fehr and Gächter (2000a) show that the degree to which a person contributes to the common pool depends significantly on the mean contribution of the other participants. If the degree of contribution of the other people is rather high, a high contribution is associated with strong social approval. However, if the contribution is medium, a high contribution results in lower social approval. Thus, social approval modulates both the degree to which people participate toward the common pool, and their motive for behavior.

To summarize, social preferences often influence behavior to a strong degree. By integrating social preferences into its approach, economic theory has made significant progress toward understanding incentives, contracts, and organizations. Including social and intrinsic incentives into the theories to explain motivated behavior improved ecologic validity, and has shown that more motives exist than those based on purely financial interests. Social preference theories are able to explain interactive human behavior, such as cooperation. Although social preferences are considered to be positive, monetary incentives have the ability to undermine this effect, and to be detrimental to the degree of motivation—and, ultimately, to the level of performance. In consequence, further research is needed here (see chapter "Applied Economics—The Use of Monetary Incentives to Modulate Behavior" by Strang et al).

5 ECONOMICS AND PSYCHOLOGY: DIFFERENT OBJECTIVES? DIFFERENT MOTIVES?

This chapter introduced different approaches to motivated behavior from the various academic disciplines of biology, psychology, and economics. Motivation is defined by the directness and intensity of behavior and poses questions about how goals emerge and how they are sustained. Although this approach is common across disciplines, classical economic theories have largely ignored psychological theories and findings on motivation. Until the emergence of behavioral economics, psychologists and economists mainly worked in parallel, but separated on research about motivation. This might partly be due to differences in their research focus. While economists traditionally focus more on group or market levels in their theories, psychologists attempt to explain individual behavior. Furthermore, economists are interested in the behavioral outcomes of motivation, and in the ways in which behavior adapts to changes in incentives, whereas psychologists are more interested in the drivers and motives underlying the emergence of motivated behavior. These different perspectives have long hampered integrative theories.

In general, modern economic approaches to motivation are strongly tied to the concepts of biology and learning theories. Both rely on the assumption that there is a direct connection between a trigger and the resulting action. Thus, while biologic motives highlight the association of a specific behavior with an incentive, economists often assume that people perform at their maximum level or at a satisfactory level when there is the prospect of a financial reward. Both strains of theory rely on the simple association of desired behavior and a resulting consequence.

Advantages of classical economic theories are that they are applicable across contexts, and that they allow for clear predictions about human behavior—implying that they can be used to give more general and larger-scale advice on how to increase motivation. According to traditional economic theories, an increase in extrinsic incentives will always result in an increase in performance, meaning that an increase in monetary incentives will enhance both employee performance and cooperative behavior. Based on this assumption, motivation schemes have been launched in the corporate world. Workers and managers receive bonuses, stock options, and other monetary incentives to encourage them to perform better at their jobs (Camerer and Hogarth, 1999).

In contrast, psychological theories on motivation do not allow, and are not intended to make, such general and large-scale predictions about the outcome of motivated behavior. Psychological theories offer a collection of different motives and explanations for the emergence of motivated behavior in order to account for individual differences and the origins of motivation. An increase in performance, therefore, depends on the person, on the context and the form of initial motivation (extrinsic or intrinsic). Psychologists have challenged the classical economic view of a generally positive effect of incentives by providing compelling evidence against the corresponding assumptions. Contrary to economic theory, monetary incentives were shown to have a negative influence on motivation in specific contexts (Ariely et al., 2009), and people were shown to be influenced by factors other than solely monetary incentives. For example, intrinsic motivation has been shown to modulate motivation to a large degree (Deci et al., 1999; Fehr and Falk, 2002b). Thus, even in the absence of financial or other nontangible rewards, people will sometimes engage in a task.

Behavioral economists adapted economic theories on motivation in order to account for some of these "deviant" behaviors, and for the first time acknowledged intrinsic motives as well as personality and social preferences as variables that influence motivation. However, despite recognizable convergences among disciplines, a unifying theory is not yet in sight. The development of such a universal theory that integrates findings from all branches of disciplines seems impossible, although some researchers in the field on neuroeconomics make a claim for such (Glimcher and Rustichini, 2004). Strengthening the exchanges between disciplines might be a first step toward a unified approach.

The main task in motivation research is to make sense of the current knowledge that has been gathered in the various disciplines, especially the modulatory interaction of intrinsic, social, and extrinsic incentives. Motives are often unconscious, however, which makes it difficult to measure them. For that reason, monetary incentives as motives are very useful, because they allow an objective measure of the motivator itself. Also, long-term effects of motives need to be studied in order to develop a clearer image of the underlying processes. Long-term effects have been generally neglected in both psychology and in economics, although both areas of study could determine behavior to a great extent (Crockett et al., 2013; McClure et al., 2004).

Thus, while converging knowledge and findings from different disciplines and schools within disciplines has resulted in significant progress toward understanding motives underlying human behavior, more (interdisciplinary) research is necessary in order to formulate a unifying theory—or at least a more comprehensive theory—on human motivation.

ACKNOWLEDGMENTS

This work was supported by Deutsche Forschungsgemeinschaft (DFG) Grants INST 392/125-1 and PA 2682/1-1 (to S.Q.P.).

REFERENCES

Akerlof, G.A., 1970. The market for lemons: quality uncertainty and the market mechanism. Q. J. Econ. 84, 488–500.

Albrecht, K., Abeler, J., Weber, B., Falk, A., 2014. The brain correlates of the effects of monetary and verbal rewards on intrinsic motivation. Front. Neurosci. 8, 1–10.

Ariely, D., Gneezy, U., Loewenstein, G., Mazar, N., 2009. Large stakes and big mistakes. Rev. Econ. Stud. 76, 451–469.

Baker, G., 2000. The use of performance measures in incentive contracting. Am. Econ. Rev. 90, 415–420.

Baker, G.P., Jensen, M.C., Murphy, K.J., 1988. Compensation and incentives: practice vs. theory. J. Financ. 43, 593–616.

Baker, T.B., Piper, M.E., McCarthy, D.E., Majeskie, M.R., Fiore, M.C., 2004. Addiction motivation reformulated: an affective processing model of negative reinforcement. Psychol. Rev. 111, 33–51.

Barmettler, F., Fehr, E., Zehnder, C., 2012. Big experimenter is watching you! Anonymity and prosocial behavior in the laboratory. Games Econ. Behav. 75, 17–34.

Baumeister, R.F., Leary, M.R., 1995. The need to belong: desire for interpersonal attachments as a fundamental human motivation. Psychol. Bull. 117, 497–529.

Benabou, R., Tirole, J., 2003. Intrinsic and extrinsic motivation. Rev. Econ. Stud. 70, 489–520.

Broadhurst, P., 1959. The interaction of task difficulty and motivation: the Yerkes–Dodson Law revived. Acta Psychol. 16, 321–338.

Camerer, C., Fehr, E., 2006. When does "economic man" dominate social behavior? Science 311, 47–52.

Camerer, C., Hogarth, R.M., 1999. The effects of financial incentives in experiments: a review and capital-labor-production framework. J. Risk Uncertain. 19, 7–42.

Camerer, C., Loewenstein, G., Prelec, D., 2005. Neuroeconomics: how neuroscience can inform economics. J. Econ. Lit. XLIII, 9–64.

Cellura, A.R., 1969. The application of psychological theory in educational settings: an overview. Am. Educ. Res. J. 6, 349–382.

Chomsky, N., 1959. A review of BF skinner's verbal behavior. Language 35, 26–58.

Clark, L., Lawrence, A.J., Astley-Jones, F., Gray, N., 2009. Gambling near-misses enhance motivation to gamble and recruit win-related brain circuitry. Neuron 61, 481–490.

Crockett, M.J., Braams, B.R., Clark, L., Tobler, P.N., Robbins, T.W., Kalenscher, T., 2013. Restricting temptations: neural mechanisms of precommitment. Neuron 79, 391–401.
Davis, H.D., Sears, R.R., Miller, H.C., Brodbeck, A.J., 1948. Effects of cup, bottle and breast feeding on oral activity of newborn infants. Pediatrics 2, 549–558.
Deci, E.L., 1971. Effects of externally mediated rewards on intrinsic motivation. J. Pers. Soc. Psychol. 18, 105–115.
Deci, E.L., Ryan, R.M., 1985. Intrinsic Motivation and Self-Determination in Human Behavior. Plenum Press, New York, 17, 253.
Deci, E.L., Ryan, R.M., 2000. The "what" and "why" of goal pursuits: human needs and the self-determination of behavior. Psychol. Inq. 11, 227–268.
Deci, E.L., Ryan, R., 2002. Handbook of Self-Determination Research. The University of Rochester Press, New York.
Deci, E.L., Koestner, R., Ryan, R.M., 1999. A meta-analytic review of experiments examining the effects of extrinsic rewards on intrinsic motivation. Psychol. Bull. 125, 627–668.
Elster, J., 1989. Social norms and economic theory. J. Econ. Perspect. 3, 99–117.
Falk, A., Fischbacher, U., 2006. A theory of reciprocity. Games Econ. Behav. 54, 293–315.
Falk, A., Gächter, S., Kovács, J., 1999. Intrinsic motivation and extrinsic incentives in a repeated game with incomplete contracts. J. Econ. Psychol. 20, 251–284.
Falk, A., Fehr, E., Fischbacher, U., 2003. On the nature of fair behavior. Econ. Inq. 41, 20–26.
Fehr, E., Falk, A., 2002a. Psychological foundations of incentives. Eur. Econ. Rev. 46, 687–724.
Fehr, E., Falk, A., 2002b. Reciprocal fairness, cooperation and limits to competition. In: Fullbrook, E. (Ed.), Intersubjectivity in Economics: Agents and Structures. Tayler & Francis Group, Bury St Edmunds, pp. 28–42.
Fehr, E., Fischbacher, U., 2002. Why social preferences matter—the impact of non-selfish motives on competition, cooperation and incentives. Econ. J. 112, C1–C33.
Fehr, E., Gächter, S., 1998. Reciprocity and economics: the economic implications of Homo Reciprocans. Eur. Econ. Rev. 42, 845–859.
Fehr, E., Gächter, S., 2000a. Cooperation and punishment in public goods experiments. Am. Econ. Rev. 90 (4), 980–994.
Fehr, E., Gächter, S., 2000b. Fairness and retaliation: the economics of reciprocity. J. Econ. Perspect. 14 (3), 159–181.
Fehr, E., Gächter, S., 2002. Altruistic punishment in humans. Nature 415, 137–140.
Fehr, E., Schmidt, K.M., 1999. A theory of fairness, competition, and cooperation. Q. J. Econ. 114, 817–868.
Fehr, E., Tougareva, E., Fischbacher, U., 2014. Do high stakes and competition undermine fair behaviour? Evidence from Russia. J. Econ. Behav. Organ. 108, 354–363.
Fischbacher, U., Gächter, S., Fehr, E., 2001. Are people conditionally cooperative? Evidence from a public goods experiment. Econ. Lett. 71, 397–404.
Flora, S.R., 2004. The Power of Reinforcement. State University of New York Press, Albany.
Freud, A., 1952. The mutual influences in the development of the ego and the id: introduction to the discussion. Psychoanal. Stud. Child 7, 42–50.
Freud, S., 1961. The Ego and the Id. W.W. Norton, New York.
Frey, B.S., Jegen, R., 2001. Motivation crowding theory. J. Econ. Surv. 15 (5), 589–611.
Gibbons, R., 1997. An introduction to applicable game theory. J. Econ. Perspect. 11, 127–149.

Gleitman, H., Fridlund, A.J., Reisberg, D., 2004. Psychology, sixth ed. W.W. Norton, New York.

Glimcher, P.W., Rustichini, A., 2004. Neuroeconomics: the consilience of brain and decision. Science 306, 447–452.

Gneezy, U., Meier, S., Rey-Biel, P., 2011. When and why incentives (don't) work to modify behavior. J. Econ. Perspect. 25, 191–210.

Goldstein, K., 1939. The Organism: A Holistic Approach to Biology Derived from Pathological Data in Man. American Book Publishing, Salt Lake City.

Hau, R., Martini, U., 2012. PONS Wörterbuch für Schule und Studium Latein-Deutsch. PONS GmbH, Stuttgart.

Heckhausen, J., Heckhausen, H., 2006. Motivation und Handeln: Einführung und Überblick. Springer, Berlin, Heidelberg.

Herkner, W., 1986. Psychologie. Springer, Wien.

Hirschman, C.E., 1980. Innovativeness, novelty seeking and consumer creativity. J. Consum. Res. 7, 283–295.

Hsee, C.K., Yu, F., Zhang, J., Zhang, Y., 2003. Medium maximization. J. Consum. Res. 30, 1–14.

Hull, C.L., 1943. Principles of Behavior: An Introduction to Behavior Theory. Appleton Century, Oxford.

Hull, C.L., 1952. A Behavior System: An Introduction to Behavior Theory Concerning the Individual Organism. Yale University Press, New Haven.

James, W., 1890. The Principles of Psychology. H. Holt and Company, New York.

Jenkins Jr., G.D., Mitra, A., Gupta, N., Shaw, J.D., 1998. Are financial incentives related to performance? A meta-analytic review of empirical research. J. Appl. Psychol. 83, 777–787.

Kalivas, P.W., Volkow, N.D., 2005. The neural basis of addiction: a pathology of motivation and choice. Am. J. Psychiatry 162, 1403–1413.

Keller, J.A., 1981. Grundlagen der Motivation. Urban & Schwarzenberg, München.

Kenning, P., Plassmann, H., 2005. NeuroEconomics: an overview from an economic perspective. Brain Res. Bull. 67, 343–354.

Koob, G.F., Le Moal, M., 2001. Drug addiction, dysregulation of reward, and allostasis. Neuropsychopharmacology 24, 97–129.

Kubie, L.S., 1948. Instincts and homoeostasis. Psychosom. Med. 10, 15–30.

Kunz, A.H., Pfaff, D., 2002. Agency theory, performance evaluation, and the hypothetical construct of intrinsic motivation. Account. Org. Soc. 27, 275–295.

Lawler, E.E., Porter, L.W., 1967. Antecedent attitudes of effective managerial performance. Organ. Behav. Hum. Perform. 2, 122–142.

Markland, D., 1999. Self-determination moderates the effects of perceived competence on intrinsic motivation in an exercise setting. J. Sport Exerc. Psychol. 21, 351–361.

Maslow, A.H., 1943. A theory of human motivation. Psychol. Rev. 50, 370–396.

Maslow, A.H., 1954. The instinctoid nature of basic needs. J. Pers. 22, 326–347.

McClelland, D.C., 1987. Human Motivation. Cambridge University Press, Cambridge.

McClure, S.M., Laibson, D.I., Loewenstein, G., Cohen, J.D., 2004. Separate neural systems value immediate and delayed monetary rewards. Science 306, 503–507.

Mitchell, T.R., 1982. Motivation: new directions for theory, research, and practice. Acad. Manage. Rev. 7, 80–88.

Modell, A.H., 1993. The Private Self. Harvard University Press, Cambridge.

Morgan, C.L., 1912. Instincts and Experience. The Macmillian Company, New York.
Mowrer, O.H., 1951. Two-factor learning theory: summary and comment. Psychol. Rev. 58, 350–354.
Mullainathan, S., Thaler, R.H., 2001. Behavioral economics. Int. Encycl. Soc. Behav. Sci. 1094–1100.
Neher, A., 1991. Maslow's theory of motivation a critique. J. Humanist. Psychol. 31, 89–112.
Nevid, J.S., 2013. Psychology: Concepts and Applications, fourth ed. Wadsworth Cengage Learning, Belmont.
Olds, J., 1953. The influence of practice on the strength of secondary approach drives. J. Exp. Psychol. 46, 232–236.
Opsahl, R.L., Dunnette, M.D., 1966. Role of financial compensation in industrial motivation. Psychol. Bull. 66, 94–118.
Pavlov, I.P., 1941. Lectures on Conditioned Reflexes. Conditioned Reflexes and Psychiatry, vol. 2 Lawrence & Wishart, Wishart, London.
Porter, L.W., Lawler, E.E., 1982. What Job Attitudes Tell About Motivation. Harvard Business Review Reprint Service, Boston.
Ryan, R.M., 2012. The Oxford Handbook of Human Motivation. Oxford University Press, New York.
Ryan, R.M., Deci, E.L., 2000. Self-determination theory and the facilitation of intrinsic motivation, social development, and well-being. Am. Psychol. 55, 68–78.
Sanfey, A.G., 2007. Social decision-making: insights from game theory and neuroscience. Science 318, 598–602.
Sanfey, A.G., Rilling, J., Aronson, J., Nystrom, L., Cohen, J., 2003. The neural basis of economic decision-making in the Ultimatum Game. Science 300, 1755–1758.
Schuster, J.R., Clark, B., Rogers, M., 1971. Testing portions of the Porter and Lawler model regarding the motivational role of pay. J. Appl. Psychol. 55, 187–195.
Sherrington, C.S., 1916. The Integrative Action of the Nervous System. Cambridge University Press Archive, Cambridge.
Skinner, B.F., 1938. The Behavior of Organisms: An Experimental Analysis. Appleton-Century, Oxford.
Skinner, B.F., 2011. About Behaviorism. Vintage, New York.
Skinner, B.F., 2014. Contingencies of Reinforcement: A Theoretical Analysis, third ed. The B. F. Skinner Foundation, Cambridge.
Steers, R.M., Porter, L.W., Bigley, G.A., 1996. Motivation and Leadership at Work, sixth ed. McGraw-Hill, New York.
Strang, S., Gross, J., Schuhmann, T., Riedl, A., Weber, B., Sack, A., 2015. Be nice if you have to—the neurobiological roots of strategic fairness. Soc. Cogn. Affect. Neurosci. 10, 790–796.
Strombach, T., et al., 2015. Social discounting involves modulation of neural value signals by temporoparietal junction. Proc. Natl. Acad. Sci. 112 (5), 1619–1624.
Trigg, A.B., 2004. Deriving the Engel curve: Pierre Bourdieu and the social critique of Maslow's hierarchy of needs. Rev. Soc. Econ. 62, 393–406.
Watson, J.B., 1913. Psychology as the behaviorist views it. Psychol. Rev. 20, 158–177.
Watson, J.B., 1930. Behaviorism (Rev. ed.), Norton, New York.
White, R.W., 1959. Motivation reconsidered: the concept of competence. Psychol. Rev. 66, 297–333.

White, K., Lehman, D.R., 2005. Culture and social comparison seeking: the role of self-motives. Pers. Soc. Psychol. Bull. 31, 232–242.

Wiest, W.M., 1967. Some recent criticisms of behaviorism and learning theory: with special reference to Breger and McGaugh and to Chomsky. Psychol. Bull. 67, 214–225.

Wike, E.L., Barrientos, G., 1958. Secondary reinforcement and multiple drive reduction. J. Comp. Physiol. Psychol. 51, 640–643.

Wobker, I., 2015. The price of envy—an experimental investigation of spiteful behavior. Manag. Decis. Econ. 35, 326–335.

Yerkes, R.M., Dodson, J.D., 1908. The relation of strength of stimulus to rapidity of habit-formation. J. Comp. Neurol. Psychol. 18, 459–482.

Young, P.T., 1959. The role of affective processes in learning and motivation. Psychol. Rev. 66, 104–125.

Zimbardo, P.G., 2007. The Lucifer Effect: Understanding Why People Turn Evil. Random House, New York.

CHAPTER

A benefit–cost framework of motivation for a specific activity

2

B. Studer[*,†,1], **S. Knecht**[*,†]

[*]*Mauritius Hospital, Meerbusch, Germany*

[†]*Institute of Clinical Neuroscience and Medical Psychology, Medical Faculty, Heinrich-Heine-University Düsseldorf, Düsseldorf, Germany*

[1]*Corresponding author: Tel.: +49-2159-679-5114; Fax: +49-2159-679-1535, e-mail address: bettina.studer@stmtk.de*

Abstract

How can an individual be motivated to perform a target exercise or activity? This question arises in training, therapeutic, and education settings alike, yet despite—or even because of—the large range of extant motivation theories, finding a clear answer to this question can be challenging. Here we propose an application-friendly framework of motivation for a specific activity or exercise that incorporates core concepts from several well-regarded psychological and economic theories of motivation. The key assumption of this framework is that motivation for performing a given activity is determined by the *expected benefits* and the *expected costs* of (performance of) the activity. Benefits comprise positive feelings, gains, and rewards experienced during performance of the activity (*intrinsic benefits*) or achieved through the activity (*extrinsic benefits*). Costs entail effort requirements, time demands, and other expenditure (*intrinsic costs*) as well as unwanted associated outcomes and missing out on alternative activities (*extrinsic costs*). The expected benefits and costs of a given exercise are subjective and state dependent. We discuss convergence of the proposed framework with a selection of extant motivation theories and briefly outline neurobiological correlates of its main components and assumptions. One particular strength of our framework is that it allows to specify five pathways to increasing motivation for a target exercise, which we illustrate and discuss with reference to previous empirical data.

Keywords

Motivation, Benefit, Costs, Exercise, Effort, Value

1 INTRODUCTION

How can a child be motivated to do homework or chores? How can an employee be motivated to work hard? How can a stroke patient be enticed to perform a demanding training to regain lost physical or cognitive functions? In short, how can an individual be motivated to carry out a given activity, and do so with high effort and persistence? Given the range of extant theories in the scientific literature and the large variance in the focus, scope, and terminology of different models, finding an answer to this question can be a struggle. Our goal is to address this challenge by formulating a convergent, application-friendly framework of motivation for a specific exercise or activity. The core assumption underlying our framework is that motivation for performing a given activity is the result of a comparison of the anticipated benefits vs the anticipated costs associated with (performance of) the activity. We highlight that our model is not intended as a comprehensive theory of motivation. Rather, it aims to serve as a focused framework that incorporates and unifies core concepts from a range of extant psychological and economic theories of motivation and can help structure and guide the development of interventions targeting motivation in therapeutic, educational, or sports settings.

In the first section of this chapter, we will describe the framework, its assumptions, components, and terminology. In the second section, we will discuss in more detail, how a selection of well-regarded psychological and economic theories of motivation, namely Self-Determination Theory (SDT) (Deci, 1980; Ryan and Deci, 2000b), Expectancy Value Theory (Vroom, 1964), Temporal Motivation Theory (TMT) (Steel and König, 2006), and Effort-Discounting Theory (eg, Botvinick et al., 2009; Hartmann et al., 2013; Kivetz, 2003), fit into the proposed framework and in which aspects they differ from it. We will also briefly outline extant knowledge about neurobiological correlates of the main assumptions and components of the proposed framework. The third and final section of this chapter will present some examples on how the framework might be applied to training and therapy programs, using both hypothetical scenarios and previously published empirical data.

2 THE PROPOSED BENEFIT–COST FRAMEWORK OF MOTIVATION

The core assumption of the proposed framework is that an individual's motivation to perform a specific exercise or activity is determined by the *expected benefits* of and the *expected costs* associated with the exercise. In short, the overall *expected benefit* comprises anticipated positive feelings, experiences, and gains arising during performance of the activity or achieved through the activity. The *overall expected cost*, on the other hand, entails effort requirements, time demands or other necessary expenditure, unwanted associated outcomes, and the cost of missing out on alternative activities. Our framework assumes that benefits and costs work antagonistically on motivation, such that motivation to perform a given exercise will be high if the overall expected benefit clearly outweighs the overall expected cost, but low if the expected benefit and expected cost are of similar magnitude. Importantly, our framework defines both the overall expected benefit and the overall expected cost as being:

(1) *Multifactorial*, meaning that the overall benefit of a given exercise or activity is determined by multiple benefits of different natures, for instance positive affect, self-confirmation, feeling of progress, increase in social status, and more tangible benefits, such as learning and performance gains or financial gains. However, similar to the concept of subjective utility in economic theory (Bernoulli, 1954; Edwards, 1961, 1962; Karmarkar, 1978), our framework assumes that these different benefits and dimensions can be integrated into an overall subjective benefit quantifiable on a single internal scale. The same is assumed for the overall cost of an exercise or activity. Again, the overall expected cost reflects an integration of multiple costs of various natures (for instance required physical effort, mental effort, financial investments) into an internal overall measure.

Further, our framework assumes that the expected benefits and expected costs of an exercise or activity are:

(2) *Subjective*. That is to say, the anticipated benefits and costs of a given activity or exercise are not constant across individuals, but rather codetermined by an individual's personality, capabilities, goals, attitudes, social reference, and past experiences. As a simplified example, consider an outgoing extravert student and a shy, introvert student who are asked to a give a public talk. We would expect that the extravert student will enjoy public speaking more, and thus the subjective anticipated benefit of this activity would be higher for this students compared to the introvert student. As another example, the perceived benefit of carrying out a difficult work assignment is expected to be higher if one's coworker is paid equally for the same work than if they are paid a lot more than oneself. The same is true for costs. For instance, climbing the same set of stairs would require higher physical and mental effort for a stroke patient with deficits in balance and walking functions than for a healthy individual, and thus subjective expected costs of climbing the stairs would be higher for the stroke patient.

(3) *State dependent*. That is to say, the expected benefits and expected costs of a given activity and for a given individual are not constant across time. For instance, the subjective costs of the same cycling exercise are expected to be higher when one is fatigued than when one is well rested, and the perceived benefit of eating an apple is higher when hungry than when saturated.

2.1 SUBJECTIVE BENEFIT

Building upon the distinction between intrinsic and extrinsic motivation psychological theories of motivation (eg, Deci, 1980; Eccles and Wigfield, 2002; Harackiewicz, 2000; Ryan and Deci, 2000a,b; Vallerand, 2007), the proposed framework differentiates two main classes of benefits which determine the overall *subjective expected benefit* of an exercise or activity (Fig. 1): (i) anticipated direct benefits of the exercise per se (*intrinsic benefits*) and (ii) anticipated benefits of instrumental

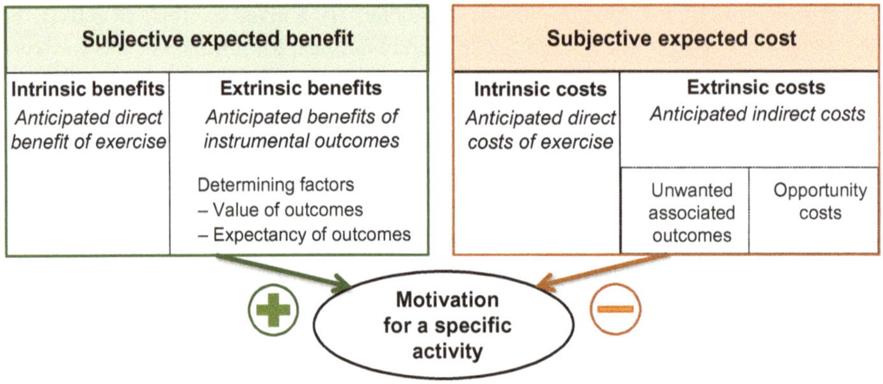

FIG. 1

Motivation as the net result of a benefit–cost evaluation. The degree of motivation for a specific exercise is determined by overall subjective expected benefit and the overall subjective expected cost of the exercise. See text for further explanation.

outcomes achieved through the exercise (*extrinsic benefits*). *Intrinsic benefits* are positive feelings that an individual experiences during the performance of the exercise itself, such as enjoyment, pleasure, satisfaction, feeling of accomplishment, competence or mastery, and, in the case of a group activity, sense of belonging (for more elaboration on intrinsic benefits, see also Oudeyer et al., 2016). Meanwhile, *extrinsic benefits* contain gains, positive feelings, rewards, and goals one wants to achieve through the exercise or activity (instrumental outcomes). Examples would be health gains, performance gains, social recognition, or financial rewards. Following Subjective Expected Utility Theory (eg, Bernoulli, 1954; Edwards, 1962; Steel and König, 2006), Expectancy Value Theories (Atkinson, 1957; Eccles and Wigfield, 2002; Lawler and Porter, 1967; Vroom, 1964), and Self-Efficacy Theory (Bandura and Locke, 2003), our framework postulates that the magnitude of an extrinsic benefit is determined by two factors: The *value* of the instrumental outcome and the *expectancy* of the instrumental outcome. Value entails the personal attractiveness and degree of importance of the instrumental outcome. *Expectancy* means the perceived likelihood that the instrumental outcome will be achieved. Let us for instance assume that an individual aims to lose weight through exercising. This individuals' motivation for treadmill running would be expected to be high if they strongly believe that treadmill running is an effective way to achieve weight loss, but small if the individual considers treadmill running to be unlikely to positively impact body weight. In addition to the effectiveness of the exercise or activity itself, beliefs about the *personal* ability to achieve a certain outcome also impact expectancy. Going back to the treadmill example, expectancy of achieving weight loss would be small if an individual strongly doubts that they will be able to persist with the exercise long enough for it to become effective.

In line with economic theories (eg, Bernoulli, 1954; Edwards, 1962; Kahneman and Tversky, 1979; Steel and König, 2006), our framework assumes that all expected *intrinsic* and *extrinsic benefits* of a given activity are aggregated into an *overall subjective expected benefit*. The integration formula however is not specified. In other words, our framework assumes *that* various *extrinsic benefits* and *intrinsic benefits* are integrated, but makes no assumptions about *how* they are combined. Indeed, the relationship between intrinsic and extrinsic benefits (or motivators) is a topic of active debate in the field (see Strang et al., 2016) and might not be constant but rather vary across situations.

2.2 SUBJECTIVE COST

Analogue to benefits, our framework differentiates two main classes of expected subjective costs: (i) expected *intrinsic costs* and (ii) expected *extrinsic costs. Intrinsic costs* are integral to the performance of the activity or exercise itself, for instance required physical work, negative feelings or affect, mental effort, or pain. *Extrinsic costs* are those arising as an indirect result of performing an exercise. *Extrinsic costs* include unwanted associated outcomes (eg, injury or social disapproval for an activity that is negatively regarded by others) and the cost of missing out on alternative activities, termed "*opportunity cost*" in economic theory (eg, Buchanan, 1979, 2008) and our framework. The opportunity cost of an exercise can be quantified as the motivational value of the best alternative activity that is available simultaneously to and has to be given up for the target activity. The motivational value of that alternative in turn is again determined by the subjective expected benefit and the subjective expected cost of that alternative activity. As a consequence, our framework predicts that motivation to perform a given exercise is also dependent on the availability and subjective valuation of alternative activities (see also Engelmann and Hein, 2013 for a discussion on how availability of alternatives influences valuation and choice). As a hypothetical example: imagine you want to go to the gym with a friend. Both you and your friend enjoy working out and believe that exercising is good for your health. However, your friend also likes sunbathing in the park. Our framework would predict that your friend would be more motivated to accompany you to the gym on a rainy day than on a sunny day (assuming all other factors and circumstances have remained the same).

Our framework's assumptions regarding the integration of different *intrinsic* and *extrinsic costs* of a given activity mirror those described for benefit integration. That is to say, our framework again postulates that all intrinsic and extrinsic expected subjective costs are integrated into one internal quantity, but makes no assumptions about the precise manner of this integration.

2.3 MOTIVATION AS THE RESULT OF BENEFIT–COST COMPARISON[a]

Our framework postulates that the degree of motivation for performance of a given activity is determined through (implicit or explicit) comparison of the overall *expected benefit* and the overall *expected cost* of the activity. How exactly this comparison is

[a]We chose the term "benefit–cost" rather than the more conventional "cost–benefit" comparison/framework to emphasize the positive dimension in this evaluation, in line with the conceptualization of motivation as the driving force behind (goal-directed) behavior (see also Section 2.4).

computed, and in particular whether benefit and cost are compared linearly (ie, through subtraction) or nonlinearly (ie, through hyperbolic or exponential discounting), is still unclear and left unspecified in our framework, because contradicting findings and postulations have been made in extant empirical and theoretical work (see, eg, Luhmann, 2013 vs Ray and Bossaerts, 2011; or Hartmann et al., 2013 vs Bonnelle et al., 2015). However, independently of the precise computation, the proposed framework predicts that motivation is large, when the overall expected benefit clearly outweighs the overall expected cost, and small when perceived benefit and cost are close to each other. Further, when the subjective expected cost outweighs the subjective expected benefit, lack of motivation is predicted, and the degree of this lack of motivation is expected to scale with the relative dominance of costs.

2.4 MOTIVATION–BEHAVIOR RELATIONSHIP

Many definitions of motivation highlight the close coupling between motivation and behavior, characterizing motivation as the force that activates, energizes, and directs behavior (see Kleinginna and Kleinginna, 1981). For instance, Hebb (1955, p. 244) states "'motivation' refers here in a rather general sense to the energizing of behavior, and especially to the sources of energy in a particular set of responses that keep them temporarily dominant over others and account for continuity and direction in behavior." Steers and Porter (1987, pp. 5–6) write "When we discuss motivation, we are primarily concerned with (1) what energizes human behavior; (2) what directs or channels such behavior; and (3) how this behavior is maintained or sustained." And, Petri and Govern (2012, p. 4) define motivation as "the concept we use when describing the forces acting on or within an organism to initiate and direct behavior." In line with these definitions, our framework postulates that motivation for a given exercise determines the probability that the individual will carry out the target exercise, the exercise amount or intensity, and how long an individual persists (exercise duration) (see Fig. 2). At the same time, our framework recognizes that a necessary prerequisite of this translation of motivation into behavior at a given time point is that performance of the target activity is possible. Therefore, our framework specifically predicts that motivation determines behavior in a dose-related manner when performance of the target activity is possible in the current environment and situation.

2.5 THE CHALLENGE OF SUBJECTIVITY AND STATE DEPENDENCY

The proposed framework assumes that *expected benefits* and *costs* of a given activity are subjective and state dependent. These assumptions (which are in fact shared by most motivation theories, although expressed in variant terminology) can pose a challenge for real-life application: If something that is perceived as an important benefit by one individual may not be acknowledged by another at all, and if an outcome or a factor only influences motivation in a certain state, how can effective motivation

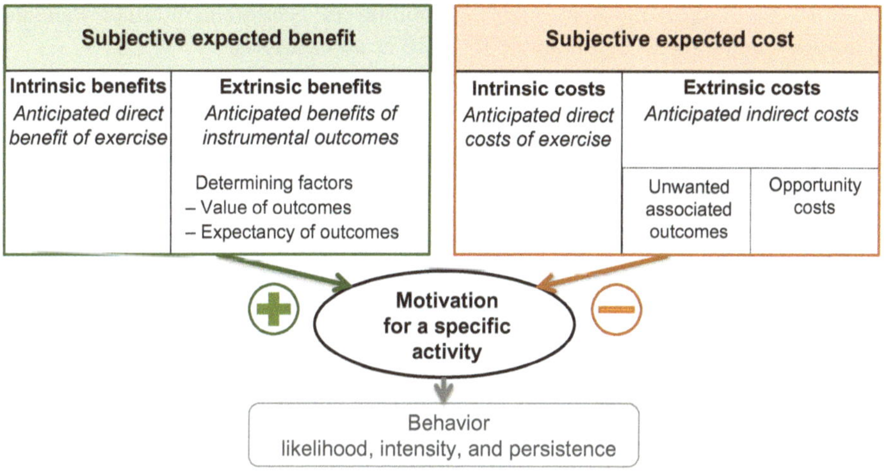

FIG. 2

Final proposed benefit–cost framework of motivation. The graph shows the final proposed framework including the link to behavior. See text for further explanation.

enhancement strategies be found? One approach would be to first examine the personality and state factors that significantly influence subjective evaluation of benefits and costs of the target activity (for instance through questionnaire assessments and systematic observation of state-related fluctuations or experimental manipulation of state), and then use this knowledge to design individual- and state-tailored interventions. At the same time, subjectivity and state dependency are most likely not unlimited, since some experiences and outcomes appear to be consistently perceived as positive by most individuals, including primary and secondary rewards [eg, food, erotic images, or monetary gains (Berridge, 2009; Rogers and Hardman, 2015; Sescousse et al., 2013)] and more abstract experiences such as autonomy, competence, personal control, learning progress, and social approval (see, eg, Deci and Ryan, 1987; Izuma et al., 2008; Leotti and Delgado, 2011; Oudeyer et al., 2007; Rademacher et al., 2010). Anticipation of such benefits should thus nearly always have a positive effect upon motivation, albeit with (inter- and intraindividually) varying effect *strength* is. Likewise, previous research indicates that pain (externally set) requirements for physical or mental effort, financial losses, and social disapproval/punishment are typically perceived as negative or aversive (eg, Bonnelle et al., 2016; Brooks and Berns, 2013; Fields, 1999; Friman and Poling, 1995; Kohls et al., 2013; Prevost et al., 2010; Seymour et al., 2007). Anticipation of such costs should thus nearly always have a reducing effect upon motivation (with some variability in effect strength). A second potential approach to the development of motivation enhancement tools would therefore be to aim to identify and use manipulable factors that robustly affect motivation in most individuals (see for instance our study reported in "Increasing Self-Directed Training in Neurorehabilitation Patients Through Competition" by Studer et al., 2016 as an example).

3 CONVERGENCE AND DIFFERENCES WITH EXTANT MOTIVATION THEORIES

In the following, we discuss how four well-regarded psychological and economic theories of motivation fit into our framework, and in what aspects they diverge from the just described framework. In addition, Table 1 provides an overview over influencing factors of the main components in our framework that can be extracted from these theories.

3.1 SELF-DETERMINATION THEORY

SDT (Deci, 1980; Deci and Ryan, 2000; Ryan and Deci, 2000b, 2007) is a relatively complex macro-theory of motivation. It is built on the core assumption that humans have innate *needs for competence*, *autonomy*, and *relatedness* to others, and seek out activities that satisfy these needs. According to SDT, motivation for a given activity is determined by the (perceived) degree to which the activity provides feelings of competence, autonomy, and relatedness, as well as by the current strength of these needs (subject to individual and state differences). A second assumption of SDT is that *intrinsic motivation* should be differentiated from *extrinsic motivation*, with intrinsic motivation being seen as the better type of motivation for securing personal well-being and advancing personal growth. Further, SDT postulates that extrinsic motivation can be divided into four subtypes, characterized by a varying degree of internalization of the benefit of a target activity and how the behavior is regulated. On one end of this four division spectrum are activities that are performed purely to satisfy an external demand (external regulation). On the other end of the spectrum are activities that are performed to achieve fully internalized instrumental outcomes and that are integrated into the repertoires of behaviors that satisfy psychological needs (integrated regulation). The two remaining subtypes, termed "introjected regulation" and "identified regulation" lie in between these two poles. SDT postulates that perception of autonomy, and thereby also the "quality" or height of motivation, increases from conditions of "external regulation" through to activities under "integrated regulation." A related assumption of SDT is that there is a degree of antagonism between extrinsic and intrinsic motivation, and that adding externally controlled incentives to an activity (for instance monetary rewards) will hamper intrinsic motivation.

While SDT has a different focus than our framework and diverges in some assumptions, many of its components and described influencing factors can be reconciled with our proposed model. For instance, the differentiation between intrinsic and extrinsic motivation can be found in our framework in the distinction between intrinsic and extrinsic benefits and costs. The assumption that motivation is affected by the degree of internalization and perceived autonomy is also broadly compatible with the two determining factors of *extrinsic benefits* in our framework: SDT defines activities under integrated regulation and high autonomy as those that are perceived as both valuable to and under personal control of the individual. These two characteristics roughly correspond to a high personal *value* of and high personal *expectancy* of instrumental outcomes. One point of divergence is that SDT (implicitly) assumes that intrinsic motivation beats extrinsic motivation, or in the terminology of our framework, that *intrinsic benefits* contribute more strongly to the overall *expected benefit* of an activity than *extrinsic benefits*. Given that integration relationships are unspecified in our framework, such an outweighing of *intrinsic benefits* is not incompatible with our proposition, but other constant or situation-dependent weighting functions and integration formulas are equally permitted by our framework.

3.2 EXPECTANCY VALUE THEORY

Expectancy Value Theory (Vroom, 1964) postulates that motivation for a given behavior or action is determined by two factors: (i) *expectancy*, ie, how probable it is that a wanted (instrumental) outcome is achieved through the behavior or action; (ii) *value*, ie, how much the individual values the desired outcome. These two core factors are integrated through multiplication, such that motivation = expectancy × value. Motivation is large when both expectancy and value are high, but disappears when one of these factors equals zero. Vroom further differentiates two subcomponents of the factor expectancy. The first subcomponent relates to an individual's belief about their personal ability to perform a given activity at a required level, in other words, the perceived relationship between effort and performance. This subcomponent is termed "expectancy" (just like the overall factor). The second subcomponent relates to (an individual's belief about) the probabilistic association between a performed activity and the wanted outcome (termed "instrumentality"). These two subcomponents are again integrated through multiplication, such that overall expectancy is high when an individual both beliefs that they will be personally able to perform a given activity and that successful performance of this activity will likely lead to the wanted outcome.

Eccles et al. (Eccles, 1983; Eccles and Wigfield, 2002; Wigfield and Eccles, 2000) and Lawler and Porter (1967) extended Vrooms model and define influencing factors of expectancy and value. For instance, Lawler and Porter (1967) state that value is determined by the degree to which an outcome is believed to satisfy needs for security, esteem, autonomy, and self-actualization. Eccles and colleagues (Eccles, 1983; Eccles and Wigfield, 2002; Wigfield and Eccles, 2000) argue that expectancy and value are affected by task-specific beliefs (ie, perceived difficulty) and individuals' self-schema and goals, which in turn are influenced by other peoples' beliefs, socialization, and personal past achievement experiences. These authors further listed four components of task value: (i) degree of enjoyment (intrinsic value), (ii) personal importance of doing well in a given task (attainment value), (iii) the degree of fit with current goals (utility value), and (iv) "relative cost," including required effort, lost alternative opportunities, and negative affect. Finally, and in direct alignment with our framework, Eccles and colleagues state that expectancy and value directly influence performance, persistence, and choice.

All three described variants of Expectancy Value Theory are broadly consistent with our framework, and the two core components *expectancy* and *value* have been incorporated into our model as determining factors of expected *extrinsic benefits*. There are however some differences in the precise understanding of these factors. For instance, in Eccles and colleagues' model, expected costs are directly integrated into value estimation, rather than represented as a separate factor (as in our framework). Meanwhile, Lawler and Porter's model does not consider costs at all, and neither their model nor Vroom's theory explicitly differentiate between intrinsic and extrinsic benefits.

3.3 TEMPORAL MOTIVATION THEORY

TMT is a utility-based model proposed by Steel and König (2006) that focuses on how the attractiveness of an activity or choice option is affected by the temporal distance of the realization of associated outcomes. TMT assumes that the subjective expected utility of an activity is determined by the summed utility of the possible gains minus the summed (negative utility) of possible losses associated with the activity. The utility of each possible gain (and each possible loss) is determined by the anticipated *value* of the gain multiplied by the *expectancy* of the gain, divided by the temporal *delay* of the gain. Comparable to our framework, value is understood as the amount of satisfaction an outcome is believed to bring (a given subject and in a given situation), and expectancy is defined as the perceived probability that an outcome will occur (also influenced by individual and situational factors). Temporal delay refers to how far away in the future the realization of a gain lies, and discounts the value of gain and loss outcomes. The further in the future a gain is, the smaller its perceived value. Finally, drawing on Cumulative Prospect Theory (Tversky and Kahneman, 1992), TMT assumes that gains and losses have different value and expectancy weighting functions.

Many of the basic assumptions of TMT are matching those of our framework, and therefore TMT can easily be reconciled with our model. Specifically, TMT covers how the *value* of *extrinsic benefits* (gain) and *extrinsic costs* (losses) is calculated, and underlines that temporal distance of an instrumental outcome, or "temporal discounting" in economics terminology, is an important factor in these calculations. Thus, TMT allows us to specify the temporal distance of an instrumental or associated outcome as one factor influencing *extrinsic benefits* and *extrinsic costs* (see Table 1). Features of our framework that are not explicitly mentioned in TMT are *intrinsic benefits* and *intrinsic costs*, although one could argue that the broad definition of gains and losses in TMT includes both instrumental and direct (intrinsic) gains/losses. Furthermore, TMT differs from our framework in how opportunity costs affect motivation and behavior. In our framework, opportunity costs directly affect motivation for a given activity. In TMT, opportunity costs are not considered

Table 1 Influencing Factors of *Subjective Expected Benefit*, *Subjective Expected Cost*, and Their Subcomponents Extracted from Previous Motivation Theories (Nonexhaustive List)

Component		Dimension	Influencing Factors Extracted from Previous Motivation Theories
Subjective expected benefit	Intrinsic benefits	—	Intensity of need for autonomy (SDT) Intensity of need for competence (SDT) Intensity of need for relatedness (SDT) Perceived degree of satisfaction of need for autonomy (SDT) Perceived degree of satisfaction of need for competence (SDT) Perceived degree of satisfaction of need for relatedness (SDT) Degree of enjoyment (ET+) Degree of interest (ET+)
	Extrinsic benefits	Value of instrumental outcomes	Personal importance of outcome (ET; ET+; ET*; TMT, EDT); degree of internalization (SDT) Personal attractiveness/desirability of outcome (ET; TMT, EDT) Intensity of need for autonomy (SDT) Intensity of need for competence (SDT) Intensity of need for relatedness (SDT) Perceived degree of satisfaction of need for autonomy (SDT; ET*) Perceived degree of satisfaction of need for competence (SDT) Perceived degree of satisfaction of need for security (ET*) Perceived degree of satisfaction of need for esteem (ET*) Perceived degree of satisfaction of need for self-actualization (ET*) Degree of fit with short-term and long-term goals (ET; ET+) Societal/other's beliefs about importance (ET+) Self-schemata, personal, and social identities (ET+) Temporal delay (TMT) Delay weighting function(TMT) Value weighting function (TMT) Reference point; current state (TMT)
		Expectancy of instrumental outcomes	Instrumentality (ET) Perceived probability of outcome (TMT) Probability weighting function (TMT) Self-efficacy (ET)

Continued

Table 1 Influencing Factors of *Subjective Expected Benefit*, *Subjective Expected Cost*, and Their Subcomponents Extracted from Previous Motivation Theories (Nonexhaustive List)—cont'd

Component		Dimension	Influencing Factors Extracted from Previous Motivation Theories
Subjective expected cost	Intrinsic costs	—	Perceived personal control, competence/ability (ET; ET+) Perceived task difficulty (ET; ET+) Previous achievement experience (ET; ET+) Society's/others' belief about personal competence (ET+) Perceived level of effort (EDT) Effort weighting function (EDT) Performance anxiety (ET+) Fear of failure (ET+)
	Extrinsic costs	Unwanted associated outcomes	Reference point; current state (TMT) Loss value (TMT) Loss weighting function (TMT) Loss probability (TMT) Loss probability weighting function (TMT)
		Opportunity cost	Perception of forgone opportunities/alternatives (ET+)

Note. *Abbreviations in brackets indicate the theory of which the listed influencing factor was extracted from:* EDT, *Effort-Discounting Theory (Hartmann et al., 2013; Kivetz, 2003; Prevost et al., 2010; and others);* ET, *Vroom's Expectancy Value Theory (Vroom, 1964);* ET*, *Lawler and Porter's Expectancy Value Theory (Lawler and Porter, 1967);* ET+, *Eccles's Expectancy Value Theory (Eccles, 1983; Eccles and Wigfield, 2002);* SDT, *Self-Determination Theory (Ryan and Deci, 2000b, 2007);* TMT, *Temporal Motivation Theory (Steel and Konig, 2006).*

in the initial evaluation of the utility of a given activity. Instead, subjective utilities of all available activities are first independently assessed and then compared in order to guide choice toward the activity with the highest utility.

3.4 EFFORT-DISCOUNTING THEORY

Effort-Discounting Theory (EDT) (eg, Bonnelle et al., 2015; Botvinick et al., 2009; Hartmann et al., 2013; Kivetz, 2003; Prevost et al., 2010) also builds on Utility Theory and focused on how effort requirements affect subjective utility of an activity. EDT postulates that the utility of a given activity is determined by two factors, expected gain [determined by the value and probability of (rewarding) outcomes] and effort requirements. Both factors are defined as subjective and thus vary across individuals. Different formulas of how expected reward and effort requirements are integrated into subjective utility have been postulated, including hyperbolic (Prevost et al., 2010) and parabolic (Hartmann et al., 2013) discounting of expected reward by physical effort. However, the underlying core assumption is identical for all formulations of EDT, namely that effort diminishes utility, in other words, motivation.

EDT's assumptions are fully compatible with our framework, which also describes motivation as the result of a benefit–cost evaluation. The main difference between our framework and EDT is that the scope of our model is wider. Classical EDT concerns itself primarily with (physical) effort costs and extrinsic benefits (although see Kivetz, 2003), whereas our framework also considers other types of intrinsic costs, as well as *extrinsic costs* and *intrinsic benefits.*

4 CONVERGENCE WITH FINDINGS FROM NEUROECONOMIC RESEARCH

The neurobiological underpinnings of motivation are discussed in details in Section 3 of this volume (see Bernacer et al., 2016; Hegerl and Ulke, 2016; Kroemer et al., 2016; Losecaat Vermeer et al., 2016; Morales et al., 2016; Umemoto and Holroyd, 2016) and fall outside the scope of this chapter. However, we note that findings from recent neuroimaging and electrophysiological studies on decision-making and choice behavior align with several of the assumptions of the proposed benefit–cost framework. For instance, a large body of neuroimaging studies showed that the aspects determining *expectancy* (eg, probability) and *value* (eg, magnitude, risk, and temporal delay) of (extrinsic) rewards are reflected in activation patterns of a network of brain regions, including midbrain, striatum, orbitofrontal, ventromedial and lateral prefrontal cortex, anterior insula, anterior cingulate cortex, and inferior parietal cortex, during evaluation and selection of decision options (eg, Berns and Bell, 2012; Huettel et al., 2005; Hutcherson et al., 2012; Kim et al., 2008; Smith et al., 2009; Studer et al., 2012; Symmonds et al., 2010; Tobler et al., 2009). Furthermore, convergent evidence from functional magnetic resonance studies in humans and single-cell recordings in nonhuman animals indicates that midbrain dopaminergic neurons, striatum, orbitofrontal cortex, and ventromedial prefrontal cortex support the integration of different aspects of and multiple types of expected benefits into a singular internal measure (eg, Bartra et al., 2013; Lak et al., 2014; Levy and Glimcher, 2012; Montague and Berns, 2002; Raghuraman and Padoa-Schioppa, 2014).

Neurobiological correlates of (subjective) *intrinsic costs*, in particular physical effort requirements, have also been observed in activation patterns of midcingulate cortex, anterior insula, and dorsal and ventral striatum (Bonnelle et al., 2016; Day et al., 2011; Prevost et al., 2010), and pharmacological manipulation of dopaminergic signaling (in the nucleus accumbens) affects perceived effort as well as willingness to exert effort for a given reward (Denk et al., 2005; Hamid et al., 2016; Salamone et al., 2007). Moreover, two recent neuroimaging studies using laboratory decision-making paradigms showed that the brain encodes not only the overall subjective expected benefit of the chosen option or action but also that of the (best) alternative, forgone choice option or action [represented in the frontopolar cortex (Boorman et al., 2009, 2011)], which could conceivably serve to signal opportunity costs. Tonic levels of dopamine in the nucleus accumbens have also been suggested to signal opportunity costs (Niv et al., 2007).

Finally, there is accumulating evidence that the dorsal anterior cingulate cortex and ventral striatum might support the evaluative comparison of expected benefits and expected costs inherent to our framework (eg, Bonnelle et al., 2016; Croxson et al., 2009; Schouppe et al., 2014; Shenhav et al., 2013; Walton et al., 2006).

5 APPLICATION EXAMPLES

Our framework allows specifying a number of different pathways to increasing motivation for a target activity. In the following, we present these pathways with the help of previously published studies, as far as available. We hope that this example-based elaboration will provide further understanding of our framework, but also inspire and help to direct development of future applications targeting motivation in therapeutic, training, and educational settings.

5.1 PATHWAY #1: BOOSTING THE INTRINSIC BENEFIT OF THE ACTIVITY

One approach to boost motivation for a given exercise could be to increase its *intrinsic benefit*, in other words, augment the "fun factor" of the exercise, or boost the sense of achievement an individual experiences during exercise performance. An increasingly popular strategy that can be counted in this category is "gamification," where game elements and design techniques are applied to training and learning programs. For instance, a current trend in rehabilitation is to substitute or complement traditional motor exercises with video games entailing similar body movements. The underlying assumption of this approach is that video games are more enjoyable and fun than traditional exercises and thus associated with higher motivation, exercise frequency, and intensity (see Lohse et al., 2013). Case studies, feasibility studies, and first clinical trials have provided encouraging results in the form of high enjoyment ratings and compliance (Galna et al., 2014; Joo et al., 2010; McNulty et al., 2015); however, further randomized, placebo-controlled clinical trials are warranted in order to assert the effectiveness of video game use in enhancing rehabilitation outcome (Barry et al., 2014; Lohse et al., 2013). Another implementation of gamification is to build motivation-boosting elements of games, for instance choice (Wulf and Adams, 2014), competition (Studer et al., 2016), or monetary rewards (Goodman et al., 2014), into the exercise program without changing the exercise format itself. While gamification is usually discussed in the context of intrinsic motivation, we note that in some cases, such motivation-boosting game elements could also serve as new instrumental outcomes (see Section 5.2), rather than (exclusively) modulate *intrinsic benefit*.

5.2 PATHWAY #2: ADDING NEW EXTRINSIC BENEFITS TO AN ACTIVITY

A second approach to increasing motivation for a target activity that can be derived from our framework is to add new performance-based incentives—or *extrinsic benefits* to the target activity. There are several published studies using this approach. For instance, Jeffery et al. (1998) tested whether attendance at supervised walking sessions offered as part of a weight loss intervention for obese adults could be increased through monetary incentives. Each time an individual attended an exercise session, they received a small payment. This approach was effective: Attendance at the exercise sessions was twice as high in the treatment group receiving monetary rewards than in the control group (no added incentives). Similarly, Markham et al. (2002) designed a motivational intervention for absenteeism in manufacturing employees, which consisted of public recognition and awards/personalized gifts for good attendance. This intervention had an impressive effect: Absenteeism decreased by approximately 37% compared to before intervention implementation. The public recognition intervention was further about twice as effective as a control intervention, in which individuals were simply informed about their rate of absence but not awarded for good attendance. Our model would explain the findings of these two studies as follows: Monetary gains, awards, and public recognition carry positive value. Since these rewards were contingent upon performance of the target activity (attending the exercise session and attending work, respectively), they constitute new instrumental outcomes, and their addition thus increased the overall subjective expected benefit of the target activity. Further, in line with the observed raise in attendance rates, our framework would predict that the resulting enhancement of motivation would translate into a higher likelihood of attendance.

5.3 PATHWAY #3: INCREASING VALUE AND EXPECTANCY OF EXTRINSIC BENEFITS

A third potential approach to increase motivation would be to augment the perceived *value* of instrumental outcomes. For instance, a recent study by Hulleman et al. (2010, Study 2) tested whether students' motivation for majoring in psychology could be enhanced through manipulation of perceived personal relevance. Psychology undergraduate students were asked to write either an essay on the relevance of a course topic to their personal life (intervention group) or a factual paper on a course topic (control group). Both postintervention grade performance and reported interest in majoring in psychology were higher in the intervention group than in the control group, and this effect could be related to an higher ratings of personal relevance of the course material. Our framework would interpret these findings as follows: The writing exercise in the intervention group drew students' attention to the personal relevance of their currently studied course material. This led to an increase in the personal *value* of acquiring psychological knowledge (ie, an instrumental outcome of studying psychology) and thus augmented the subjective *expected benefit* of studying psychology. And again, our framework would predict that the resulting motivation enhancement would manifest as more study-related behavior (eg, more time spent reading course material), which in turn could explain the higher achieved grades.

A related approach would be to boost the *expectancy* of an instrumental outcome of a target activity. For instance, Hsee et al. (2003) suggested that motivation for an activity where the desired instrumental outcome is temporally distant could be increased by providing tokens or points (termed a "medium") immediately after performance of the target activity. Such tokens are assumed to enhance motivation through two mechanisms: (i) by providing the individual with a new immediate extrinsic benefit (see Section 5.2), and (ii) by illustrating and highlighting progress toward the distant goal (likely increasing expectancy). A recent study by Van Voorhees et al. (2013) offers another example of how expectancy can be augmented: This randomized clinical trial tested the effectiveness of three different information brochures in motivating primary care patients with depression to participate in an Internet support group. Patients were either given (i) a generic referral card, (ii) a brochure containing testimonials of other patients highlighting how helpful the support group is, or (iii) a recommendation letter from the treating physician. The authors expected that the physician's recommendation letter would be most effective; however, the results revealed that sign-up rate and engagement measures were highest in the group given the testimonial brochure. Applying our model, the success of the testimonial brochure intervention could be explained through being effective in augmenting *expectancy* of the self-help group reducing suffering (as an instrumental outcome). The positive testimonials of other patients in the same situation may have positively affected patients' belief about the effectiveness of the Internet self-help group and thereby raised the subjective expected benefit of taking part. A third empirical example of this approach was recently published by Brown et al. (2015), who ran a series of experiments in which students were given one of two descriptions of a biomedical research project. The 'intervention group' description highlighted the societal and communal impact of the research project (for instance, that the developed technology would help improve the lives of babies and injured soldiers); whereas the 'control group' description did not. Subsequently, the students were questioned about their willingness to study biomedical sciences and work in biomedical research. Willingness to enter biomedical research was higher in the intervention group than in the control group, and this effect could be explained by perceived societal impact of biomedical research (assessed through ratings) being higher in the intervention group. Again, this effect could be explained in terms of modification of the expectancy of the instrumental outcome (improving lives of vulnerable) of conducting biomedical research (the target activity): Reading an explicit example of a life-changing biomedical innovation might have increased the students' *expectancy* that such societal benefits are reached through biomedical research.

5.4 PATHWAY #4: REDUCING PERCEIVED INTRINSIC COSTS

Our framework predicts that motivation could also be heightened by lessening the overall *subjective expected cost* of the target activity. A fourth pathway to increase motivation would thus be to reduce *intrinsic costs*, for instance physical effort. Unfortunately, empirical examples on how subjective effort could be modulated are still rare. However, given our framework's assumption that intrinsic costs are state dependent, one conceivable way to increase motivation for an exercise would be to ensure that the exercise is planned for a time point where the subject is well rested, for instance at the beginning rather than at the end of a training or therapy session. Intriguingly, a recent study found that a mood manipulation (through subliminal priming with happy or sad faces) during cycling exercising affected both performance and ratings of perceived exertion, with performance being higher and perceived exertion being lower when positive mood was induced (Blanchfield et al., 2014). Other manipulations of the exercise environment [for instance, playing of music (Fritz et al., 2013; Lin and Lu, 2013)] could also positively affect perception of effort requirements.

5.5 PATHWAY #5: REDUCING EXTRINSIC COSTS BY ELIMINATING ATTRACTIVE ALTERNATIVES

A second approach to reduce overall *subjective expected cost*, and thus a fifth pathway to increasing motivation, would be to lessen the opportunity costs of a target exercise by making attractive alternative activities unavailable for a given time window. In neurorehabilitation, which relies heavily on active training and sustained intensive training efforts by the patient, removing alternatives to rehabilitative training, or more generally physical activity, belong to the "tricks of the trade." For example, rehabilitation patients can be motivated to be physically active by removing the alternative to relax in bed through a mechanical bed blockage during daytime. As another example, imagine a rehabilitation patient who has regained some walking abilities, but for whom walking still requires a high level of effort. When wanting to reach a different location in the hospital, this patient might often choose use of a wheelchair over walking, given that both activities allow reaching the location target (instrumental benefit) and wheelchair use requires a lot less effort. Removing the wheelchair can thus be a fruitful way to entice the patient to walk more, as it eliminates the low effort alternative activity and thus reduces the opportunity costs of walking (= target activity). Note that such strategies are not usually forced on patients, but rather are proposed, explained, and implemented only with a patient's consent. Therefore, the employment of restrictions on the availability of attractive options is self-controlled. In behavioral economics, such self-controlled elimination of alternatives in anticipation of (later) lapses in motivation for a target activity is also referred to as "precommitment" (eg, Crockett et al., 2013; Kurth-Nelson and Redish, 2012).

6 CONCLUDING REMARKS

In this chapter, we have introduced a benefit–cost framework of motivation for a specific activity or exercise, discussed how this framework builds upon and converges with influential previous motivation theories, and outlined five strategies how motivation could be increased that were derived from this framework. Most of presented examples for these pathways to motivation enhancement entailed physical activities or exercise, but the outlined strategies would be equally transferable to other contexts, such as cognitive training, job performance, and so on. While the proposed framework is not intended as a comprehensive theory of human motivation, we believe that it can support future development of effective motivation enhancement tools for educational, training, and therapeutic settings, particularly when combined with emerging knowledge about the neuronal underpinnings of motivation and goal-directed behavior.

REFERENCES

Atkinson, J.W., 1957. Motivational determinants of risk-taking behavior. Psychol. Rev. 64, 359.

Bandura, A., Locke, E.A., 2003. Negative self-efficacy and goal effects revisited. J. Appl. Psychol. 88, 87–99.

Barry, G., Galna, B., Rochester, L., 2014. The role of exergaming in Parkinson's disease rehabilitation: a systematic review of the evidence. J. Neuroeng. Rehabil. 11, 1–10.

Bartra, O., Mcguire, J.T., Kable, J.W., 2013. The valuation system: a coordinate-based meta-analysis of bold FMRI experiments examining neural correlates of subjective value. Neuroimage 76, 412–427.

Bernacer, J., Martinez-Valbuena, I., Martinez, M., Pujol, N., Luis, E., Ramirez-Castillo, D., Pastor, M.A., 2016. Chapter 5—Brain correlates of the intrinsic subjective cost of effort in sedentary volunteers. In: Studer, B., Knecht, S. (Eds.), Progress in Brain Research, vol. 229. Elsevier, Amsterdam, pp. 103–123.

Bernoulli, D., 1954. Exposition of a new theory on the measurement of risk. Econometrica 22, 23–36.

Berns, G.S., Bell, E., 2012. Striatal topography of probability and magnitude information for decisions under uncertainty. Neuroimage 59, 3166–3172.

Berridge, K.C., 2009. 'Liking' and 'wanting' food rewards: brain substrates and roles in eating disorders. Physiol. Behav. 97, 537–550.

Blanchfield, A.W., Hardy, J., Marcora, S.M., 2014. Non-conscious visual cues related to affect and action alter perception of effort and endurance performance. Front. Hum. Neurosci. 8, 1–16.

Bonnelle, V., Veromann, K.R., Burnett Heyes, S., Lo Sterzo, E., Manohar, S., Husain, M., 2015. Characterization of reward and effort mechanisms in apathy. J. Physiol. Paris 109, 16–26.

Bonnelle, V., Manohar, S., Behrens, T., Husain, M., 2016. Individual differences in premotor brain systems underlie behavioral apathy. Cereb. Cortex 26, 807–819.

Boorman, E.D., Behrens, T.E., Woolrich, M.W., Rushworth, M.F., 2009. How green is the grass on the other side? Frontopolar cortex and the evidence in favor of alternative courses of action. Neuron 62, 733–743.

Boorman, E.D., Behrens, T.E., Rushworth, M.F., 2011. Counterfactual choice and learning in a neural network centered on human lateral frontopolar cortex. PLoS Biol. 9, e1001093.

Botvinick, M.M., Huffstetler, S., Mcguire, J.T., 2009. Effort discounting in human nucleus accumbens. Cogn. Affect. Behav. Neurosci. 9, 16–27.

Brooks, A.M., Berns, G.S., 2013. Aversive stimuli and loss in the mesocorticolimbic dopamine system. Trends Cogn. Sci. 17, 281–286.

Brown, E.R., Smith, J.L., Thoman, D.B., Allen, J.M., Muragishi, G., 2015. From bench to bedside: a communal utility value intervention to enhance students' biomedical science motivation. J. Educ. Psychol. 107, 1116–1135.

Buchanan, J.M., 1979. Cost and Choice: An Inquiry in Economic Theory. University of Chicago Press, Chicago.

Buchanan, J.M., 2008. Opportunity cost. In: Durlauf, S.N., Blume, L.E. (Eds.), The New Palgrave Dictionary of Economics. Palgrave Macmillan, Basingstoke, http://www.dictionaryofeconomics.com/article?id=pde2008_O000029, doi:http://dx.doi.org/10.1057/9780230226203.1222.

Crockett, M.J., Braams, B.R., Clark, L., Tobler, P.N., Robbins, T.W., Kalenscher, T., 2013. Restricting temptations: neural mechanisms of precommitment. Neuron 79, 391–401.

Croxson, P.L., Walton, M.E., O'reilly, J.X., Behrens, T.E.J., Rushworth, M.F.S., 2009. Effort-based cost–benefit valuation and the human brain. J. Neurosci. 29, 4531–4541.

Day, J.J., Jones, J.L., Carelli, R.M., 2011. Nucleus accumbens neurons encode predicted and ongoing reward costs in rats. Eur. J. Neurosci. 33, 308–321.

Deci, E.L., 1980. The Psychology of Self-Determination. Heath, Lexington, MA.

Deci, E.L., Ryan, R.M., 1987. The support of autonomy and the control of behavior. J. Pers. Soc. Psychol. 53, 1024–1037.

Deci, E.L., Ryan, R.M., 2000. The "what" and "why" of goal pursuits: human needs and the self-determination of behavior. Psychol. Inq. 11, 227–268.

Denk, F., Walton, M.E., Jennings, K.A., Sharp, T., Rushworth, M.F., Bannerman, D.M., 2005. Differential involvement of serotonin and dopamine systems in cost-benefit decisions about delay or effort. Psychopharmacology (Berl.) 179, 587–596.

Eccles, J.S., 1983. Expectancies, values, and academic behaviors. In: Spence, J.T. (Ed.), Achievement and Achievement Motives—Psychological and Sociological Approaches. W.H. Freeman and Company, San Francisco, pp. 75–146.

Eccles, J.S., Wigfield, A., 2002. Motivational beliefs, values, and goals. Annu. Rev. Psychol. 53, 109–132.

Edwards, W., 1961. Behavioral decision theory. Annu. Rev. Psychol. 12, 473–498.

Edwards, W., 1962. Utility, subjective probability, their interaction, and variance preferences. J. Confl. Resolut. 6, 42–51.

Engelmann, J.B., Hein, G., 2013. Contextual and social influences on valuation and choice. Prog. Brain Res. 202, 215–237.

Fields, H.L., 1999. Pain: an unpleasant topic. Pain (Suppl. 6), S61–S69.

Friman, P.C., Poling, A., 1995. Making life easier with effort: basic findings and applied research on response effort. J. Appl. Behav. Anal. 28, 583–590.

Fritz, T.H., Hardikar, S., Demoucron, M., Niessen, M., Demey, M., Giot, O., Li, Y., Haynes, J.-D., Villringer, A., Leman, M., 2013. Musical agency reduces perceived exertion during strenuous physical performance. Proc. Natl. Acad. Sci. U.S.A. 110, 17784–17789.

Galna, B., Jackson, D., Schofield, G., Mcnaney, R., Webster, M., Barry, G., Mhiripiri, D., Balaam, M., Olivier, P., Rochester, L., 2014. Retraining function in people with Parkinson's disease using the microsoft kinect: game design and pilot testing. J. Neuroeng. Rehabil. 11, 1–12.

Goodman, R.N., Rietschel, J.C., Roy, A., Jung, B.C., Macko, R.F., Forrester, L.W., 2014. Increased reward in ankle robotics training enhances motor control and cortical efficiency in stroke. J. Rehabil. Res. Dev. 51, 213–228.

Hamid, A.A., Pettibone, J.R., Mabrouk, O.S., Hetrick, V.L., Schmidt, R., Vander Weele, C.M., Kennedy, R.T., Aragona, B.J., Berke, J.D., 2016. Mesolimbic dopamine signals the value of work. Nat. Neurosci. 19, 117–126.

Harackiewicz, J.M., 2000. Intrinsic and Extrinsic Motivation: The Search for Optimal Motivation and Performance. Academic Press, San Diego.

Hartmann, M.N., Hager, O.M., Tobler, P.N., Kaiser, S., 2013. Parabolic discounting of monetary rewards by physical effort. Behav. Process. 100, 192–196.

Hebb, D.O., 1955. Drives and the CNS (conceptual nervous system). Psychol. Rev. 62, 243.

Hegerl, U., Ulke, C., 2016. Chapter 10—Fatigue with up- vs downregulated brain arousal should not be confused. In: Studer, B., Knecht, S. (Eds.), Progress in Brain Research, vol. 229. Elsevier, Amsterdam, pp. 239–254.

Hsee, C.K., Yu, F., Zhang, J., Zhang, Y., 2003. Medium maximization. J. Consum. Res. 30, 1–14.

Huettel, S.A., Song, A.W., Mccarthy, G., 2005. Decisions under uncertainty: probabilistic context influences activation of prefrontal and parietal cortices. J. Neurosci. 25, 3304–3311.

Hulleman, C.S., Godes, O., Hendricks, B.L., Harackiewicz, J.M., 2010. Enhancing interest and performance with a utility value intervention. J. Educ. Psychol. 102, 880.

Hutcherson, C.A., Plassmann, H., Gross, J.J., Rangel, A., 2012. Cognitive regulation during decision making shifts behavioral control between ventromedial and dorsolateral prefrontal value systems. J. Neurosci. 32, 13543–13554.

Izuma, K., Saito, D.N., Sadato, N., 2008. Processing of social and monetary rewards in the human striatum. Neuron 58, 284–294.

Jeffery, R.W., Wing, R.R., Thorson, C., Burton, L.R., 1998. Use of personal trainers and financial incentives to increase exercise in a behavioral weight-loss program. J. Consult. Clin. Psychol. 66, 777.

Joo, L.Y., Yin, T.S., Xu, D., Thia, E., Fen, C.P., Kuah, C.W., Kong, K.H., 2010. A feasibility study using interactive commercial off-the-shelf computer gaming in upper limb rehabilitation in patients after stroke. J. Rehabil. Med. 42, 437–441.

Kahneman, D., Tversky, A., 1979. Prospect theory: an analysis of decision under risk. Econometrica 47, 263–291.

Karmarkar, U.S., 1978. Subjectively weighted utility: a descriptive extension of the expected utility model. Organ. Behav. Hum. Perform. 21, 61–72.

Kim, S., Hwang, J., Lee, D., 2008. Prefrontal coding of temporally discounted values during intertemporal choice. Neuron 59, 161–172.

Kivetz, R., 2003. The effects of effort and intrinsic motivation on risky choice. Mark. Sci. 22, 477–502.

Kleinginna Jr., P.R., Kleinginna, A.M., 1981. A categorized list of motivation definitions, with a suggestion for a consensual definition. Motiv. Emot. 5, 263–291.

Kohls, G., Perino, M.T., Taylor, J.M., Madva, E.N., Cayless, S.J., Troiani, V., Price, E., Faja, S., Herrington, J.D., Schultz, R.T., 2013. The nucleus accumbens is involved in both the pursuit of social reward and the avoidance of social punishment. Neuropsychologia 51, 2062–2069.

Kroemer, N.B., Burrasch, C., Hellrung, L., 2016. Chapter 6—To work or not to work: Neural representation of cost andbenefit of instrumental action. In: Studer, B., Knecht, S. (Eds.), Progress in Brain Research, vol. 229. Elsevier, Amsterdam, pp. 125–157.

Kurth-Nelson, Z., Redish, A.D., 2012. Don't let me do that!—models of precommitment. Front. Neurosci. 6, 138.

Lak, A., Stauffer, W.R., Schultz, W., 2014. Dopamine prediction error responses integrate subjective value from different reward dimensions. Proc. Natl. Acad. Sci. U.S.A. 111, 2343–2348.

Lawler, E.E., Porter, L.W., 1967. Antecedent attitudes of effective managerial performance. Organ. Behav. Hum. Perform. 2, 122–142.

Leotti, L.A., Delgado, M.R., 2011. The inherent reward of choice. Psychol. Sci. 22, 1310–1318.

Levy, D.J., Glimcher, P.W., 2012. The root of all value: a neural common currency for choice. Curr. Opin. Neurobiol. 22, 1027–1038.

Lin, J.-H., Lu, F.J.-H., 2013. Interactive effects of visual and auditory intervention on physical performance and perceived effort. J. Sports Sci. Med. 12, 388–393.

Lohse, K., Shirzad, N., Verster, A., Hodges, N., Van Der Loos, H.F.M., 2013. Video games and rehabilitation: using design principles to enhance engagement in physical therapy. J. Neurol. Phys. Ther. 37, 166–175.

Losecaat Vermeer, A.B., Riečanský, I., Eisenegger, C., 2016. Chapter 9—Competition, testosterone, and adult neurobehavioral plasticity. In: Studer, B., Knecht, S (Eds.), Progress in Brain Research, vol. 229. Elsevier, Amsterdam, pp. 213–238.

Luhmann, C.C., 2013. Discounting of delayed rewards is not hyperbolic. J. Exp. Psychol. Learn. Mem. Cogn. 39, 1274–1279.

Markham, S.E., Scott, K.D., Mckee, G.H., 2002. Recognizing good attendance: a longitudinal, quasi-experimental field study. Pers. Psychol. 55, 639–660.

Mcnulty, P.A., Thompson-Butel, A.G., Faux, S.G., Lin, G., Katrak, P.H., Harris, L.R., Shiner, C.T., 2015. The efficacy of Wii-based movement therapy for upper limb rehabilitation in the chronic poststroke period: a randomized controlled trial. Int. J. Stroke 10, 1253–1260.

Montague, P.R., Berns, G.S., 2002. Neural economics and the biological substrates of valuation. Neuron 36, 265–284.

Morales, I., Font, L., Currie, P.J., Pastor, R., 2016. Chapter 7—Involvement of opioid signaling in food preference and motivation: studies in laboratory animals. In: Studer, B., Knecht, S. (Eds.), Progress in Brain Research, vol. 229. Elsevier, Amsterdam, pp. 159–187.

Niv, Y., Daw, N.D., Joel, D., Dayan, P., 2007. Tonic dopamine: opportunity costs and the control of response vigor. Psychopharmacology 191, 507–520.
Oudeyer, P.-Y., Kaplan, F., Hafner, V.V., 2007. Intrinsic motivation systems for autonomous mental development. IEEE Trans. Evol. Comput. 11, 265–286.
Oudeyer, P.-Y., Gottlieb, J., Lopes, M., 2016. Chapter 11—Intrinsic motivation, curiosity, and learning: theory and applications in educational technologies. In: Studer, B., Knecht, S (Eds.), Progress in Brain Research, vol, 229. Elsevier, Amsterdam, pp. 257–284.
Petri, H., Govern, J., 2012. Motivation: Theory, Research, and Application. Wadsworth Publishing, Belmont, CA.
Prevost, C., Pessiglione, M., Metereau, E., Clery-Melin, M.-L., Dreher, J.-C., 2010. Separate valuation subsystems for delay and effort decision costs. J. Neurosci. 30, 14080–14090.
Rademacher, L., Krach, S., Kohls, G., Irmak, A., Grunder, G., Spreckelmeyer, K.N., 2010. Dissociation of neural networks for anticipation and consumption of monetary and social rewards. Neuroimage 49, 3276–3285.
Raghuraman, A.P., Padoa-Schioppa, C., 2014. Integration of multiple determinants in the neuronal computation of economic values. J. Neurosci. 34, 11583–11603.
Ray, D., Bossaerts, P., 2011. Positive temporal dependence of the biological clock implies hyperbolic discounting. Front. Neurosci. 5, 2.
Rogers, P.J., Hardman, C.A., 2015. Food reward. What it is and how to measure it. Appetite 90, 1–15.
Ryan, R.M., Deci, E.L., 2000a. Intrinsic and extrinsic motivations: classic definitions and new directions. Contemp. Educ. Psychol. 25, 54–67.
Ryan, R.M., Deci, E.L., 2000b. Self-determination theory and the facilitation of intrinsic motivation, social development, and well-being. Am. Psychol. 55, 68.
Ryan, R.M., Deci, E.L., 2007. Active human nature: self-determination theory and the promotion and maintenance of sport, exercise, and health. In: Hagger, M.S., Chatzisarantis, N.L. D. (Eds.), Intrinsic Motivation and Self-Determination in Exercise and Sport. Human Kinetics, Champaign, IL, pp. 1–19.
Salamone, J.D., Correa, M., Farrar, A., Mingote, S.M., 2007. Effort-related functions of nucleus accumbens dopamine and associated forebrain circuits. Psychopharmacology 191, 461–482.
Schouppe, N., Demanet, J., Boehler, C.N., Ridderinkhof, K.R., Notebaert, W., 2014. The role of the striatum in effort-based decision-making in the absence of reward. J. Neurosci. 34, 2148–2154.
Sescousse, G., Caldu, X., Segura, B., Dreher, J.C., 2013. Processing of primary and secondary rewards: a quantitative meta-analysis and review of human functional neuroimaging studies. Neurosci. Biobehav. Rev. 37, 681–696.
Seymour, B., Daw, N., Dayan, P., Singer, T., Dolan, R., 2007. Differential encoding of losses and gains in the human striatum. J. Neurosci. 27, 4826–4831.
Shenhav, A., Botvinick, M.M., Cohen, J.D., 2013. The expected value of control: an integrative theory of anterior cingulate cortex function. Neuron 79, 217–240.
Smith, B.W., Mitchell, D.G.V., Hardin, M.G., Jazbec, S., Fridberg, D., Blair, R.J.R., Ernst, M., 2009. Neural substrates of reward magnitude, probability, and risk during a wheel of fortune decision-making task. Neuroimage 44, 600–609.
Steel, P., König, C.J., 2006. Integrating theories of motivation. Acad. Manag. Rev. 31, 889–913.
Steers, R.M., Porter, L.W., 1987. Motivation and Work Behaviour. McGraw-Hill, New York.

Strang, S., Park, S., Strombach, T., Kenning, P., 2016. Chapter 12—Applied economics: the use of monetary incentives to modulate behavior. In: Studer, B., Knecht, S (Eds.), Progress in Brain Research, vol. 229. Elsevier, Amsterdam, pp. 285–301.

Studer, B., Apergis-Schoute, A.M., Robbins, T.W., Clark, L., 2012. What are the odds? The neural correlates of active choice during gambling. Front. Neurosci. 6, 1–16.

Studer, B., Van Dijk, H., Handermann, R., Knecht, S., 2016. Chapter 16—Increasing self-directed training in neurorehabilitation patients through competition. In: Studer, B., Knecht, S (Eds.), Progress in Brain Research, vol. 229. Elsevier, Amsterdam, pp. 367–388.

Symmonds, M., Bossaerts, P., Dolan, R.J., 2010. A behavioral and neural evaluation of prospective decision-making under risk. J. Neurosci. 30, 14380–14389.

Tobler, P.N., Christopoulos, G.I., O'doherty, J.P., Dolan, R.J., Schultz, W., 2009. Risk-dependent reward value signal in human prefrontal cortex. Proc. Natl. Acad. Sci. U.S.A. 106, 7185–7190.

Tversky, A., Kahneman, D., 1992. Advances in prospect theory: cumulative representation of uncertainty. J. Risk Uncertain. 5, 297–323.

Umemoto, A., Holroyd, C.B., 2016. Chapter 8—Exploring individual differences in task switching: persistence and other personality traits related to anterior cingulate cortex function. In: Studer, B., Knecht, S (Eds.), Progress in Brain Research, vol. 229. Elsevier, Amsterdam, pp. 189–212.

Vallerand, J., 2007. A hierarchical model of intrinsic and extrinsic motivation for sport and physical activity. In: Hagger, M.S., Chatzisarantis, N.L.D. (Eds.), Intrinsic Motivation and Self-Determination in Exercise and Sport. Human Kinetics, Champaign, IL, pp. 255–279.

Van Voorhees, B.W., Hsiung, R.C., Marko-Holguin, M., Houston, T.K., Fogel, J., Lee, R., Ford, D.E., 2013. Internal versus external motivation in referral of primary care patients with depression to an internet support group: randomized controlled trial. J. Med. Internet Res. 15, e42.

Vroom, V.H., 1964. Work and Motivation. Wiley, Oxford, England.

Walton, M.E., Kennerley, S.W., Bannerman, D.M., Phillips, P.E.M., Rushworth, M.F.S., 2006. Weighing up the benefits of work: behavioral and neural analyses of effort-related decision making. Neural Netw. 19, 1302–1314.

Wigfield, A., Eccles, J.S., 2000. Expectancy–value theory of achievement motivation. Contemp. Educ. Psychol. 25, 68–81.

Wulf, G., Adams, N., 2014. Small choices can enhance balance learning. Hum. Mov. Sci. 38, 235–240.

CHAPTER

Control feedback as the motivational force behind habitual behavior

3

O. Nafcha*,[1], E.T. Higgins[†], B. Eitam*,[1]

**University of Haifa, Haifa, Israel*

[†]Columbia University, New York, NY, United States

[1]Corresponding authors: Tel.: 054-6734574; Fax: 972 (4) 8240966 (O.N.); Office Tel.: 972 (4) 8249666; Fax: 972 (4) 8240966 (B.E.), e-mail address: ornafcha@gmail.com; beitam@psy.haifa.ac.il

Abstract

Motivated behavior is considered to be a product of integration of a behavior's subjective benefits and costs. As such, it is unclear what motivates "habitual behavior" which occurs, by definition, after the outcome's value has diminished. One possible answer is that habitual behavior continues to be selected due to its "intrinsic" worth. Such an explanation, however, highlights the need to specify the motivational system for which the behavior has intrinsic worth. Another key question is how does an activity attain such intrinsically rewarding properties. In an attempt to answer both questions, we suggest that habitual behavior is motivated by the influence it brings over the environment—by the control motivation system, including "control feedback." Thus, when referring to intrinsic worth, we refer to a representation of an activity that has been reinforced due to it being effective in controlling the environment, managing to make something happen. As an answer to when does an activity attain such rewarding properties, we propose that this occurs when the estimated instrumental outcome expectancy of an activity is positive, but the *precision* of this expectancy is low. This lack of precision overcomes the chronic dominance of outcome feedback over control feedback in determining action selection by increasing the relative weight of the control feedback. Such a state of affairs will lead to repeated selection of control relevant behavior and entails insensitivity to outcome devaluation, thereby producing a habit.

Keywords

Control, Habit, Motivation, Sense of agency, Goal-directed, Action selection, Anorexia, Comparator, Cybernetic models

This chapter explores the relations between control feedback and habitual behavior. Control feedback is the information about the degree of control an organism has over the environment (Eitam et al., 2013). We propose that control feedback will, under certain conditions, induce habitual behavior

Progress in Brain Research, Volume 229, ISSN 0079-6123, http://dx.doi.org/10.1016/bs.pbr.2016.06.008

The chapter is divided into two major sections. The first selectively reviews existing computational models of action selection and regulation, starting with cybernetic models (eg, Carver and Scheier, 1981; Miller et al., 1960; Powers, 1973a) and then models focusing on more elementary actions (eg, the comparator model). This section also discusses the role of control feedback as implemented in these frameworks. The second section focuses on habitual- vs goal-directed behavior and outlines our conceptual framework for how habitual behavior is acquired and maintained through control feedback. Finally, we discuss some practical implementations that arise from the proposed model, such as eating disorders.

1 COMPUTATIONAL MODELS OF ACTION SELECTION AND REGULATION

Much of our time is invested in the pursuit of goals. Accordingly, the literature on goal pursuit is huge and rife with definitions of goals crossing different levels of analysis (see De Houwer and Moors, 2015; Higgins and Scholer, 2015; Marr, 1982). For instance, a goal is what one is trying to accomplish—the object or aim of an action (Locke et al., 1981); it is a "a cognitive representation of a desired end point that impacts evaluations, emotions, and behaviors" (Fishbach and Ferguson, 2007, p. 491; Kruglanski, 1996) or a "cognitive representation linking means or actions with desired outcomes" (Mustafić and Freund, 2012, p. 493). Definitions aside, the control of behavior in light of one's current standing in relation to a goal is required in order to pursue goals successfully. One class of models that has been applied to this process is cybernetic models.

1.1 CYBERNETIC MODELS OF GOAL-DIRECTED BEHAVIOR

According to a cybernetic control model, the overarching objective is to reduce perceived discrepancy between a current state and a desired goal state by relying on feedback processes. The concept of cybernetic control is derived from engineering (Wiener, 1948) and was also inspired by physiology (eg, homeostasis, Cannon, 1932). Wiener (1948) coined the term "cybernetic" from the Greek word "steersman" as is proper to the function that this model is designed to achieve (Powers, 1978).

The term "self-regulation," developed in this context by Carver and Scheier (1982, 2011), refers to the "sense of purposive processes, the sense that self-corrective adjustments are taking place as needed to stay on track for the purpose being served" (Carver and Scheier, 2011, p. 3). The key cybernetic unit is "the negative feedback loop" (Carver and Scheier, 1982). The "negative" refers to its function to *reduce* discrepancy between the current state and the desired end-state. The loop is comprised of four functional elements: a reference point, a comparator, input, and output functions. A goal within a negative feedback loop is the reference point one desires or intends to achieve (Carver and Scheier, 1982). The role of the input function is to identify one's current state in respect to that goal. Finally, the comparator continuously compares (monitors) the input function and the reference value. The result of the comparison determines the output function—the behavior that seems appropriate to reduce the gap between the current state and the desired end-state. The output function—through the selected behavior—affects the environment and consequently the perceived input changes until the gap is nullified (Carver and Scheier, 1982, 2011; Miller et al., 1960). See Fig. 1 for illustration.

1.2 A COMPUTATIONAL MODEL FOR MOTOR ACTION SELECTION

Internal models and comparators also play an important theoretical role in the literature on motor control. Internal models can be distinguished into two types. The first is a "forward model" that predicts the sensory consequences given a current state and a motor command (Wolpert et al., 1995). This sensory prediction is available due to the simulation of the movement driven by an "efference copy" of the motor command (Holst and Mittelstaedt, 1950; Sperry, 1950). The second type of model is the "inverse model," which uses an outcome to infer the motor command that could have produced it (Wolpert et al., 1995).

One of the most influential models of motor control based on the principle of cybernetic control is the comparator model (Blakemore et al., 1999; Frith, 1992; Frith et al., 2000; Wolpert et al., 1995). The comparator model itself includes both forward and inverse models, and was initially conceived to explain motor execution, learning, and control. The comparator units in the model rely on probabilistic estimation, comparison, and inference, and enable quantifying the fit between the desired effects (motor goals), motor commands, and environmental results (Körding and Wolpert, 2006; Wolpert et al., 2003). A first comparator compares the current state and the desired state. A second comparator compares the desired state and the forward model related to the motor command (ie, the predicted state of the world given execution of the

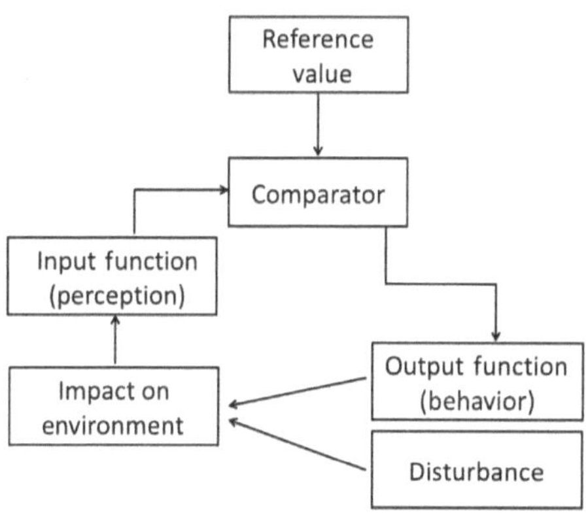

FIG. 1

An illustration of cybernetic model's elements and dynamics (as proposed by Carver and Scheier, 1982, 1990, 2011; Powers, 1973a,b). The desired goal/drive serves as the *reference value*; the current state is the *input function*; the *comparator* contrasts the current state with the desired one; *the output function* aims to reduce this gap; the effect of behavior +noise leads to the update of the input function.

command). A third comparator compares the current state and the predicted state. The model was extended to explain the "self-other distinction," such as explaining why, when, and how are perceptual sensory effects of self-generated actions vs other-generated actions attenuated (Blakemore et al., 1999, 2000), and how is the estimation of the timing of a self-caused, voluntary action vs involuntary action and its effect shifted one toward the other (Intentional Binding, Haggard et al., 2002).

In particular, the comparator model was expanded to explain the "sense of agency," the experience one has of controlling one's own actions and the external world, as well as distinguishing when it is one's own action that is responsible for an environmental change (Haggard and Tsakiris, 2009; but see Synofzik et al., 2008). The typical application of the comparator model to the sense of agency includes the second comparator and, especially, the third comparator. An error signal from the first comparator indicates a discrepancy between the current state and the desired state, and the need to reselect or modify the motor plan to reduce the error; a process that mirrors a change within the negative feedback unit (Carver and Scheier, 1982; Miller et al., 1960). The lack of an error signal will result in the smooth selection of the intended behavior until goal completion (Carver and Scheier, 1982; or an "exit" signal Miller et al., 1960).

An error signal produced by the third comparator (actual vs own action predicted state) is directly related to the sense of agency; when an error signal exists, self-causality and control are reduced (Pacherie, 2001, 2007, 2008, but see Synofzik et al., 2008 for limitations). Conversely, when no such error signal is detected—the effect is estimated to be self-generated and this estimation feeds in to downstream processes; for example, evidence from our lab suggests that the motor plan that is responsible for an own action effect is rewarded (see further elaboration on this issue in the section later). This is manifested in both faster (Eitam et al., 2013; Karsh and Eitam, 2015a) and more frequent selection of the action (Karsh and Eitam, 2015a).

Although this latter (third comparator) comparison is absent in the negative feedback loops, which involve the assessment of desired states or outcomes, we propose that control (ie, self-causality) information could have a similar regulatory function, and especially when the information regarding goal or current (goal relevant) state is lacking or imprecise (cf. White, 1959). Regarding mechanism, we suggest adding a similar negative feedback loop to the (existing) third comparator by which the system strives to minimize the discrepancy between the current actual state (striving for agency) and the predicted state. Such an addition would, for example, enable persistence, even when the output of the (outcome-concerned) negative feedback loop is imprecise (noisy) as long as the outcome expectancy is positive. The persistence would be driven by the control-driven negative feedback loop.

1.3 MOTIVATION FROM CONTROL

The behaviorist's emphasis on reward and punishment (eg, Skinner, 1953) is still the basis of many models of motivation in psychology and neuroscience (Steels, 2004). The key assumptions of this framework are as follows: first, the main goal of the organism is to maintain bodily homeostasis (eg, body temperature); second, this goal is met through the organism's tendency to seek reward and avoid punishment (Beck, 2000; Steels, 2004).

In the book *Beyond Pleasure and Pain*, Higgins (2012) reviews the substantial evidence in the psychological literature that people want (ie, are motivated by) more than just desired results. Another important source of motivation is "control" (managing what happens) and the relation between control and what he termed "value" (having desired results). Applying this perspective to information processing, Eitam et al. (2013) differentiated between types of information pertaining to different motivations, referring to the information about our standing in relation to a desired outcome as constituting "outcome feedback," and the information about the degree of control the organism has over the environment as constituting "control feedback." Outcome feedback is the information about progressing toward a goal as discussed earlier and control feedback is the information that is relevant for decisions of agency. It was assumed that both types of information could motivate action.

Early empirical support for the notion that information about one's control can be motivating appears in Stephens' (1934) largely overlooked paper that documented that, when "something" happens after a response, it strengthens the corresponding response. And this is even the case for feedback about negative outcomes (see also Thorndike, 1927). Later on, reviewing evidence that animals are seemingly motivated by outcome-neutral events, White (1959) coined the term "effectance" for the motivation to influence or interact with the environment. An important precursor to our current hypothesis is White's proposal that the hypothesized effectance "drive" influences behavior even when it does not promise the satisfaction of a current homeostatic need or obtain a tangible reward (ie, no obtained outcome).[a] Also resonating with the motivating force of control, deCharms (1968) suggested that personal causation is reinforcing; thus when behavior is perceived as stemming from "the person's choices" it will be valued more than behavior judged to stem from an external force (see also Deci and Ryan, 1985a,b). Similarly, Nuttin (1973) proposed a "causality pleasure" that is the result of the perception of being the initiator of the action.

Drawing on an analogy with the established motivating effects of outcome feedback (and more generally, of tangible rewards), Eitam et al. (2013) tested whether control feedback also motivates independent of outcomes. As we briefly mentioned earlier, their research showed that trivial and valence-neutral control feedback (a "flash" following a key press) motivates behavior. In their study, participants were instructed to press one of four keys that corresponded to one of four target stimuli. In one condition (the "Immediate Effect" condition), immediately after participants

[a]Another key insight of White's was that the relationship between control and outcome motivation is hierarchical and the latter will control behavior only when the influence of outcome motivation is weakened.

pressed a key, the circle changed its color and disappeared. Conversely, for a "No Effect" condition, the circle merely continued in its downward path, regardless of the key press (participants were assured beforehand that "the game is working properly"). Since, multiple replications showed that participants in the Immediate Effect condition were on average ~30 ms faster compared to those in the No Effect condition. Recently, Karsh and Eitam (2015a) generalized this finding by using a free choice version of the earlier paradigm (the EMFC task, see also Karsh and Eitam, 2015b). One of the key contributions of their research was to replicate the earlier pattern under conditions in which control motivation actually damaged participants' overall task performance because they were asked to "respond randomly." This is because counter to what counted as successful performance of the task (ie, what counted as positive outcome success), participants' responses were biased toward keys that were associated with a higher probability to deliver effects (ie, were more likely to deliver positive control feedback) and away from ones with a low probability to deliver effects. Specifically, participants tended to select the key that was associated with the highest chance to deliver an effect with a higher frequency than they tended to select the key associated with the lowest probability to deliver control feedback—despite this lowering their outcome performance given the task instructions.

This research also found evidence suggesting that the degree of contingency between actions and effects is to some degree accessible to consciousness, and that such awareness is associated with a preference for selecting the key associated with the highest probability of leading to positive control feedback (Karsh and Eitam, 2015a). Conversely, response speed, which Karsh and Eitam (2015b) argued to be more sensitive to the completion of a lower level of response selection (the parameters specifying how a movement is to be performed) was not associated with awareness of action–effect contingency. The modification of these "low-level" action parameters of the action is apparently related to implicit decisions of agency (Eitam et al., 2013; Karsh and Eitam, 2015a,b).

Returning to the comparator model (Blakemore et al., 1999; Frith et al., 2000; Wolpert et al., 1995) with the above in mind, it is possible to draw an analogy between the information generated by the comparator model's first comparator (current state vs motor goal) with what we called outcome feedback (cf. Carver and Scheier, 1982; Powers, 1973a,b).[b] In contrast, the source of motivation from control is the (lack of) error signal coming from the third comparator (current vs predicted state)—one that has no counterpart in the classic cybernetic models of goal pursuit, which dealt solely with outcome feedback.

[b]More speculatively, the second comparator may be loosely equated with what Higgins (2012) called "truth" effectance, or "truth feedback" in the informational language of Eitam et al. (2013). Here, we argue that for control feedback to control behavior this assessment of whether a simulated action vis-a-vis a goal should generate a "in the right direction" output.

1.4 HIERARCHICAL ORGANIZATION OF GOALS, INTENTION, AND MEANS

Let us now consider how behavior is represented hierarchically in order to substantiate a later claim that, like outcome feedback, control feedback can also target a specific level of abstraction. Goals can be represented at very different levels of abstraction (eg, Carver and Scheier, 2011; Trope and Liberman, 2010) from "wanting to be a decent person," to "donating money to the needy," to "calling the bank to transfer the money," and so on. Powers (1973a,b) suggested that control systems, which underlie the self-regulation of behavior, are hierarchically organized as superordinate and subordinate goal loops. The more abstract goals (eg, to be a decent person) residing at the top of the hierarchy, below them abstract principles (eg, specifying what "decent" means) followed by specific action programs that are intended to meet the concrete goals that operationalize these abstract principles (Carver and Scheier, 1981, 1982, 2011). Concrete goals may be associated with sequences of actions, which are in turn attained by even lower parameters that operate as low-level goals (ie, configuration, sensation, and intensity goals; see Carver and Scheier, 1982; Powers, 1973a). Thus, both very abstract and very concrete goals can serve as reference points for self-regulation.

Behavioral output is determined by monitoring the input information at the adequate level of abstraction and by comparing it to the reference value that is transferred from the level above. To repeat, the behavioral output of a given level serves as the reference value for the next (lower) level (Carver and Scheier, 1982, 1990, 2011; Powers, 1973a). In addition, during the execution of the lower level action, the activation of the higher level action representation is required (Botvinick, 2008).

Similarly, Searle (1983) distinguishes between a prior *intention* (a goal or reference state) that is independent from the execution of the intended action and an *intention in action* (a lower level implementation) that is sensitive to the internal and external context. Pacherie (2006, 2007, 2008) develops Searle's classification and defines three stages of intention specification. There are F(uture) intentions that refer to future-directed intentions. Similar to Searle's prior intention, the F intention always will precede (and is orthogonal to) the action itself.

The intention in action is divided into P(resent) and M(otor) intentions. The P intention is still a relatively abstract intention—the program (Powers, 1973a) or script (Schank and Abelson, 1977) that follows from the F intention. It serves to guide and monitor the ongoing action with sensitivity to the target of the action, to its timing, context, and perceptual characteristics. It may be consciously accessed and thus influences one's conscious experience. Lastly, the M (motor) intention is the lowest level or most concrete intention. It translates the perceptual contents of the P intention into a sensorimotor representation "through a precise specification of the spatial and temporal characteristics of the constituent elements of the selected motor program" (Pacherie, 2007, p. 3). Conscious access to this type of intention is considered to be limited as it is connected to the details about how the action is performed (Pacherie, 2006). Pacherie (2007) further proposes that the earlier differentiation between three levels of goals (intentions) parallels a similar differentiation among three levels of means specificity. The means which serve the most abstract F intentions are represented as subgoals, the means which serve P intentions are represented as specific actions, and the means that serve the M intentions are represented as specific movements.

Another theory that emphasizes on hierarchical representation of goals is action identification theory (AIT; Vallacher and Wegner, 1985, 1987). According to this theory, people tend to construe their actions at one of two levels of abstraction: a "low level" of identification, which refers to how the action (or what action) is to be performed (ie, the concrete yet verbalizable aspects of action execution); and a "high level" of identification in which the action is construed in relation to the goal or the reason for, the why of, performing the action (Wegner et al., 1989).

1.5 CONTROL FEEDBACK IS DIRECTED TOWARD DIFFERENT LEVELS OF THE ACTION HIERARCHY

There is considerable support for the notion that people represent or frame their actions hierarchically (in addition to the earlier review, see also Badre, 2008). The abstractness of the goal representation is associated with the process of action selection (Badre et al., 2010). Specifically, most of the models that involve hierarchical loops respect the means-ends hierarchy, such that the type of outcome feedback that is relevant for self-regulation differs according to the abstractness of the corresponding goal (Powers, 1973a,b). Here, we propose that the type of *control* feedback also differs according to the abstractness of the goal. de Vignemont and Fourneret (2004), for example, distinguished between a sense of agency about an action's execution (I am the initiator of the action) and the exact manner in which the action is performed (I am the cause of the action's performance). Similarly, Pacherie (2006) distinguished between the F intention and the experience of intentional causativeness—the P intention and the sense of initiation vs the M intention and the sense of control.

Recently, Karsh and Eitam (2015b) suggested that conscious knowledge of one's agency (eg, knowledge of the best effector to attain control over the environment) was associated with the selection of an effector (a subgoal or specific action according to Pacherie, 2006). In contrast, the implicit decision of agency (another form of control feedback) influenced the selection of low-level motor parameters (the specific nature of the movement).

Thus, similarly to cybernetic models of goal pursuit (ie, based on outcome feedback), control feedback may also target different levels of abstraction of the action representation. Using Pacherie's (2007) terms, it is possible that the relevant control feedback for the M intention is the third comparator of the comparator model, and hence is sensitive to what is relevant to that comparator (eg, temporal and spatial contiguity; Karsh and Eitam, 2015b). Such "low-level" control feedback informs the system that it was the one that performed the observed movement (independent of monitoring the attainment of the movement's goal). Similarly, it is possible that different control feedback is associated with more abstract goals (corresponding to Pacherie's P/F intentions).

In the next section, we consider how control motivation relates to habitual behavior. We first review some differences between habitual- and goal-directed behavior. We then outline our framework for proposing that control feedback is a key mechanism underlying habitual behavior.

2 OUTCOME VS CONTROL MOTIVATION AND FEEDBACK

Motivation is a theoretical construct that refers to the reasons (or forces) for why people and other animals choose particular actions at particular times and places (Beck, 2000; Lewin, 1935) and persist in performing them in the face of obstacles (Deci et al., 1999; Sansone and Thoman, 2006). In other words, to be motivated is "to have preferences that will direct choices" (Higgins, 2012; p. 24). Studer and Knecht ("A cost-benefit model of motivation for activity", this volume) suggest that motivation results from an integration of subjective benefits and costs of an activity. In other words, motivated behavior is seemingly a product of integration between the value of the reward (objective and subjective) and on its expected demand on resources (eg, the effort required to attain it, Bijleveld et al., 2012; Kool et al., 2013; Silvestrini and Gendolla, 2013).

Until recently, only outcomes were considered in the computation of subjective reward but based on our exposition earlier, we propose that reward from control is a second, independent source of value to take into account. Motivation's influence on behavior is classically parsed into two distinct influences: one that refers to the "direction" of behavior and corresponds to action selection processes; and another, "energizing" effect, that refers to processes underlying effort allocation, such as the amount of resources that the organism should invest in a behavior (Dickinson and Balleine, 2002; Niv et al., 2006). In this chapter, we focus mostly on action selection and how control feedback influences them as an answer to what motivates habits—instrumental behavior that continues to be performed even when the relevant external outcome (for which it was the means) has lost its value.

Tackling a related question, Higgins (2012) describes two classic answers to the question of what motivates people to continue working when goal accomplishment is not immediate (ie, "distant outcomes"). The first explanation, the incentives approach, is consistent with the behaviorist framework mentioned earlier (Beck, 2000; Hull, 1943; Rachlin, 1976; Skinner, 1953). According to this framework, people engage in activities instrumentally; with activities construed as a sequence of means to external ends. By this (incentives) approach, we do things because we want/need to have the outcomes that we have learned that these activities may bring, or because they can help reduce the probability of unwanted outcomes.

A second possibility is that people continue pursuing an activity due to rewarding properties of the activity itself. By this approach we do things because we like/enjoy/are interested in the "activities themselves." Famously, Deci and Ryan (1985a,b, 2000) highlight the distinction between "intrinsic" and "extrinsic" motivation, with extrinsic motivation referring to external outcomes that control behavior (eg, money, praise) and intrinsic motivation referring to behaviors that are performed due to their "inherently satisfying" nature (eg, are fun or challenging).

A timely question is *who or what is intrinsically motivated.* Is it the "organism" (eg, organismic integration theory; Deci and Ryan, 1985a,b)? Is it the conscious perceiver? Is it a subsystem? Or rather is it a specific representation of an action as is proposed in current models of outcome-based action selection (Redgrave et al., 1999). If the latter, one may further ask at what level of abstraction of the action representation does intrinsic motivation have its effect? A final key question is through what mechanism does an activity *itself* attain rewarding properties?

Relatedly, Higgins (2012) subscribes to a third, hybrid answer to the question of what motivates people when goal accomplishment is not immediate. The hybrid is that incentives initiate an activity, but once the action has started, valued intrinsic properties are discovered and these "take over" and lead to persistence. By this version, an activity can be at different times extrinsically and intrinsically motivated. What begins as a means to an end takes is no longer tied to the original goal-what Allport (1937) described as becoming "Functionally Autonomous."

Here, we define an "intrinsically motivated activity" narrowly: as a representation of an activity that has been rewarded due to it being effective in controlling the environment, in making something happen, independent of goal attainment (ie, by receiving control feedback rather than leading to the attainment of a valued outcome or outcome feedback; Eitam et al., 2013; Karsh and Eitam, 2015a,b). Note that we are not arguing that this exhausts the concept of "intrinsic motivation," but rather that control is a nonoutcome-dependent motivation, which can to some degree be explained mechanistically.

As we alluded to earlier, one immediate result of adopting such a mechanistic perspective is that we can offer an explanation of why "intrinsic motivation" so defined may be hampered by so called "extrinsic motivation." It is because outcome feedback (and hence reward from outcomes) will generally trump control feedback (cf. White, 1959). We can also predict when this will *not* be the case, as we describe later.

2.1 HABITUAL- VS GOAL-DIRECTED BEHAVIOR

The distinction between goal-directed or purposive and habitual behavior is older than modern (20th century) psychology (eg, James, 1890). While goal-directed behavior is argued to be preplanned and flexible, habitual behavior is considered to be reactive and inflexible (Gillan et al., 2015; Wood and Rünger, 2016). Operationally, assessing whether a behavior is goal-directed or habitual is accomplished using a variety of experimental procedures that quantify the sensitivity of the behavior to outcome devaluation (Adams, 1982; Adams and Dickinson, 1981; Balleine, 2005; Balleine and Dickinson, 1998a,b; Colwill and Rescorla, 1985; Gillan et al., 2015; Klossek et al., 2008). Such procedures typically include two phases. In the first, an animal learns to select and execute an action that leads to a specific desired outcome. Then, in second phase, the value of the outcome is reduced, such as by using the "specific satiety" procedure (eg, Balleine and Dickinson, 1998b) or by inducing an aversion to a food reward (eg, Adams and Dickinson, 1981; Colwill and Rescorla, 1985). When such interventions lead to a reduction in the frequency of the response that was instrumentally associated with the outcome, the response is said to be goal-directed. Thus, goal-directed behavior is operationally defined as one that disappears after outcome devaluation. Conversely, behavior that continues to be performed at basically the same rate after outcome devaluation is considered to be habitual.

Another common operationalization for classifying goal-directed vs habitual behavior is through testing the behavior's sensitivity to degradation of the (causal) contingency between the behavior and the outcome. Here in the second phase, the desired outcome is given regardless of whether the learned instrumental behavior is performed. Once again, a reduction in the frequency of the behavior is taken as evidence that it is goal-directed (Colwill and Rescorla, 1986; Dickinson and Mulatero, 1989), whereas persistence of the behavior at the same basic rate is evidence for the behavior having become habitual.

2.2 HOW ARE HABITS FORMED AND MAINTAINED?

Previous studies suggest several possible answers to the question of why are behaviors still performed even though they have lost their goal instrumentality. One answer is that habitual behavior is nonmotivated behavior and is the residual behavior following devaluation of a desired goal (Adams, 1982; Adams and Dickinson, 1981; Balleine, 2005; Balleine and Dickinson, 1998b; Bargh, 1994; Wood and Neal, 2007). This is not a satisfying answer. Given that behavior does not typically unfold in a vacuum, it is difficult to understand why a behavior would persist without being motivated in some way. In classic terms, why would it not "extinguish"? Thus, it is more plausible to argue that the habitual behavior continues to be motivated by some source. But what source? According to the "motivated cueing" approach (Wood and Neal, 2007, 2009), habitual behavior is a "motivated response disposition" that is activated *directly* through the context cue because that *cue* was associated with positive reinforcement from past performance. This activation can occur without a mediating goal because the goal's reward value has previously conditioned the cue.

Another possible answer is that habitual behavior is a form of goal dependent, yet automatic, behavior operating even when the goal it serves is itself unconscious or automatic (Aarts and Dijksterhuis, 2000). In this case, the context cue activates the goal and the goal automatically activates the corresponding habitual behavior. Importantly, both sources of motivation (the motivated cueing and the goal dependent automaticity) stem from goals (either past or current automatic). In other words, these answers continue to argue that *habits are motivated by outcomes*. And this holds despite the worth of the outcomes being devalued.

Alternatively, one could consider that the habitual behaviors persist despite the worth of the outcomes being devalued because the worth of the habitual behaviors no longer derives from outcomes and, instead, derives from a different motivational system. We propose that the habitual behavior is motivated by an outcome independent source—by the *degree of control it affords over the environment*, as signaled by control feedback. A unique prediction from this perspective is that analogous to goal-directed behavior being sensitive to outcome devaluation, habitual behavior should be sensitive to control devaluation (eg, a decrease in control contingency or the worth of having an effect). If supported, this prediction could be a key to future intervention programs for extinguishing unwanted habits. But before considering this, we now consider how an activity might attain such control-related rewarding properties.

2.3 THE BIRTH OF A HABIT

Our starting assumption is that a hierarchical relation exists between reward from outcome feedback and reward from control feedback (cf. White, 1959). Specifically, as long as outcome feedback is sufficiently precise (in the Bayesian sense of the inverse of the standard deviation), there is a tendency to rely on that information alone to select which action to take. As an example of precise outcome information, when my goal is 100 steps away and I know that I have already walked 60 steps, then I know that 40 more steps will bring me to my goal. Control information has little relevance in such a case.

Given this assumption, we propose that one route for inducing habitual behavior is by reducing the precision of the output of the outcome feedback process. Such a reduction in precision can occur when the outcome feedback (ie, the input to this comparator) is insufficiently precise, as when it is vague, unreliable, or altogether absent. Alternatively, this reduction in precision can occur by setting continuous, abstract, or "infinite" goals (eg, a "do your best" goal, see Campion and Lord, 1982; Touré-Tillery and Fishbach, 2011). This unreliability will lead to lowering the weight of the outcome feedback output for any process that uses it as input, including action selection. Assuming that the weighting of outcome and control feedback in action selection is relative and that these influences compete for action selection, lowering the weighting of the outcome feedback will increase the (relative) weight of control feedback. In other words, such unreliability of the outcome comparator's output "releases" action selection from the dominance of outcome feedback.

A necessary condition for habit formation is that an action be performed and, typically, repeatedly so. To that end, an action must be deemed relevant and connected to goal attainment (ie, it must be perceived to be goal relevant). In other words, people need to know about the goal pursuit process, they need to know that they are "moving in the right direction" (Higgins, 2015). This goal relevancy could be derived from either "top-down" information from social learning or other prior knowledge or through "bottom-up" learning due to repeated rewarding of the response (Thorndike, 1927; Wood and Neal, 2007). Thus, when outcome feedback is imprecise, more attention will be paid to the goal pursuit process itself, to the manner of the goal pursuit, including how an action is executed (or the fact it is executed). That is, they will pay attention to control feedback. This would mean paying less attention to outcomes such as the outcome devaluation that might be occurring, which would lead to habitual behavior.

Let us return to the earlier walking example. If we do not know how much we still have to go, we at least need to believe that every step is a step in the right direction toward the goal. And, if we continue walking, we will *eventually* reach our goal. The lack of precision enables focusing on the execution of the action and leads to positive ongoing *control* feedback in reference to the goal, which simultaneously reinforces the current action—one step at a time.

2.4 EMPIRICAL RESULTS

Recent results from our lab provide initial support for the above proposals. In two experiments, we tested the proposal that a decrease in the precision of outcome feedback will increase the weight of control feedback, and thereby lead to the formation of habitual behavior. The experiments included two phases: an induction phase and a testing phase. In the induction phase, participants performed a bogus creativity task that allowed us to independently manipulate the precision of the outcome performance feedback and the existence of (vs lack of) control feedback given to them (see Table 1). In this phase, participants were told that "the more people are creative, the more they base their judgments on their intuition" and that "in the present task we ask them to tap into their intuitive–subliminal perception skills and guess which letter (S, D, H, J) was subliminally flashed on the computer screen." None actually were, but all but one participant believed that letters were presented. Participants were further told that their goal was to attain 350 creativity points. The probability of receiving "correct" feedback was manipulated so that each key (subliminal letter) was associated with a different probability. For example, for one participant pressing the "S" key led to "correct" feedback 90% of the time; pressing "D" 60% of the time; "H" 30%; and "J" never led to "correct" feedback. This assignment was counterbalanced between participants.

In order to quantify the strength of habitual behavior, the second phase was essentially an outcome devaluation procedure in which the goal of the task was changed to participants being instructed to respond randomly (the EMFC task, Karsh and Eitam, 2015a; "random" here meaning probability matching, Bar-Hillel, and Wagenaar, 1991). Now, selecting the instrumental action from the induction phase would actually *damage* performance of the new task (as it would bias specific responses).

Before the testing phase began participants were informed that they are now going to take part in a second task that is also related to creativity but one that will not involve any guessing of subliminal letters. In this second task they were required on every trial to *randomly* select one of the four letters (S, D, H, J). No (outcome) feedback was given on success in being random

In the test, participants received own action effects (white flashes) only in the third (of four) block (a "saving" block). To test for extinction of the responses from

Table 1 The Conditions Differed in *the Precision of the Outcome Feedback* and the *Existence of Control Feedback*

	Induction Phase			Testing Phase	
Condition	**Clear Outcome Feedback**	**Control Feedback**	**Goal Relevance**	**Outcome Devaluation**	**Control Devaluation**
1	Running score	Effect (white "flash")	Yes A flash equals 1, 2, or 3 points "creativity points"	Blocks 1–4	Blocks 1–2 Block 4
2	Running score	None	None		
3	None	Effect (white "flash")	None		
4	None	None	None		
5	None	Effect (white "flash")	Yes A flash equals 1, 2, or 3 points "creativity points"		

Participants in Condition 1 had complete information. Each time they were "correct" they received a white flash (control feedback) and the score was (randomly) raised by 1, 2, or 3 "creativity points." In Condition 2 participants saw the updating score (without an effect). In Condition 3 participants also saw flashes (following key presses) but they were also informed that these were in no way related to their performance, but instead are a test of one version of a computer–human interface. Participants of Condition 4 (a control group) did not receive any feedback. Finally, participants of Condition 5 (the "habit inducing" condition) received a perceptual effect (white flash) every time they pressed a "correct" key. But, they were also informed that a white flash might reflect 1, 2 points, or 3 points. This inserted imprecision in the outcome feedback and hence. In the current standing vis-a-vis the goal.

the induction (first) phase the probability by which a key press led to an effect corresponded to the probabilities for receiving (outcome, control, or both) feedback in the induction phase. Thus, the key which led to the highest probability to obtain creativity points in the induction phase (an outcome which was now devalued) was associated in the testing phase with the higher probability to deliver control feedback (an action contingent perceptual effect).

To test for our hypothesis that habitual behavior would be sensitive to *control devaluation* (analogous to the sensitivity of instrumental behavior to outcome devaluation), in the first 120 trials of the test phase, we also devaluated control by eliminating the perceptual effect (a white flash), As stated earlier control (but not value), feedback was reinstated in the next 60 trials in order to examine "savings"—note, throughout the testing phase participants' goal was to be as random as possible and there was no feedback on the randomness of performance (see Table 1).

The key finding was that, in the savings block of the testing phase, participants who received imprecise but positive outcome feedback combined with control feedback (a flash) at the induction phase (Condition 5, see Table 1) showed the strongest evidence for habitual behavior. These participants responses in the saving block were the most biased toward the (habitual) highest probability for effect key from the induction phase when we reinstated the control feedback (the white flashes). This pattern of results was replicated in a second experiment.

The results also provided preliminary support for the existence of a hierarchical relationship between outcome and control feedback. During the induction phase, when participants received control feedback but were also explicitly told that it was irrelevant to their goal of "attaining creativity points" (Condition 3) their pattern of performance was identical to that of the control group which did not receive any feedback at all. Additionally, these participants did not show any indication of having acquired a habit of pressing the "high probability key" in the saving block in the testing phase.

3 CONCLUDING REMARKS

On the one hand, relying on habits is useful because of their automatic, relatively effortless character (ie, efficiency; James, 1890; Wood and Rünger, 2016). On the other, the same stability makes it difficult to rid ourselves of bad habits. In the present chapter, we tried to shed new light on the motivational force behind habitual behavior and to consider how and when an action attains such rewarding properties.

Several burning questions arise in regard to the proposed framework. To what extent does control-driven habit formation explain dysfunctional habits? For example, might this framework explain some "addictive" behaviors (eg, email checking)? Can malfunctioning of the hypothesized processes underlie disorders such as obsessive compulsive disorder and impulsive behavior?

One area to which the present framework could be applied is eating disorders, such as anorexia nervosa. The lack of perceived/actual control was associated with engagement in abnormal eating behaviors (Shapiro, 1981; Shapiro et al., 1996) and Strauss and Ryan (1987) have proposed that various "autonomy-related" issues exist in anorexia nervosa. Anorexia could be construed as habitual control over food intake. The creation of such a habit from the perspective of control motivation is as follows: one has a goal to be attractive, to be as thin as you ought to be in order to be attractive. "Eating less" is the dominant means to achieve this goal. The vagueness and open endedness of this "being attractive" goal leads to the output from outcome feedback being constantly imprecise. This increases the relative weight of control motivation and control feedback, which makes the means of "eating less," and constantly checking on its effects (control feedback) more worthwhile and habitual—independent of any success in becoming more attractive. A possible intervention could be to reduce the worth of control motivation and control feedback by introducing a more precise attractiveness goal and clear outcome feedback, such as tying attractiveness to having a specific weight window determined by height and body type.

To conclude, we have suggested in this chapter that control motivation with control feedback is the motivational force that preserves habitual behavior. Accordingly, we offer a new perspective on habitual behavior. From our perspective, habit is a case where behavior that originates in goal pursuit becomes continuously motivated by control feedback, independent of outcome motivation and feedback. An activity attains such control rewarding properties when the link between it and the goal pursuit outcomes it produces has been weakened. When the output of monitoring one's outcome attainment becomes imprecise but still considered to be "positive," the relative weight of control motivation and control feedback increases, and control relevant behavior is selected. This, in turn, leads to insensitivity to outcome devaluation and the creation (or manifestation) of habitual behavior.

Earlier we defined "intrinsically motivated activity" as a representation of an activity that has been rewarded due to it's being effective in controlling the environment (ie, by receiving control feedback). We further argued that habitual activity is a behavior that is reinforced by control feedback. Is it possible to reverse our argument and also claim that the shift from goal-directed behavior to habitual before reflects the shift from extrinsically to intrinsically motivated behavior? Our speculative and tentative answer is "no"—simply because there are other sources which may underlie such a shift (eg, extensive practice). In fact, it may be the case that further research may differentiate between motivated and nonmotivated "habitual behavior."

Further, we are used to use the term "habitual behavior" in the context of the operation of outcome devaluation but our findings suggest, to some degree, that outcome devaluation may merely create the conditions for *revealing* habits. Specifically, but overcoming the default dominance of outcomes in action selection and enabling other forces (eg, control) to assert themselves.

Conceptually, we are also used to name repetitive behavior as being habitual; this however, again raises the question of what exactly makes this behavior habitual? And the common answer will be: "repetition." Without a better definition we risk circularity. Our proposal of control-motivated habits is one way to circumvent circularity. Further research will show how much of "habitual behavior" can be explained by adopting it.

REFERENCES

Aarts, H., Dijksterhuis, A., 2000. Habits as knowledge structures: automaticity in goal-directed behavior. J. Pers. Soc. Psychol. 78 (1), 53.

Adams, C., 1982. Variations in the sensitivity of instrumental responding to reinforcer devaluation. Q. J. Exp. Psychol. B 34 (B), 77–98.

Adams, C.D., Dickinson, A., 1981. Instrumental responding following reinforcer devaluation. Q. J. Exp. Psychol. B 33 (B), 109–121.

Allport, G.W., 1937. The functional autonomy of motives. Am. J. Psychol. 50, 141–156.

Badre, D., 2008. Cognitive control hierarchy and the rostro-caudal organization of the frontal lobes. Trends Cogn. Sci. 12 (5), 193–200.
Badre, D., Kayser, A.S., D'Esposito, M., 2010. Frontal cortex and the discovery of abstract action rules. Neuron 66 (2), 315–326.
Balleine, B.W., 2005. Neural bases of food-seeking: affect arousal and reward in corticostriato limbic circuits. Physiol. Behav. 86, 717–730.
Balleine, B.W., Dickinson, A., 1998a. Goal-directed instrumental action: contingency and incentive learning and their cortical substrates. Neuropharmacology 37, 407–419.
Balleine, B.W., Dickinson, A., 1998b. The role of incentive learning in instrumental outcome revaluation by sensory-specific satiety. Anim. Learn. Behav. 26, 46–59.
Bargh, J.A., 1994. The four horsemen of automaticity: awareness, intention, efficiency, and control in social cognition. In: Wyer, R.S., Srull, T.K. (Eds.), Handbook of Social Cognition. Lawrence Erlbaum, Hillsdale, NJ, pp. 1–40.
Bar-Hillel, M., Wagenaar, W.A., 1991. The perception of randomness. Adv. Appl. Math. 12, 428–454.
Beck, R.C., 2000. Motivation: Theory and Principles. Prentice Hall, New Jersey.
Bijleveld, E., Custers, R., Aarts, H., 2012. Adaptive reward pursuit: how effort requirements affect unconscious reward responses and conscious reward decisions. J. Exp. Psychol. Gen. 141 (4), 728.
Blakemore, S.J., Frith, C.D., Wolpert, D.M., 1999. Spatio-temporal prediction modulates the perception of self-produced stimuli. J. Cogn. Neurosci. 11, 551–559.
Blakemore, S.J., Wolpert, D., Frith, C., 2000. Why can't you tickle yourself? Neuroreport 11 (11), R11–R16.
Botvinick, M.M., 2008. Hierarchical models of behavior and prefrontal function. Trends Cogn. Sci. 12 (5), 201–208.
Campion, M.A., Lord, R.G., 1982. A control systems conceptualization of the goal-setting and changing process. Organ. Behav. Hum. Perform. 30 (2), 265–287.
Cannon, W.B., 1932. The Wisdom of the Body. Norton, New York, NY.
Carver, C.S., Scheier, M.F., 1981. The self-attention-induced feedback loop and social facilitation. J. Exp. Soc. Psychol. 17 (6), 545–568.
Carver, C.S., Scheier, M.F., 1982. Control theory: a useful conceptual framework for personality–social, clinical and health psychology. Psychol. Bull. 92 (1), 111.
Carver, C.S., Scheier, M.F., 1990. Origins and functions of positive and negative affect: a control-process view. Psychol. Rev. 97 (1), 19.
Carver, C.S., Scheier, M.F., 2011. Self-regulation of action and affect. In: Vohsand, K.D., Baumeister, R.F. (Eds.), Handbook of Self-Regulation. Guilford Press, New York, NY, pp. 3–21.
Colwill, R.M., Rescorla, R.A., 1985. Postconditioning devaluation of reinforcer affects instrumental responding. J. Exp. Psychol. Anim. Behav. Process 11, 120–132.
Colwill, R.M., Rescorla, R.A., 1986. Associative structures in instrumental learning. In: Bower, G.H. (Ed.), The Psychology of Learning and Motivation, vol. 20. Academic Press, San Diego, CA, pp. 55–104.
De Houwer, J., Moors, A., 2015. Levels of analysis in social psychology. Theory and Explanation in Social Psychology. Guilford Press, New York, London, 24–40.
de Vignemont, F., Fourneret, P., 2004. The sense of agency: a philosophical and empirical review of the "who" system. Conscious. Cogn. 13 (1), 1–19.
DeCharms, R., 1968. Personal Causation: The Internal Affective Determinants of Behavior. Academic Press, New York, NY.

Deci, E.L., Ryan, R.M., 1985a. The general causality orientations scale: self-determination in personality. J. Res. Pers. 19 (2), 109–134.
Deci, E.L., Ryan, R.M., 1985b. Intrinsic Motivation and Self-Determination in Human Behavior. Plenum Press, New York, NY.
Deci, E.L., Ryan, R.M., 2000. The "what" and "why" of goal pursuits: human needs and the self-determination of behavior. Psychol. Inq. 11 (4), 227–268.
Deci, E.L., Koestner, R., Ryan, R.M., 1999. A meta-analytic review of experiments examining the effects of extrinsic rewards on intrinsic motivation. Psychol. Bull. 125 (6), 627.
Dickinson, A., Balleine, B., 2002. The role of learning in the operation of motivational systems. In: Gallistel, C.R. (Ed.), Learning, Motivation and Emotion, vol. 3. 497–533.
Dickinson, A., Mulatero, C.W., 1989. Reinforcer specificity of the suppression of instrumental performance on a non-contingent schedule. Behav. Process. 19, 167–180.
Eitam, B., Kennedy, P.M., Higgins, E.T., 2013. Motivation from control. Exp. Brain Res. 229 (3), 475–484.
Fishbach, A., Ferguson, M.J., 2007. The goal construct in social psychology. In: Kruglanski, A.W.,Higgins, E.T. (Eds.), Social Psychology: Handbook of Basic Principles, vol. II. Guilford Press, New York, NY, pp. 490–515.
Frith, C.D., 1992. The Cognitive Neuropsychology of Schizophrenia. Lawrence Erlbaum Associates, Hillsdale, NJ.
Frith, C.D., Blakemore, S.J., Wolpert, D.M., 2000. Abnormalities in the awareness and control of action. Philos. Trans. R. Soc. Lond. B 355, 1771–1788.
Gillan, C.M., Otto, A.R., Phelps, E.A., Daw, N.D., 2015. Model based learning protects against forming habits. Cogn. Affect. Behav. Neurosci. 15 (3), 523–536.
Haggard, P., Tsakiris, M., 2009. The experience of agency feelings, judgments, and responsibility. Curr. Dir. Psychol. Sci. 18 (4), 242–246.
Haggard, P., Clark, S., Kalogeras, J., 2002. Voluntary action and conscious awareness. Nat. Neurosci. 5 (4), 382–385.
Higgins, E.T., 2012. Beyond Pleasure and Pain: How Motivation Works. Oxford University Press, New York, NY.
Higgins, E.T., 2015. Control and truth working together: the agentic experience of "Going in the right direction" In: Haggard, P., Eitam, B. (Eds.), The Sense of Agency. Oxford University Press, New York, NY, pp. 327–343.
Higgins, E.T., Scholer, A.A., 2015. Goal pursuit functions: working together. In: Bargh, J.A., Borgida, G. (Eds.), American Psychological Association Handbook of Social Psychology. American Psychological Association, Washington, DC.
Holst, E.V., Mittelstaedt, H., 1950. Das Reafferenzprincip (Wcchselwirkungen zwischen 'Zentralnervensystern und Peripherie'). Nnturwissettschqfirn 37, 464–476.
Hull, C.L., 1943. Principles of Behavior: An Introduction to Behavior Theory. Appleton-Century-Crofts, New York, NY.
James, W., 1890. The Principles of Psychology. Dover Publications, New York, NY.
Karsh, N., Eitam, B., 2015a. I control therefore I do: judgments of agency influence action selection. Cognition 138, 122–131.
Karsh, N., Eitam, B., 2015b. Motivation from control: a response selection framework. In: Haggard, P., Eitam, B. (Eds.), The Sense of Agency. Oxford University Press, New York, NY, pp. 265–286.
Klossek, U.M.H., Russell, J., Dickinson, A., 2008. The control of instrumental action following outcome devaluation in young children aged between 1 and 4 years. J. Exp. Psychol. Gen. 137, 39–51.

Kool, W., McGuire, J.T., Wang, G.J., Botvinick, M.M., 2013. Neural and behavioral evidence for an intrinsic cost of self control. PLoS One 8 (8), e72626.
Körding, K.P., Wolpert, D.M., 2006. Bayesian decision theory in sensorimotor control. Trends Cogn. Sci. 10 (7), 319–326.
Kruglanski, A.W., 1996. Goals as knowledge structures. In: Gollwitzer, P.M., Bargh, J.A. (Eds.), The Psychology of Action: Linking Cognition and Motivation to Behavior. Guilford Press, New York, NY, pp. 599–619.
Lewin, K., 1935. A Dynamic Theory of Personality: Selected Papers. McGraw Hill, New York, NY.
Locke, E.A., Shaw, K.N., Saari, L.M., Latham, G.P., 1981. Goal setting and task performance: 1969–1980. Psychol. Bull. 90 (1), 125.
Marr, D., 1982. Vision. W.H. Freeman, San Francisco, CA.
Miller, G.A., Galanter, E., Pribram, K.H., 1960. Plans and the Structure of Behavior. Holt, Rinehart and Winston, Inc.,, New York, NY.
Mustafić, M., Freund, A.M., 2012. Means or outcomes? Goal orientation predicts process and outcome focus. Eur. J. Dev. Psychol. 9 (4), 493–499.
Niv, Y., Joel, D., Dayan, P., 2006. A normative perspective on motivation. Trends Cogn. Sci. 10 (8), 375–381.
Nuttin, J.R., 1973. Pleasure and reward in human motivation and learning. In: Berlyne, D.E., Madsen, K.B. (Eds.), Pleasure, Reward, Preference: Their Nature, Determinants, and Role in Behavior. Academic Press, New York, NY, pp. 243–273.
Pacherie, E., 2001. Agency lost and found. Philos. Psychiatr. Psychol. 8 (2–3), 173–176.
Pacherie, E., 2006. Towards a dynamic theory of intentions. In: Pockett, S., Banks, W.P., Gallagher, S. (Eds.), Does Consciousness Cause Behavior? An Investigation of the Nature of Volition. MIT Press, Cambridge, MA, pp. 145–167.
Pacherie, E., 2007. The sense of control and the sense of agency. Psyche 13 (1), 1–30.
Pacherie, E., 2008. The phenomenology of action: a conceptual framework. Cognition 107 (1), 179–217.
Powers, W.T., 1973a. Behavior: The Control of Perception. Aldine, Chicago, (p. ix).
Powers, W.T., 1973b. Feedback: beyond behaviorism. Science 179 (4071), 351–356.
Powers, W.T., 1978. Quantitative analysis of purposive systems: some spadework at the foundations of scientific psychology. Psychol. Rev. 85 (5), 417.
Rachlin, H., 1976. Behavior and Learning. Freeman, San Francisco, CA.
Redgrave, P., Prescott, T.J., Gurney, K., 1999. The basal ganglia: a vertebrate solution to the selection problem. Neuroscience 89, 1009–1023.
Sansone, C., Thoman, D.B., 2006. Maintaining activity engagement: individual difference in the process of self-regulating motivation. J. Pers. 74 (6), 1697–1720.
Schank, R.C., Abelson, R.P., 1977. Scripts, Plans, and Understanding. Lawrence Erlbaum Associates, Hillsdale, NJ.
Searle, J.R., 1983. Intentionality: An Essay in the Philosophy of Mind. Cambridge University Press.
Shapiro, D., 1981. Autonomy and Rigid Character. Basic Books, New York, NY.
Shapiro Jr., D.H., Schwartz, C.E., Astin, J.A., 1996. Controlling ourselves, controlling our world: psychology's role in understanding positive and negative consequences of seeking and gaining control. Am. Psychol. 51 (12), 1213.
Silvestrini, N., Gendolla, G.H., 2013. Automatic effort mobilization and the principle resource conservation: one can only prime the possible and justified. J. Pers. Soc. Psychol. 104 (5), 803.
Skinner, B.F., 1953. Science and Human Behavior. Macmillan, New York, NY.

Sperry, R.W., 1950. Neural basis of spontaneous optokinetic response produced by visual inversion. J. Comp. Physiol. Psychol. 43, 482–489.
Steels, L., 2004. The autotelic principle. In: Lida, F., Pfeifer, R., Steels, L., Kuniyoshi, Y. (Eds.), Embodied Artificial Intelligence. Springer-Verlag, Berlin Heidelberg, Germany, pp. 231–242.
Stephens, J.M., 1934. The influence of punishment on learning. J. Exp. Psychol. 17, 536–555.
Strauss, J., Ryan, R.M., 1987. Autonomy disturbances in subtypes of anorexia nervosa. J. Abnorm. Psychol. 96 (3), 254.
Synofzik, M., Vosgerau, G., Newen, A., 2008. Beyond the comparator model: a multifactorial two-step account of agency. Conscious. Cogn. 17 (1), 219–239.
Thorndike, E.L., 1927. The law of effect. Am. J. Psychol., 212–222.
Touré-Tillery, M., Fishbach, A., 2011. The course of motivation. J. Consum. Psychol. 21 (4), 414–423.
Trope, Y., Liberman, N., 2010. Construal-level theory of psychological distance. Psychol. Rev. 117, 440–463.
Vallacher, R.R., Wegner, D.M., 1985. A Theory of Action Identification. Erlbaum, Hillsdale, NJ.
Vallacher, R.R., Wegner, D.M., 1987. What do people think they're doing? Action identification and human behavior. Psychol. Rev. 94 (1), 3.
Wegner, D.M., Vallacher, R.R., Dizadji, D., 1989. Do alcoholics know what they're doing? Identifications of the act of drinking. Basic Appl. Soc. Psychol. 10 (3), 197–210.
White, R.W., 1959. Motivation reconsidered: the concept of competence. Psychol. Rev. 66, 297–333.
Wiener, N., 1948. Cybernetics: Control and Communication in the Animal and the Machine. MIT Press, Cambridge, MA.
Wolpert, D.M., Ghahramani, Z., Jordan, M.I., 1995. An internal model for sensorimotor integration. Science 269 (5232), 1880.
Wolpert, D.M., Doya, K., Kawato, M., 2003. A unifying computational framework for motor control and social interaction. Philos. Trans. R. Soc. Lond. B Biol. Sci. 358 (1431), 593–602.
Wood, W., Neal, D.T., 2007. A new look at habits and the habit-goal interface. Psychol. Rev. 114 (4), 843.
Wood, W., Neal, D.T., 2009. The habitual consumer. J. Consum. Psychol. 19, 579–592.
Wood, W., Rünger, D., 2016. Psychology of habit. Annu. Rev. Psychol. 67, 289–314.

SECTION

2

Assessing Motivation

CHAPTER

Quantifying motivation with effort-based decision-making paradigms in health and disease

4

T.T.-J. Chong[*,†,‡,1], **V. Bonnelle**[§], **M. Husain**[§,¶]

[*]*Macquarie University, Sydney, NSW, Australia*

[†]*ARC Centre of Excellence in Cognition and its Disorders, Macquarie University, Sydney, NSW, Australia*

[‡]*Monash Institute of Cognitive and Clinical Neurosciences, Monash University, Clayton, VIC, Australia*

[§]*University of Oxford, Oxford, United Kingdom*

[¶]*John Radcliffe Hospital, Oxford, United Kingdom*

[1]*Corresponding author: Tel.: +61-2-9850-2980; Fax: +61-2-9850-6059, e-mail address: trevor.chong@mq.edu.au*

Abstract

Motivation can be characterized as a series of cost–benefit valuations, in which we weigh the amount of effort we are willing to expend (the cost of an action) in return for particular rewards (its benefits). Human motivation has traditionally been measured with self-report and questionnaire-based tools, but an inherent limitation of these methods is that they are unable to provide a mechanistic explanation of the processes underlying motivated behavior. A major goal of current research is to quantify motivation objectively with effort-based decision-making paradigms, by drawing on a rich literature from nonhuman animals. Here, we review this approach by considering the development of these paradigms in the laboratory setting over the last three decades, and their more recent translation to understanding choice behavior in humans. A strength of this effort-based approach to motivation is that it is capable of capturing the wide range of individual differences, and offers the potential to dissect motivation into its component elements, thus providing the basis for more accurate taxonomic classifications. Clinically, modeling approaches might provide greater sensitivity and specificity to diagnosing disorders of motivation, for example, in being able to detect subclinical disorders of motivation, or distinguish a disorder of motivation from related but separate syndromes, such as depression. Despite the great potential in applying effort-based paradigms to index human motivation, we discuss several caveats to interpreting current and future studies, and the challenges in translating these approaches to the clinical setting.

Progress in Brain Research, Volume 229, ISSN 0079-6123, http://dx.doi.org/10.1016/bs.pbr.2016.05.002

Keywords

Motivation, Decision-making, Effort, Reward, Apathy

1 WHAT IS MOTIVATION?

Life is replete with instances in which we must weigh the potential benefits of a course of action against the associated amount of effort. Students must decide how intensively to study for an exam based on its importance. Employees decide how much effort to put into their jobs given their wage. Motivation is that process which facilitates *overcoming the cost of an effortful action to achieve the desired outcome*. It is a complex and multifaceted phenomenon, operating in several different domains: motivation to take a course of action, or to engage in cognitive effort, or to engage in emotional interaction. It is also influenced by many developmental, cultural, and environmental factors. A further challenge in studying motivation across individuals is that there is significant interindividual variability, ranging from healthy individuals who are highly motivated, to patients with disorders of motivation who suffer from debilitating disorders of diminished motivation, such as apathy.

Our current understanding of motivation has been shaped by the prescient observations of early philosophers and psychologists. In the 19th century, Jeremy Bentham cataloged a table of the "springs of action" that operate on the will to motivate one to act (Bentham, 1817). Shortly after this, William James, inspired by Darwin's recently published *Theory of Natural Selection* (Darwin, 1859), favored a more biological approach. He suggested that motivation comprised genetically programmed "instincts," which maintained or varied behavior in the face of changing circumstances to promote survival (James, 1890). Developing this idea, William McDougall outlined the instinct theory of motivation, in which he attributed all human behavior to 18 "instincts," or motivational dispositions (McDougall, 1908). He proposed that these instincts were important in driving goal-oriented behavior, which requires one to first attend to certain objects (the perceptual or cognitive component); experience an emotional excitement when perceiving that object (the emotional component); and initiate an act toward that object (the volitional component). This idea of fixed instincts later evolved to the concept of "needs" or "drives" giving rise to motivated behavior (Hull, 1943; Maslow, 1943).

More recently, motivation has been conceptualized as the behaviorally relevant processes that enable an organism to regulate its external and/or internal environments (Ryan and Deci, 2000; Salamone, 1992). These processes typically involve sensory, motor, cognitive, and emotional functions working together (Pezzulo and Castelfranchi, 2009; Salamone, 2010). However, only in the last few decades has attention turned to uncovering the precise mechanisms underlying motivated behavior in humans. Traditionally, studies on human motivation have been qualitative, or relied on subjective self-report or questionnaire-based measures (Table 1). The limitation of a questionnaire-based approach is that it is necessarily limited in its ability

Table 1 Questionnaires in Common Use to Measure Motivation in Healthy Individuals and Patients with Disorders of Diminished Motivation (eg, Apathy)

Healthy Individuals[a]	
Academic Amotivation Inventory	Legault et al. (2006)
Academic Motivation Scale	Vallerand et al. (1992)
Intrinsic Motivation Inventory	Choi et al. (2009) and Ryan (1982)
Sports Motivation Scale	Pelletier et al. (1995)
Patients[b]	
Apathy Evaluation Scale	Marin et al. (1991)
Apathy Inventory	Robert et al. (2002)
Apathy Scale	Starkstein et al. (1992) and Starkstein et al. (2001)
Behavioral Assessment of Dysexecutive Syndrome	Norris and Tate (2000)
Brief Psychiatric Rating Scale	Overall and Gorham (1962)
Dementia Apathy Interview and Rating	Strauss and Sperry (2002)
Dimensional Apathy Scale	Radakovic and Abrahams (2014)
Frontal Systems Behavior Scale	Grace and Malloy (2001)
Irritability Apathy Scale	Burns et al. (1990)
Key Behavior Change Inventory	Belanger et al. (2002)
Lille Apathy Rating Scale	Sockeel et al. (2006)
Neuropsychiatric Inventory	Cummings et al. (1994)
Positive and Negative Syndrome Scale	Kay et al. (1987)
Scale for the Assessment of Negative Symptoms	Andreasen (1984)

[a]Questionnaires validated for healthy individuals do not contain defined cut-offs for lack of motivation (eg, Pelletier et al., 1995; Vallerand et al., 1992).
[b]Patient questionnaires either focus entirely on apathy, or include questions on apathy as one or more items within their inventory.

to provide a mechanistic account of the processes underlying motivated behavior. Curiously, the questionnaires that are in use today have either been validated for use in the healthy population, or in patients (see Weiser and Garibaldi, 2015, for an extensive review), but few are in common use to measure motivation in both populations. This is likely to reflect historical trends, as current evidence suggests that motivation in health and disease is likely to be on a continuum (Chong and Husain, 2016).

The importance of being able to objectively characterize the cost–benefit processes that underlie motivated behavior is especially important in the clinical domain. Disorders of motivation, such as apathy, are common in several neurological and psychiatric disorders, such as Parkinson's disease (PD), stroke, depression, and schizophrenia. However, apathy is often under-recognized and under-treated, with one of the reasons being that we lack of a sensitive means to classify these disorders, and track their response to treatment. Questionnaires rely on patients having sufficient insight to respond to the questions that are posed, which is often not the case (de Medeiros et al., 2010; Njomboro and Deb, 2012; Starkstein et al., 2001). Although several questionnaires attempt to take this into account by providing alternative versions based on information provided by a caregiver, some other informant, or the clinician, responses to these multiple versions often only marginally concur (Chase, 2011).

Ultimately, therefore, there is a significant need to develop more objective methods to better characterize the mechanisms underlying human motivation, in both health and disease. Here, we discuss the utility of translating effort-based decision-making paradigms from the literature on nonhuman animals to index human motivation. For this reason, we do not consider emotional motivation, but focus on studies of effort operationalized in the physical and cognitive domains. This review primarily aims to summarize the potential and the limitations of the numerous methodologies that have been reported; a more detailed discussion of the underlying neurobiology of motivation is presented separately (Chong and Husain, 2016).

2 MOTIVATION AS EFFORT FOR REWARD

Recently, there has been a surge of interest in developing a mechanistic account of the neural and computational processes underlying motivated behavior in human health and disease. The vast majority of studies on the neurobiology of decision-making have inferred an animal's motivation by observing its response to rewarding outcomes. For example, a large corpus of studies has examined the effect of varying the delay—*temporal discounting*—or uncertainty of an outcome—*risk aversion* and *probability discounting* (Cardinal, 2006). In the language of more contemporary behavioral studies of motivation, animals must compute the perceived value (or "utility") of the motivational stimulus vs the costs (such as delay or uncertainty) involved in obtaining it (Salamone and Correa, 2012). Motivation has therefore been conceptualized in neuroeconomic terms as a cost–benefit trade-off, in which the animal seeks to maximize utility while minimizing the associated cost.

Effort Is Costly: In the last 5 years, particular interest has focused on another important component of motivation—namely, the amount of effort that an animal must be prepared to invest for a given reward. Effort, like delay and uncertainty, is usually perceived as a cost. It is particularly salient and aversive—so much so that a consistent finding across species is that animals will seek to minimize the amount of effort that they exert in pursuit of a given reward (Hull, 1943). Consequently, effort has the effect of *devaluing* the reward associated with it, such that the greater amount of effort that is required, the less the subjective value of the reward to the individual. This phenomenon is known as, "effort discounting."

This recent interest in human effort-related processes is grounded in a rich and substantial history of similar research in nonhuman animals, led predominantly by the pioneering work of John Salamone and his colleagues (Salamone and Correa,

2012; Salamone et al., 2006, 2007). These approaches have been extremely useful in capturing individual differences in animals, and providing an insight into the neural activity that underlies the trade-off between effort and reward. The many effort-based decision-making paradigms that have been developed in animals therefore offer a solid foundation on which to construct models of motivated behavior and motivational dysfunction in humans.

Effort-Based Decision-Making Is Useful to Capture Individual Differences: Motivation has been conceptualized as comprising two distinct phases. Both are usually driven by the presence of a target object that is typically a reward or highly valued reinforcer to the organism (eg, a preferred food). Usually, however, these rewards are not immediately available, and the organism must first overcome any distances or barriers between it and the target object (Pezzulo and Castelfranchi, 2009; Ryan and Deci, 2000; Salamone, 2010; Salamone and Correa, 2012). The first phase of motivated behavior therefore requires the organism to initiate behaviors that bring it in close proximity of the reward (the *approach phase*, also sometimes referred to as the preparatory/appetitive/seeking phase), before the reward can ultimately be consumed (the *consummatory phase*) (Craig, 1917; Markou et al., 2013).

The animal's behavior during the approach phase, therefore, represents the amount of effort that it is willing to exert in return for the reward on offer. It reflects behavior that is highly adaptive, as it enables the organism to exert effort to overcome the costs separating it from its rewards (Salamone and Correa, 2012). Importantly, however, although animals in general will seek to minimize effort, *individual animals* will differ in terms of the minimum amount of effort they are willing to invest for a given reward. Observing choice behavior during this approach phase of a decision-making task is therefore a particularly useful means to index the individual variability in motivation.

Effort Can Be Operationalized in Different Domains: One factor that influences the way in which effort interacts with reward to constrain choice behavior relates to the domain in which effort must be exerted (Fig. 1). Effort is often operationalized in terms of some form of physical requirement. In nonhuman animals, for example, it has been defined in terms of the height of a barrier to scale; the weight of a lever press; the number of handle turns; or the number of nose-pokes. Given that much of the research on effort-based decision-making has emerged from the animal literature, it is unsurprising that effort in human studies is also often defined physically—for example, as the number of button presses on a keyboard (Porat et al., 2014; Treadway et al., 2009), or the amount of force delivered to a hand-held dynamometer (Bonnelle et al., 2016; Chong, 2015; Chong et al., 2015; Cléry-Melin et al., 2011; Kurniawan et al., 2010; Prévost et al., 2010; Zénon et al., 2015).

However, effort can be perceived not only physically, but in the cognitive domain as well. Studies examining cognitive effort-based decisions in nonhuman animals are extremely rare, due to the associated challenges in training the animals to perform the task. One of the few attempts to do so was reported recently, and required rodents to identify in which one of five locations a target stimulus appeared, with cognitive effort being manipulated as the duration for which the target stimulus remained

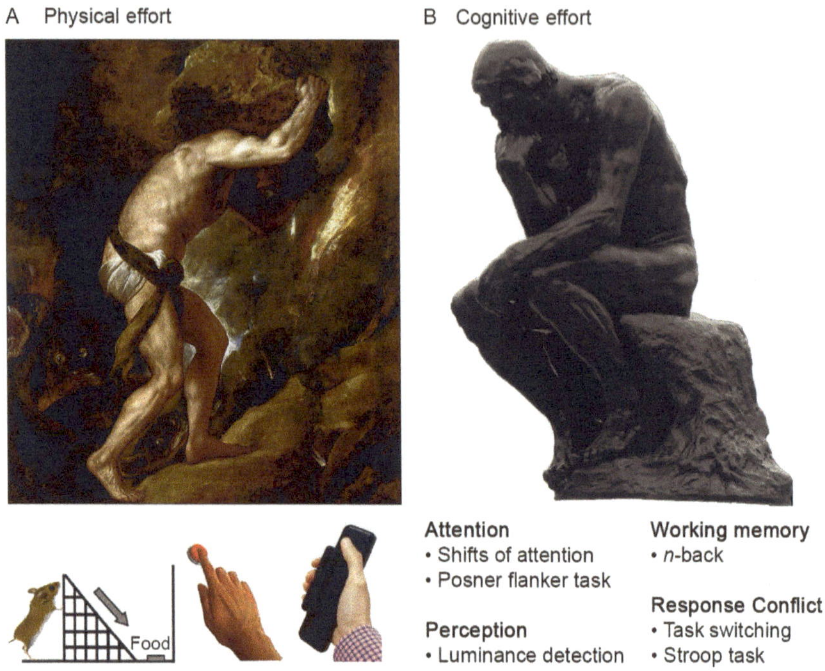

FIG. 1

Effort is typically operationalized in the physical and cognitive domains. (A) Physical effort has been manipulated in terms of the height or steepness of a barrier that an animal must overcome in pursuit of reward, or, in humans, as the number of button presses, or the amount of force applied to a hand-held dynamometer. (B) Cognitive effort in humans has been manipulated across several cognitive faculties. Note that many effortful tasks are aversive, not only because of the associated physical or cognitive demand, but also because of the greater amount of time it takes to complete the task, and the lower likelihood of completing it. For example, pushing a boulder up a mountain is aversive, not only because of the physical demand involved, but also because of the amount of time it would take, and the low probability of successfully accomplishing the task. In the case of Sisyphus, the effort involved in pushing the boulder up the mountain is considerable; the time it would take for him to do so and successfully maintain it at the peak is an eternity; and the probability of him completing the task is zero, thus infinitely reducing the subjective value of this course of action (and vindicating it as a suitable form of divine retribution). The distinction between effort, temporal, and probability discounting is discussed in Section 3.5.

Image credits: Left—*Titian, 1549,* Sisyphus, *Oil on canvas, 217 × 216 cm, Museo del Prado, Madrid.* Right—*Rodin, c1904,* Le Penseur, *Bronze, Musée Rodin, Paris.*

on (Hosking et al., 2014, 2015). In humans, there has been growing interest in the neural mechanisms that underlie cognitive effort-based decisions. Typically in these studies, cognitive load is manipulated in paradigms involving spatial attention (Apps et al., 2015), task switching (Kool et al., 2010; McGuire and Botvinick, 2010), conflict (eg, the Stroop effect (Schmidt et al., 2012)), working memory (eg, as an *n*-back task (Westbrook et al., 2013)), and perceptual effort tasks similar to those described previously (Reddy et al., 2015). These studies confirm that, like physical effort, cognitive demands carry an intrinsic effort cost (Dixon and Christoff, 2012; Kool et al., 2010; McGuire and Botvinick, 2010; Westbrook et al., 2013).

In summary, organisms must be sensitive to effort-related response costs, and make decisions based upon cost/benefit analyses. Today, we have a great deal of knowledge on the neural circuits that process information about the value of motivational stimuli, the value and selection of actions, and the regulation of cost/benefit decision-making processes that integrate this information to guide behavior (Croxson et al., 2009; Guitart-Masip et al., 2014; Kable and Glimcher, 2009; Phillips et al., 2007; Roesch et al., 2009). Much of this knowledge on the neurobiological determinants of decision-making has been gleaned from paradigms in nonhuman animals, involving operant procedures requiring responses on ratio schedules for preferred rewards, or dual-alternative tasks in the form of T-maze barrier procedures. In the following section, we survey the development of these different paradigms in effort-based decision-making in nonhuman animals, prior to considering their utility in human studies of motivated decision-making (Fig. 2).

3 EXPERIMENTAL APPROACHES TO EFFORT DISCOUNTING

3.1 FIXED AND PROGRESSIVE RATIO PARADIGMS

Operant conditioning paradigms are a commonly used approach to determining the willingness of an animal to work for reward (Fig. 2A) (Randall et al., 2012; Salamone et al., 1991, 2002; Schweimer and Hauber, 2005). Typically, the animal is first trained to perform an action in return for a reward (Hodos, 1961). In a *fixed ratio* (FR) study, a predefined number of operant responses are required to receive one unit of reinforcer (eg, five lever-presses for one unit of reward) (Salamone et al., 1991). In a *progressive ratio* (PR) paradigm, the number of operant responses required to obtain one unit of reward gradually increases over sequential trials—for example, in an exponential design, the number of nose-pokes required for the delivery of successive rewards might be 2, 4, 8, 16, 32, etc. (Beeler et al., 2012; Randall et al., 2012).

Relative to FR paradigms, PR paradigms have been found to generate greater response variability, which has been useful to study individual differences in behavior (Randall et al., 2012, 2014). By requiring the animal to repeatedly make choices between effort and reward under conditions in which the ratio requirement gradually increases, PR paradigms use the *break-point* as the key metric of motivation. The break-point is the last ratio that the animal is willing to complete for the reward on offer, and therefore represents the maximum amount of effort that it is willing to execute for that reward (Richardson and Roberts, 1996).

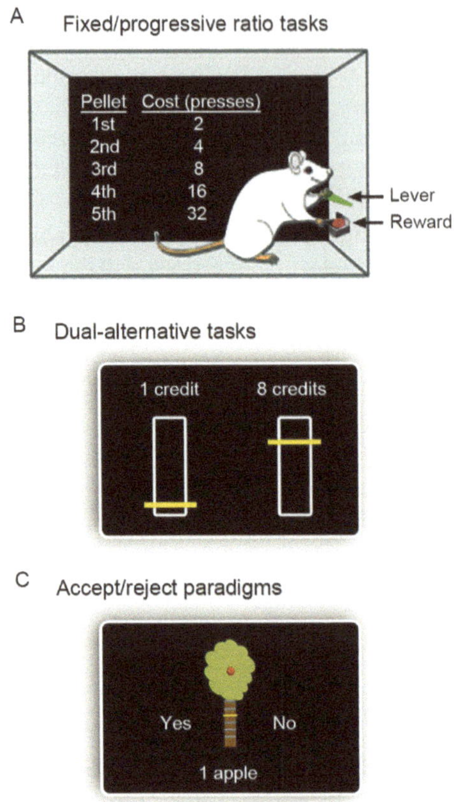

FIG. 2

Different approaches to effort-based decision-making. (A) In an operant paradigm, the subject decides how much effort to invest for a given reward. Illustrated is a progressive ratio paradigm. (B) In a dual-alternative paradigm, participants choose between two options—for example, a fixed baseline option vs a variable, more valuable, offer. In the example, participants choose whether they prefer to exert the lowest level of effort for 1 credit, or a higher level of effort for 8 credits. (C) In an accept/reject paradigm, participants are offered a single combination of effort and reward, and they decide to accept or reject the given offer. Here, participants choose whether they are willing to exert a high level of effort (indicated by the *yellow* bar) for the given reward (1 apple).

Panel B: After Apps, M., Grima, L., Manohar, S., Husain, M., 2015. The role of cognitive effort in subjective reward devaluation and risky decision-making. Sci. Rep. 5, 16880. Panel C: Adapted from Chong, T.T.-J., Bonnelle, V., Manohar, S., Veromann, K.-R., Muhammed, K., Tofaris, G., Hu, M., Husain, M., 2015. Dopamine enhances willingness to exert effort for reward in Parkinson's disease. Cortex 69, 40–46.

PR paradigms have been used for decades, primarily to study the reinforcing effects of psychostimulants and drug-seeking behavior in rodents (Richardson and Roberts, 1996; Stoops, 2008). More recently, several groups have used these tasks in humans to index motivation. For example, studies in children have used lever-press responses in return for monetary rewards, and found that break-points vary as a function of age and gender (Chelonis et al., 2011a). Similar investigations have shown that break-points can be increased following administration of psychostimulants such as methylphenidate, which increase levels of monoamines including dopamine (Chelonis et al., 2011b). In contrast, acute phenylalanine/tyrosine depletion, which reduces dopamine levels, has the effect of lowering break-points (Venugopalan et al., 2011). Such reports link parsimoniously with the literature in animals, by showing the importance of dopamine in increasing the motivation to work for reward (Chong and Husain, 2016).

In attempting to understand the mechanisms of motivated decision-making, it is particularly important to disentangle *choices* from the associated instrumental responses. A limitation of PR paradigms is that they are unable to do so unambiguously. Specifically, the break-points determined in a PR paradigm represent both the amount of effort that an animal is willing to invest for a particular reward, as well as the amount of effort that it is physically capable of performing for that reward. Thus, they are a function, not only of the animal's preferences, but also motor parameters that may be secondarily and nonspecifically affected by the experimental manipulation. This may be particularly important in the case of dopaminergic manipulations, as dopamine is known to augment the vigor with which physical responses are made (Niv et al., 2007), and the task would therefore be unable to disentangle the effect of dopamine on motivation vs its motor effects. In sum, a potential difficulty with operant conditioning paradigms in motivation research is that a lower break-point can be viewed as either a reduced willingness to expend effort, or due to a reduction in motor activity.

3.2 DUAL-ALTERNATIVE DESIGNS IN NONHUMAN ANIMALS

One paradigm that has been used to examine effort-based choices involves providing animals with a choice between a highly valued reinforcer (eg, a greater amount of food or a preferred food such as Bioserve pellets) and a less-valued reinforcer (eg, a smaller amount of food or lab chow) that is concurrently available. The key manipulation is that the rodent is required to exert a particular amount of effort (eg, climbing a barrier) to obtain the more valued reward. At baseline, most rodents will be willing to exert a greater amount of effort in exchange for the more valuable reward (Salamone et al., 1991).

The classic design in rodents involves the animal having to make a choice between the two offers in a T-maze procedure (Cousins et al., 1996; Salamone et al., 1994; Walton et al., 2002). It is first trained to learn the locations of the less- and more highly valued reinforcer, which are placed in opposite arms of the T-maze. Then, after an experimental intervention (a lesion or pharmacological manipulation), a physical barrier is added to the high-reward arm, which the animal must now overcome to obtain the more lucrative offer. The rate at which the high-effort/high-reward offer is chosen can be taken as a proxy of the animal's motivation, and one can then compare differences in these rates as a function of the experimental manipulation.

An advantage of this paradigm over the PR paradigm is that here it is possible to separate choice (the progression of a rodent down one arm of the T-maze) from motor execution (climbing the barrier). However, it remains important to ensure that the animal's choices are not influenced by the probability that they will succeed in overcoming that barrier to reach the reward. In addition, one potential limitation of this design is that the reinforcement magnitude for each arm typically remains the same on each trial. Thus, as the rodents become satiated after repeated visits to the large-reward arm, choice behavior may be more variable during later trials, which may in turn reduce the sensitivity of the task to different manipulations (Denk et al., 2005).

To overcome this reservation, the paradigm subsequently evolved to vary the amount of reward on offer in what has been termed an *effort-discounting* paradigm (Bardgett et al., 2009; Floresco et al., 2008). In this version, after a rodent chooses a high-reward option, the total reward available on that arm is reduced by one unit prior to the subsequent trial. By repeating this procedure until the rodent chooses the small-reward arm, it is possible to derive the *indifference points* between two choices to calculate sensitivities to different costs and reward amounts (Richards et al., 1997). This may be a more sensitive approach to determining the neurobiological substrates of effort-based decision-making (Green et al., 2004; Richards et al., 1997).

Over the last 35 years, these dual-alternative tasks have been of great utility in identifying the distributed circuit that regulates motivated decision-making in rodents. By systematically inactivating or lesioning specific components of the putative reward network, T-maze procedures have revealed that dopamine depletion in the nucleus accumbens biases rats toward the low-effort/low-reward option (Cousins et al., 1996; Salamone et al., 1994). Using similar procedures, lesions of the rodent medial prefrontal cortex, including the anterior cingulate cortex, led to fewer effortful choices, in contrast to lesions of the prelimbic/infralimbic and orbitofrontal cortices, which did not (Rudebeck et al., 2006; Walton et al., 2002, 2003). A final important example of the utility of the T-maze procedure is that bilateral inactivation of the basolateral amygdala, or unilateral inactivation of the basolateral amygdala concurrent with inactivation of the contralateral anterior cingulate cortex, decreases effortful behavior driven by food reward (Floresco and Ghods-Sharifi, 2007).

In summary, much of the knowledge that we have now of the neural regions responsible for effort-based decision-making has been based on applying these simple effort-discounting paradigms (Font et al., 2008; Ghods-Sharifi and Floresco, 2010; Hauber and Sommer, 2009; Mingote et al., 2008; Nunes et al., 2013a,b; Salamone and Correa, 2012; Salamone et al., 2007).

3.3 DUAL-ALTERNATIVE DESIGNS IN HUMANS

Given the utility of dual-alternative paradigms in animals, several tasks have been designed to translate these effort-discounting paradigms to humans (Fig. 2B). One example of a task that was inspired by the T-maze procedures in rodents is the effort expenditure for rewards task (Treadway et al., 2009; Wardle et al., 2011). In this task, effort is operationalized as the number of button presses delivered in a fixed period of time. The high-effort condition typically requires 100 button presses using the non-dominant fifth digit within 21 s, whereas the low-effort condition requires 30 button presses using the dominant index finger within 7 s. The reward for successfully completing the low-effort task was fixed at $1.00, but that for the high-effort task was varied between $1.24 and $4.30. This experiment also included a probabilistic component to the reward outcome, such that successful completion of each trial was rewarded with either high (88%), medium (50%), or low (12%) probability, and participants were informed of this prior to the beginning of the trial.

The most straightforward approach to analysing such data is to define motivation as the proportion of trials in which participants opt for the high-effort/high-reward option relative to the low-effort/low-reward option. This simple *ratio measure* has been used to characterize effort-based decision-making in several patient populations, including depression (Treadway et al., 2012a), schizophrenia (Barch et al., 2014), and autism (Damiano et al., 2012). For example, patients with major depressive disorder are typically less willing to choose the high-effort/high-reward option than healthy controls (Treadway et al., 2012a), as are patients with schizophrenia with a high degree of negative symptoms (Gold et al., 2013). In contrast, patients with autism spectrum disorder were more willing to expend effort than controls, regardless of the reward contingencies (Damiano et al., 2012).

In addition to ratio analyses, data from dual-alternative paradigms can also be subject to *computational modeling approaches*, to quantify effort discounting within individual subjects. For example, a recent study aimed to model effort discounting in a physical effort task (Klein-Flügge et al., 2015). Participants were required to exert sustained contractions on a hand-held dynamometer for a fixed duration of time, and at varying levels of force. The levels of force for each subject were independently calibrated to their maximal voluntary contraction (MVC). They were then required to choose between a low-effort/low-reward option and a high-effort/high-reward offer, with the magnitude of the effort and reward varied from trial to trial.

The authors then fitted several models of effort discounting—including linear, quadratic, hyperbolic, and sigmoidal functions—which differ in their predictions of how effort should subjectively devalue the reward on offer (Fig. 3). For example, linear models would predict constant discounting of value with increasing effort, such that an additional fixed cost devalues reward by the same amount. These linear models have been suggested in the context of effort-based choice behavior when persistent effort has to be made over time (eg, repeated lever presses). In contrast, concave models (eg, parabolic) would predict that changes in effort at higher levels would have greater impact on subjective value than changes at lower levels, and

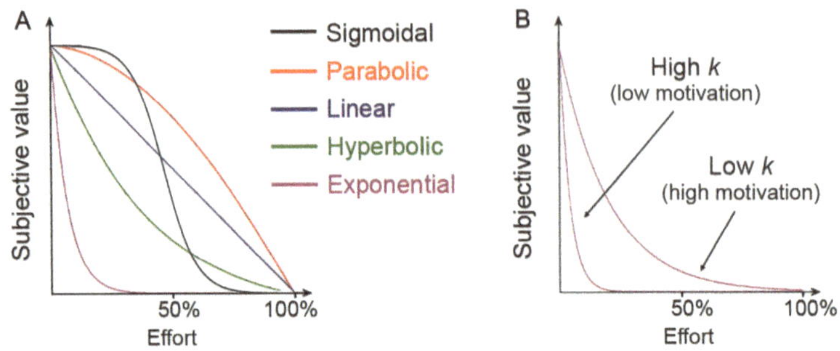

FIG. 3

Effort-discounting functions are useful to quantify individual differences in motivated decision-making. (A) Classes of function that have been used to computationally model effort-discounting behavior. These functions differ in their predictions of how effort should subjectively devalue the reward on offer. (B) An example of the utility of modeling effort discounting to capture individual differences. Two hypothetical participants are illustrated here in the context of a task in which effort discounting is exponential. The less motivated individual has a steeper discounting function, as indexed by a higher discounting parameter (k). These parameters can then be used to compare individual differences in motivation.

convex models (eg, hyperbolic) would predict the opposite. With Bayesian model comparisons, the authors found that a sigmoidal model, incorporating characteristics of both the concave and convex functions, appeared to best describe effort-discounting behavior.

By fitting sigmoidal functions to individual participants, it was possible to derive unique, subject-specific parameters that describe each individual's effort discounting. In this specific instance, the parameters fitted included the steepness of the curve and the turning point of the sigmoid. Although deriving these parameters was not the principal aim of this study (which was to compare effort and temporal discounting), the approach demonstrates the potential utility of deriving specific parameters which may then be used to index individuals' motivation, and to follow it over the course of a disease or of treatment.

A third approach to quantify effort-based decisions in individuals is to use *staircase paradigms* in order to derive subject-specific effort indifference points (Klein-Flügge et al., 2015; Westbrook et al., 2013). This approach typically involves holding the value of the low-effort/low-reward option constant, while titrating the high-effort/high-reward option incrementally as a function of participants' responses. Thus, if the high-effort/high-reward offer is rejected, then participants on a subsequent trial will be presented with an offer that has an incrementally lower effort requirement or higher reward value. Repeating this procedure then leads to a point at which participants are indifferent between the baseline option and each of the higher effort levels. These indifferent point values can thus be used as an objective metric to characterize how costly individuals perceive increasing amounts of effort, in an identical manner to that described for the apple-gathering task described next (Chong et al., 2015).

3.4 ACCEPT/REJECT TASKS IN HUMANS

Another approach inspired by effort-discounting paradigms in animals has been to present participants with a single combination of effort and reward on individual trials and have them decide whether to accept or reject each of the combinations on offer (Fig. 2C) (Bonnelle et al., 2015, 2016; Chong et al., 2015). A potential advantage of this approach, relative to the dual-alternative designs predominantly used in animals, is that it involves simpler displays, which may be more suitable to testing patient populations who might have impaired information processing (Bonnelle et al., 2015).

Here, we provide an illustrative example of an effort-based decision-making task we recently developed, which demonstrates the utility of such paradigms to index human motivation (Bonnelle et al., 2015, 2016; Chong et al., 2015). In this task, participants were presented with cartoons of apple trees and were instructed to accumulate as many apples as possible based on the combinations of stake and effort that were presented (Fig. 4A). Effort was operationalized as the amount of force delivered to a pair of hand-held dynamometers and was indexed to each participant's MVC, as determined at the beginning of each experiment. By referencing the effort levels to each individual's maximum force, we were able to normalize the difficulty of each level across individuals.

Potential rewards were indicated by the number of apples on the tree, while the associated effort was indicated by the height of a yellow bar positioned on the tree trunk, and ranged over six levels as a function of each participant's MVC. On each trial, participants decided whether they were willing to exert the specified level of effort for the specified stake. If they judged the particular combination of stake and effort to be "not worth it," they selected the "No" response and the next trial would commence. If, however, they decided to engage in that trial, they selected the "Yes" option and began squeezing the dynamometer in order to receive the apples on offer.

Dissecting the Components of Motivation: One of the advantages of this paradigm is that it is possible to separate different components of motivated behavior. Specifically, by parametrically manipulating effort and reward in an accept/reject context, this task was able to differentially examine the effect of effort and reward on individuals' choices (Bonnelle et al., 2015). In one set of analyses, we applied logistic regression techniques to derive the effort indifference points for each participant—that is, the effort level at which each reward was accepted and rejected on 50% of occasions (Bonnelle et al., 2015; Chong et al., 2015). The converse analysis was undertaken to determine reward indifference points as a function of effort level.

The power of this approach is that it achieves a quantifiable point of equivalence between increasing amounts of effort and reward. This allowed us then to examine reward and effort indifference points separately, and use these points to define a preference function for each subject, characterized by a subject-specific slope and intercept. We found that apathy ratings were correlated with the intercept of individuals'

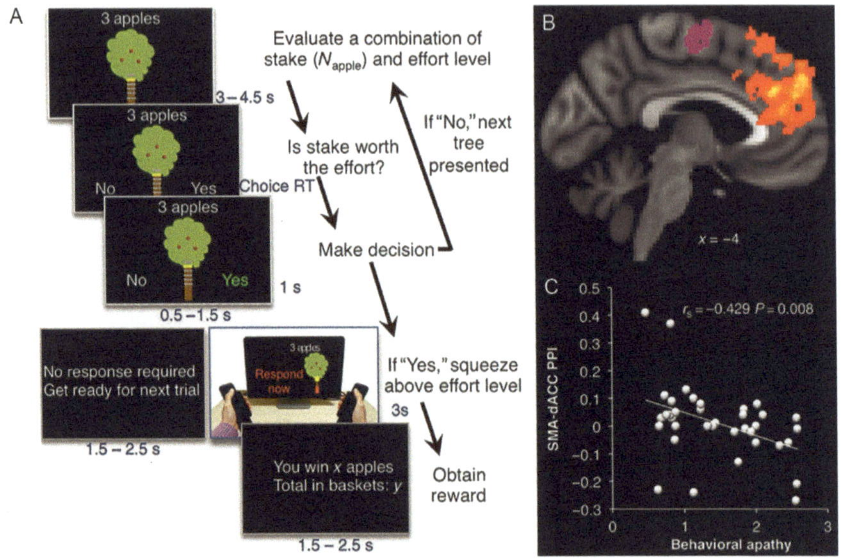

FIG. 4

(A) In the apple-gathering task, each trial started with an apple tree showing the stake (number of apples) and effort level required to win a fraction of this stake (trunk height) (Bonnelle et al., 2016). Rewards were indicated by the number of apples in the tree and effort was indicated by the height of a *yellow* bar on the tree trunk. Effort was operationalized as the amount of force to be delivered to hand-held dynamometers as a function of each individual's maximum voluntary contraction (MVC). Participants made an accept/reject decision as to whether to engage in an effortful response for the apples on offer. To control for fatigue, the accept option was followed by a screen indicating that no response was required on 50% of trials. (B) Relation between the supplementary motor area (SMA) functional connectivity and apathy traits. *Yellow–orange* voxels depict regions in which activity during the decision period on "accept" trials was more strongly correlated with activity in the SMA (*purple*) in more motivated individuals. (C) Correlation between behavioral apathy scores and the strength of the correlation (or functional connectivity) between the SMA and the dorsal anterior cingulate cortex.

Adapted from Bonnelle, V., Manohar, S., Behrens, T., Husain M., 2016. Individual differences in premotor brain systems underlie behavioral apathy. Cereb. Cortex 26 (2), 2016, 807–819.

effort indifference lines, which was a measure of the spontaneous level of effort that individuals were willing to engage for the smallest possible reward. In contrast, there was no relationship between apathy scores and the slope of the effort indifference line, which represented how much reward influenced the subjective cost associated with effort. These results demonstrate how a task can explain apathetic traits more sensitively than questionnaire-based measures and may be utilized to examine impairments in motivation in patient populations (Bonnelle et al., 2015).

Characterizing the Neural Substrates of Motivation: This paradigm has also been applied to determine the neural correlates of lowered motivation (apathy) in healthy individuals (Bonnelle et al., 2016). Using functional magnetic resonance imaging (fMRI), individuals who had higher subjective apathy ratings were found to be more sensitive to physical effort and had greater activity in areas associated with effort discounting, such as the nucleus accumbens. Interestingly, however, lower motivation was associated with increased activity in areas involved in action anticipation, such as the supplementary motor area (SMA) and cingulate motor zones. Furthermore, these less motivated individuals had decreased structural and functional connectivity between the SMA and anterior cingulate cortex (Fig. 4B). This led to the hypothesis that decreased structural integrity of the anterior cingulum might be associated with suboptimal communication between key nodes involved in action energization and preparation, leading to increased physiological cost, and increased effort sensitivity, to initiate action. This speculation remains to be confirmed, but serves to illustrate the utility of applying effort-based paradigms to capture the range of interindividual differences in motivation, even within healthy individuals, and to reveal their functional and structural markers.

Detecting Subclinical Deficits in Motivation: In addition to characterizing motivation in healthy individuals, a further useful role for effort-based paradigms is in detecting subclinical deficits in motivation within patient populations. Disorders of diminished motivation are currently diagnosed based on questionnaire-based measures of motivation, which may be insufficiently sensitive to detect more subtle motivational deficits. Using the apple-gathering task, we were able to show that patients with PD, regardless of their medication status, were willing to invest less effort for low rewards, as revealed by their lower effort indifference points (Fig. 5) (Chong et al., 2015). Importantly, none of these patients were clinically apathetic as assessed with the Lille Apathy Rating Scale (LARS), suggesting that deficits in motivation may nevertheless be present in individuals who are not clinically apathetic, but that these deficits are detectable with a sufficiently sensitive measure. Thus, the utility of these paradigms is being able to quantify components of effort-based decisions that may lead to earlier diagnosis and institution of therapy than would be otherwise possible with conventional self-report-based questionnaires. Furthermore, given the potential sensitivity of these techniques, they may offer us a more objective means of diagnosis and monitoring responses to treatment (Chong and Husain, 2016).

Distinguishing Apathy from Related Symptoms: Although it is conventionally established that apathy is separate from depression (Kirsch-Darrow et al., 2006; Levy et al., 1998; Starkstein et al., 2009), it is clear that these two disorders share several overlapping features, which may sometimes be difficult to distinguish. The utility of effort-based decision-making paradigms is in their potential to dissociate the two. For example, in the apple-gathering task, there was no relationship between effort indifference point measures and responses on a depression scale (the depression, anxiety, and stress scale, DASS) (Chong et al., 2015). This is similar to other studies that have shown that effort discounting is strongly correlated with apathy, but not with related symptoms such as diminished expression in

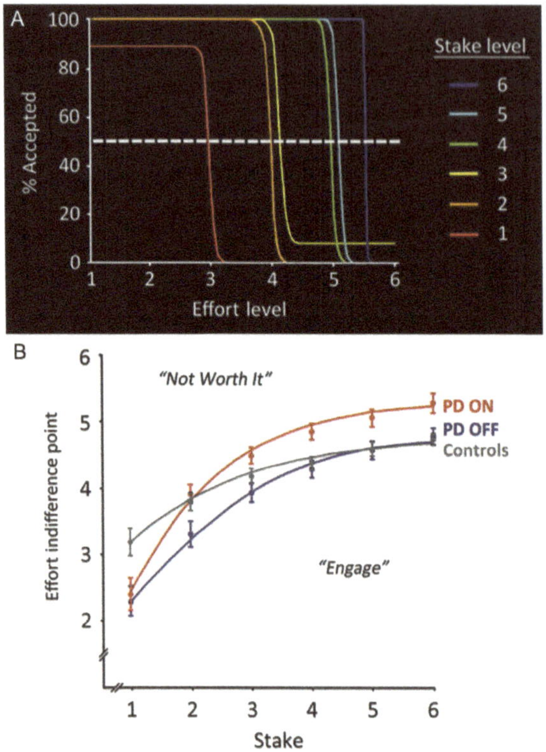

FIG. 5

We recently applied the apple-gathering task to patients with Parkinson's disease (Chong et al., 2015). (A) An example of the fitted probability functions for a representative participant. Logistic functions were used to plot the probability of engaging in a trial as a function of the effort level for each of the six stakes. Each participant's effort indifference points—the effort level at which the probability of engaging in a trial for a given stake is 50% (indicated by the *dashed line*)—were then computed. (B) Effort indifference points were then plotted as a function of stake for patients and controls. Regardless of medication status, patients had significantly lower effort indifference points than controls for the lowest reward. However, for high rewards, effort indifference points were significantly higher for patients when they were ON medication, relative not only to when they were OFF medication, but even compared to healthy controls. *Error bars* indicate ± 1 SEM.

Adapted from Chong, T.T.-J., Bonnelle, V., Manohar, S., Veromann, K.-R., Muhammed, K., Tofaris, G., Hu, M., Husain, M., 2015. Dopamine enhances willingness to exert effort for reward in Parkinson's disease. Cortex 69, 40–46.

schizophrenia (Hartmann et al., 2015). Effort-based tasks may therefore offer an objective means to quantifiably distinguish apathy from other symptoms of neurologic and psychiatric disease, which bear some surface resemblance to apathy, but which may have potentially different underlying mechanisms.

3.5 THE CHALLENGES OF EFFORT-DISCOUNTING TASKS

The preceding discussion highlights the range of effort-discounting paradigms that have been applied, using different methodologies and different methods of analysis. A challenge in isolating effort as a unique cost is that it is often associated with other costs, such as risk or temporal delay. In designing and applying effort-based paradigms, it is critical to consider and account for other factors that might impact on individuals' decision-making. To illustrate the measures that we have taken to control for these other costs, here we consider a cognitive effort task that we recently applied to measure motivation in healthy individuals (Apps et al., 2015).

In this cognitive effort study (Fig. 6), we manipulated effort as the number of switches of attention from one spatial location to another. We used an rapid serial visual presentation (RSVP) paradigm, in which participants had to attend to one of two peripheral target streams, to the left and right of fixation, for a target number "7." Each of these peripheral target streams was surrounded by three, task-irrelevant, distractor streams. Simultaneously, they had to fixate on a central stream of characters for a number "3," which was a cue to switch their attention to the opposite stream. We operationalized effort as the number of times attention had to be switched from one stream to the other (1–6), and verified that this corresponded to subjective increases in perceived cognitive effort.

Each experimental session commenced with an extensive training session, in which participants became practiced at each of the six different effort levels. After the training phase, participants undertook the critical choice phase, which required them to choose between a fixed, low-effort/low-reward baseline option, and a variable, high-effort/high-reward offer. The baseline option involved performing the lowest level of effort (one attentional switch) for 1 credit, and the offer varied from 2 to 6 attentional switches for 2 to 10 credits. Participants were instructed that each credit would be converted to monetary reward at the conclusion of the experiment.

Controlling for Probability Discounting: Choice data showed that, as predicted, participants chose the higher effort option less frequently with increasing effort levels, which would be consistent with the considerable literature on effort discounting summarized previously. However, this raises a challenging issue in the effort-discounting literature, which is how to control for probability costs. A well-established finding in economics is that humans are risk-averse and prefer a certain outcome over one that is associated with a degree of risk (probability discounting). In the context of an effort-based decision-making paradigm, it is therefore important to ensure that individuals' aversion to the higher effort levels is not due to the relatively lower likelihood that they will be able to successfully perform them (see Fig. 1).

Indeed, on this cognitive effort task, we found that individuals' performance did decline as a function of effort. Critically, however, we took a methodological approach to minimize the effect of probability discounting as a potential factor in our results. During the preliminary training phase, participants were rewarded a credit for every trial performed adequately. We set the requirements for a successful (rewarded) trial at a level that every participant was *able to achieve* on almost every trial. Thus,

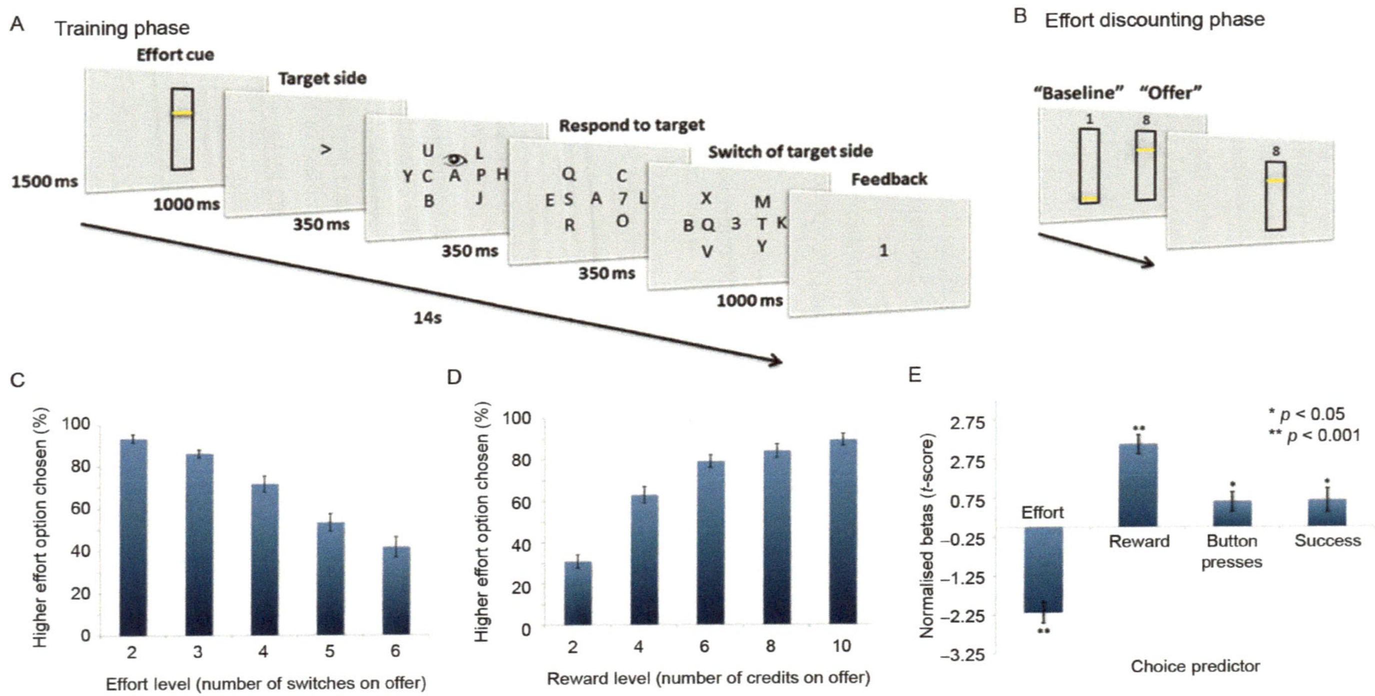

FIG. 6

See figure legend on opposite page.

even though performance declined with increasing effort, the rates at which participants were reinforced were very similar across effort levels. In a subsequent logistic regression analysis, we found that, even though the ability to complete a given effort level did influence individuals' preferences, effort was a significantly better predictor of choice behavior than success rates. These procedures therefore allowed us to minimize and account for the effect of probability discounting in a cognitive effort-discounting task.

Controlling for Temporal Discounting: Most effortful tasks take longer to complete than those that are less effortful (see Fig. 1). For example, a commonly employed procedure involves manipulating effort as the number of presses of a button or a lever (Treadway et al., 2009). An advantage of this procedure is that it draws from a rich tradition in research on nonhuman animals, and is simple to implement in the laboratory. However, although it is intuitive that a higher number of presses is more effortful, such a manipulation is also associated with a greater time cost. A very well-established finding in humans is that temporal delays are discounted hyperbolically, such that we tend to prefer smaller amounts sooner, rather than larger amounts later. Thus, another challenge in designing effort-based tasks is therefore to be able to ensure that any apparent effort discounting is not being driven by an element of temporal discounting.

FIG. 6

In a recent cognitive effort task, we manipulated cognitive effort as the number of shifts of attention in a rapid serial visual presentation task (Apps et al., 2015). (A) In a preliminary training phase, participants maintained central fixation as an array of letters changed rapidly and attend to a target stream presented horizontally to the left or right of a central stream, in order to detect targets (the number "7"). The initial target side was indicated at the beginning of the trial by an *arrow*. During each trial, a cue in the center of the screen (anumber "3") indicated that the target side was switching, requiring participants to make a peripheral shift of attention. Effort was manipulated as the number of attentional shifts, which varied from one to six. In the training session feedback was provided in the form of credits (1 credit or 0) at the end of each trial if participants successfully detected a sufficient number of targets. (B) Effort-discounting task. Choices were made between a fixed "baseline" and a variable "offer." The baseline was fixed at the lowest effort and reward (1 credit, 1 shift). The offer varied in terms of reward and effort (2, 4, 6, 8, 10 credits and 2, 3, 4, 5, 6 shifts). Choices on this task indexed the extent to which rewards were devalued by shifts of attention. (C) Results showed that shifts of attention were effortful and devalued rewards. As the number of attentional shifts increased, the less likely it was that the offer was chosen. (D) Similarly, as the amount of reward offered increased, the more likely it was that the offer was chosen. (E) Results of a logistic regression analysis, showing that effort was a significantly better predictor of choice than task success and the number of button presses for each effort level. The *y*-axis shows mean normalized betas for predictors of choosing the higher effort/higher reward offer.

Adapted from Apps, M., Grima, L., Manohar, S., Husain, M., 2015. The role of cognitive effort in subjective reward devaluation and risky decision-making. Sci. Rep. 5, 16880.

In the case of the cognitive effort task described earlier, controlling the temporal profile of each effort level was relatively straightforward. We set each trial to last a fixed duration of 14 s, and participants had to sustain their attention on the task for that entire period, with effort being manipulated simply as the number of spatial shifts of attention (Apps et al., 2015). This ensured that the temporal parameters of every trial at every effort level were identical. In the physical effort tasks that we have employed, we have attempted to overcome the issue of temporal discounting through the use of hand-held dynamometers (Bonnelle et al., 2015, 2016; Chong et al., 2015), which are an effective means to minimize the temporal difference between low- (eg, 40% MVC) and high-effort trials (eg, 80% MVC). This difference is further minimized by holding the actual duration of each trial constant.

The Effect of Fatigue on Effort Discounting: An important feature of effort as a cost is that it accumulates over time. Thus, with increasing time-on-task, individuals are likely to fatigue, which will have an obvious effect on their choice preferences later in the experiment. In all of the traditional tasks described in animals, the animal must actually execute their chosen course of action. Thus, it is possible that decisions in the later parts of the experiment might be affected by the accumulation of effort in the form of fatigue.

In humans, several approaches have been adopted to eliminate the effect of fatigue on participants' responses. The main approach has been to require participants to perform only a random subset of their revealed preferences. In the case of our cognitive effort task, these random trials were deferred until the conclusion of the experiment (Apps et al., 2015), whereas other tasks have required the choices to be executed immediately after the response is provided (Bonnelle et al., 2015, 2016; Klein-Flügge et al., 2015). In studies that have required participants to execute choices on every trial, it is important to verify that increasing failures to complete the high-effort trials cannot account for any preference shifts (eg, with regression techniques) (Treadway et al., 2012a).

Few studies have explicitly attempted to model the effect of fatigue on choice decision-making (Meyniel et al., 2012, 2014). More recently, however, fatigue has become the subject of increasing neuroscientific interest (Kurzban et al., 2013). For example, there have been recent attempts to computationally model a labor/leisure trade-off in describing when the brain decides to rest (Kool and Botvinick, 2014). A closer integration between the effects of fatigue on effort discounting should be an important focus of future studies.

4 FUTURE CHALLENGES AND APPLICATIONS

The preceding sections surveyed the different techniques that have been applied to quantify effort-based decision-making in human and nonhuman animals. Applying these techniques in humans has given us great insight into the mechanisms of effort-based motivation in healthy individuals and has provided us with an understanding of the neural circuitry involved in reward valuation and effort discounting.

Given the volume of research that will surely follow in the next few years, a challenge will be to parse the wealth of data from disparate paradigms across, and within, species. For example, the decision-making process in a dual-alternative design is necessarily different from that of an accept/reject design, which differs again from decision-making in a foraging context. Tasks also differ according to the degree to which they account for such factors as probability discounting, temporal discounting and fatigue, and reinforcement can occur with varying magnitudes and schedules. Furthermore, various domains of effort have been examined across the species—including perceptual, cognitive, and physical effort. Given this heterogeneity, perhaps it is all the more impressive that, despite the wide range of methodologies employed, most findings in studies of effort-based decisions have been relatively consistent—pointing, for example, to the importance of dopamine within the mesocorticolimbic system as being critical in overcoming effort for reward (Chong and Husain, 2016; Salamone and Correa, 2012).

However, future research will need to clarify the precise effect of varying task parameters on choice. For example, one distinction that is yet to be clarified is the difference in the way the brain processes costs associated with different types of effort (eg, cognitive vs physical). Phenomenologically, cognitive and physical effort are perceived as distinct entities. Furthermore, physical effort has the advantage of being relatively straightforward to manipulate in animals; being easily characterized objectively (eg, as force); and having demonstrable physiological and metabolic correlates. In contrast, cognitive effort is more difficult to conceptualize; cannot be defined in metabolic terms; and may be experienced differently depending on the cognitive faculty that is being loaded (attention, working memory, etc.).

This distinction between cognitive and physical effort processing is an example of a question that is not only relevant to understanding the basic neuroscience of motivation—of how the brain processes different effort costs—but also one that is clinically relevant. For example, at present there is a somewhat arbitrary distinction between constructs such as "mental" or "physical" apathy, which is intuitive, and based primarily on questionnaire data. This distinction suggests that the domains are separate, but the extent to which they rely on shared vs independent mechanisms has not been thoroughly investigated. Studies in animals suggest potentially dissociable neural substrates (Cocker et al., 2012; Hosking et al., 2014, 2015), but the neural correlates underlying the subjective valuation of cognitive and physical effort in humans remains to be defined (but see Schmidt et al., 2012).

The natural extension of the literature on effort-based decisions is its applications to diagnosing and monitoring disorders of diminished motivation in patients (Chong and Husain, 2016). Several authors have suggested that effort-based decision-making paradigms could be useful for modeling the motivational dysfunction seen in multiple neurological and psychiatric conditions (Salamone and Correa, 2012; Salamone et al., 2006, 2007; Yohn et al., 2015). Effort is a particularly salient variable in individuals with apathy who lack the ability to initiate simple day-to-day activities (Levy and Dubois, 2006; van Reekum et al., 2005). This lack of internally generated actions may stem from impaired incentive motivation: the ability to convert basic valuation of reward into action execution (Schmidt et al., 2008). Only relatively recently, however, have researchers started to apply effort-based decision-making paradigms to assess patients with clinical disorders of motivation.

Despite studies of effort-based decisions in patients being a relatively recent undertaking, several populations have already been tested. The broad conclusion from many of these studies is similar, with apathetic individuals being inclined to exert less effort for reward: patients with PD are willing to apply less force to a dynamometer for low rewards than age-matched controls (Chong et al., 2015; Porat et al., 2014); patients with major depression fail to modulate the amount of effort they exert in return for primary or secondary rewards (Cléry-Melin et al., 2011; Sherdell et al., 2012; Treadway et al., 2012a); patients with schizophrenia are less inclined to perform a perceptually, cognitively, or physically demanding task for monetary reward than controls (Reddy et al., 2015). Collectively, these studies show that deficits in effort-based decision-making are not unique to any one disease entity (Barch et al., 2014; Dantzer et al., 2012; Fervaha et al., 2013a,b; Gold et al., 2013; Treadway et al., 2012b).

On the one hand, this may be taken as evidence that apathy, as a common thread between these conditions, is associated with damage to a mesocorticolimbic system that generates internal association between action and its consequences. This would be consistent with preclinical studies, suggesting a key involvement of medial prefrontal areas and the pallido–striatal complex in the anticipation and execution of effortful actions. However, the question arises as to why different pathologies leading to different brain disorders give rise to the identical phenotype of reduced motivation to exert effort. Do the behavioral manifestations of higher effort indifference points or higher break-points in apathetic patients simply represent the same surface phenotype of some common underlying neural dysfunction? Or are there distinguishing features to the impairments of effort-based decisions within these populations that may be dissociable with sufficiently sensitive measures? A focus of future research will be to identify the specific components of effort-based decision-making that are affected in these populations (eg, the evaluation of the effort costs vs the costs of having to act).

Although the translation of effort-based tasks from animals to patients holds great promise, a practical challenge will be to precisely identify the parameters and paradigms which maximize the sensitivity and specificity of detecting any potential decision-making impairments in a population of interest. In deciding on an approach, it is worth acknowledging the advantages and limitations of the aforementioned paradigms, and their ability to capture the putative motivational deficit in the population of interest. For example, patients whose motivational deficits are more likely to be physical rather than cognitive would be more apt to be tested with a task involving effort in the former domain. However, due to the nascency of this field, extant data do now allow us to unequivocally advocate one approach over another in exploring specific motivational deficits in a given patient population. The difficulty of choosing an appropriate paradigm is exemplified by a recent study in patients with schizophrenia, who were administered several effort-based decision-making tasks in order to measure motivated behavior (Reddy et al., 2015). The tests were all essentially dual-alternative paradigms, but involved different forms of effort—namely, perceptual effort, task switching, grip force, and button presses. Although these tasks were useful in capturing some of the differences in motivation in patients with schizophrenia, they were each found to have different psychometric properties. Thus, prior to translating such effort-based paradigms for wide-spread clinical use, it remains for us to determine and standardize the parameters and constraints of these tasks to maximize the probability of detecting any motivational deficits.

In conclusion, the rich history of effort-based decision-making tasks in animals provides us with a large corpus of basic neuroscience data on which to draw. Through these paradigms, we have gained a deep understanding of the neural networks that are involved in encoding cost–benefit trade-offs. Extending these studies to humans therefore holds great potential in allowing us to understand the process of healthy motivation, and develop parsimonious models of motivation across species. A key advantage of these paradigms is their ability to sensitively capture individual differences. Furthermore, these tasks offer multiple metrics that may be more objective, sensitive, and specific to the identification of disorders of motivation than traditional self-report and questionnaire-based measures. The availability of such metrics should act as an incentive to develop new treatments, and to determine the efficacy of existing drugs. Ultimately, it is hoped that we may be able to combine different metrics of decision-making to devise a useful index of motivational impairments in disease, which will allow us to more accurately diagnose, monitor, and treat disorders of motivation.

ACKNOWLEDGMENTS

T.C. is funded by the National Health and Medical Research Council (NH & MRC) of Australia (1053226). M.H. is funded by a grant from the Wellcome Trust (098282).

REFERENCES

Andreasen, N., 1984. Scale for the Assessment of Negative Symptoms (SANS). College of Medicine, University of Iowa, Iowa City.

Apps, M., Grima, L., Manohar, S., Husain, M., 2015. The role of cognitive effort in subjective reward devaluation and risky decision-making. Sci. Rep. 5, 16880.

Barch, D.M., Treadway, M.T., Schoen, N., 2014. Effort, anhedonia, and function in schizophrenia: reduced effort allocation predicts amotivation and functional impairment. J. Abnorm. Psychol. 123, 387.

Bardgett, M., Depenbrock, M., Downs, N., Points, M., Green, L., 2009. Dopamine modulates effort-based decision-making in rats. Behav. Neurosci. 123, 242.

Beeler, J.A., McCutcheon, J.E., Cao, Z.F., Murakami, M., Alexander, E., Roitman, M.F., Zhuang, X., 2012. Taste uncoupled from nutrition fails to sustain the reinforcing properties of food. Eur. J. Neurosci. 36, 2533–2546.

Belanger, H.G., Brown, L.M., Crowell, T.A., Vanderploeg, R.D., Curtiss, G., 2002. The Key Behaviors Change Inventory and executive functioning in an elderly clinic sample. Clin. Neuropsychol. 16, 251–257.

Bentham, J., 1817. A table of the springs of action. R Hunter, London.

Bonnelle, V., Veromann, K.-R., Burnett Heyes, S., Sterzo, E., Manohar, S., Husain, M., 2015. Characterization of reward and effort mechanisms in apathy. J. Physiol. Paris 109, 16–26.

Bonnelle, V., Manohar, S., Behrens, T., Husain, M., 2016. Individual differences in premotor brain systems underlie behavioral apathy. Cereb. Cortex 26 (2), 807–819.

Burns, A., Folstein, S., Brandt, J., Folstein, M., 1990. Clinical assessment of irritability, aggression, and apathy in Huntington and Alzheimer disease. J. Nerv. Ment. Dis. 178, 20–26.

Cardinal, R.N., 2006. Neural systems implicated in delayed and probabilistic reinforcement. Neural Netw. 19, 1277–1301.

Chase, T., 2011. Apathy in neuropsychiatric disease: diagnosis, pathophysiology, and treatment. Neurotox. Res. 19, 266–278.

Chelonis, J.J., Gravelin, C.R., Paule, M.G., 2011a. Assessing motivation in children using a progressive ratio task. Behav. Processes 87, 203–209.

Chelonis, J.J., Johnson, T.A., Ferguson, S.A., Berry, K.J., Kubacak, B., Edwards, M.C., Paule, M.G., 2011b. Effect of methylphenidate on motivation in children with attention-deficit/hyperactivity disorder. Exp. Clin. Psychopharmacol. 19, 145–153.

Choi, J., Mogami, T., Medalia, A., 2009. Intrinsic motivation inventory: an adapted measure for schizophrenia research. Schizophr. Bull. 36, 966–976.

Chong, T.T.-J., 2015. Disrupting the perception of effort with continuous theta burst stimulation. J. Neurosci. 35, 13269–13271.

Chong, T.T.-J., Husain, M., 2016. Chapter 17—The role of dopamine in the pathophysiology and treatment of apathy. In: Studer, B., Knecht, S. (Eds.), Progress in Brain Research, vol. 229, Elsevier, Amsterdam, pp. 389–426.

Chong, T.T.-J., Bonnelle, V., Manohar, S., Veromann, K.-R., Muhammed, K., Tofaris, G., Hu, M., Husain, M., 2015. Dopamine enhances willingness to exert effort for reward in Parkinson's disease. Cortex 69, 40–46.

Cléry-Melin, M.L., Schmidt, L., Lafargue, G., Baup, N., Fossati, P., Pessiglione, M., 2011. Why don't you try harder? An investigation of effort production in major depression. PLoS One 6, e23178.

Cocker, P.J., Hosking, J.G., Benoit, J., Winstanley, C.A., 2012. Sensitivity to cognitive effort mediates psychostimulant effects on a novel rodent cost/benefit decision-making task. Neuropsychopharmacology 37, 1825–1837.

Cousins, M.S., Atherton, A., Turner, L., Salamone, J.D., 1996. Nucleus accumbens dopamine depletions alter relative response allocation in a T-maze cost/benefit task. Behav. Brain Res. 74, 189–197.

Craig, W., 1917. Appetites and aversions as constituents of instincts. Proc. Natl. Acad. Sci. U.S.A. 3, 685–688.

Croxson, P., Walton, M., O'Reilly, J., Behrens, T., Rushworth, M., 2009. Effort-based cost-benefit valuation and the human brain. J. Neurosci. 29, 4531–4541.

Cummings, J.L., Mega, M., Gray, K., Rosenberg-Thompson, S., Carusi, D.A., Gornbein, J., 1994. The Neuropsychiatric Inventory comprehensive assessment of psychopathology in dementia. Neurology 44, 2308–2314.

Damiano, C.R., Aloi, J., Treadway, M., Bodfish, J.W., Dichter, G.S., 2012. Adults with autism spectrum disorders exhibit decreased sensitivity to reward parameters when making effort-based decisions. J. Neurodev. Disord. 4, 13.

Dantzer, R., Meagher, M.W., Cleeland, C.S., 2012. Translational approaches to treatment-induced symptoms in cancer patients. Nat. Rev. Clin. Oncol. 9, 414–426.

Darwin, C., 1859. On the origin of species by means of natural selection, or the preservation of favoured races in the struggle for life. John Murray, London.

de Medeiros, K., Robert, P., Gauthier, S., Stella, F., Politis, A., Leoutsakos, J., Taragano, F., Kremer, J., Brugnolo, A., Porsteinsson, A.P., Geda, Y.E., 2010. The Neuropsychiatric Inventory-Clinician rating scale (NPI-C): reliability and validity of a revised assessment of neuropsychiatric symptoms in dementia. Int. Psychogeriatr. 22, 984–994.

Denk, F., Walton, M.E., Jennings, K.A., Sharp, T., Rushworth, M.F., Bannerman, D.M., 2005. Differential involvement of serotonin and dopamine systems in cost-benefit decisions about delay or effort. Psychopharmacology (Berl.) 179, 587–596.

Dixon, M.L., Christoff, K., 2012. The decision to engage cognitive control is driven by expected reward-value: neural and behavioral evidence. PLoS One 7, e51637.

Fervaha, G., Foussias, G., Agid, O., Remington, G., 2013a. Neural substrates underlying effort computation in schizophrenia. Neurosci. Biobehav. Rev. 37, 2649–2665.

Fervaha, G., Graff-Guerrero, A., Zakzanis, K.K., Foussias, G., Agid, O., Remington, G., 2013b. Incentive motivation deficits in schizophrenia reflect effort computation impairments during cost-benefit decision-making. J. Psychiatr. Res. 47, 1590–1596.

Floresco, S.B., Ghods-Sharifi, S., 2007. Amygdala-prefrontal cortical circuitry regulates effort-based decision making. Cereb. Cortex 17, 251–260.

Floresco, S.B., Tse, M.T.L., Ghods-Sharifi, S., 2008. Dopaminergic and glutamatergic regulation of effort- and delay-based decision making. Neuropsychopharmacology 33, 1966–1979.

Font, L., Mingote, S., Farrar, A.M., Pereira, M., Worden, L., Stopper, C., Port, R.G., Salamone, J.D., 2008. Intra-accumbens injections of the adenosine A2A agonist CGS 21680 affect effort-related choice behavior in rats. Psychopharmacology (Berl.) 199, 515–526.

Ghods-Sharifi, S., Floresco, S.B., 2010. Differential effects on effort discounting induced by inactivations of the nucleus accumbens core or shell. Behav. Neurosci. 124, 179–191.

Gold, J.M., Strauss, G.P., Waltz, J.A., Robinson, B.M., Brown, J.K., Frank, M.J., 2013. Negative symptoms of schizophrenia are associated with abnormal effort-cost computations. Biol. Psychiatry 74, 130–136.

Grace, J., Malloy, P.F., 2001. Frontal Systems Behavior Scale (FrSBe): Professional Manual. Psychological Assessment Resources, Lutz, Florida.

Green, L., Myerson, J., Holt, D.D., Slevin, J.R., Estle, S.J., 2004. Discounting of delayed food rewards in pigeons and rats: is there a magnitude effect? J. Exp. Anal. Behav. 81, 39–50.

Guitart-Masip, M., Duzel, E., Dolan, R., Dayan, P., 2014. Action versus valence in decision making. Trends Cogn. Sci. 18, 194–202.

Hartmann, M.N., Hager, O.M., Reimann, A.V., Chumbley, J.R., Kirschner, M., Seifritz, E., Tobler, P.N., Kaiser, S., 2015. Apathy but not diminished expression in schizophrenia is associated with discounting of monetary rewards by physical effort. Schizophr. Bull. 41, 503–512.

Hauber, W., Sommer, S., 2009. Prefrontostriatal circuitry regulates effort-related decision making. Cereb. Cortex 19, 2240–2247.

Hodos, W., 1961. Progressive ratio as a measure of reward strength. Science 134, 943–944.

Hosking, J., Cocker, P., Winstanley, C., 2014. Dissociable contributions of anterior cingulate cortex and basolateral amygdala on a rodent cost/benefit decision-making task of cognitive effort. Neuropsychopharmacology 39, 1558–1567.

Hosking, J., Floresco, S., Winstanley, C., 2015. Dopamine antagonism decreases willingness to expend physical, but not cognitive, effort: a comparison of two rodent cost/benefit decision-making tasks. Neuropsychopharmacology 40, 1005–1015.

Hull, C., 1943. Principles of Behavior: An Introduction to Behavior Theory. Appleton-Century, New York.

James, W., 1890. The Principles of Psychology. Henry Holt, Boston.

Kable, J.W., Glimcher, P.W., 2009. The neurobiology of decision: consensus and controversy. Neuron 63, 733–745.

Kay, S.R., Fiszbein, A., Opfer, L.A., 1987. The positive and negative syndrome scale (PANSS) for schizophrenia. Schizophr. Bull. 13, 261–276.

Kirsch-Darrow, L., Fernandez, H.F., Marsiske, M., Okun, M.S., Bowers, D., 2006. Dissociating apathy and depression in Parkinson disease. Neurology 67, 33–38.

Klein-Flügge, M.C., Kennerley, S.W., Saraiva, A.C., Penny, W.D., Bestmann, S., 2015. Behavioral modeling of human choices reveals dissociable effects of physical effort and temporal delay on reward devaluation. PLoS Comput. Biol. 11, e1004116.

Kool, W., Botvinick, M.M., 2014. A labor/leisure tradeoff in cognitive control. J. Exp. Psychol. Gen. 143, 131–141.

Kool, W., McGuire, J.T., Rosen, Z.B., Botvinick, M.M., 2010. Decision making and the avoidance of cognitive demand. J. Exp. Psychol. Gen. 139, 665–682.

Kurniawan, I., Seymour, B., Talmi, D., Yoshida, W., Chater, N., Dolan, R., 2010. Choosing to make an effort: the role of striatum in signaling physical effort of a chosen action. J. Neurophysiol. 104, 313–321.

Kurzban, R., Duckworth, A., Kable, J., Myers, J., 2013. An opportunity cost model of subjective effort and task performance. Behav. Brain Sci. 36, 661–679.

Legault, L., Green-Demers, I., Pelletier, L.G., 2006. Why do high school students lack motivation in the classroom? Toward an understanding of academic amotivation and the role of social support. J. Educ. Psychol. 98, 567–582.

Levy, R., Dubois, B., 2006. Apathy and the functional anatomy of the prefrontal cortex-basal ganglia circuits. Cereb. Cortex 16, 916–928.

Levy, M.L., Cummings, J.L., Fairbanks, L.A., Masterman, D., Miller, B.L., Craig, A.H., Paulsen, J.S., Litvan, I., 1998. Apathy is not depression. J. Neuropsychiatry Clin. Neurosci. 10, 314–319.

Marin, R.S., Biedrzycki, R.C., Firinciogullari, S., 1991. Reliability and validity of the Apathy Evaluation Scale. Psychiatry Res. 38, 143–162.

Markou, A., Salamone, J., Bussey, T., Mar, A., Brunner, D., Gilmour, G., Balsam, P., 2013. Measuring reinforcement learning and motivation constructs in experimental animals: relevance to the negative symptoms of schizophrenia. Neurosci. Biobehav. Rev. 37, 2149–2165.

Maslow, A.H., 1943. A theory of human motivation. Psychol. Rev. 50, 370–396.

McDougall, W., 1908. An Introduction to Social Psychology. Methuen, London.

McGuire, J.T., Botvinick, M.M., 2010. Prefrontal cortex, cognitive control, and the registration of decision costs. Proc. Natl. Acad. Sci. U.S.A. 107, 7922–7926.

Meyniel, F., Sergent, C., Rigoux, L., Daunizeau, J., Pessiglione, M., 2012. Neurocomputational account of how the human brain decides when to have a break. Proc. Natl. Acad. Sci. U.S.A. 110, 2641–2646.

Meyniel, F., Safra, L., Pessiglione, M., 2014. How the brain decides when to work and when to rest: dissociation of implicit-reactive from explicit-predictive computational processes. PLoS Comput. Biol. 10, e1003584.

Mingote, S., Font, L., Farrar, A.M., Vontell, R., Worden, L.T., Stopper, C.M., Port, R.G., Sink, K.S., Bunce, J.G., Chrobak, J.J., Salamone, J.D., 2008. Nucleus accumbens adenosine A2A receptors regulate exertion of effort by acting on the ventral striatopallidal pathway. J. Neurosci. 28, 9037–9046.

Niv, Y., Daw, N., Joel, D., Dayan, P., 2007. Tonic dopamine: opportunity costs and the control of response vigor. Psychopharmacology (Berl.) 191, 507–520.

Njomboro, P., Deb, S., 2012. Poor dissociation of patient-evaluated apathy and depressive symptoms. Curr. Gerontol. Geriatr. Res. 2012, 1–8.

Norris, G., Tate, R.L., 2000. The Behavioural Assessment of the Dysexecutive Syndrome (BADS): ecological, concurrent and construct validity. Neuropsychol. Rehabil. 10, 33–45.

Nunes, E.J., Randall, P.A., Hart, E.E., Freeland, C., Yohn, S.E., Baqi, Y., Müller, C.E., López-Cruz, L., Correa, M., Salamone, J.D., 2013a. Effort-related motivational effects of the VMAT-2 inhibitor tetrabenazine: implications for animal models of the motivational symptoms of depression. J. Neurosci. 33, 19120–19130.

Nunes, E.J., Randall, P.A., Podurgiel, S., Correa, M., Salamone, J.D., 2013b. Nucleus accumbens neurotransmission and effort-related choice behavior in food motivation: effects of drugs acting on dopamine, adenosine, and muscarinic acetylcholine receptors. Neurosci. Biobehav. Rev. 37, 2015–2025.

Overall, J.E., Gorham, D.R., 1962. The brief psychiatric rating scale. Psychol. Rep. 10, 799–812.

Pelletier, L.G., Fortier, M.S., Vallerand, R.J., Tuson, K.M., Briere, N.M., Blais, M.R., 1995. Toward a new measure of intrinsic motivation, extrinsic motivation, and amotivation in sports: the sport motivation scale (SMS). J. Sport Exerc. Psychol. 17, 35–53.

Pezzulo, G., Castelfranchi, C., 2009. Intentional action: from anticipation to goal-directed behavior. Psychol. Res. 73, 437–440.

Phillips, P.E., Walton, M.E., Jhou, T.C., 2007. Calculating utility: preclinical evidence for cost-benefit analysis by mesolimbic dopamine. Psychopharmacology 191, 483–495.

Porat, O., Hassin-Baer, S., Cohen, O.S., Markus, A., Tomer, R., 2014. Asymmetric dopamine loss differentially affects effort to maximize gain or minimize loss. Cortex 51, 82–91.

Prévost, C., Pessiglione, M., Météreau, E., Cléry-Melin, M.-L., Dreher, J.-C., 2010. Separate valuation subsystems for delay and effort decision costs. J. Neurosci. 30, 14080–14090.

Radakovic, R., Abrahams, S., 2014. Developing a new apathy measurement scale: dimensional apathy scale. Psychiatry Res. 219, 658–663.

Randall, P.A., Pardo, M., Nunes, E.J., Lopez Cruz, L., Vemuri, V.K., Makriyannis, A., Baqi, Y., Muller, C.E., Correa, M., Salamone, J.D., 2012. Dopaminergic modulation of effort-related choice behavior as assessed by a progressive ratio chow feeding choice task: pharmacological studies and the role of individual differences. PLoS One 7, e47934.

Randall, P.A., Lee, C.A., Nunes, E.J., Yohn, S.E., Nowak, V., Khan, B., Shah, P., Pandit, S., Vemuri, V.K., Makriyannis, A., Baqi, Y., 2014. The VMAT-2 inhibitor tetrabenazine affects effort-related decision making in a progressive ratio/chow feeding choice task: reversal with antidepressant drugs. PLoS One 9, e99320.

Reddy, L.F., Horan, W.P., Barch, D.M., Buchanan, R.W., Dunayevich, E., Gold, J.M., Lyons, N., Marder, S.R., Treadway, M.T., Wynn, J.K., Young, J.W., Green, M.F., 2015. Effort-based decision-making paradigms for clinical trials in schizophrenia: part 1—psychometric characteristics of 5 paradigms. Schizophr. Bull. 41, sbv089.

Richards, J.B., Mitchell, S.H., Wit, H., Seiden, L.S., 1997. Determination of discount functions in rats with an adjusting-amount procedure. J. Exp. Anal. Behav. 67, 353–366.
Richardson, N.R., Roberts, D.C., 1996. Progressive ratio schedules in drug self-administration studies in rats: a method to evaluate reinforcing efficacy. J. Neurosci. Methods 66, 1–11.
Robert, P.H., Clairet, S., Benoit, M., Koutaich, J., Bertogliati, C., Tible, O., Caci, H., Borg, M., Brocker, P., Bedoucha, P., 2002. The apathy inventory: assessment of apathy and awareness in Alzheimer's disease, Parkinson's disease and mild cognitive impairment. Int. J. Geriatr. Psychiatry 17, 1099–1105.
Roesch, M.R., Singh, T., Brown, P.L., Mullins, S.E., Schoenbaum, G., 2009. Ventral striatal neurons encode the value of the chosen action in rats deciding between differently delayed or sized rewards. J. Neurosci. 29, 13365–13376.
Rudebeck, P.H., Walton, M.E., Smyth, A.N., Bannerman, D.M., Rushworth, M.F., 2006. Separate neural pathways process different decision costs. Nat. Neurosci. 9, 1161–1168.
Ryan, R.M., 1982. Control and information in the intrapersonal sphere: an extension of cognitive evaluation theory. J. Pers. Soc. Psychol. 43, 450–461.
Ryan, R., Deci, E., 2000. Self-determination theory and the facilitation of intrinsic motivation, social development, and well-being. Am. Psychol. 55, 68–78.
Salamone, J., 1992. Complex motor and sensorimotor functions of striatal and accumbens dopamine: involvement in instrumental behavior processes. Psychopharmacology (Berl.) 107, 160–174.
Salamone, J., 2010. Motor function and motivation. In: Koob, G., Le Moal, M., Thompson, R. (Eds.), Encyclopedia of Behavioral Neuroscience. Academic Press, Oxford.
Salamone, J.D., Correa, M., 2012. The mysterious motivational functions of mesolimbic dopamine. Neuron 76, 470–485.
Salamone, J.D., Steinpreis, R.E., McCullough, L.D., Smith, P., Grebel, D., Mahan, K., 1991. Haloperidol and nucleus accumbens dopamine depletion suppress lever pressing for food but increase free food consumption in a novel food choice procedure. Psychopharmacology (Berl.) 104, 515–521.
Salamone, J.D., Cousins, M.S., Bucher, S., 1994. Anhedonia or anergia? Effects of haloperidol and nucleus accumbens dopamine depletion on instrumental response selection in a T-maze cost/benefit procedure. Behav. Brain Res. 65, 221–229.
Salamone, J., Arizzi, M., Sandoval, M., Cervone, K., Aberman, J., 2002. Dopamine antagonists alter response allocation but do not suppress appetite for food in rats: contrast between the effects of SKF 83566, raclopride, and fenfluramine on a concurrent choice task. Psychopharmacology 160 (4), 371–380.
Salamone, J.D., Correa, M., Mingote, S., Weber, S.M., Farrar, A.M., 2006. Nucleus accumbens dopamine and the forebrain circuitry involved in behavioral activation and effort related decision making: implications for understanding anergia and psychomotor slowing in depression. Curr. Psychiatr. Rev. 2, 267–280.
Salamone, J., Correa, M., Farrar, A., Mingote, S., 2007. Effort-related functions of nucleus accumbens dopamine and associated forebrain circuits. Psychopharmacology (Berl.) 191, 461–482.
Schmidt, L., d'Arc, B.F., Lafargue, G., Galanaud, D., Czernecki, V., Grabli, D., Schüpbach, M., Hartmann, A., Lévy, R., Dubois, B., Pessiglione, M., 2008. Disconnecting force from money: effects of basal ganglia damage on incentive motivation. Brain 131, 1303–1310.

Schmidt, L., Lebreton, M., Cléry-Melin, M.-L., Daunizeau, J., Pessiglione, M., 2012. Neural mechanisms underlying motivation of mental versus physical effort. PLoS Biol. 10, e1001266.

Schweimer, J., Hauber, W., 2005. Involvement of the rat anterior cingulate cortex in control of instrumental responses guided by reward expectancy. Learn. Mem. 12, 334–342.

Sherdell, L., Waugh, C.E., Gotlib, I.H., 2012. Anticipatory pleasure predicts motivation for reward in major depression. J. Abnorm. Psychol. 121, 51.

Sockeel, P., Dujardin, K., Devos, D., Deneve, C., Destee, A., Defebvre, L., 2006. The Lille apathy rating scale (LARS), a new instrument for detecting and quantifying apathy: validation in Parkinson's disease. J. Neurol. Neurosurg. Psychiatry 77, 579–584.

Starkstein, S.E., Mayberg, H.S., Preziosi, T.J., Andrezejewski, P., Leiguarda, R., Robinson, R.G., 1992. Reliability, validity, and clinical correlates of apathy in Parkinson's disease. J. Neuropsychiatry Clin. Neurosci. 4, 134–139.

Starkstein, S.E., Petracca, G., Chemerinski, E., Kremer, J., 2001. Syndromic validity of apathy in Alzheimer's disease. Am. J. Psychiatry 158, 872–877.

Starkstein, S.E., Merello, M., Jorge, R., Brockman, S., Bruce, D., Power, B., 2009. The syndromal validity and nosological position of apathy in Parkinson's disease. Mov. Disord. 24, 1211–1216.

Stoops, W.W., 2008. Reinforcing effects of stimulants in humans: sensitivity of progressive-ratio schedules. Exp. Clin. Psychopharmacol. 16, 503.

Strauss, M.E., Sperry, S.D., 2002. An informant-based assessment of apathy in Alzheimer disease. Cogn. Behav. Neurol. 15, 176–183.

Treadway, M.T., Buckholtz, J.W., Schwartzman, A.N., Lambert, W.E., Zald, D.H., 2009. Worth the 'EEfRT'? The effort expenditure for rewards task as an objective measure of motivation and anhedonia. PLoS One 4, e6598.

Treadway, M.T., Bossaller, N.A., Shelton, R.C., Zald, D.H., 2012a. Effort-based decision-making in major depressive disorder: a translational model of motivational anhedonia. J. Abnorm. Psychol. 121, 553.

Treadway, M.T., Buckholtz, J.W., Cowan, R.L., Woodward, N.D., Li, R., Ansari, M.S., Baldwin, R.M., Schwartzman, A.N., Kessler, R.M., Zald, D.H., 2012b. Dopaminergic mechanisms of individual differences in human effort-based decision-making. J. Neurosci. 32, 6170–6176.

Vallerand, R.J., Pelletier, L.G., Blais, M.R., Briere, N.M., Senecal, C., Vallieres, E.F., 1992. The academic motivation scale: a measure of intrinsic, extrinsic, and amotivation in education. Educ. Psychol. Meas. 52, 1003–1017.

van Reekum, R., Stuss, D., Ostrander, L., 2005. Apathy: why care? J. Neuropsychiatry Clin. Neurosci. 17, 7–19.

Venugopalan, V.V., Casey, K.F., O'Hara, C., O'Loughlin, J., Benkelfat, C., Fellows, L.K., Leyton, M., 2011. Acute phenylalanine/tyrosine depletion reduces motivation to smoke cigarettes across stages of addiction. Neuropsychopharmacology 36, 2469–2476.

Walton, M.E., Bannerman, D.M., Rushworth, M.F., 2002. The role of rat medial frontal cortex in effort-based decision making. J. Neurosci. 22, 10996–11003.

Walton, M.E., Bannerman, D.M., Alterescu, K., Rushworth, M.F., 2003. Functional specialization within medial frontal cortex of the anterior cingulate for evaluating effort-related decisions. J. Neurosci. 23, 6475–6479.

Wardle, M.C., Treadway, M.T., Mayo, L.M., Zald, D.H., de Wit, H., 2011. Amping up effort: effects of d-amphetamine on human effort-based decision-making. J. Neurosci. 31, 16597–16602.

Weiser, M., Garibaldi, G., 2015. Quantifying motivational deficits and apathy: a review of the literature. Eur. Neuropsychopharmacol. 25, 1060–1081.
Westbrook, A., Kester, D., Braver, T., 2013. What is the subjective cost of cognitive effort? Load, trait, and aging effects revealed by economic preference. PLoS One 8, e68210.
Yohn, S.E., Santerre, J.L., Nunes, E.J., Kozak, R., Podurgiel, S.J., Correa, M., Salamone, J.D., 2015. The role of dopamine D1 receptor transmission in effort-related choice behavior: effects of D1 agonists. Pharmacol. Biochem. Behav. 135, 217–226.
Zénon, A., Sidibé, M., Olivier, E., 2015. Disrupting the supplementary motor area makes physical effort appear less effortful. J. Neurosci. 35, 8737–8744.

SECTION

The Neurobiology of Motivation

3

CHAPTER

Brain correlates of the intrinsic subjective cost of effort in sedentary volunteers

5

J. Bernacer*,[1], I. Martinez-Valbuena*, M. Martinez[†], N. Pujol[‡], E. Luis[†], D. Ramirez-Castillo*, M.A. Pastor*,[†,‡]

**Mind-Brain Group (Institute for Culture and Society, ICS), University of Navarra, Pamplona, Spain*

[†]Neuroimaging Laboratory, Center for Applied Medical Research (CIMA), University of Navarra, Pamplona, Spain

[‡]Clínica Universidad de Navarra, University of Navarra, Pamplona, Spain

[1]Corresponding author: Tel.: +34-948425600; Fax: +34-948425619, e-mail address: jbernacer@unav.es

Abstract

One key aspect of motivation is the ability of agents to overcome excessive weighting of intrinsic subjective costs. This contribution aims to analyze the subjective cost of effort and assess its neural correlates in sedentary volunteers. We recruited a sample of 57 subjects who underwent a decision-making task using a prospective, moderate, and sustained physical effort as devaluating factor. Effort discounting followed a hyperbolic function, and individual discounting constants correlated with an indicator of sedentary lifestyle (global physical activity questionnaire; $R=-0.302$, $P=0.033$). A subsample of 24 sedentary volunteers received a functional magnetic resonance imaging scan while performing a similar effort-discounting task. BOLD signal of a cluster located in the dorsomedial prefrontal cortex correlated with the subjective value of the pair of options under consideration ($Z>2.3$, $P<0.05$; cluster corrected for multiple comparisons for the whole brain). Furthermore, effort-related discounting of reward correlated with the signal of a cluster in the ventrolateral prefrontal cortex ($Z>2.3$, $P<0.05$; small volume cluster corrected for a region of interest including the ventral prefrontal cortex and striatum). This study offers empirical data about the intrinsic subjective cost of effort and its neural correlates in sedentary individuals.

Keywords

Decision making, Effort discounting, GPAQ, Risk discounting, Sedentary lifestyle, Subjective value, Utility

Progress in Brain Research, Volume 229, ISSN 0079-6123, http://dx.doi.org/10.1016/bs.pbr.2016.05.003

1 INTRODUCTION

Decision making and action performance depend on an evaluation of the balance of costs and benefits. As explained in chapter "A Cost-Benefit Model of Motivation" by Studer and Knecht (Studer and Knecht, 2016), both factors have a dual contribution, namely, intrinsic and extrinsic. Let us consider the case of a 1-h jogging session for a usual runner. On the side of benefits, there is the intrinsic value of physical exercise stemming from the positive feelings that it causes in the runner. In addition, extrinsic subjective benefits may include, for example, an increase of the runner's probabilities to win an upcoming race and thereby achieve an economic reward. On the side of costs, there is an obvious intrinsic cost due to the energy expenditure that physical exercise requires. Additional intrinsic factors might include the temporal cost related to achieving an expected reward (eg, improving performance, winning a race, etc.), or the expense of running apparel. Extrinsic costs mainly refer to the loss of putative benefits that alternative activities (such as going out with friends or watching TV at home) may entail. Regarding these factors, we can assume that a regular runner is motivated for a particular running session because subjective benefits overcome subjective costs. However, if we consider instead the case of a beginner, subjective benefits are likely to be lower because the intrinsic value of exercise and extrinsic value of instrumental outcomes are less familiar. Furthermore, the intrinsic cost of effort, as well as the cost associated to forgoing alternative activities, might be extremely high. Thus, it should not come as a surprise that the beginner is poorly motivated for each running session.

This chapter summarizes our study of the intrinsic subjective cost of effort at both behavioral and neural levels. We were particularly interested in learning how the subjective weighing of effort depends on whether physical exercise is habitual for the agent. For this purpose, we analyzed effort discounting in a sample of volunteers with various levels of physical activity, from sedentary to highly active. We then studied the brain correlates of effort weighing in a subsample of sedentary volunteers.

Peters and Büchel (2010) describe a brief taxonomy of value types in decision making, including outcome, goal, decision, and action values. Whereas outcome and goal values are unrelated to costs, decision value depends on the subjective discounting of the objective value of a reward. Action value reflects the pairing of an action with any of the other types of values, and thus it could be either related or unrelated to costs. Therefore, decision value is the only type of value that is strictly related to subjective costs. In general terms, as it is described by prospect theory, subjective value (SV) is the expected objective outcome of the actions discounted by various factors of risk, time, and effort (see, for example, Kable and Glimcher, 2007; Prévost et al., 2010; Weber and Huettel, 2008). This theoretical and experimental framework was first described in the field of economics (Kahneman and Tversky, 1979), was later extrapolated to behavioral psychology (Green and Myerson, 2004) and, most recently, has become a productive field of research in neuroscience. In keeping with the focus of this chapter, we concentrate on literature in neuroscience to explain the background of our topic.

A primary goal of neuroscientific studies of value-based decision making is to describe the brain correlates of SV, ie, the brain area that encodes the subjective discounting of a reward. Thirty euros are objectively better than 10 euros, but they could be perceived as less valuable if: (1) they are not immediately available; (2) we are not sure about obtaining them; or (3) we have to exert some effort to obtain them. The actual weight of these discounting factors is subjective and state dependent, but there is clear evidence that they share a common neural correlate in humans. Based on a meta-analysis of functional magnetic resonance imaging (fMRI) studies, Levy and Glimcher propose that the ventromedial prefrontal cortex (VMPFC) encodes SV irrespective of the nature of the reward (Levy and Glimcher, 2012). This valuation is carried out by integrating sensory inputs (from parietal and occipital cortices), information about the internal state of the agent (subcortical inputs), and personal preferences in terms of discounting factors (from other regions of the prefrontal cortex). Then the value signal is conveyed to motor-related cortical areas which, in association with the basal ganglia, produce the behavioral output. The engagement of VMPFC in value coding has been verified by extensive research (see, for example, Bartra et al., 2013; Dreher, 2013; Montague et al., 2006; O'Doherty, 2011). Pharmacologically, this valuation seems to depend on monoaminergic signaling (Arrondo et al., 2015; Bernacer et al., 2013; Jocham et al., 2011). In the following paragraphs, we briefly summarize the main findings about intrinsic subjective costs in decision making in the fields of psychology and neuroscience.

As mentioned earlier, the main discounting factors in decision making (ie, factors that determine intrinsic costs) are time, risk, and effort. In 2004, Green and Myerson published an integrative review on temporal and probabilistic discounting in human behavior (Green and Myerson, 2004). As reported in this review, the intrinsic cost of temporal delay is usually assessed experimentally with a very simple task, for which volunteers are asked to choose between a relatively small immediate reward and a larger delayed reward (for example, $150 now vs $1000 in 6 months). Using the responses of each volunteer, a discounting curve is calculated that shows the subjective devaluation of a reward (*Y* axis) with increasing delays (*X* axis). As Green and Myerson explain, even though temporal discounting curves were first described as exponential, they seem to follow a hyperbola-like shape. This shape has been extensively replicated in psychology and neuroscience (see, for example, Estle et al., 2006; Kable and Glimcher, 2007; Kobayashi and Schultz, 2008; McKerchar et al., 2009; Peters and Büchel, 2009; Pine et al., 2010; Wittmann et al., 2007). Concerning risk discounting, the procedure and results are very similar. In this case the experimenter offers two options that differ in probability of obtaining a reward (for example, $150 guaranteed vs 30% probability of obtaining $1000). Once again, a hyperbola-like function produces the best fit to the experimental data (Estle et al., 2006; Green and Myerson, 2004; Weber and Huettel, 2008), with the *X* axis representing the odds against winning the reward. Finally, the characterization of the effort discounting curve is quite recent (Hartmann et al., 2013). Hartmann and collaborators report that effort discounting is best defined as a parabolic curve as opposed to the hyperbolic curve suggested by other authors (Mitchell, 2004; Prévost et al., 2010). In these studies, the task consists of choosing between a small, noneffortful reward, and a larger reward that requires squeezing a handle with variable intensity. Thus, the value of the reward is discounted by increasing levels of effort.

At the neural level, the main brain area whose activity correlates with discounting functions is the ventral prefrontal cortex. Kable and Glimcher followed the behavioral approach explained earlier to calculate the SV of the option that volunteers chose while inside an fMRI scanner (Kable and Glimcher, 2007). For example, if a volunteer chooses \$30 with a temporal delay of 30 days, the SV is the objective value (30) multiplied by the subjective intrinsic cost (or individual temporal discounting factor) of waiting for 30 days (say, for example, 0.25). Thus, an objective reward of \$30 is reduced to 7.5. These authors found that the BOLD signal of VMPFC and ventral striatum correlated with the SV of the chosen option. These results have been replicated by others (Gregorios-Pippas, 2009; Prévost et al., 2010; Wittmann et al., 2007). SV discounted by probability has been described to have similar brain correlates, although other areas such as the intraparietal sulcus have also been included (Peters and Büchel, 2009). With respect to physical effort discounting, the main brain areas involved in SV are the striatum, supplementary motor area, anterior cingulate, VMPFC, and motor cortex (Burke et al., 2013; Croxson et al., 2009; Kurniawan et al., 2010, 2011; Prévost et al., 2010; Treadway et al., 2012). In the next paragraph we discuss in more detail the tasks employed in these effort-discounting experiments in order to highlight the novelty of the research that we present later in this chapter.

Investigations of the brain correlates of effort discounting have attracted increasing interest in recent years. For instance, fMRI experiments have sought to assess the brain areas associated with effort-based decision making, their interactions with other discounting factors, and the influence of dopamine on this process. The theoretical background of these experiments comes from Salamone's research on rats (see Salamone, 2009 for a review). The cornerstone of this research is the relationship between effort, decision making, dopamine, and nucleus accumbens. To our knowledge, one of the first translational studies that attempted to assess the neural correlates of effort discounting in humans was the work by Botvinick et al. (2009). However, the type of effort involved in their task was mental effort. Previously, although in a different context, Pessiglione et al. (2007) studied the motivational role of subliminal images and its influence on brain activity associated with decision-making processes. Remarkably, they measured motivation as the grip force exerted when squeezing a handle, and reported that the ventral pallidum encoded both conscious and subliminal motivation. This type of task (hand grip) is adopted by most of the subsequent studies on physical effort discounting (Bonnelle et al., 2016; Burke et al., 2013; Kurniawan et al., 2010; Meyniel et al., 2013; Prévost et al., 2010; Skvortsova et al., 2014), although some have used different paradigms involving button presses per time unit (Kroemer et al., 2014; Scholl et al., 2015; Treadway et al., 2009). What is important to emphasize at this point is that all of these experiments involve a decision about an *immediate* effort. Also, both hand gripping and button pressing might not be optimal for evaluating the willingness of a subject to make an effort in real life: whereas everyday decisions are often discounted by strong efforts (ie, driving a car instead of walking or using the elevator instead of the stairs), within experimental settings subjects might be more highly motivated and thus more willing to make a brief and relatively small effort.

For these reasons, we decided to adopt a different paradigm for which the effort under consideration is prospective and sustained, and therefore of potentially greater ecological validity: namely, running on a treadmill. To implement our study, we first recruited a large sample of volunteers who undertook a decision-making task for which they had to decide between a small, noneffortful reward, and a larger reward that required running for a certain period of time on a treadmill. We collected information about their lifestyle (ie, daily level of activity) with the intention of testing the ecological validity of our task, that is, the correlation between effort discounting and the level of physical activity in a normal week. We then recruited a subsample of sedentary volunteers who received an fMRI scan while doing a similar decision-making task. Using neurocomputational methods, we investigated brain activity to determine which areas are correlated with effort discounting-related signals. In the following sections we describe these two experiments in detail and then discuss the implications of our results for the understanding of motivation.

2 METHODS

In this chapter we report the results from two experiments. The first aimed to calculate individual and group effort-discounting curves when the effort at stake is prospective, moderate, and sustained. In addition, we aimed to test whether the decaying constants of individual curves correlated with a lifestyle indicator, assessed by administration of the Global Physical Activity Questionnaire (GPAQ) published by the WHO (http://www.who.int/chp/steps/resources/GPAQ_Analysis_Guide.pdf). The second experiment aimed to assess brain activity in sedentary subjects when effort is the main devaluating factor in a decision-making task. We used neurocomputational methods to evaluate the neural correlates of SV and effort discounting. These two parameters were estimated from the individual curves obtained in the first experiment.

2.1 SUBJECTS

The protocol of the experiment was approved by the Committee of Ethics for Research of the University of Navarra. A sample of 57 subjects (age 18–25, 26 females) was recruited within the environment of the university. Hence, they all had a similar profile in terms of age, income, and educational level; however, they were not asked to fulfill any special requirements in terms of sedentary lifestyle prior to the study in order to ensure a certain degree of diversity to facilitate correlation of the data with effort-discounting constants. A subsample of volunteers ($N=24$, 14 female) was recruited from this initial sample for the second experiment. Inclusion criteria were: (1) a low score in GPAQ together with no past history of habitual running (this criterion is explained in detail in Section 2.2); (2) no fMRI scan incompatibilities; (3) ability to follow a physical exercise program for the following 3 months; (4) no neurological or psychiatric disorders, as assessed by the Mini International Neuropsychiatric Interview (Cummings et al., 1994). The third criterion was part of an additional project not reported here. All participants provided signed informed consent before the scan.

2.2 GLOBAL PHYSICAL ACTIVITY QUESTIONNAIRE

We estimated the active lifestyle of the volunteers with the Spanish version of the GPAQ. This test queries the volunteers about their physical activity during a normal week. It is divided into four sections: work, everyday movement between places, recreational activities, and sedentary behavior. In each of the first three sections they have to disclose the amount of time (in hours and minutes) they spend doing moderate or vigorous physical activity. In the last section, they have to report the number of hours they spend sitting or reclining in a typical day. The dependent variable is the number of METs (metabolic equivalents), which is the ratio of a person's working metabolic rate relative to the resting metabolic rate. One MET corresponds to a consumption of 1 kcal/kg/h. According to WHO guidelines, four METs are assigned to time spent in moderate activities, and eight METs to time spent in vigorous activities. Time spent traveling between places is considered moderate activity.

With regards to the subsample of subjects included in the fMRI experiment ($N=24$), volunteers were interviewed to verify that they had never done habitual running before. Their mean GPAQ score was 1023.3 (standard error of the mean, SEM = 192.1), ranging from 0 to 3360. Eleven participants scored lower than 600, considered to be extremely sedentary by the WHO. Concerning the remaining 13 participants included in the fMRI study, most of their GPAQ score (77.2% on average) was due to walking between home and campus. Only 2 participants had a score higher than 600 due to recreational activities, in particular team sports. Since they reported that their engagement in such activities was occasional, and not habitual, they were finally included in the experiment. The GPAQ score of these 24 subjects was significantly lower than that of the remaining participants (Mann–Whitney $U=249.5$, $N_{\text{fmri}}=24$, $N_{\text{no\_fmri}}=32$, $P=0.026$, two-tailed).

2.3 TASKS

The tasks of both experiments were coded in Cogent 2000 (Wellcome Department of Imaging Neuroscience, UCL, London, UK) and Matlab (Mathworks, Natick, MA). For the first experiment, we used a modified version of the most common task used for temporal and risk discounting (Kable and Glimcher, 2007), which has also been employed to assess effort discounting (Hartmann et al., 2013) (Fig. 1). Subjects were

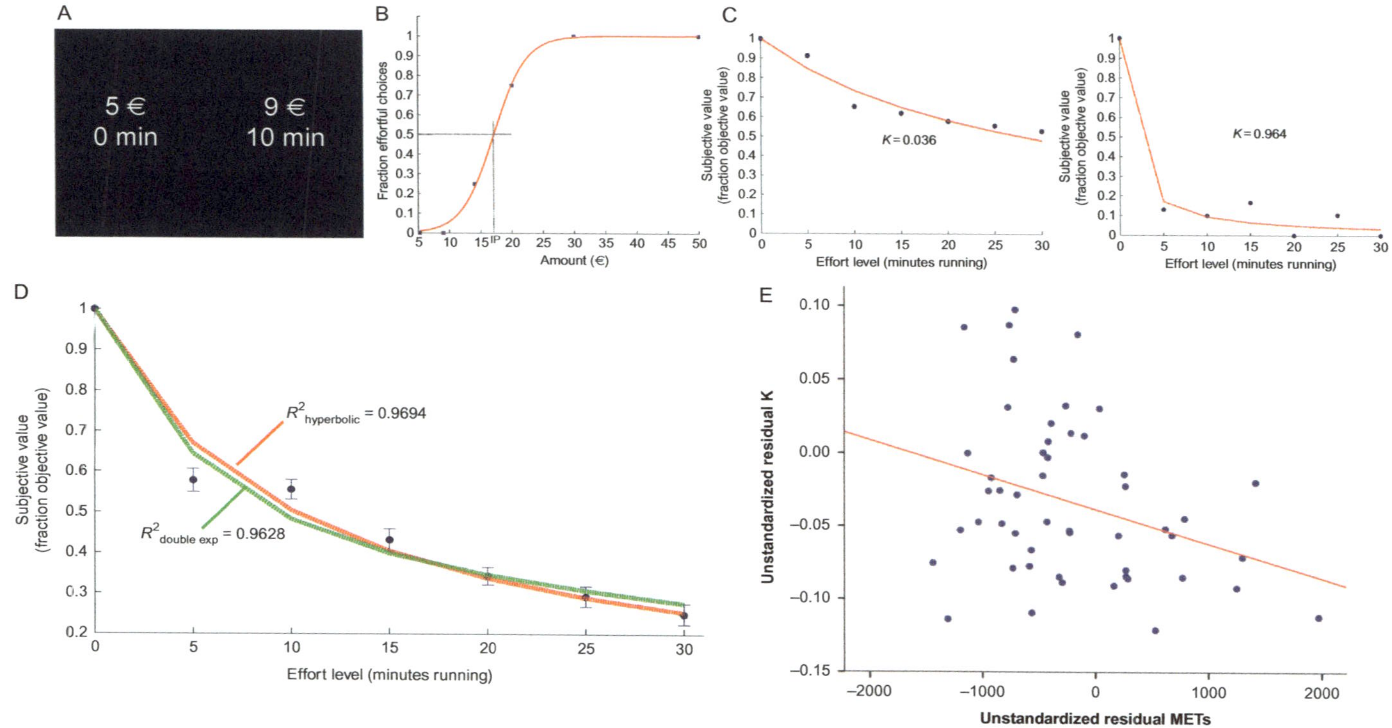

FIG. 1

Behavioral task and main results of the first experiment. (A) Task used to assess effort discounting in the whole sample (N=57). A fixed option (winning 5 € without effort) was presented simultaneously with an effortful option that entailed a larger reward together with different levels of effort. See Section 2.3 for details. (B) Example of logistic fitting to the actual behavior of one participant for 30 min running in the treadmill. The X axis represents money (in €), and the Y axis is a fraction of the effortful choice. The intersection of the *dashed line* with the X axis represents the indifference point (IP). (C) Two examples of hyperbolic effort-discounting curves for two individuals, showing low (*left*) and high (*right*) effortdiscounting. (D) Group hyperbolic and double exponential fitting to effort discounting. *Data points* represent the median, and *error bars* indicate the SEM. R^2 indicates goodness of fit after sum of least squares, adjusted for the number of constants in each formula. (E) Scatterplot to illustrate the partial correlation of individual hyperbolic K and habitual physical activity (METs), controlling for the individual R^2 values. Unstandardized residuals are calculated by a linear regression considering K (or METs) as a dependent variable, and R^2 as an independent variable.

instructed about the general framework of the project, and they were presented sequentially several pairs of options from which they had to choose one: one of the options (randomly presented on the left or right side of the screen) was always present and involved a 5 € reward in exchange for no effort. The other option entailed a higher amount of money (5.25, 9, 14, 20, 30, or 50 €) together with different required efforts (5, 10, 15, 20, 25, and 30 min periods of running on a treadmill). Therefore, there were 36 different pairs of options presented, and each of them was randomly displayed four times (144 trials in total, divided into 2 sessions of 72 trials). Subjects had to respond by pressing the left or right arrow of the keyboard. They were not informed about the structure of the task, and they were told that both reward and effort were hypothetical (see Section 4). We used a similar task to calculate risk discounting, another devaluating factor used in the fMRI task (see later). The task and data analysis were identical to the effort discounting task, substituting effort levels for probability of winning the reward (90%, 75%, 50%, 33%, 10%, and 5%).

The fMRI task was similar to the one used in the behavioral study just described, although there were key differences (Fig. 2). Again, two options were presented at the same time, and volunteers had to choose one of them by pressing a left or right button with the index or middle finger (respectively) of their right hand. In this case, both options entailed the possibility of winning 30 € (fixed reward). In addition, each option included a certain probability of winning the reward (30%, 40%, 50%, 60%,

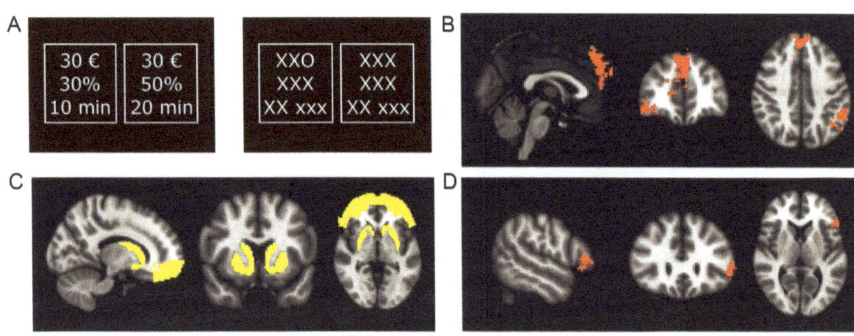

FIG. 2

fMRI task and neuroimaging results. (A) *Left*, The decision-making task includes pairs of options involving the probability (30–70%) of winning a fixed reward (30 €) in exchange for some effort (10–30 min running in a treadmill) "task pairs." *Right*, Display of the motor control used in the task. Subjects were instructed to select the option with the "O." (B) Clusters surviving the statistical threshold ($Z>2.3$, $P<0.05$ whole-brain cluster correction) for the comparison of difference of subjective value vs motor control. (C) Region of interest used to assess the neural correlates of effort-related subjective value, including the striatum and ventral prefrontal cortex. (D) Clusters surviving the statistical threshold ($Z>2.3$, $P<0.05$ small volume cluster correction) for the comparison of difference of effort discounting vs motor control. Right side of the brain is displayed on the left side of the image for coronal and axial views.

or 70%) together with a required effort (10, 15, 20, 25, or 30 min running on a treadmill). Subjects were explained that after the scan one of the trials would be picked at random and the chosen option would be recorded. Then, they entered a lottery determined by the probability of the chosen option, and if they won they were asked to do the required physical exercise in exchange for the money during the following week. If they lost the lottery, they would not get any money nor do any exercise. Payments were given as vouchers for the university's book shop.

Pairs of options were selected individually for each volunteer, guaranteeing seven "difficult" pairs (SV of both options were nearly identical), six "easy" pairs (SV were very different), and seven pairs of medium "difficulty" (SV were similar). Therefore, in total, 20 different pairs of options (task pairs) were presented. As explained earlier, SV corresponds to the actual reward (30 €) multiplied by the discounting factors of effort and risk, which were obtained in the first experiment.

Each of the 20 task pairs were presented nine times. In addition to these 180 trials, 45 motor control trials were included (Fig. 2). There were also 45 trials in which subjects could choose a certain "noneffortful reward" (30 €, 100%, 0 min vs 30 €, 0%, 0 min), and 45 additional trials involving a certain reward together with "maximum effort" (30 €, 100%, 35 min vs 30 €, 0%, 35 min). In total, 315 trials were presented to each volunteer, divided into 3 sessions of 105 trials each (about 12 min). The options stayed on the screen up to 4 s or until the subject responded. The order and position of the options (left or right) were randomly arranged. Trials were separated by a fixation cross of random variable duration (2–6 s).

2.4 BEHAVIORAL DATA ANALYSIS AND CURVE FITTING

Data processing and curve fitting were performed using Matlab, and statistical analyses were carried out using SPSS 15.0 (SPSS Inc., Chicago). We first calculated the function that best describes the behavior of each participant. To do so, for each subject and effort level we looked for the situation in which the SV of the effortless option was equal to the SV of a particular effort level (ie, indifference point). This was inferred by plotting for each effort level the number of times (out of four) that each reward was preferred instead of the 5 € (effortless option). For example, for the effort level of 10 min running, one particular subject may have the following behavior: 5.25, 0 times chosen (0/4); 9 €, 1/4; 14 and 20 €, 3/4; 30 and 50 €, 4/4. This data were then fitted to a logistic function (Eq. 1) to calculate which amount of money corresponded to a 2/2 behavior, that is, the indifference point (Fig. 1):

$$y(\text{Money}) = \frac{k}{1 + \mathrm{e}^{-G(\text{Money} - r_0)}} \tag{1}$$

Curve fitting was performed by a script that tested all the possible combinations of 100 different values of the constants in the logistic function [k(0.5,1.5); G(0.1,10); r_0(1100)]. The best fitting was the maximum value after calculating the sum of least squares for each combination. After this, the discounting factor of each effort level was calculated by dividing each indifference point by the money corresponding to the effortless option (5 €). Finally, these discounting factors were plotted and four different fittings were evaluated according to the literature: hyperbolic (Eq. 2), exponential (Eq. 3), double exponential (Eq. 4) (Green and Myerson, 2004; Prévost et al., 2010), and parabolic (Eq. 5) (Hartmann et al., 2013):

$$y(\text{Effort}) = \frac{1}{1 + K * \text{Effort}} \tag{2}$$

$$y(\text{Effort}) = e^{-c * \text{Effort}} \tag{3}$$

$$y(\text{Effort}) = \frac{e^{-\beta * \text{Effort}} + e^{-\partial * \text{Effort}}}{2} \tag{4}$$

$$y(\text{Effort}) = A - H * \text{Effort}^2 \tag{5}$$

Again, curve fitting was carried out by a script that tested different combinations of the constants included in each formula, and the best fitting was chosen by sum of least squares. In this case, 1000 different values of each constant were tested.

In order to evaluate the best fitting for the whole sample, we calculated the median of the indifference points for each effort level, obtained the discounting factors as before, plotted them and assessed the same fitting functions.

2.5 fMRI SETTING

We used a 3T fMRI scanner (Siemens TRIO, Erlangen, Germany) and a 32-channel head coil. Between 170 and 274 volumes (depending on the subjects' reaction times) were acquired in each of the 3 sessions, using an echo-planar imaging sequence to measure BOLD contrast (or "activity") (resolution = $3 \times 3 \times 3$ mm^3; TR/TE = 3000/30 ms; FOV = 192×192 mm^2, Flip angle = 90 degree; 64, 48, and 48 volumes acquired in the coronal, sagittal, and axial planes, respectively). The first five volumes were discarded for T1 equilibration effects. An anatomical T1 MPRAGE image was also collected (TR = 1620 ms; TE = 3.09 ms; inversion time (TI) = 950 ms; FOV = $256 \times 192 \times 160$ mm^3; flip angle = 15 degree; image resolution = 1 mm isotropic).

fMRI data were analyzed with FSL (created by the Analysis Group, FMRIB, Oxford, UK, http://fmrib.ox.ac.uk/fsl) (Jenkinson et al., 2012). Prior to any data processing, the skull was removed from all T1 images using the BET tool included in FSL package. Individual T2* images were processed with FEAT (FMRI Expert Analysis Tool). They were realigned, motion corrected, and spatially smoothed with a Gaussian kernel of 5 mm (full-width half maximum). Each time series was high-pass filtered (100 s cutoff). Images were registered to the corresponding T1 image and finally normalized to MNI template.

2.5.1 General linear model for the fMRI data

Each individual time series was fitted to a general linear model (GLM) with 10 explanatory variables (EVs). The model was mainly intended to assess the effect of subjective effort discounting in decision making, considering also the effect of risk discounting and the interaction between both. Thus, the EVs included in the model are as follows (Table 1):

The general appearance of task pairs and the motor control is shown in Fig. 2A. Maximum reward/maximum effort pairs correspond to the presentation of the following options: {30 €, 100%, 35 min} vs {30 €, 0%, 35 min}. Maximum reward/no effort corresponds to pairs {30 €, 100%, 0 min} vs {30 €, 0%, 0 min}.

All regressors were convolved with a canonical double gamma hemodynamic response function (HRF). The average reaction time of all participants was 1.905 s. For that reason, the duration of all events was set to 2 s. Due to the slow nature of the fMRI HRF, we did not expect significant differences between a fixed (2 s) or a variable (linked to event-related reaction times) duration of events, since both figures were very similar.

EV1, EV8, EV9, and EV10 account for brain activity during the presentation of task pairs, pairs involving a maximum effort (35 min), pairs involving no effort (0 min), and motor control pairs, respectively. For the present report, these are variables of no interest except for EV10, which was subtracted from the EVs of interest. EV2 (Shannon's entropy) explains brain activity in relation to a behavioral measure of uncertainty; it is also excluded from the present report.

EV3 accounts for brain activity associated with the SV of the pair or "difference SV." The SV of each option was calculated by multiplying the actual reward (30 €) by the discounting factors of effort and risk, which were estimated in the first

Table 1 Structure of the General Linear Model (GLM) Used to Analyze the fMRI Data

	Explanatory Variable	Onset	Parametric Modulator
EV1	Task pairs	Task pair on screen	No (boxcar)
EV2	Uncertainty	Task pair on screen	Shannon's entropy
EV3	SV of the pair	Task pair on screen	$\lvert SV_{chosen} - SV_{not\_chosen} \rvert$
EV4	ED factor of chosen option	Task pair on screen	ED factor
EV5	RD factor of chosen option	Task pair on screen	RF factor
EV6	ED of the pair	Task pair on screen	$\lvert ED_{chosen} - ED_{not\_chosen} \rvert$
EV7	RD of the pair	Task pair on screen	$\lvert RD_{chosen} - RD_{not\_chosen} \rvert$
EV8	MR/NE pair	MR/NE on screen	No (boxcar)
EV9	MR/ME effort pair	MR/ME effort on screen	No (boxcar)
EV10	Motor control	Motor control on screen	No (boxcar)

ED, *effort discounting;* EV, *explanatory variable;* MR/ME, *maximum reward/maximum effort;* MR/NE, *maximum reward/no effort;* RD, *risk discounting;* SV, *subjective value.*

experiment. Since both options were simultaneously presented on screen, we calculated the SV of the pair, that is, the difference between the SV of both options. We used absolute value rather than signed differences because, as suggested by other authors (FitzGerald et al., 2009), it fits with the idea that agents weigh the values of different options, and then select between them stochastically according to probabilities derived from a nonlinear choice distribution. This assumes that agents can select the option with the lowest SV; hence, the difference between SVs might be negative. In this case, the correlation of a negative parameter value with a negative neurophysiological signal would be difficult to interpret. Instead, using absolute values allows us to search for the brain areas whose BOLD signal correlates with the net SV of the pair, which could be a better indicator of deliberation itself, irrespective of the chosen action.

EVs 4–7 explain the contributions of effort and risk in the subjective valuation of the options. In detail, EVs 4 and 5 account for the effect of effort and risk, respectively, on the selected option of the task pair. Further, EVs 6 and 7 explain the contribution of effort and risk (respectively) to the overall weighing of the task pair, irrespective of the selected option, and they were computed as the absolute value of the difference between the discounting factors of both options included in the pair. Therefore, whereas EVs 4 and 5 are linked to the actual selection of one of the options, EVs 6 and 7 are associated with the deliberation process. Effort- and risk-discounting factors were calculated from the discounting curves of the first experiment. Note that discounting factors close to 1 involve low discounting, that is, a SV close to the objective reward; when discounting factors are close to 0, they have a maximum effect in reducing SV.

The main interest of this experiment was to assess the brain correlates of effort discounting in SV. For that reason, the comparisons of interest that are presented here are EV3 vs EV10, and EV6 vs EV10. The former comparison reveals those brain areas whose activity correlated specifically with the difference between SVs (the net SV of the pair under consideration), considering effort and risk as devaluators. The absolute value of this difference can be understood as an index of decision difficulty (Shenhav et al., 2014). According to our model, brain areas revealed by this contrast would have a boosted BOLD signal when both options of the pair had a disparate SV (ie, "easy decisions"), and a reduced BOLD signals when both options had a similar SV (ie, difficult decisions). The latter comparison is similar, but it is intended to expose those brain areas whose activity correlates with the SV of the pair considering only the effect of effort on the deliberation process, excluding risk devaluation. In this case, the brain areas revealed by this comparison would have an increased BOLD signal just in those pairs whose options have dissimilar degrees of subjective effort discounting (in general, low effort vs high effort pairs), and a baseline activity in trials with similar levels of demanded effort (irrespective to the effort intensity). Once the individual statistical parametric maps were calculated for each session, a second level analysis was performed to average all three individual sessions; then, the whole sample statistical map was calculated in a third-level analysis. We corrected for multiple comparisons by thresholding these group maps at $Z > 2.3$, with cluster correction of $P < 0.05$ (Worsley et al., 1992). The analysis for the first contrast (difference SV) was carried out for the whole brain. Based on previous literature concerning the role of effort in SV discounting (discussed earlier), we restricted our analysis of the neural correlates of effort discounting to a large region of interest including the ventral prefrontal cortex and striatum (12,186 voxels in total) (Fig. 2C).

3 RESULTS

3.1 EXPERIMENT 1: EFFORT DISCOUNTING AND CORRELATION WITH LIFESTYLE

GPAQ data were not collected from one volunteer (male). As expected, the sample ($N=56$) showed high variability in terms of physical activity measured in METs: mean = 1395, SEM = 183.7, min = 0, max = 9200). Median values differed between male (1360 METs) and female (840 METs), and this difference was statistically significant (Mann–Whitney $U=249$; $N_{male}=29$; $N_{female}=27$; $P=0.019$, two-tailed).

With regards to effort discounting, the behavior of the whole sample is best described by a hyperbolic function according to the following adjustment values (R^2 adjusted for the number of variables in each function): hyperbolic = 0.9694; exponential = 0.9024; double exponential = 0.9628; parabolic = 0.5297) (Fig. 1). Note that the double exponential curve is also a good predictor of the sample's behavior, while the parabolic fitting is the poorest. Interestingly, in terms of individual fitting, the hyperbolic curve is the best predictor for the same number of subjects as the double exponential ($N=20$). The behavior of 16 subjects approximates an exponential curve, whereas the parabolic function is optimal for only 1. Since the best fitting for the sample is hyperbolic, subsequent analyses take the individual constants (K) from the hyperbola-like discounting function. When comparing male and female participants, there are no statistical differences in hyperbolic K (Mann–Whitney $U=398.5$; $N_{male}=30$; $N_{female}=27$; $P=0.917$, two-tailed) or R^2 goodness of fit (Mann–Whitney $U=328$; $N_{male}=30$; $N_{female}=27$; $P=0.218$, two-tailed).

Having achieved the goal of the first part of the study, we then focused on the task of building an ecological model for effort discounting. For this we correlated the individual hyperbolic decaying constants with the individual METs value, controlling for the individual adjustment (R^2) to the hyperbolic curve. This partial (instead of a bivariate) correlation was carried out in order to consider the fact that the hyperbolic function was not the best fit for all subjects. Since the correlated variables followed a normal distribution ($P > 0.05$ after Kolmogorov–Smirnov test), we performed a Pearson's partial correlation test. Statistical analyses revealed a significant correlation between both variables: $r = -0.302$, $P=0.033$ ($N=51$ after discarding outliers, that is, extreme values higher or lower than three times interquartile range). As predicted, this means that the effort discounting is higher (higher values of K) for subjects with a sedentary lifestyle (lower METs values).

In conclusion, this first experiment demonstrates that: (1) effort discounting in a sample of university students is described by a hyperbolic function; and (2) a task including a prospective effort as devaluating factor is a proper indicator of the active or sedentary lifestyle of the subjects.

3.2 EXPERIMENT 2: BRAIN CORRELATES OF EFFORT DISCOUNTING IN SEDENTARY SUBJECTS

3.2.1 Behavioral results of the fMRI task

The fMRI task was designed so that the effect of effort as a discounting factor is revealed by the selection of the lower probability option of the pair (ie, if effort were not a discounting factor for the participants they would always choose the high probability option, which is obviously more advantageous). The sedentary lifestyle of the volunteers was reflected in their choices during the fMRI task. Focusing on the nine times that each "difficult" and "medium" pair was presented, subjects chose on average the high probability option 5.7 ($\pm$0.56) times more often, irrespective of the demanded effort. "Easy" pairs were discarded from this analysis because some of them involved a high probability/low effort option vs low probability/high effort option. Even though all participants of the second experiment did not do physical exercise regularly, there was some variability in the degree of habitual physical activity and the individual value of the hyperbolic decaying constants. Interestingly, we found a negative correlation between the number of times that the high probability option was chosen and the hyperbolic decaying constant (Spearman's rho $= -0.414$, $P = 0.05$, $N = 23$ after discarding one outlier). This result confirms that subjects with a higher effort discounting (higher K) tended to prefer the low probability/low effort option.

3.2.2 Imaging results

In this section we report those areas whose activity correlated with (1) SV of the pair or "difference SV" and (2) effort-related discounting factor of the pair when contrasted with the motor control.

With respect to the SV of the pair, we performed a whole-brain analysis that revealed a large cluster in the dorsomedial prefrontal cortex as well as a cluster located in the right ventrolateral prefrontal cortex (VLPFC) and different aspects of the parietal cortex (Fig. 2; Table 2). In other words, these brain areas had a higher BOLD signal for those pairs of options with a high difference SV (ie, difference of SV between option A and B), and a low BOLD signal for those options with a low difference SV.

Finally, following the literature discussed earlier, we restricted our analysis of the neural correlates of effort discounting to a large region of interest including the ventral prefrontal cortex and striatum (12,186 voxels in total) (Fig. 2). The analysis revealed a cluster located in the left VLPFC (Fig. 2; Table 2). Therefore, BOLD signal in this area correlated with effort discounting of the pair, which can be treated as

Table 2 Clusters Surviving the Statistical Threshold ($Z>2.3$, $P<0.05$ Corrected) for the Two Comparisons of Interest

Cluster	Voxels	*Z* max	*P*	Coordinates (*X*, *Y*, *Z*)	Area
Difference subjective value vs motor control (whole brain)					
1	1174	3.75	1.91×10^{-6}	0, 44, 40	DMPFC
2	768	3.98	0.00016	−20, −38, 12	L parietal
3	551	3.8	0.00236	36, 26, −18	R VLPFC
4	364	3.45	0.0321	−52, −64, 38	L angular gyrus
Difference effort discounting vs motor control (ROI)					
1	158	3.72	0.0358	−54, 28, 6	L VLPFC

Coordinates are given in standard space. See text for details about the region of interest. DMPFC, *dorsomedial prefrontal cortex;* L, *left;* R, *right;* ROI, *region of interest;* VLPFC, *ventrolateral prefrontal cortex.*

equivalent to the effort-based difference SV (ie, SV excluding the effect of risk discounting).

In summary, our neurocomputational imaging results suggest that the DMPFC is associated with the SV of the pair of options under consideration, taking into account both effort and risk discounting, and the VLPFC is related to effort discounting in decision making.

4 DISCUSSION

In this section we discuss the implications of our two experiments, whose main results can be summarized as follows. First, we have described the hyperbola-like discounting function of effort, using for the first time a prospective, moderate, and sustained form of physical exercise. We have demonstrated the ecological validity of our approach by proving the association between the decaying constant and the level of physical activity of the volunteers. Second, we have evidence that indicates the neural correlates of two different effort-related neurocomputational parameters, namely, SV and effort discounting of the pair: DMPFC and VLPFC, respectively.

Even though the role of effort in decision making at behavioral and neural levels has been the focus of a large number of studies in recent years, these studies are limited by the fact that the demanded effort of their chosen task is immediate and brief (see, for example, Bonnelle et al., 2016; Burke et al., 2013; Croxson et al., 2009; Hartmann et al., 2013; Kurniawan et al., 2011; Prévost et al., 2010; Skvortsova et al., 2014; Treadway et al., 2012). Because of this limitation, the relationship between the experimental intrinsic cost of effort and the active or sedentary lifestyle of subjects has not been analyzed previously. Thus, we decided to adapt a task commonly used in this kind of experiments by including an exercise that could inform us about the weight of effort on the participants' daily lives. In our opinion, the validity of our approach is confirmed by the correlation between the individual effort-discounting constant and the metabolic consumption of the participants measured as METs, as recommended by the WHO. Next we comment on the implications of these functions for understanding the intrinsic cost of effort.

The hyperbolic curve has been reported to describe the role of other discounting factors, such as temporal delay and risk (Green and Myerson, 2004; Kable and Glimcher, 2007). Considering the shape of this curve, we see that a mild initial contribution of the devaluator rapidly lowers the SV of the expected reward; this steep decrease gradually lessens, as increases in the intensity of the devaluator (longer temporal delays, higher odds against winning, or higher effort) have diminishing impact on the subjective discounting of the reward. The dynamics of the hyperbolic function is mainly explained by the constant K: a high K involves a very steep decrease of value, whereas K values closer to zero yield milder curves. Thus, the intrinsic cost of effort changes in parallel with K values. We found that the decisions of a large amount of the participants, as well as the whole sample's behavior, were also described by a double exponential discounting function, as proposed by some authors in temporal discounting (Mcclure et al., 2007). In this case, the utility function is decomposed into two processes, each accounted by a different constant: β and δ. Depending on the actual values, the former usually relates to a quicker and abrupt decay of the function, whereas the latter relates to a more harmonic exponential trend for higher amounts of the devaluator. In the context of temporal discounting and primary reward, McClure and collaborators termed β the "impatient" component, whereas they related the δ component with planning and deliberation (Mcclure et al., 2007). Applying this analogy to our task, β may be understood as the "passive" component, as it accounts for the initially sharp decline of the SV with low levels of effort. In our opinion, this initial strong devaluating effect of physical effort is the reason why the parabolic function provided the worst fit, contrary to recent research (Hartmann et al., 2013). The work by Hartmann and collaborators involved a handgrip task, where an initial low effort does not have such as a strong intrinsic cost as a prospective sustained exercise. Their approach, however, may provide useful information about actual immediate efforts and reward.

Our neuroimaging analyses indicate the brain correlates of effort discounting in decision making. To our knowledge, this is the first time that a prospective moderate sustained effort has been used in this kind of experiment. One of the main advantages of our task is to remove the effects of motor preparation and immediate feeling of vigor from the decision itself. When assessing the neural correlates of effort-discounted decision making with an immediate intense effort, brain activity may be associated with the decision, preparation of the movement or immediate motivation, among other factors. Another possible strategy to overcome this limitation is to separate choice and execution periods during the handgrip task (Kurniawan et al., 2010). Instead, we decided to use an ecologically valid and generalizable task, as shown in the first experiment. The key brain areas tagged by our neurocomputational analyses are the DMPFC and VLPFC.

Whole-brain analysis of the SV of the pair revealed a significant cluster in the DMPFC. The neurocomputational methods that assess the neural bases of SV in decision making allow for two different approaches, depending on the task. On the one hand, if only one option is displayed on the screen (the other being fixed and implicit), the variable of interest is usually the SV of the chosen option (for example, Kable and Glimcher, 2007). On the other hand, if both options are displayed on the screen, the best strategy is to model the absolute value of the pair (FitzGerald et al., 2009). This reflects more accurately the subject's weighing of both options. These authors report a cluster in the VMPFC (or subgenual area) as the neural correlate of difference value. In our study, the brain correlates include the DMPFC. The discrepancy between FitzGerald et al.'s study and ours may be due to the absence or presence of discounting factors in the decision-making process. Whereas their task is a direct valuation of items, we asked our volunteers to employ more resources in evaluating their willingness to make an effort in exchange for a higher probability to win. According to a recent meta-analysis carried out on over 200 neuroimaging articles about SV, the DMPFC seems to be part of a network whose activity correlates with the salience of SV rather than SV itself (Bartra et al., 2013). Thus, BOLD signal would increase with both subjective reward and punishments, and would decrease with neutral values. In light of our results, the interpretation could be similar: DMPFC's BOLD signal is higher when the difference value of the choice is large and lower when it is small. Depending on the task, a high difference value may be a consequence of either a reward (vs neutral) or a punishment (vs neutral). The meta-analysis by Bartra et al. includes several different tasks and the foci in DMPFC could be understood as the difference value when two options are presented simultaneously as well as a value-based salience signal.

Another intriguing result of our experiment is the indication of the VLPFC as a neural correlate of differential effort discounting: its activity tracks the effort-discounted value of the pair, as it is very active for pairs with disparate values of effort discounting and weakly active for pairs with similar effort discounting. The involvement of this brain region in effort-related processing has been suggested by other authors. Schmidt et al. (2009) presented a series of arousing pictures prior to effort exertion in exchange for a reward. They found that activity in VLPFC correlated with the level of arousal, interpreting VLPFC function as a "motivating" signal which facilitates effort exertion to obtain a reward. Although we did not include any motivating stimulus in our task, pairs with a higher difference of effort discounting might require extra motivation to overcome the negative effect of high effort. It should be taken into account that a high difference of effort discounting always means a comparatively high effort level in our task. However, a low difference of effort discounting could be due to similar effort levels, irrespective to the magnitude of the demanded effort. In this case the motivation signal could be irrelevant, as choosing either option does not make a big difference in terms of effort exertion. With respect to the literature on decision making, a recent experiment suggests the role of VLPFC in temporal discounting: in this case, it is thought to process a state-dependent cognitive control signal in order to determine the SV of waiting for delayed reward (Wierenga et al., 2015). The authors of this study found that VLPFC was especially active in sated volunteers and interpreted this activity as a cognitive control signal that helps them to wait for larger reward. Applying this to our results, pairs with a high difference of effort discounting would require a control signal to evaluate whether the more effortful option is really worth the effort when considered next to the other, much easier option.

One of the possible limitations of our first experiment is that the reward and prospective efforts are hypothetical for the subjects. However, a within-subject experiment on temporal discounting including hypothetical vs real reward revealed that both approximations account for the subject's behavior in a similar way (Johnson and Bickel, 2002). Within-subjects experiments in this context have been criticized because they do not consider the fact that volunteers may remember their responses to the previous condition of the task, although the key results (no differences between real and hypothetical reward) have been replicated with other methods (Lagorio and Madden, 2005; Madden et al., 2004). In addition, many behavioral studies on temporal and risk discounting have used hypothetical instead of real reward (Estle et al., 2006; Green and Myerson, 2004; Green et al., 2013; McKerchar et al., 2009 among others). In any case, this potential limitation does not affect our second experiment, where subjects were informed about the random selection of one of the presented pairs and the possibility of actually winning a reward in exchange for demanded effort. Another possible limitation of our task is ambiguity concerning whether we are assessing effort or temporal discounting, since effort load is measured as time (minutes running in the treadmill). Conceptually, however, the influence of temporal delay on our task is negligible. In the first experiment subjects were instructed to imagine they were ready to start the exercise and then make the decision between the fixed option (5 € reward with no demanded effort) and the more rewarding but effortful option. Thus, other factors such as time spent going to the gym, changing clothes, etc. were attenuated, as the decision was presented as if these things had already occurred. In the second experiment, where actual efforts and reward were at stake, the effect of temporal delay was diminished by the fact that subjects were told they would receive the reward (and make the required effort) during the week following the scan. Therefore, the actual point in time of obtaining the reward did not covary with the load of the exerted effort.

5 CONCLUSIONS

In this chapter we have analyzed behaviorally and at a neural level the intrinsic cost of effort in economic decision making. This is one of the main factors that contribute negatively to motivation for a specific exercise. We have designed a task to calculate individual and group effort discounting, and we have proven its validity and generalizability in relation to the sedentary lifestyle of volunteers. Finally, we have shown that different aspects of the prefrontal cortex (dorsomedial and ventrolateral) are associated with the subjective weighing of effort in decision making. We hope these results contribute to a better understanding of the subjective costs that affect motivation.

REFERENCES

Arrondo, G., Aznarez-Sanado, M., Fernandez-Seara, M.A., Goñi, J., Loayza, F.R., Salamon-Klobut, E., Heukamp, F.H., Pastor, M.A., 2015. Dopaminergic modulation of the trade-off between probability and time in economic decision-making. Eur. Neuropsychopharmacol. 25, 817–827.

Bartra, O., McGuire, J.T., Kable, J.W., 2013. The valuation system: a coordinate-based meta-analysis of BOLD fMRI experiments examining neural correlates of subjective value. Neuroimage 76, 412–427.

Bernacer, J., Corlett, P.R., Ramachandra, P., McFarlane, B., Turner, D.C., Clark, L., Robbins, T.W., Fletcher, P.C., Murray, G.K., 2013. Methamphetamine-induced disruption of frontostriatal reward learning signals: relation to psychotic symptoms. Am. J. Psychiatry 170, 1326–1334.

Bonnelle, V., Manohar, S., Behrens, T., Husain, M., 2016. Individual differences in premotor brain systems underlie behavioral apathy. Cereb. Cortex, 26 (2), 807–819.

Botvinick, M.M., Huffstetler, S., McGuire, J.T., 2009. Effort discounting in human nucleus accumbens. Cogn. Affect. Behav. Neurosci. 9, 16–27.

Burke, C.J., Brünger, C., Kahnt, T., Park, S.Q., Tobler, P.N., 2013. Neural integration of risk and effort costs by the frontal pole: only upon request. J. Neurosci. 33, 1706–1713.

Croxson, P.L., Walton, M.E., O'Reilly, J.X., Behrens, T.E.J., Rushworth, M.F.S., 2009. Effort-based cost-benefit valuation and the human brain. J. Neurosci. 29, 4531–4541.

Cummings, J.L., Mega, M., Gray, K., Rosenberg-Thompson, S., Carusi, D., Gornbein, J., 1994. The neuropsychiatric inventory: comprehensive assessment of psychopathology in dementia. Neurology 44, 2308–2314.

Dreher, J.-C., 2013. Neural coding of computational factors affecting decision making. Prog. Brain Res. 202, 289–320.

Estle, S.J., Green, L., Myerson, J., Holt, D.D., 2006. Differential effects of amount on temporal and probability discounting of gains and losses. Mem. Cognit. 34, 914–928.

FitzGerald, T.H.B., Seymour, B., Dolan, R.J., 2009. The role of human orbitofrontal cortex in value comparison for incommensurable objects. J. Neurosci. 29, 8388–8395.

Green, L., Myerson, J., 2004. A discounting framework for choice with delayed and probabilistic rewards. Psychol. Bull. 130, 769–792.

Green, L., Myerson, J., Oliveira, L., Chang, S.E., 2013. Delay discounting of monetary rewards over a wide range of amounts. J. Exp. Anal. Behav. 100, 269–281.

Gregorios-Pippas, L., 2009. Short-term temporal discounting of reward value in human ventral striatum. J. Neurophysiol. 101, 1507–1523.

Hartmann, M.N., Hager, O.M., Tobler, P.N., Kaiser, S., 2013. Parabolic discounting of monetary rewards by physical effort. Behav. Processes 100, 192–196.

Jenkinson, M., Beckmann, C.F., Behrens, T.E.J., Woolrich, M.W., Smith, S.M., 2012. FSL. Neuroimage 62, 782–790.

Jocham, G., Klein, T.A., Ullsperger, M., 2011. Dopamine-mediated reinforcement learning signals in the striatum and ventromedial prefrontal cortex underlie value-based choices. J. Neurosci. 31, 1606–1613.

Johnson, M.W., Bickel, W.K., 2002. Within-subject comparison of real and hypothetical money rewards in delay discounting. J. Exp. Anal. Behav. 77, 129–146.

Kable, J.W., Glimcher, P.W., 2007. The neural correlates of subjective value during intertemporal choice. Nat. Neurosci. 10, 1625–1633.

Kahneman, D., Tversky, A., 1979. Prospect theory: an analysis of decision under risk. Econometrica 47, 263–291.
Kobayashi, S., Schultz, W., 2008. Influence of reward delays on responses of dopamine neurons. J. Neurosci. 28, 7837–7846.
Kroemer, N.B., Guevara, A., Ciocanea Teodorescu, I., Wuttig, F., Kobiella, A., Smolka, M.N., 2014. Balancing reward and work: anticipatory brain activation in NAcc and VTA predict effort differentially. Neuroimage 102, 510–519.
Kurniawan, I.T., Seymour, B., Talmi, D., Yoshida, W., Chater, N., Dolan, R.J., 2010. Choosing to make an effort: the role of striatum in signaling physical effort of a chosen action. J. Neurophysiol. 104, 313–321.
Kurniawan, I.T., Guitart-Masip, M., Dolan, R.J., 2011. Dopamine and effort-based decision making. Front. Neurosci. 5, 81.
Lagorio, C.H., Madden, G.J., 2005. Delay discounting of real and hypothetical rewards III: steady-state assessments, forced-choice trials, and all real rewards. Behav. Processes 69, 173–187.
Levy, D.J., Glimcher, P.W., 2012. The root of all value: a neural common currency for choice. Curr. Opin. Neurobiol. 22, 1027–1038.
Madden, G.J., Raiff, B.R., Lagorio, C.H., Begotka, A.M., Mueller, A.M., Hehli, D.J., Wegener, A.A., 2004. Delay discounting of potentially real and hypothetical rewards: II. Between- and within-subject comparisons. Exp. Clin. Psychopharmacol. 12, 251–261.
Mcclure, S.M., Ericson, K.M., Laibson, D.I., Loewenstein, G., Cohen, J.D., 2007. Time discounting for primary rewards. J. Neurosci. 27, 5796–5804.
McKerchar, T.L., Green, L., Myerson, J., Pickford, T.S., Hill, J.C., Stout, S.C., 2009. A comparison of four models of delay discounting in humans. Behav. Processes 81, 256–259.
Meyniel, F., Sergent, C., Rigoux, L., Daunizeau, J., Pessiglione, M., 2013. Neurocomputational account of how the human brain decides when to have a break. Proc. Natl. Acad. Sci. U.S.A. 110, 2641–2646.
Mitchell, S.H., 2004. Effects of short-term nicotine deprivation on decision-making: delay, uncertainty and effort discounting. Nicotine Tob. Res. 6, 819–828.
Montague, P.R., King-Casas, B., Cohen, J.D., 2006. Imaging valuation models in human choice. Annu. Rev. Neurosci. 29, 417–448.
O'Doherty, J.P., 2011. Contributions of the ventromedial prefrontal cortex to goal-directed action selection. Ann. N. Y. Acad. Sci. 1239, 118–129.
Pessiglione, M., Schmidt, L., Draganski, B., Kalisch, R., Lau, H., Dolan, R.J., Frith, C.D., 2007. How the brain translates money into force: a neuroimaging study of subliminal motivation. Science 316 (5826), 904–906.
Peters, J., Büchel, C., 2009. Overlapping and distinct neural systems code for subjective value during intertemporal and risky decision making. J. Neurosci. 29, 15727–15734.
Peters, J., Büchel, C., 2010. Neural representations of subjective reward value. Behav. Brain Res. 213, 135–141.
Pine, A., Shiner, T., Seymour, B., Dolan, R.J., 2010. Dopamine, time, and impulsivity in humans. J. Neurosci. 30, 8888–8896.
Prévost, C., Pessiglione, M., Météreau, E., Cléry-Melin, M.-L., Dreher, J.-C., 2010. Separate valuation subsystems for delay and effort decision costs. J. Neurosci. 30, 14080–14090.
Salamone, J.D., 2009. Dopamine, behavioral economics, and effort. Front. Behav. Neurosci. 3, 1–12.

Schmidt, L., Clery-Melin, M.-L., Lafargue, G., Valabregue, R., Fossati, P., Dubois, B., Pessiglione, M., 2009. Get aroused and be stronger: emotional facilitation of physical effort in the human brain. J. Neurosci. 29, 9450–9457.

Scholl, J., Kolling, N., Nelissen, N., Wittmann, M.K., Harmer, C.J., Rushworth, M.F.S., 2015. The good, the bad, and the irrelevant: neural mechanisms of learning real and hypothetical rewards and effort. J. Neurosci. 35, 11233–11251.

Shenhav, A., Straccia, M.A., Cohen, J.D., Botvinick, M.M., 2014. Anterior cingulate engagement in a foraging context reflects choice difficulty, not foraging value. Nat. Neurosci. 17, 1249–1254.

Skvortsova, V., Palminteri, S., Pessiglione, M., 2014. Learning to minimize efforts versus maximizing rewards: computational principles and neural correlates. J. Neurosci. 34, 15621–15630.

Studer, B., Knecht, S., 2016. Chapter 2—A benefit–cost framework of motivation for a specific activity. In: Studer, B., Knecht, S. (Eds.), Progress in Brain Research, vol. 229. Elsevier, Amsterdam, pp. 25–47.

Treadway, M.T., Buckholtz, J.W., Schwartzman, A.N., Lambert, W.E., Zald, D.H., 2009. Worth the "EEfRT"? The effort expenditure for rewards task as an objective measure of motivation and anhedonia. PLoS One 4, 1–9.

Treadway, M.T., Buckholtz, J.W., Cowan, R.L., Woodward, N.D., Li, R., Ansari, M.S., Baldwin, R.M., Schwartzman, A.N., Kessler, R.M., Zald, D.H., 2012. Dopaminergic mechanisms of individual differences in human effort-based decision-making. J. Neurosci. 32, 6170–6176.

Weber, B.J., Huettel, S.A., 2008. The neural substrates of probabilistic and intertemporal decision making. Brain Res. 1234, 104–115.

Wierenga, C.E., Bischoff-Grethe, A., Melrose, A.J., Irvine, Z., Torres, L., Bailer, U.F., Simmons, A., Fudge, J.L., McClure, S.M., Ely, A., Kaye, W.H., 2015. Hunger does not motivate reward in women remitted from anorexia nervosa. Biol. Psychiatry 77, 642–652.

Wittmann, M., Leland, D.S., Paulus, M.P., 2007. Time and decision making: differential contribution of the posterior insular cortex and the striatum during a delay discounting task. Exp. Brain Res. 179, 643–653.

Worsley, K.J., Evans, A.C., Marrett, S., Neelin, P., 1992. A three-dimensional statistical analysis for CBF activation studies in human brain. J. Cereb. Blood Flow Metab. 12, 900–918.

CHAPTER

To work or not to work: Neural representation of cost and benefit of instrumental action

6

N.B. Kroemer*[,1], C. Burrasch*[,†], L. Hellrung*

**Technische Universität Dresden, Dresden, Germany*

[†]University of Lübeck, Lübeck, Germany

[1]Corresponding author: Tel.: +49-351-463-42206; Fax: + 49-351-463-42202, e-mail address: nils.kroemer@tu-dresden.de

Abstract

By definition, instrumental actions are performed in order to obtain certain goals. Nevertheless, the attainment of goals typically implies obstacles, and response vigor is known to reflect an integration of subjective benefit and cost. Whereas several brain regions have been associated with cost/benefit ratio decision-making, trial-by-trial fluctuations in motivation are not well understood. We review recent evidence supporting the motivational implications of signal fluctuations in the mesocorticolimbic system. As an extension of "set-point" theories of instrumental action, we propose that response vigor is determined by a rapid integration of brain signals that reflect value and cost on a trial-by-trial basis giving rise to an online estimate of utility. Critically, we posit that fluctuations in key nodes of the network can predict deviations in response vigor and that variability in instrumental behavior can be accounted for by models devised from optimal control theory, which incorporate the effortful control of noise. Notwithstanding, the post hoc analysis of signaling dynamics has caveats that can effectively be addressed in future research with the help of two novel fMRI imaging techniques. First, adaptive fMRI paradigms can be used to establish a time–order relationship, which is a prerequisite for causality, by using observed signal fluctuations as triggers for stimulus presentation. Second, real-time fMRI neurofeedback can be employed to induce predefined brain states that may facilitate benefit or cost aspects of instrumental actions. Ultimately, understanding temporal dynamics in brain networks subserving response vigor holds the promise for targeted interventions that could help to readjust the motivational balance of behavior.

Keywords

Response vigor, Striatum, Effort, Action control, Motivation, fMRI, Dopamine, Utility, Reward

Progress in Brain Research, Volume 229, ISSN 0079-6123, http://dx.doi.org/10.1016/bs.pbr.2016.06.009

1 INTRODUCTION

Despite good intentions, we do not always manage to give our best. Even when the action required to obtain a desirable goal is seemingly simple such as repeated button presses (BP), the behavioral output is characterized by an inherent variability. This variability in response to the same goal is typically treated as "noise" and handled by averaging of behavioral responses across a sequence of trials. However, if we suppose that actions are realized because a brain signal is translated into behavioral output, this noise might be indicative of the neural processes that give rise to vigor, not only qualitatively, but quantitatively. As a result, shared trial-by-trial differences in behavioral or neural responses can help to identify the underlying processes of motivation (Kroemer et al., 2014). As an illustrative example, we may consider a group of workers of a company. The management defines the goals for the workers' productivity, a set level that has to be met. Nevertheless, the workers typically differ relative to the set level in their average productivity (interindividual differences) as they do in their productivity from minute to minute, hour to hour, or even day to day (intraindividual variability). In the past decades, substantial progress has been made in identifying brain regions that correspond with the behavioral output *on average*. Whereas interindividual differences have received considerable attention in research, little is known about intraindividual variability, mainly because variability in response vigor after accounting for the incentive at stake and its subjective value was treated as uninformative noise (or residual variance error term, ε). Nevertheless, such intraindividual variability may entail information on which other motivational factors drive response vigor beyond the prospective incentive.

In this review, we will address the intriguing question why performance varies given the same incentive. We posit that variability can be partially accounted for by trial-by-trial fluctuations in the anticipation of costs and benefits of action. In other words, we propose that some of the variability in behavior occurs because our perception of costs and benefits is not constant and does not correspond to a true, yet unobservable, "subjective" value, which is merely corrupted by noise. Instead, valuation signals in response to the same incentive might be better characterized in terms of value distributions (Kroemer et al., 2016), where stronger signals are indicative of higher online estimates of subjective value (ie, higher anticipated benefit or lower cost). In turn, cue-induced reinforcement signals reflecting utility could support the invigoration of instrumental behavior (Kroemer et al., 2014). Arguably, there is also uninformative noise on top of the observed variability at the level of behavior and brain response, but emerging evidence suggest that brain response variability is an important and reproducible characteristic influencing behavior (Dinstein et al., 2015; Garrett et al., 2013, 2014; Kroemer et al., 2016). Such intraindividual variability can help us to fundamentally improve our understanding of the brain processes that subserve motivated behavior because it enables us to test strong hypotheses about the translation of brain response to action. Notably, this probes complementary information to the common parametric analysis based on subjective value, because an implicit assumption is often the global stability of such estimates across a sequence of trials. Temporal dynamics can thus provide additional insights into the adaptive transfer of value to action. By exploiting information contained in signaling dynamics of brain and behavior, this approach sheds light on the differences between brain regions that "set the tone" for work (eg, by tracking the expected value) and helps to dissociate it from other task-positive regions that actually put the demand to work (eg, by supporting faster motor responses). Moreover, we will describe how such a framework can be put to the test by employing recent advances in real-time fMRI (rt-fMRI), which enable the detection and utilization of current brain states (as in adaptive paradigms) or the feedback-based volitional induction of specified brain states (as in neurofeedback).

2 NEUROECONOMIC PERSPECTIVE ON EFFORT

The expenditure of effort implicates costs to an individual. On the one hand, motor effort incurs metabolic costs, which have often been considered to be negligible because of their small magnitude relative to the overall metabolic rate at rest. However, recent evidence derived from physiology research challenges this conclusion since our body actively seeks to minimize energy expenditure even when the potential gain is low in terms of total calories (Selinger et al., 2015; Zadra et al., 2016; for a full discussion, see Section 5). On the other hand, the investment of effort incurs opportunity costs simply because the individual cannot commit any other desired activity to the same degree at the same time (eg, Kurzban et al., 2013; Niv et al., 2007; Westbrook and Braver, 2016). In this perspective, dynamics arise in search of the optimal allocation of effort and a sense of effort can in turn encourage shifts in task allocation in order to optimize the division of limited processing capacities between the two tasks (Kurzban et al., 2013). In other words, opportunity costs of effort can induce soft constraints on processing capacities similar to memory load effects. As a result, according to neuroeconomic theories, effort requirements will lead to the discounting of a reward at stake thereby reducing its overall utility (Kivetz, 2003; Phillips et al., 2007). The reduced utility will shift choices to offers that require less effort or it will lead to less investment of effort in return for the potential benefit. Hence, the neuroeconomics framework offers a benchmark of an optimal decision-maker that we will use to describe how such optimality is approximated in many situations by individuals when it comes to estimating costs and benefits of instrumental action.

So how does *homo economicus* decide to work or rest? One particularly promising explanation is provided by cost evidence accumulation models (Meyniel et al., 2013, 2014). In these models, cost evidence accumulates during the exertion of effort and dissipates during extended rest, triggering effort cessation (with "exhaustion") and resumption (with "recovery"; Meyniel et al., 2013, 2014). Our proposal is complementary to cost evidence accumulation because one might regard it as a generalization of the idea. First, we argue that anticipated benefit corresponds to a distributed value signal, rather than to a fixed "true" value. In other words, analogous to costs, benefits are more actively inferred which introduces (partially shared) trial-by-trial fluctuations in brain response and behavior. Second, we hypothesize that a complementary cost signal may fluctuate also independently of fatigue, while fatigue is one likely key contributor to such variability. Third, instead of a model-based approach derived from behavioral data, we will focus on a more "model-free" approach where model parameters are not mapped onto brain regions, but brain response is used to constrain the feature/parameter space. This change in emphasis is mainly employed to demonstrate that these approaches do complement each other and may eventually help in building a more coherent understanding of cost/benefit analyses in the brain.

3 THE NEUROMODULATION OF EFFORT

An agent's "wanting" of a reward is commonly measured in terms of instrumental responses such as repeated lever presses. To estimate utility, reward schedules can be systematically manipulated to determine indifference points (ie, two options have approximately the same subjective value making it difficult to predict the exact choice), response thresholds (ie, responses occur when stimulus intensity exceeds a given threshold), or breakpoints. Breakpoints are operationalized by gradually increasing the response requirement on progressive ratio schedules of rewards until an individual ceases to respond (eg, Wanat et al., 2010). In contrast to wanting, "liking" is characterized by specific hedonic orofacial expressions and associated with separable neuroanatomical circuits (Berridge, 1996; Berridge and Kringelbach, 2015).

A considerable body of evidence has conclusively demonstrated that dopamine function is necessary for an animal's wanting to invest effort for the prospect of receiving reward in return. For example, lesions of the nucleus accumbens (NAcc) or the basolateral amygdala (BLA), dopamine depletion in NAcc or anterior cingulate cortex (ACC), and administration of dopamine antagonists lead to marked increases of the response threshold (Denk et al., 2005; Floresco and Ghods-Sharifi, 2007; Ostrander et al., 2011; Phillips et al., 2007; Salamone et al., 2007). Conversely, low doses of D-amphetamine (which increase dopamine transmission) improve the tolerance of animals to increased response costs (Floresco et al., 2008). However, these effects might be attributable to an improved tolerance to delay, which is a potential confound of effort requirements when more work also leads to increases in the delay to the reward receipt. Similarly, reduced cue-induced dopamine release in the NAcc with escalating response requirements on progressive ratio schedules might also be driven by increases in the delay to reward, but not increasing effort requirements per se (Wanat et al., 2010).

Dopaminergic effects on effort can also be dissociated from the modulatory effects of other neurotransmitters. Whereas serotonergic (Denk et al., 2005) or glutamatergic (Floresco et al., 2008) interventions affect an individual's sensitivity to delay, the representation of effort costs does not appear to be altered. Likewise, the depletion of serotonin brain levels impairs reversal learning while effort discounting remains unaffected (Izquierdo et al., 2012). In contrast, activation of GABAergic neurons in the ventral pallidum increases effort discounting (Farrar et al., 2008), and their input is modulated by a subpopulation of striatal neurons that coexpress adenosine (Mingote et al., 2008). Adenosine receptor modulation has been repeatedly shown to affect effort expenditure in concert with dopaminergic neuromodulation (Font et al., 2008; Worden et al., 2009). Collectively, these results indicate that dopamine consistently improves the tolerance to response costs in animal studies.

Notably, dopamine acts via two distinct neural pathways in the striatum, namely the D1 "go" circuit and the D2 "no-go" circuit (eg, Frank and Hutchison, 2009; Frank et al., 2004). Whereas response vigor maps intuitively onto the D1 "go" circuit, which is critically involved in learning from positive outcomes, response costs are thought to be encoded by the D2 "no-go" circuit, which is critically involved in learning from negative outcomes (Frank and Hutchison, 2009; Frank et al., 2004). Lower levels of D2 receptors in the striatum are considered to be one of the hallmarks of addiction (Volkow et al., 2011), which is associated with marked differences in the subjective value of work for drug vs monetary reward (eg, Bühler et al., 2010). Furthermore, initial evidence suggested that D2 receptor availability is also reduced in obese individuals (Wang et al., 2001), flanked by animal studies indicating that this deficit could be diet-induced (Johnson and Kenny, 2010). However, this finding has not been consistently replicated to date, which might be due to a nonlinear association with BMI (Horstmann et al., 2015). Notably, recent animal studies demonstrate that a D2 receptor knockdown strongly reduces physical activity in an environment enriched with voluntary exercise opportunities, facilitating the development of obesity (Beeler et al., 2015). Therefore, it has been argued that alterations in dopaminergic neurotransmission could potentially explain the observed differences in motivation and learning in obesity (Kroemer and Small, 2016).

Human studies targeting the dopaminergic system have corroborated the importance of dopamine in effort expenditure and effort discounting. Using [^{18}F]fallypride positron emission tomography (PET), which shows high affinity for D2/D3 receptors, Treadway et al. (2012) demonstrated that high-effort choices during low-probability trials (ie, very high opportunity costs) in a reward task were associated with stronger D-amphetamine-induced dopamine release in the caudate and ventromedial prefrontal cortex (vmPFC). Furthermore, they found a negative correlation with high-effort choices over all trials and D-amphetamine-induced dopamine release in the left and right insula (Treadway et al., 2012) suggesting that the effects of dopamine release in the insula are orthogonal to the effect in the mesocorticolimbic system. Beierholm et al. (2013) showed that the administration of L-DOPA, which increases tonic levels of dopamine, enhances the modulatory effect of the average reward rate (supposedly reflected in tonic dopamine levels; Niv et al., 2007) on response vigor. This modulatory effect was specific to L-DOPA as the administration of citalopram, a selective serotonin reuptake inhibitor, did not affect response vigor.

Further evidence is provided by studies on the loss of dopaminergic neurotransmission as in Parkinson's disease (PD). For example, selective deterioration of substantia nigra (SN) dopamine neurons occurs in aging and is known as one of the risk factors for the development of PD, and this deterioration leads to a loss in signal fidelity (Branch et al., 2014), which possibly increases the noise in the representation of stimuli and the motor output in response to them. Higher variability and lower accuracy of motor behavior were commonly interpreted as consequences of generalized movement slowing (bradykinesia) in PD, but recent studies showed that PD patients can be as accurate as healthy participants. However, PD patients show much steeper discounting of reward as a function of motor effort (Mazzoni et al., 2007), an effect that is attenuated by dopaminergic medication (Chong et al., 2015). Intriguingly, Manohar et al. (2015) showed that the shift in the cost/benefit ratio of effort expenditure observed in PD patients might be explained by an effortful mechanism of noise control, which can be employed to improve the precision if the costs appear to be justified by the incentives at stake. This model provides a parsimonious framework to integrate how online estimates of utility, literally, may go hand in hand with motor control policies that implement action with an optimal balance of force and precision (for details, see Section 6).

To summarize, a mounting body of evidence has demonstrated that dopamine is critically involved in action control and the invigoration of behavior. These observations in animals and humans are flanked by studies in mice, where the absence of dopamine signaling causes severe apathy and, ultimately, starvation (Palmiter, 2007, 2008). Whereas the importance of dopamine signaling in the allocation of effort is well established, we are only beginning to unravel the exact functional contributions of brain regions within the motivation network to response vigor.

4 BRAIN REGIONS SUBSERVING THE ALLOCATION OF EFFORT

In the past decades, "valuation," "reward," or "action control" networks have been well characterized along anatomical (eg, Haber and Knutson, 2010) and functional axes (eg, Liu et al., 2011; Peters and Büchel, 2011). Since a full review is beyond the scope of this chapter, we will focus here on the implications of these insights for reward-related action control and trial-by-trial estimation of utility in several candidate brain regions (Fig. 1) evaluated in Kroemer et al. (2014). The study's design and findings are schematically summarized in Figs. 1 and 2. Briefly, four reinforcement levels were indicated by cues before the onset of the motor-response phase. In this 3 s response interval, vigor was instrumental that is each button press was rewarded by an individualized reward unit, which was multiplied with the reinforcement level [0,1,10,100] and feedback on the reward obtained was displayed after each trial. The key result of the full mixed-effects model of brain response and behavior was that anticipatory responses at the stage of the reinforcement cue were predictive of subsequent response vigor (Kroemer et al., 2014) and that the analysis of trial-by-trial dynamics revealed a network of "shared labor" that we will briefly review in light of recent results.

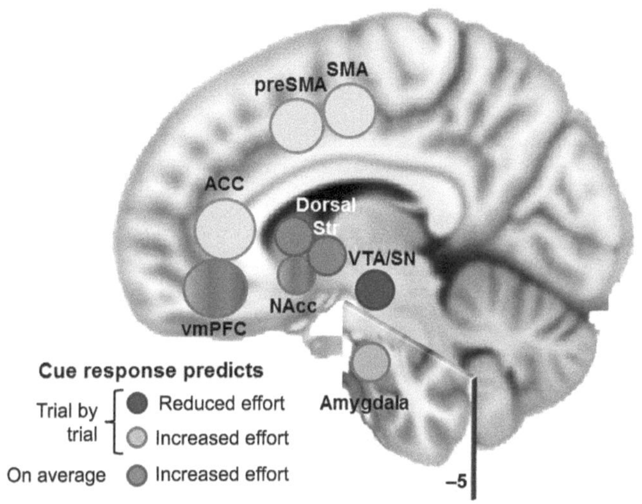

FIG. 1

Schematic summary of the results by Kroemer et al. (2014). All regions-of-interest show evidence for encoding the reward level information and except for the VTA/SN they showed a positive association with effort (trend level for ACC and preSMA). Using a full-mixed effects analysis, the contribution to trial-by-trial fluctuations in effort on the one hand and average effort on the other hand could be disentangled. Average effort was predicted by increased cue-induced activation in the NAcc, dorsal striatum, and vmPFC. Above-average effort, however, could be predicted by increased cue-induced signals in the amygdala, NAcc, and vmPFC and *decreased* cue-induced signals in the VTA/SN. These results point to a dissociation between NAcc and VTA/SN (work more vs less) as well as the dorsal and ventral striatum (set level vs online estimate of utility).

4.1 VENTRAL STRIATUM/NAccs

The ventral striatum has long been hypothesized to support the invigoration of behavior, in concert with its role in reinforcement learning (Collins and Frank, 2015; Mannella et al., 2013). Recent studies in rodents demonstrate that the expression of actions is essential for dopamine signaling according to reinforcement-learning principles (Syed et al., 2016) and that minute-by-minute changes in dopamine levels reflect the willingness to work for reward (Hamid et al., 2016). In humans, stronger anticipatory reward-cue responses predict stronger subsequent expenditure of motor effort and this is also observed when the cue signals the absence of reward (Kroemer et al., 2014). Notably, striatal responses are attenuated when a high-effort choice has to be made, but when participants voluntarily choose to exert effort, stronger brain responses are again observed (Schouppe et al., 2014). The signal in the NAcc may therefore also impinge on rational behavior leading to disadvantageous decisions and costly errors, suggesting that it reflects a tendency to approach (Chumbley et al., 2014). Furthermore, the variability of the NAcc response to food reward is predictive

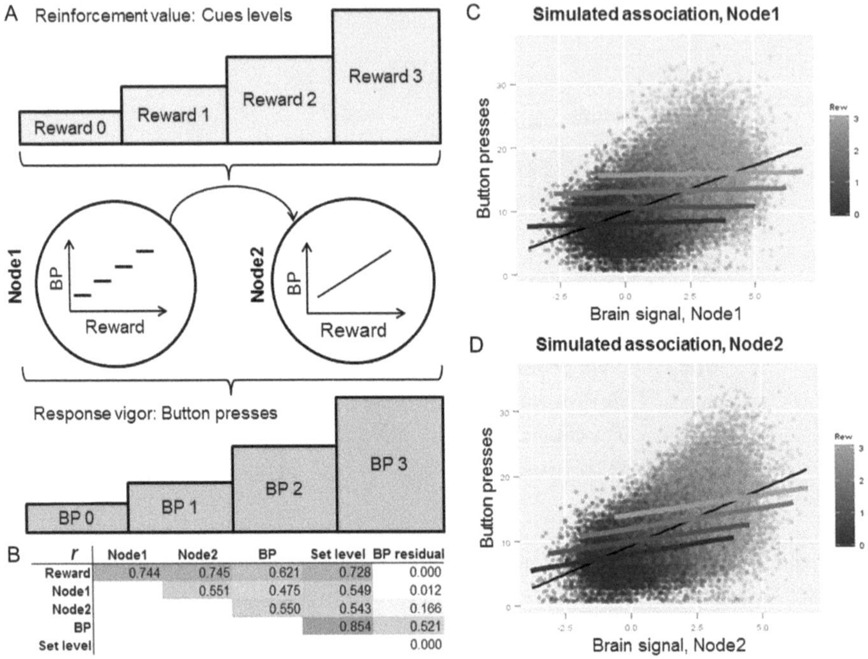

B

r	Node1	Node2	BP	Set level	BP residual
Reward	0.744	0.745	0.621	0.728	0.000
Node1		0.551	0.475	0.549	0.012
Node2			0.550	0.543	0.166
BP				0.854	0.521
Set level					0.000

FIG. 2

Correlations of brain signal and button presses (BP) can be driven by two complementary processes as demonstrated by simulations. The simulation resembles the design used in Kroemer et al. (2014) and involves four reward levels (RLs; coded as [0, 1, 2, 3]), 96 trials in total, and 500 agents. Signal strength of the nodes is simulated in accordance with single-trial betas. (A) Within the small network, Node1 represents the difference between RLs, which is translated into more BP on average (ie, modulation of the intercept for each RL; resembling the putative role of the dorsal striatum). The "value" information stored in Node1 is also used to set the target amplitude of the brain response in Node2. In Node2, brain responses are actively sampled from a Gaussian distribution set to the average of Node1's response for each RL (and with the same "noise level" as Node1) and then probabilistically translated into BP (resembling the putative role of the ventral striatum). (B) While the overall correlations between reward, BP, and the signal in Node1 and Node2 are highly similar, only the signal in Node2 (see panel D vs C depicting Node1) is associated with trial-by-trial fluctuations in BP (BP residual). (C) and (D) The *thin black regression lines* depict the correspondence between vigor and brain signal across RLs whereas the *thick gray-scaled regression lines* depict the correspondence between vigor and brain signal within each RL (color coded in *gray* shades).

of the variability in food intake and the reinforcement value of food (Kroemer et al., 2016). Since variability of the reinforcement signal was about as reproducible across different sessions as its amplitude, it suggests that brain responses in the NAcc should be characterized not only in terms of their average amplitude but rather as "value distributions" (Kroemer et al., 2016). When the reward cannot be increased by voluntarily spending more effort, the ventral striatum (and dopaminergic midbrain) tracks the net value of an option that is the reward discounted by the required effort needed to obtain it (Botvinick et al., 2009; Croxson et al., 2009; Kurniawan et al., 2013). Collectively, these results point to the ventral striatum/NAcc as a prime candidate brain region representing an integrated online estimate of utility for a given action policy.

4.2 DORSAL STRIATUM (CAUDATE/PUTAMEN)

The dorsal striatum mainly receives projections from the SN and is hypothesized to be involved in planning, execution, and automatization of motor behavior. Evidently, these processes are critically involved in the formation of habits and the implementation of habitual behavior. Hence, the dorsal striatum has been dubbed "the actor" of the reward-signaling pathways (O'Doherty et al., 2004) and distinguished from "the critic": the ventral striatum. With regard to the invigoration of behavior, it has been hypothesized that dopamine, particularly in the dorsal striatum (Wang et al., 2013), provides an energy budget for action (Beeler, 2012; Beeler et al., 2012, 2015), which is also supported by recent studies in rodents demonstrating that the dorsal striatum represents nutritional value, not the hedonic value, of sugar (Tellez et al., 2016). Likewise, the dorsal striatum is sensitive to the availability of food beyond caloric content in humans (Blechert et al., 2016). Thus, the dorsal striatum may be sensitive to alterations in metabolism and metabolic state (Kroemer and Small, 2016).

There is conclusive evidence in animals (Wang et al., 2013) and humans demonstrating that the dorsal striatum is involved in encoding effort requirements (Kurniawan et al., 2010, 2013) or average effort spend for a given reward (Kroemer et al., 2014). The most pervasive evidence comes from a series of studies involving mice that were genetically engineered to lack tyrosine hydroxylase in dopamine neurons. Tyrosine hydroxylase is the rate-limiting enzyme in the synthesis of dopamine, and its lack causes hypoactivation, aphagy, and, ultimately starvation unless feeding is rescued by the treatment with L-DOPA (Palmiter, 2007, 2008). Since food hedonics and spatial learning of food rewards remain functional, these alterations can be attributed to a lack of motivation to engage in behavior. Crucially, restoration of dopamine signaling in the dorsal striatum is sufficient to restore feeding and locomotion. This illustrates how essential dopamine function within the dorsal striatum is to instrumental behavior (Palmiter, 2007, 2008).

4.3 DOPAMINERGIC MIDBRAIN (VTA/SN)

The ventral tegmental area (VTA), SN, and the retrorubral cell groups constitute the dopaminergic midbrain, and the functional connectivity between the dopaminergic midbrain and the striatum resembles a feedback loop (Haber and Knutson, 2010). Strong reward-related increases in the BOLD response can be reliably observed in the VTA (http://www.neurosynth.org/analyses/terms/reward/), and multimodal imaging studies suggest that this brain response in the VTA/SN and ventral striatum is correlated with dopamine release as measured using [11C]raclopride PET in the ventral striatum (Schott et al., 2008).

Neurophysiology research in animals has demonstrated that response costs attenuate the neural response in the VTA/SN, which indicates that the value of rewards is discounted by the delay to its receipt (Kobayashi and Schultz, 2008), the risk (Stauffer et al., 2014), or the effort needed to obtain it (Pasquereau and Turner, 2013; Varazzani et al., 2015). Critically, dopamine neurons in the SN pars compacta reflected upcoming effort cost during anticipation, which was associated with the negative influence of effort on action selection (Varazzani et al., 2015). This observation may explain why stronger anticipatory cue-responses were associated with *reduced* effort expenditure in humans, in contrast to cue signals in the amygdala, NAcc, or vmPFC (Kroemer et al., 2014). Notably, if the reward value cannot be increased by spending more effort, the VTA/SN tracks the net value of an option in conjunction with the ventral striatum (Croxson et al., 2009). Thus, the VTA/SN possibly encodes the (average) value of the reward at stake discounted by the effort, which is going to be invested in order to obtain the desired reward.

4.4 VENTROMEDIAL PREFRONTAL CORTEX

Human neuroimaging research has conclusively shown that the vmPFC is important in transferring subjective value to action (Grabenhorst and Rolls, 2011; Levy and Glimcher, 2012). In addition to subjective-value information, the vmPFC may forward a second-order valuation signal reflecting the confidence in a decision based on value judgments (Lebreton et al., 2015). Opportunity costs of action such as delay are reliably encoded in the vmPFC (Peters and Büchel, 2011), and it has been shown to integrate cost and benefit information (Basten et al., 2010), but there is also evidence that effort costs may recruit a different functional network (Prevost et al., 2010; Rudebeck et al., 2006). Nevertheless, Kroemer et al. (2014) found that above-average cue responses in the vmPFC predicted above-average effort expenditure, in concert with the amygdala and the NAcc, and D-amphetamine-induced dopamine release in the vmPFC is associated with effort discounting (Treadway et al., 2012). Hence, while the exact contribution to effort-based decision-making still remains largely elusive, the extensive body of evidence on subjective value, choice, and subsequent implementation of behavior strongly suggests that the vmPFC is involved in the online estimation of utility.

4.5 AMYGDALA

Despite the classical focus of amygdala research on the processing of emotions and fear conditioning, the amygdala appears to be generally involved in encoding relevance (in concert with the ventral striatum; Ousdal et al., 2012) and salience (eg, Anderson and Phelps, 2001), exerting a bottom-up "priority bias" on other regions within the mesocorticolimbic circuit (Mannella et al., 2013). The strong structural connections between the amygdala and the ventral striatum are ideally suited to subserve rapid encoding of stimulus–outcome associations and conditioning in general (Haber and Knutson, 2010). Accordingly, cue-induced dopamine release in the NAcc is modulated by one of the distinct cores within the amygdala, the BLA. In rodents, inactivation of the BLA reduces cue-induced dopamine release in the NAcc, which attenuates cue-induced conditioned approach behavior (Jones et al., 2010). Moreover, the transfer of information between the BLA and the prefrontal cortex (ie, the ACC) affects effort discounting since inactivation (Floresco and Ghods-Sharifi, 2007) or lesions (Ostrander et al., 2011) of the BLA make animals avoid high-effort requirements to obtain high-reward options.

Furthermore, human imaging studies have provided compelling evidence that the amygdala is involved in the cost/benefit trade-off. For example, Basten et al. (2010) showed that the amygdala encodes the costs associated with specific stimulus–outcome associations. While the ventral striatum provides an estimate of the benefits, the amygdala forwards the representation of the implied costs to the "comparator region" vmPFC. In this region, costs and benefits are integrated and the evidence for a given option is accumulated by the interconnected intraparietal sulcus, which, ultimately, gives rise to the decision (Basten et al., 2010). However, when behavior needs to be invigorated, fluctuations in the amygdala may reflect the effectiveness of the induction of behavioral approach (Kroemer et al., 2014). To summarize, the amygdala appears to be critically involved in the cost/benefit trade-off, which is essential to adaptive action control, and the BLA in particular might regulate the invigoration of behavior by the prospect of reward.

4.6 SUPPLEMENTARY MOTOR AREA

In order to perceive physical effort, a motor signal needs to carry information about the intensity of muscle contraction, and the supplementary motor area (SMA) is known to represent this important component (Zenon et al., 2015). For example, brain activation in the SMA correlates with the exerted force to obtain rewards on a grip device (Pessiglione et al., 2007), and cue-induced signals indicating reward correlate with effort expenditure (Kroemer et al., 2014). Intriguingly, disruption of the SMA signal by the application of repeated transmagnetic stimulation (rTMS) increases grip force (White et al., 2013) and reduces perceived effort (Zenon et al., 2015). Likewise, activation within the SMA may correspond to a brain state of vigilance, which could contribute to particularly vigorous responding (Hinds et al., 2013). Notably, recent work has suggested that behavioral apathy, a trait-like characteristic characterized by lack of motivation to initiate behavior or responses associated with increased effort sensitivity, is associated with greater recruitment of SMA and cingulate motor regions as well as decreased structural and functional connectivity between the SMA and ACC (Bonnelle et al., 2016). Taken together, this evidence suggests that the SMA is a promising target within the circuit to modify response vigor, possibly by affecting the subjective perception of effort expenditure.

4.7 ANTERIOR CINGULATE CORTEX

The ACC has drawn a lot of attention in neuroscientific research, leading to many influential theories that seek to explain the diverse set of published results. However, a detailed discussion is beyond the scope of this review. By summarizing popular theories of ACC function and their evidence, Holroyd and Yeung (2011) proposed a unifying framework suggesting that the ACC is critically involved in setting high-level plans. As a result, performance and valence monitoring occurs in order to keep track of the implementation of goals by hierarchical reinforcement learning. Striking evidence for the involvement of the ACC comes from animal research. For example, rats with ACC lesions cease to work for high reward if it requires high effort (Rudebeck et al., 2006; Walton et al., 2003). In humans, the ACC has been shown to be involved in cognitive (Botvinick et al., 2009; Westbrook and Braver, 2016) and physical effort discounting (Croxson et al., 2009). Consequently, it has been suggested that signals from ACC to NAcc or the dopaminergic midbrain act top-down to support an agent in overcoming effort-related response costs (Walton et al., 2006). One potential mechanism may be the active control of a "gain rate" that determines the signal-to-noise (SNR) ratio within the motivation network (Verguts et al., 2015). In the neurocomputational framework by Verguts et al. (2015), reward and cost feedback from the dopaminergic midbrain provides input into ACC, which supports learning of action policies to allocate effort. Collectively, these studies suggest that the ACC is involved in learned and, possibly, strategic aspects of effort allocation.

5 METABOLIC COSTS AS A CONSTRAINT IN EFFORT EXPENDITURE

Any type of physical effort expended comes at the cost of metabolism (O'Dwyer and Neilson, 2000). Given a choice, animals usually choose the less effortful option to pursue an objective in order to avoid unnecessary metabolic costs (Salamone et al., 2007; Walton et al., 2006). As a principal law of survival, all energy expenditure must be compensated by intake of energy; and therefore, it is imperative to maximize the cost/benefit ratio according to economic principles. Consequently, the anticipated benefit of the reward needs to surpass the perceived costs of action for a reward to motivate its approach (cf. Proffitt, 2006). Although the potential to save energy by means of optimizing motor behavior appears to be low relative to the overall energy expenditure, research in the past decades has demonstrated that individuals prefer to move in energetically optimal ways (Selinger et al., 2015). The timescale of such an optimization process has been often debated along evolutionary or developmental timescales, but recent theories of motor control have emphasized the potential of the continuous and dynamic optimization of energetic cost. For example, humans rapidly adapt their walking speed to different levels of difficulty (as dynamically defined via an exoskeleton), and this change in preference is associated with an optimization of energy expenditure (Selinger et al., 2015).

Moreover, metabolic state influences estimates of walkable distances, which can be considered as subjective estimates of effort. When energy is readily available, individuals are more inclined to exert effort suggesting that the perception of effort is bioenergetically scaled (Zadra et al., 2016). Likewise, the recently proposed "thrift theory" of dopamine proposes a conserve–expense axis where the availability of energy determines exploratory vs exploitatory behavior (Beeler et al., 2012).

Consequently, a better understanding of the physiological factors underlying effort allocation can be gained from populations with aberrant metabolism or altered body composition. For example, BMI has been negatively associated with the willingness to exert effort in order to obtain snack foods (Mathar et al., 2015) but is positively associated with their liking (Goldfield et al., 2011). This inversion of preference appears to trace back to the individual metabolic costs, which are recruited by instrumental behavior. Since the energy expended by the skeletal muscles increases along with the weight they have to support, movement becomes more costly with increasing BMI (Browning et al., 2007; Leibel et al., 1995), and low physical activity is a predisposing factor for obesity (Fogelholm et al., 2007; Pietilainen et al., 2008). Impaired glucose metabolism, which is associated with obesity (Chan et al., 1994; Colditz et al., 1990; Mokdad et al., 2003), could further contribute to an avoidance of effortful behaviors. The effectiveness of glucose metabolism determines how fast and efficiently energy can be allocated. Thereby, reduced insulin sensitivity may inflate the perception of effort (McArdle et al., 2010). These results corroborate a wealth of evidence indicating that BMI is positively associated with an increased fraction of time spent on sedentary activities, which require little physical activity (eg, watching TV) in everyday life (Beunza et al., 2007; Kaleta and Jegier, 2007; Matthews et al., 2008; Mitchell et al., 2013; Rhodes et al., 2012). In line with these results, in obese women, eating away from home and consumption of instant meals (vs self-prepared food) is associated with increased impulsivity and excess caloric intake (Appelhans et al., 2012) suggesting that the perceived effort to prepare food may contribute to the maintenance of unhealthy diets.

Moreover, dopamine and obesity are closely linked via genetic mechanisms or endocrine signals of metabolic state (Kroemer and Small, 2016). Briefly, polymorphisms in the ANKK (TAQIA) and FTO gene have been associated with obesity, weight gain, and altered D2 receptor functioning (Sevgi et al., 2015; Stice et al., 2008, 2015; Sun et al., 2015). In animals, it has been shown that ghrelin increases motivation to work for food (King et al., 2011), and leptin regulates effort allocation for food and mesolimbic dopamine via the midbrain (Davis et al., 2011). Furthermore, insulin resistance was found to alter dopamine turnover (Kleinridders et al., 2015). In humans, endocrine signals have been shown to be associated with dopamine function as well (Caravaggio et al., 2015; Dunn et al., 2012), in line with their modulatory effect on motivation. Consequently, insulin, leptin, and ghrelin also modulate anticipatory (Grosshans et al., 2012; Kroemer et al., 2013, 2015; Malik et al., 2008) and consummatory responses (Kroemer et al., 2016; Sun et al., 2014) to food in the mesocorticolimbic system. Therefore, future research focusing on individuals with obesity and/or diabetes could provide insights into the role of physiology in the perception of effort. These results may in turn be utilized to develop strategies to improve public health or treatment of metabolic disorders.

Whereas the metabolic costs of action are evident, it remains under debate as to what degree cognitive effort exerts metabolic costs as well. Multiple studies suggest that cognitive performance is influenced by metabolism. For example, cognitive performance improves after glucose administration (Hall et al., 1989; Kennedy and Scholey, 2000; Manning et al., 1992, 1998; Meikle et al., 2004; Riby et al., 2004; Smith et al., 2011), and peripheral blood glucose levels are reduced after periods of sustained cognitive demand (Donohoe and Benton, 1999; Fairclough and Houston, 2004; Gailliot and Baumeister, 2007; Gailliot et al., 2007; Scholey et al., 2001), although this has not been replicated consistently (Inzlicht et al., 2014; Molden et al., 2012). Moreover, the effects may be domain-specific (Orquin and Kurzban, 2015). Nevertheless, mental workload is associated with changes in respiratory measures of metabolism that indicate increased energy expenditure (Backs and Seljos, 1994). Relatedly, mental effort is commonly experienced as aversive (Cuvo et al., 1998; Eisenberger, 1992), and humans avoid engaging in unnecessary demanding cognitive activities (Kool et al., 2010; McGuire and Botvinick, 2010), which suggests that it comes at a subjective cost. Alternatively, it has been proposed that mental effort only imposes opportunity costs (Kurzban et al., 2013), which are mediated by dopamine function as well, but that the associated metabolic costs are negligible (Westbrook and Braver, 2016). However, dopamine antagonists do not seem to affect cognitive effort as they do affect physical effort in rodents (Hosking et al., 2015) calling for future research on the neurobiological basis of potential effort-cost domains.

To conclude, the metabolic costs of action serve as constraints on energy expenditure, possibly because of the evolutionary need to optimize costs and benefit of goal-directed behavior in order to support allostasis and avoid potential starvation (Korn and Bach, 2015). While the metabolic costs of cognitive control are controversially debated, we propose that the effortful control of noise may provide a unifying framework for motor and cognitive control policies that are optimized according to the anticipated costs and benefits of behavior (Manohar et al., 2015).

6 THE EFFORTFUL CONTROL OF NOISE AS A UNIFYING FRAMEWORK

Whereas metabolic costs of action have been used to describe how organisms optimize energy expenditure (Selinger et al., 2015), this perspective is much more controversial when it comes to cognitive effort. For a long time, resource models of cognitive control have relied on metaphorical abstractions of psychological and physiological processes (willpower). The influential work on blood levels of glucose as a physiological correlate of the cognitive control resource (Gailliot and Baumeister, 2007; Gailliot et al., 2007) has helped to put the metaphors to a test, even if the concept of "ego depletion" has arguably been proven to be too simplistic

(Inzlicht et al., 2014; Kurzban et al., 2013; Lurquin et al., 2016). Instead of depleting a limited resource, metabolic state may exert its influence via shifts in the motivational balance between labor and leisure (Inzlicht et al., 2014). Hence, metabolic state may put action on a metabolic budget (Beeler et al., 2012), where costs and benefits are evaluated dynamically, which may give rise to trial-by-trial fluctuations in motivation that reflect the current motivational balance (Kroemer et al., 2014; Meyniel et al., 2013, 2014).

To this end, optimal control models of behavior can help to describe how normative improvement in behavior can be achieved according to neuroeconomic principles of utility (Manohar et al., 2015; Meyniel et al., 2013; Rigoux and Guigon, 2012; Shenhav et al., 2013). Here, we will focus on the effortful control of noise framework (Manohar et al., 2015) as a recent extension that holds the potential to integrate seemingly distinct elements of action control into the (parsimonious) challenge to adjust "noise" according to anticipated costs and benefits (Fig. 3). Within this framework, the expected value of a particular control command is determined by three elements: (1) First of all, the expected value is driven by the incentive: the reward discounted by time. The reward term takes into account that high response vigor, represented by the term cost of force u_F, leads to faster gratification. (2) The second parameter reflects noise in motor control. The noise parameter is a function of baseline variability and increases with response vigor. Crucially, the slope of this increase is reduced by the precision weight, u_P. (3) Lastly, regulation of noise is constrained by the cost term of precision and force $-(|u_P|^2+|u_F|^2)$ (Manohar et al., 2015). Within this computational framework, it is possible to optimize precision and force, which would lead to normative improvements in performance. This is achieved because higher incentives increase the "reward term" in the equation, thereby leading to a different set of parameters for the optimal balance between cost of precision and force, $\mathbf{u}$.

Similar to the "control of noise" framework by Manohar et al., Verguts et al. (2015) have suggested that the active control of a gain parameter via the ACC, which may boost the SNR within the motivation network, supports the allocation of effort according to the learned value of action policies (Verguts et al., 2015). Thus, the control of noise or gain as a challenge in instrumental action may provide a major advance in our understanding because it helps to reconcile two important neuromodulatory functions of dopamine. In addition to the rich literature on dopamine and action control, dopamine has been shown to regulate signal fidelity and noise (Garrett et al., 2013, 2015; Li and Rieckmann, 2014). Within the control of noise framework, dopamine could support a more costly mode of action control that it characterized by a better ratio of response vigor to noise (Manohar et al., 2015). Hence, performance might be improved via increases in force or increases in precision (ie, decreases in the slope between increasing force and noise) or both. This costly mode is employed according to its utility that is whenever incentives (intrinsic or extrinsic) encourage optimal performance and, thereby, pay the costs of control (Manohar et al., 2015). As a result of this framework, we can hypothesize that this costly mode of control is characterized by a specific brain state that supports such vigilance (Hinds et al., 2013) or vigor (Kroemer et al., 2014). For example, a

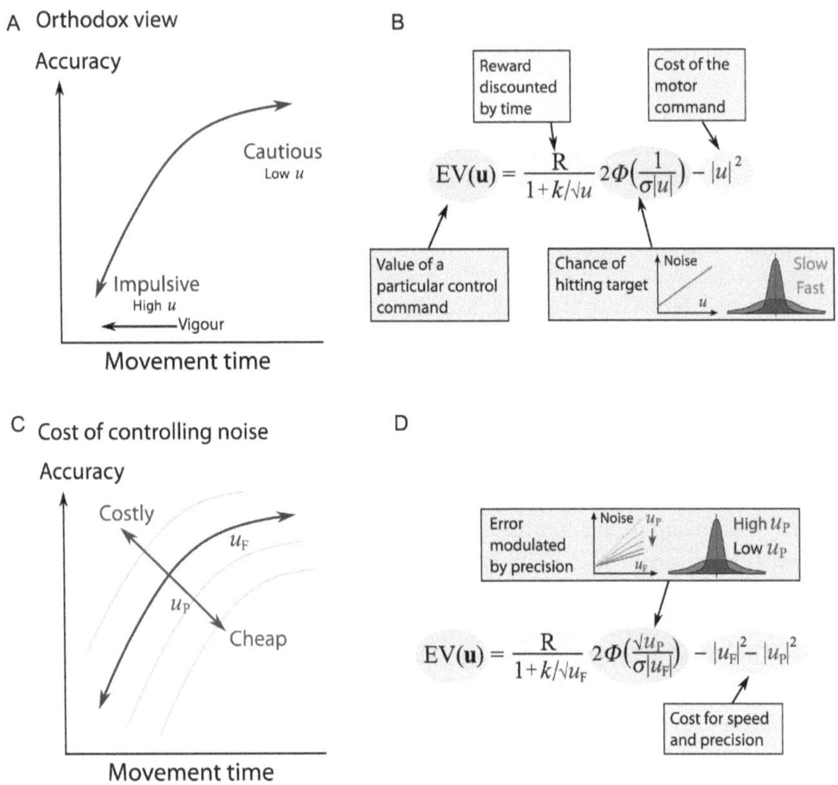

FIG. 3

Summary of the noise control framework provided by Manohar et al. (2015). According to the orthodox view of the speed–accuracy trade-off (upper panel), increases in vigor amplify noise, thereby reducing accuracy of behavior (A). This is expressed in equation (B). The introduction of a precision weight u_p that is at the same time costly allows for normative increases in performance (C) and extends the equation (D). Noise control is optimal when the potential reward exceeds the implicated costs of increased velocity and reduced variability. **u**, cost of precision (*p*) and force (*f*); *k*, discount rate of the reward; σ, noise term.

cognitive control signal forwarded by the dorsal ACC may correspond to a more costly control mode by indexing a more effortful control policy that is, nevertheless, "worth the effort" in terms of the expected value of control (cf. Shenhav et al., 2013; Verguts et al., 2015). Furthermore, it is conceivable that such a priorization will involve multiple nodes within the network such that the brain state could be probabilistically detected based on a specific spatio-temporal profile. Once we have developed a working model of what the spatial and temporal profile of a "vigorous" brain state is, we can try to translate this into an experimental setting to test, if we can actually predict effort from online signals of utility.

7 A SIMPLE SIMULATED NETWORK OF SHARED LABOR

The correlation between the amplitude of brain signals and the effort provides tentative support for an association. Notwithstanding, the exact functional contribution of a brain region to the invigoration of behavior is hard to parse simply from the observed correlation without exploiting the information provided by the signaling dynamics of brain and behavior. To illustrate this more formally, we simulated two nodes that contribute in complementary ways to instrumental behavior (Fig. 2). Yet, at the subject and group level, it is virtually impossible to decompose this contribution (ie, *black regression lines*, Fig. 2C and D) without the addition of signaling dynamics within reward levels (RLs) to the equation (*thick gray regression lines*, Fig. 2C and D). The design of the simulated study is analogous to Kroemer et al. (2014) and involves four RLs (coded as [0, 1, 2, 3]), 96 trials in total, and 500 agents.

At the neural layer, we assumed two key nodes. Node1 (representing the function of the dorsal striatum as described before) encodes the RL faithfully with random Gaussian noise added to the representation, Noise $\sim N(\mu=0,\sigma=1)$. Hence, Node1 represents the set level, which is defined by the reward within the task because it is well-known that higher incentives encourage more effort. Node2 (representing the function of the NAcc as described before) uses the input of the "set level" Node1 (ie, the average signal stratified by RL) and samples from a Gaussian distribution with the amplitude parameter μ set to the "neural" representation of the RL information, $N(\mu=\text{Node1}_{\text{RL}},\sigma=1)$. This sampling scheme produces indistinguishable distributions of brain responses for Node1 and Node2 since the only difference is how we have defined noise: In Node1, random noise is added to the representation of the RL whereas in Node2, the noise arises because the brain is actively sampling from a population of brain responses (Garrett et al., 2013; Kroemer et al., 2016). These simulated brain signals can be thought of as single-trial beta estimates that is they reflect the strength of a signal relative to a baseline (intercept), and we used this descriptive level because more extensive simulations based on neurobiological temporal characteristics would be beyond the scope of this illustration.

At the output level, we assume that both nodes contribute to the invigoration of behavior, yet in complementary ways. With Node1 representing the RL, we translate this difference into an overall shift of BP with increasing RLs. Within a mixed-effects regression framework, this would correspond to a slope that modulates the RL slope (ie, the difference between RLs in average BP) on BP by the activation in Node1, b_{Node1}. The set level is then calculated by the regression: $\text{setBP}_{\text{RL}}=\text{intercept}+b_{\text{Node1}}\times\text{Node1}$. We initialized the slope b_{Node1} by sampling the value of each simulated agent from a Gaussian distribution, $N(\mu=2,\sigma=1)$. In addition, we translated the signal in Node2, which is also dependent on the representation in Node1, into proportional increases of response vigor, b_{Node2}. This was done by, again, sampling each agent's parameter from a Gaussian distribution, $N(\mu=0.75,\sigma=1)$, to calculate BP as (rounded) output from the regression: $\text{BP}_{\text{RL}}=\text{setBP}_{\text{RL}}+b_{\text{Node2}}\times\text{Node2}+\text{Noise}\ (\sim N(\mu=0,\sigma=2))$.

A third node might be added to the simulation that tracks response costs resembling the contribution of the VTA/SN in Kroemer et al. (2014). We also simulated such an extended network and observed that hyperbolic discounting of pending response vigor could resemble the main empirical findings of (a) positive correlations among the nodes of the network, (b) positive correlations with overall vigor, and (c) negative correlations with trial-by-trial fluctuations in vigor. However, we also observed that this ensemble was much more sensitive to the choice of parameters (eg, range of neural discount rates and statistical dependence of nodes and BP) which illustrates the need to inform more complex future simulations with empirically derived constraints. Of note, the basic pattern of results in the two-node simulation was also robust to nonlinearity in gain or transfer functions. In line with the empirical results of Kroemer et al. (2014), we reduced the variability of behavioral response vigor and brain response with higher reward incentives in an alternative simulation. Critically, in both simulated and empirical cases, there was no evidence for a potential confounding effect of nonlinear gain. Furthermore, when a log-sigmoid transfer function (logsig in MATLAB) between the signals of Node1 and Node2 was used, the association between Node2 and trial-by-trial fluctuations was attenuated and such nonlinearity in transfer did not induce false-positive correlations.

To summarize, this simulation demonstrates that a correspondence between effort and brain signals can be driven by a correspondence between reward and the willingness to work for it on average. Empirically, this correspondence has been shown for the dorsal striatum in animals (Wang et al., 2013) and humans (Kroemer et al., 2014; Kurniawan et al., 2013). However, recent studies focusing on the NAcc highlight the importance of signaling dynamics in dopamine release (Hamid et al., 2016) and of action as the target of learned contingencies (Syed et al., 2016) suggesting that the translation from brain response to action might be achieved via online estimates of utility, as captured by Node2 within our simulation. The correspondence between Node2 and the NAcc is supported by evidence in humans as well (Kroemer et al., 2014, 2016; Kurniawan et al., 2013). Notably, we consider the full mixed-effects modeling approach as a consequential second step after an initial voxel-based mapping, but before more comprehensive frameworks for effective connectivity are employed, which also incur more assumptions (eg, dynamic causal modeling, DCM), since full mixed-effects models may help to effectively constrain the feature space. Preferably, in future research, the current simulations would be extended based on neurobiological constraints of signaling dynamics to mimic network interactions at a much more comprehensive level. In addition, we will describe how advanced real-time imaging techniques can be employed as a means to test hypotheses derived from simulations or observations from experimental studies to advance our mechanistic understanding.

8 TOWARDS TRIAL-BY-TRIAL BRAIN STATES AS A MEANS TO PREDICT ACTION

In order to take trial-by-trial fluctuations into account, the corresponding neural activity must be detected online and transformed into an applicable action such as the display of a feedback signal for learning. This concept called neurofeedback has been made available for all neuroimaging modalities by recent technical advances. As a result, neurofeedback based on electroencephalography (EEG) and/or rt-fMRI has been used to test scientific theories and to translate their findings into potential clinical applications. EEG-based neurofeedback offers systems that are portable and cheap while providing a very high temporal resolution. EEG-based neurofeedback has been applied to a wide range of disorders so far, such as attention deficit hyperactivity disorder or epilepsy (cf. Thibault et al., 2015), and as a means to improve cognitive performance (Gruzelier, 2014; Vernon, 2005).

In contrast, rt-fMRI offers a higher spatial resolution, which enables detecting the current brain activity at a whole-brain level. This technique supports (1) using the current brain activity as a feedback signal to learn to deliberately induce a specific brain state (Scharnowski and Weiskopf, 2015; Sulzer et al., 2013a; Weiskopf, 2012) or (2) using the current brain state as a trigger for a targeted interaction with participants during the runtime of the experiment such as the presentation of a stimulus (Fig. 4). These so-called adaptive fMRI paradigms allow establishing a time–order relationship, which is a prerequisite for causality. Once a predefined brain state is successfully detected, for example, an online estimate of utility across the previously described motivation network, the stimulus presentation can be adjusted accordingly. In other words, the online detection of brain states can be used analogous to a factor in a factorial design, which turns conventional offline correlational analyses into explicit hypothesis tests within a rigorous experimental framework. Thus, the otherwise hidden brain state becomes amenable to studying the functional implications of a given spatial and temporal profile. This can help to overcome some of the inferential limitations inherent in a merely correlative offline analysis. As a proof of principle, Hinds et al. (2013) have demonstrated that preceding activation in the SMA may correspond to a brain state of vigilance (or labor), whereas preceding activation in the default mode network may correspond to the state of leisure (cf. Inzlicht et al., 2014).

Applications of rt-fMRI take into account the intraindividual variability and, therefore, can be adapted to the neural dynamics over the runtime of the task to investigate the corresponding behavioral output as pioneered by Hinds et al. (2013). Technically, this requires unified software setups, which combine real-time analysis and stimulus presentation in a single-computer setup (eg, Hellrung et al., 2015). In general, the concept of utilizing brain states as a target for experiments has been described by Lorenz et al. (2016) and dubbed "the automatic neuroscientist." This work describes the use of rt-fMRI for individual optimization of fMRI paradigms to inform us more specifically about brain function. Nevertheless, neurofeedback and adaptive fMRI paradigms can also be combined as closed-loop systems, and it has been shown that the moment-to-moment feedback about attentional state can enhance the ability to sustain attention (deBettencourt et al., 2015).

Notwithstanding, neurofeedback and adaptive fMRI paradigms implicate several challenges for experimental design. First, technical issues such as stability of the software setup and increased sensitivity to motion need to be addressed. Second, it requires detailed a priori knowledge about brain processes to develop an experimental setup which can help to address a specific hypothesis, for example, if brain

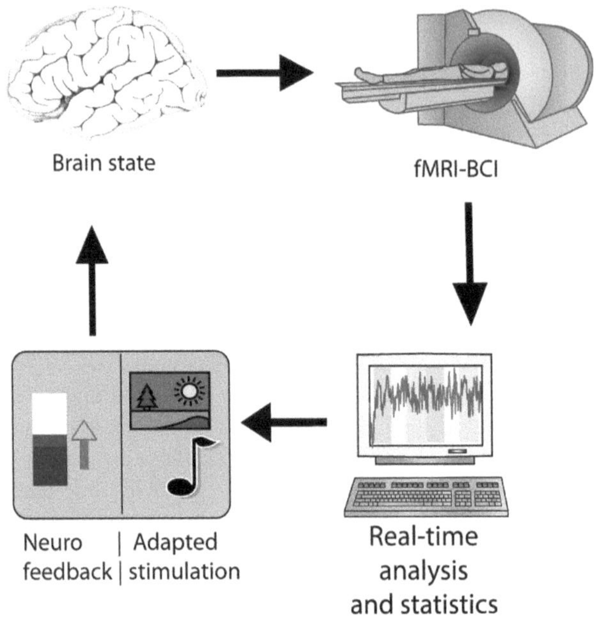

FIG. 4

Rationale of adaptive paradigms and neurofeedback as a means to study brain function. In a real-time fMRI setup, brain activity can be analyzed during the runtime of the experiment. The results of this analysis can be either used (1) to be presented as feedback signal in order to train subjects in volitional control of their brain response or (2) to adapt the paradigm. The latter approach establishes a prerequisite for stimulus presentation and enables strict neurocognitive hypothesis testing resembling factorial designs. Importantly, both methods enable online interaction with the subject, which may improve the correspondence between the investigation of brain response and behavior. Such methods can be employed in addition to conventional designs and analysis to test predictions on the functional implications of signal fluctuations within networks or single nodes. For example, based on our review, we would hypothesize that volitional up- or downregulation of the brain signal, learned via neurofeedback, would enable participants to up- and downregulate their response vigor while behavioral responses are pending (neurofeedback). Moreover, adaptive paradigms could be used to prompt behavioral responses whenever an a priori defined brain state (eg, strong activation in nucleus accumbens, weak activation in the dopaminergic midbrain nuclei) is reliably detected.

fluctuations in the NAcc and VTA/SN have opposite effects on response vigor despite the positive correlation of the time series in general. In close correspondence with the hypothesis, methods have to be carefully adapted to the question of interest in terms of the best choice of ROIs, algorithms, and presentation of the stimuli during the experiment. Third, the hemodynamic response lag imposes a neurophysiological limit for the speed of a prospective adaptation in rt-fMRI applications.

Yet, behavioral phenomena such as the invigoration of behavior by the average reward rate (Beierholm et al., 2013; Niv et al., 2007; Rigoli et al., 2016) do occur at a time resolution that is amenable to fMRI adaptation. Although such tools do not necessarily establish causality between brain signal and behavior, the inverse approach enables confirmatory tests of well-defined hypotheses on the cognitive implications of brain function and may therefore help to flank conventional offline and post hoc analyses. Moreover, the increased experimental control over the sampling of brain response and behavior can help to balance designs and maximize design efficiency.

To summarize, recent progress in imaging techniques has propelled the use of online detection of brain states as a means to study behavior by addressing more specific question about brain function. For potential future applications, this leads to the question if a given brain state can be volitionally and reliably induced by the participant.

9 CAN THE INDUCTION OF A PREDEFINED BRAIN STATE CHANGE BEHAVIOR?

In the context of instrumental action and motivation, the successful regulation within the mesocorticolimbic system has been demonstrated recently. In the first study by Sulzer et al., it has been shown that during neurofeedback training, the activation of VTA and SN can be influenced volitionally (Sulzer et al., 2013b), although there was no evidence of transfer effects in learning. A second study has shown such learning effects in a posttest after three runs of VTA neurofeedback training (MacInnes et al., 2016) in contrast to control groups. In this paradigm, participants were instructed to self-induce a state of high motivation. Within three separate training runs, participants managed to enhance their VTA activation volitionally. In addition to the successful VTA regulation, MacInnes et al. (2016) found increases in functional connectivity between the VTA and bilateral hippocampus, as well as between NAcc and hippocampus. These functional connectivity changes are in line with a third study, which investigated the regulation of NAcc neurofeedback (Greer et al., 2014). Greer et al. (2014) demonstrated that NAcc regulation is feasible, although transfer effects of such regulation remain an open empirical question. Taken together, these studies indicate that volitional self-regulation of the dopaminergic midbrain or the NAcc might allow modifying the motivational drive of behavior.

Furthermore, the successful self-regulation of the amygdala has been repeatedly shown both for upregulation with positive memories (Zotev et al., 2011) and downregulation after negative stimuli (Brühl et al., 2014; Paret et al., 2014). Although these studies focused on emotional aspects and mental disorders (eg, anxiety disorders or depression), their findings are highly relevant for motivational aspects of behavior as outlined before because the amygdala modulates mesocorticolimbic target regions such as the NAcc or vmPFC. Thus far, neurofeedback studies have mostly investigated single ROI effects, but recent findings have demonstrated the use of functional-connectivity-based feedback as a promising extension

(Koush et al., 2013; Shen et al., 2015). By extending the focus to the network level, the influence of one node such as the amygdala on the motivation network as a whole emergent feature could be investigated. Notably, a first meta-study comprising 12 neurofeedback studies from different brain regions with a total of 175 subjects and 899 neurofeedback runs suggests the existence of a neurofeedback network consisting of the anterior insula, basal ganglia, dorsal parts of the parietal lobe extending to the temporal–parietal junction, ACC, dorsolateral and ventrolateral prefrontal cortex, and visual association areas including the temporal–occipital junction (Emmert et al., 2016). Collectively, these studies provide preliminary evidence that the induction of a brain state can have effects on behavior as measured inside, but also outside the scanner environment.

10 CONCLUSIONS AND FUTURE PERSPECTIVES: TARGETING THE MOTIVATION NETWORK

In this review, we have outlined how an interconnected brain network gives rise to instrumental behavior and response vigor. This motivation network encompassing aspects of valuation, reward-related learning, and action control, which are all likely "running" on dopamine, provides a reasonable starting point for future studies on the exact contribution by specific brain regions within the network. While considerable progress has been made in the past decades, future progress will be dependent on more advanced brain imaging techniques that may provide answers to more nuanced questions about brain function, which are still open but crucial to targeted interventions. We have described how by exploiting the information contained and readily available in signaling dynamics of brain and behavior, we could potentially improve our understanding of the neurobiological processes driving motivated behavior. Furthermore, individual brain states defined by large-scale brain networks and their functional connectivity could help to elucidate the neural correlates of cost/benefit decision-making. As an example, it has been shown that neural activity from the anterior insula, ventral striatum, and lateral orbitofrontal cortex predicted the participants' decisions to accept or reject a monetary offer in the ultimatum game in real-time with an accuracy of 70% (Hollmann et al., 2011). Such dynamic classification of brain states could be used in combination with interactive experimental environments as presented by Müller et al. (2012) or Lorenz et al. (2016). As we have briefly mentioned before, such adaptation requires brain states that are temporally stable in order to be reliably detected using fMRI, which has already been shown, for example, for emotional states (Okon-Singer et al., 2014). Considering instrumental actions and the allocation of effort, it has been shown that temporal dynamics are in the range of several seconds (Meyniel et al., 2013) and invigoration of behavior by the average reward rate points to transient trial-by-trial effects that we are only beginning to unravel at the neural level (Rigoli et al., 2016). In these cases, the observation and accumulation of evidence over multiple instances can help to build a reliable classification for brain states at the level of the individual, which in turn can be used as the trigger for stimulus adaptation.

To conclude, signaling dynamics entail the potential to improve our understanding as to why we do not act perfectly reproducible all the time when we are confronted with the same goal. Our vigor to work depends on costs and benefits that have to be anticipated, integrated, and, ultimately, translated into behavior. Moreover, it is unlikely that such trade-offs in action policies do not change during the course of an experiment or within an hour of observation. Instead of treating such fluctuations as just noise, we have argued that variability in brain signals and behavior provides rich information on how we decide to work or rest. We propose that shared fluctuations in response vigor and brain response are partly due to the fact that the same incentive will vary in terms of its effectiveness to invigorate behavior via online estimates of costs and benefits. In other words, sometimes the same goal does not appear as motivating to us simply because valuation signals, which supposedly invigorate behavior, will differ in their strength from trial to trial. Our proposal is complementary to previous suggestions in the growing literature on effort expenditure, which suggest that fluctuations occur mainly because of fatigue (Meyniel et al., 2013, 2014; Rigoux and Guigon, 2012). Within our framework, we regard fatigue as one instance that could influence the perceived cost of action analogous to how changes in the average reward rate may influence the perceived benefit of action. Consequently, online estimates of utility might be used in future studies to test specific hypotheses about the functional contribution of one brain region to the allocation of effort in the pursuit of a desirable goal. We expect that recent advances in imaging techniques will help to foster this process in the development of a coherent neurobiological understanding of why we sometimes work more or less vigorously given the same incentive.

REFERENCES

Anderson, A.K., Phelps, E.A., 2001. Lesions of the human amygdala impair enhanced perception of emotionally salient events. Nature 411 (6835), 305–309.

Appelhans, B.M., Waring, M.E., Schneider, K.L., Pagoto, S.L., DeBiasse, M.A., Whited, M.C., Lynch, E.B., 2012. Delay discounting and intake of ready-to-eat and away-from-home foods in overweight and obese women. Appetite 59 (2), 576–584.

Backs, R.W., Seljos, K.A., 1994. Metabolic and cardiorespiratory measures of mental effort: the effects of level of difficulty in a working memory task. Int. J. Psychophysiol. 16, 57–68.

Basten, U., Biele, G., Heekeren, H.R., Fiebach, C.J., 2010. How the brain integrates costs and benefits during decision making. Proc. Natl. Acad. Sci. U. S. A. 107 (50), 21767–21772.

Beeler, J.A., 2012. Thorndike's law 2.0: dopamine and the regulation of thrift. Front. Neurosci. 6, 116.

Beeler, J.A., Frazier, C.R., Zhuang, X., 2012. Putting desire on a budget: dopamine and energy expenditure, reconciling reward and resources. Front. Integr. Neurosci. 6, 49.

Beeler, J.A., Faust, R.P., Turkson, S., Ye, H., Zhuang, X., 2015. Low dopamine D2 receptor increases vulnerability, to obesity via reduced physical activity, not increased appetitive motivation. Biol. Psychiatry 79 (11), 887–897.

Beierholm, U., Guitart-Masip, M., Economides, M., Chowdhury, R., Duzel, E., Dolan, R., Dayan, P., 2013. Dopamine modulates reward-related vigor. Neuropsychopharmacology 38 (8), 1495–1503.

Berridge, K.C., 1996. Food reward: brain substrates of wanting and liking. Neurosci. Biobehav. Rev. 20 (1), 1–25.

Berridge, K.C., Kringelbach, M.L., 2015. Pleasure systems in the brain. Neuron 86 (3), 646–664.

Beunza, J.J., Martinez-Gonzalez, M.A., Ebrahim, S., Bes-Rastrollo, M., Nunez, J., Martinez, J.A., Alonso, A., 2007. Sedentary behaviors and the risk of incident hypertension: the SUN Cohort. Am. J. Hypertens. 20 (11), 1156–1162.

Blechert, J., Klackl, J., Miedl, S.F., Wilhelm, F.H., 2016. To eat or not to eat: effects of food availability on reward system activity during food picture viewing. Appetite 99, 254–261.

Bonnelle, V., Manohar, S., Behrens, T., Husain, M., 2016. Individual differences in premotor brain systems underlie behavioral apathy. Cereb. Cortex 26 (2), 807–819.

Botvinick, M.M., Huffstetler, S., McGuire, J.T., 2009. Effort discounting in human nucleus accumbens. Cogn. Affect. Behav. Neurosci. 9 (1), 16–27.

Branch, S.Y., Sharma, R., Beckstead, M.J., 2014. Aging decreases L-type calcium channel currents and pacemaker firing fidelity in substantia nigra dopamine neurons. J. Neurosci. 34 (28), 9310–9318.

Browning, R.C., Modica, J.R., Kram, R., Goswami, A., 2007. The effects of adding mass to the legs on the energetics and biomechanics of walking. Med. Sci. Sports Exerc. 39 (3), 515–525.

Brühl, A.B., Scherpiet, S., Sulzer, J., Stämpfli, P., Seifritz, E., Herwig, U., 2014. Real-time neurofeedback using functional MRI could improve down-regulation of amygdala activity during emotional stimulation: a proof-of-concept study. Brain Topogr. 27 (1), 138–148.

Bühler, M., Vollstadt-Klein, S., Kobiella, A., Budde, H., Reed, L.J., Braus, D.F., Buchel, C., Smolka, M.N., 2010. Nicotine dependence is characterized by disordered reward processing in a network driving motivation. Biol. Psychiatry 67 (8), 745–752.

Caravaggio, F., Borlido, C., Hahn, M., Feng, Z., Fervaha, G., Gerretsen, P., Nakajima, S., Plitman, E., Chung, J.K., Iwata, Y., Wilson, A., Remington, G., Graff-Guerrero, A., 2015. Reduced insulin sensitivity is related to less endogenous dopamine at D2/3 receptors in the ventral striatum of healthy nonobese humans. Int. J. Neuropsychopharmacol. 18 (7), pyv014.

Chan, J.M., Rimm, E.B., Colditz, G.A., Stampfer, M.J., Willett, W.C., 1994. Obesity, fat distribution, and weight gain as risk factors for clinical diabetes in men. Diabetes Care 17 (9), 961–969.

Chong, T.T., Bonnelle, V., Manohar, S., Veromann, K.R., Muhammed, K., Tofaris, G.K., Hu, M., Husain, M., 2015. Dopamine enhances willingness to exert effort for reward in Parkinson's disease. Cortex 69, 40–46.

Chumbley, J.R., Tobler, P.N., Fehr, E., 2014. Fatal attraction: ventral striatum predicts costly choice errors in humans. Neuroimage 89, 1–9.

Colditz, G.A., Willett, W.C., Stampfer, M.J., Manson, J.E., Hennekens, C.H., Arky, R.A., Speizer, F.E., 1990. Weight as a risk factor for clinical diabetes in women. Am. J. Epidemiol. 132 (3), 501–513.

Collins, A.G., Frank, M.J., 2015. Surprise! Dopamine signals mix action, value and error. Nat. Neurosci. 19 (1), 3–5.

Croxson, P.L., Walton, M.E., O'Reilly, J.X., Behrens, T.E., Rushworth, M.F., 2009. Effort-based cost-benefit valuation and the human brain. J. Neurosci. 29 (14), 4531–4541.

Cuvo, A.J., Lerch, L.J., Leurquin, D.A., Gaffaney, T.J., Poppen, R.L., 1998. Response allocation to concurrent fixed-ratio reinforcement schedules with work requirements by adults with mental retardation and typical preschool children. J. Appl. Behav. Anal. 31 (1), 43–63.

Davis, J.F., Choi, D.L., Schurdak, J.D., Fitzgerald, M.F., Clegg, D.J., Lipton, J.W., Figlewicz, D.P., Benoit, S.C., 2011. Leptin regulates energy balance and motivation through action at distinct neural circuits. Biol. Psychiatry 69 (7), 668–674.

deBettencourt, M.T., Cohen, J.D., Lee, R.F., Norman, K.A., Turk-Browne, N.B., 2015. Closed-loop training of attention with real-time brain imaging. Nat. Neurosci. 18 (3), 470–475.

Denk, F., Walton, M.E., Jennings, K.A., Sharp, T., Rushworth, M.F., Bannerman, D.M., 2005. Differential involvement of serotonin and dopamine systems in cost-benefit decisions about delay or effort. Psychopharmacology (Berl.) 179 (3), 587–596.

Dinstein, I., Heeger, D.J., Behrmann, M., 2015. Neural variability: friend or foe? Trends Cogn. Sci. 19 (6), 322–328.

Donohoe, R.T., Benton, D., 1999. Cognitive functioning is susceptible to the level of blood glucose. Psychopharmacology (Berl.) 145 (4), 378–385.

Dunn, J.P., Kessler, R.M., Feurer, I.D., Volkow, N.D., Patterson, B.W., Ansari, M.S., Li, R., Marks-Shulman, P., Abumrad, N.N., 2012. Relationship of dopamine type 2 receptor binding potential with fasting neuroendocrine hormones and insulin sensitivity in human obesity. Diabetes Care 35 (5), 1105–1111.

Eisenberger, R., 1992. Learned industriousness. Psychol. Rev. 99 (2), 248–267.

Emmert, K., Kopel, R., Sulzer, J., Bruhl, A.B., Berman, B.D., Linden, D.E., Horovitz, S.G., Breimhorst, M., Caria, A., Frank, S., Johnston, S., Long, Z., Paret, C., Robineau, F., Veit, R., Bartsch, A., Beckmann, C.F., Van De Ville, D., Haller, S., 2016. Meta-analysis of real-time fMRI neurofeedback studies using individual participant data: how is brain regulation mediated? Neuroimage 124 (Pt. A), 806–812.

Fairclough, S.H., Houston, K., 2004. A metabolic measure of mental effort. Biol. Psychol. 66 (2), 177–190.

Farrar, A.M., Font, L., Pereira, M., Mingote, S., Bunce, J.G., Chrobak, J.J., Salamone, J.D., 2008. Forebrain circuitry involved in effort-related choice: injections of the GABAA agonist muscimol into ventral pallidum alter response allocation in food-seeking behavior. Neuroscience 152 (2), 321–330.

Floresco, S.B., Ghods-Sharifi, S., 2007. Amygdala-prefrontal cortical circuitry regulates effort-based decision making. Cereb. Cortex 17 (2), 251–260.

Floresco, S.B., Tse, M.T., Ghods-Sharifi, S., 2008. Dopaminergic and glutamatergic regulation of effort- and delay-based decision making. Neuropsychopharmacology 33 (8), 1966–1979.

Fogelholm, M., Kronholm, E., Kukkonen-Harjula, K., Partonen, T., Partinen, M., Harma, M., 2007. Sleep-related disturbances and physical inactivity are independently associated with obesity in adults. Int. J. Obes. (Lond.) 31 (11), 1713–1721.

Font, L., Mingote, S., Farrar, A.M., Pereira, M., Worden, L., Stopper, C., Port, R.G., Salamone, J.D., 2008. Intra-accumbens injections of the adenosine A2A agonist CGS

21680 affect effort-related choice behavior in rats. Psychopharmacology (Berl.) 199 (4), 515–526.
Frank, M.J., Hutchison, K., 2009. Genetic contributions to avoidance-based decisions: striatal D2 receptor polymorphisms. Neuroscience 164 (1), 131–140.
Frank, M.J., Seeberger, L.C., O'Reilly, R.C., 2004. By carrot or by stick: cognitive reinforcement learning in Parkinsonism. Science 306 (5703), 1940–1943.
Gailliot, M.T., Baumeister, R.F., 2007. The physiology of willpower: linking blood glucose to self-control. Pers. Soc. Psychol. Rev. 11 (4), 303–327.
Gailliot, M.T., Baumeister, R.F., DeWall, C.N., Maner, J.K., Plant, E.A., Tice, D.M., Brewer, L.E., Schmeichel, B.J., 2007. Self-control relies on glucose as a limited energy source: willpower is more than a metaphor. J. Pers. Soc. Psychol. 92 (2), 325–336.
Garrett, D.D., Samanez-Larkin, G.R., MacDonald, S.W., Lindenberger, U., McIntosh, A.R., Grady, C.L., 2013. Moment-to-moment brain signal variability: a next frontier in human brain mapping? Neurosci. Biobehav. Rev. 37 (4), 610–624.
Garrett, D.D., McIntosh, A.R., Grady, C.L., 2014. Brain signal variability is parametrically modifiable. Cereb. Cortex 24 (11), 2931–2940.
Garrett, D.D., Nagel, I.E., Preuschhof, C., Burzynska, A.Z., Marchner, J., Wiegert, S., Jungehulsing, G.J., Nyberg, L., Villringer, A., Li, S.C., Heekeren, H.R., Backman, L., Lindenberger, U., 2015. Amphetamine modulates brain signal variability and working memory in younger and older adults. Proc. Natl. Acad. Sci. U. S. A. 112 (24), 7593–7598.
Goldfield, G.S., Lumb, A.B., Colapinto, C.K., 2011. Relative reinforcing value of energy-dense snack foods in overweight and obese adults. Can. J. Diet. Pract. Res. 72 (4), 170–174.
Grabenhorst, F., Rolls, E.T., 2011. Value, pleasure and choice in the ventral prefrontal cortex. Trends Cogn. Sci. 15 (2), 56–67.
Greer, S.M., Trujillo, A.J., Glover, G.H., Knutson, B., 2014. Control of nucleus accumbens activity with neurofeedback. Neuroimage 96, 237–244.
Grosshans, M., Vollmert, C., Vollstadt-Klein, S., Tost, H., Leber, S., Bach, P., Buhler, M., von der Goltz, C., Mutschler, J., Loeber, S., Hermann, D., Wiedemann, K., Meyer-Lindenberg, A., Kiefer, F., 2012. Association of leptin with food cue-induced activation in human reward pathways. Arch. Gen. Psychiatry 69 (5), 529–537.
Gruzelier, J.H., 2014. EEG-neurofeedback for optimising performance. I: a review of cognitive and affective outcome in healthy participants. Neurosci. Biobehav. Rev. 44, 124–141.
Haber, S.N., Knutson, B., 2010. The reward circuit: linking primate anatomy and human imaging. Neuropsychopharmacology 35 (1), 4–26.
Hall, J.L., Gonder-Frederick, L.A., Chewning, W.W., Silveira, J., Gold, P.E., 1989. Glucose enhancement of performance on memory tests in young and aged humans. Neuropsychologia 27 (9), 1129–1138.
Hamid, A.A., Pettibone, J.R., Mabrouk, O.S., Hetrick, V.L., Schmidt, R., Vander Weele, C.M., Kennedy, R.T., Aragona, B.J., Berke, J.D., 2016. Mesolimbic dopamine signals the value of work. Nat. Neurosci. 19 (1), 117–126.
Hellrung, L., Hollmann, M., Schlumm, T., Zscheyge, O., Kalberlah, C., Roggenhofer, E., Okon-Singer, H., Villringer, A., Horstmann, A., 2015. Flexible adaptive paradigms for fMRI using a novel software package 'Brain Analysis in Real-Time' (BART). PLoS One 10 (4), e0118890.
Hinds, O., Thompson, T.W., Ghosh, S., Yoo, J.J., Whitfield-Gabrieli, S., Triantafyllou, C., Gabrieli, J.D., 2013. Roles of default-mode network and supplementary motor area in

human vigilance performance: evidence from real-time fMRI. J. Neurophysiol. 109 (5), 1250–1258.

Hollmann, M., Rieger, J.W., Baecke, S., Lützkendorf, R., Müller, C., Adolf, D., Bernarding, J., 2011. Predicting decisions in human social interactions using real-time fMRI and pattern classification. PLoS One 6, e25304.

Holroyd, C.B., Yeung, N., 2011. An integrative theory of anterior cingulate cortex function: option selection in hierarchical reinforcement learning. In: Mars, R.B., Sallet, J., Rushworth, M.F.S., Yeung, N. (Eds.), Neural Basis of Motivational and Cognitive Control. MIT Press, Cambridge, MA, pp. 333–349.

Horstmann, A., Fenske, W.K., Hankir, M.K., 2015. Argument for a non-linear relationship between severity of human obesity and dopaminergic tone. Obes. Rev. 16 (10), 821–830.

Hosking, J.G., Floresco, S.B., Winstanley, C.A., 2015. Dopamine antagonism decreases willingness to expend physical, but not cognitive, effort: a comparison of two rodent cost/benefit decision-making tasks. Neuropsychopharmacology 40 (4), 1005–1015.

Inzlicht, M., Schmeichel, B.J., Macrae, C.N., 2014. Why self-control seems (but may not be) limited. Trends Cogn. Sci. 18 (3), 127–133.

Izquierdo, A., Carlos, K., Ostrander, S., Rodriguez, D., McCall-Craddolph, A., Yagnik, G., Zhou, F., 2012. Impaired reward learning and intact motivation after serotonin depletion in rats. Behav. Brain Res. 233 (2), 494–499.

Johnson, P.M., Kenny, P.J., 2010. Dopamine D2 receptors in addiction-like reward dysfunction and compulsive eating in obese rats. Nat. Neurosci. 13 (5), 635–641.

Jones, J.L., Day, J.J., Aragona, B.J., Wheeler, R.A., Wightman, R.M., Carelli, R.M., 2010. Basolateral amygdala modulates terminal dopamine release in the nucleus accumbens and conditioned responding. Biol. Psychiatry 67 (8), 737–744.

Kaleta, D., Jegier, A., 2007. Predictors of inactivity in the working-age population. Int. J. Occup. Med. Environ. Health 20 (2), 175–182.

Kennedy, D.O., Scholey, A.B., 2000. Glucose administration, heart rate and cognitive performance: effects of increasing mental effort. Psychopharmacology (Berl.) 149 (1), 63–71.

King, S.J., Isaacs, A.M., O'Farrell, E., Abizaid, A., 2011. Motivation to obtain preferred foods is enhanced by ghrelin in the ventral tegmental area. Horm. Behav. 60 (5), 572–580.

Kivetz, R., 2003. The effects of effort and intrinsic motivation on risky choice. Mark. Sci. 22 (4), 477–502.

Kleinridders, A., Cai, W., Cappellucci, L., Ghazarian, A., Collins, W.R., Vienberg, S.G., Pothos, E.N., Kahn, C.R., 2015. Insulin resistance in brain alters dopamine turnover and causes behavioral disorders. Proc. Natl. Acad. Sci. U. S. A. 112 (11), 3463–3468.

Kobayashi, S., Schultz, W., 2008. Influence of reward delays on responses of dopamine neurons. J. Neurosci. 28 (31), 7837–7846.

Kool, W., McGuire, J.T., Rosen, Z.B., Botvinick, M.M., 2010. Decision making and the avoidance of cognitive demand. J. Exp. Psychol. Gen. 139 (4), 665–682.

Korn, C.W., Bach, D.R., 2015. Maintaining homeostasis by decision-making. PLoS Comput. Biol. 11 (5), e1004301.

Koush, Y., Rosa, M.J., Robineau, F., Heinen, K., Rieger, S., Weiskopf, N., Vuilleumier, P., Van De Ville, D., Scharnowski, F., 2013. Connectivity-based neurofeedback: dynamic causal modeling for real-time fMRI. Neuroimage 81, 422–430.

Kroemer, N.B., Small, D.M., 2016. Fuel not fun: reinterpreting attenuated brain responses to reward in obesity. Physiol. Behav. 162, 37–45.

Kroemer, N.B., Krebs, L., Kobiella, A., Grimm, O., Pilhatsch, M., Bidlingmaier, M., Zimmermann, U.S., Smolka, M.N., 2013. Fasting levels of ghrelin covary with the brain response to food pictures. Addict. Biol. 18 (5), 855–862.

Kroemer, N.B., Guevara, A., Ciocanea Teodorescu, I., Wuttig, F., Kobiella, A., Smolka, M.N., 2014. Balancing reward and work: anticipatory brain activation in NAcc and VTA predict effort differentially. Neuroimage 102 (Pt. 2), 510–519.

Kroemer, N.B., Wuttig, F., Bidlingmaier, M., Zimmermann, U.S., Smolka, M.N., 2015. Nicotine enhances modulation of food-cue reactivity by leptin and ghrelin in the ventromedial prefrontal cortex. Addict. Biol. 20 (4), 832–844.

Kroemer, N.B., Sun, X., Veldhuizen, M.G., Babbs, A.E., De Araujo, I.E., Small, D.M., 2016. Weighing the evidence: variance in brain responses to milkshake receipt is predictive of eating behavior. Neuroimage 128, 273–283.

Kurniawan, I.T., Seymour, B., Talmi, D., Yoshida, W., Chater, N., Dolan, R.J., 2010. Choosing to make an effort: the role of striatum in signaling physical effort of a chosen action. J. Neurophysiol. 104 (1), 313–321.

Kurniawan, I.T., Guitart-Masip, M., Dayan, P., Dolan, R.J., 2013. Effort and valuation in the brain: the effects of anticipation and execution. J. Neurosci. 33 (14), 6160–6169.

Kurzban, R., Duckworth, A., Kable, J.W., Myers, J., 2013. An opportunity cost model of subjective effort and task performance. Behav. Brain Sci. 36 (6), 661–679.

Lebreton, M., Abitbol, R., Daunizeau, J., Pessiglione, M., 2015. Automatic integration of confidence in the brain valuation signal. Nat. Neurosci. 18 (8), 1159–1167.

Leibel, R.L., Rosenbaum, M., Hirsch, J., 1995. Changes in energy expenditure resulting from altered body weight. N. Engl. J. Med. 332 (10), 621–628.

Levy, D.J., Glimcher, P.W., 2012. The root of all value: a neural common currency for choice. Curr. Opin. Neurobiol. 22 (6), 1027–1038.

Li, S.C., Rieckmann, A., 2014. Neuromodulation and aging: implications of aging neuronal gain control on cognition. Curr. Opin. Neurobiol. 29, 148–158.

Liu, X., Hairston, J., Schrier, M., Fan, J., 2011. Common and distinct networks underlying reward valence and processing stages: a meta-analysis of functional neuroimaging studies. Neurosci. Biobehav. Rev. 35 (5), 1219–1236.

Lorenz, R., Monti, R.P., Violante, I.R., Anagnostopoulos, C., Faisal, A.A., Montana, G., Leech, R., 2016. The automatic neuroscientist: a framework for optimizing experimental design with closed-loop real-time fMRI. Neuroimage 129, 320–334.

Lurquin, J.H., Michaelson, L.E., Barker, J.E., Gustavson, D.E., von Bastian, C.C., Carruth, N.P., Miyake, A., 2016. No evidence of the ego-depletion effect across task characteristics and individual differences: a pre-registered study. PLoS One 11 (2), e0147770.

MacInnes, J.J., Dickerson, K.C., Chen, N.K., Adcock, R.A., 2016. Cognitive neurostimulation: learning to volitionally sustain ventral tegmental area activation. Neuron 89 (6), 1331–1342.

Malik, S., McGlone, F., Bedrossian, D., Dagher, A., 2008. Ghrelin modulates brain activity in areas that control appetitive behavior. Cell Metab. 7 (5), 400–409.

Mannella, F., Gurney, K., Baldassarre, G., 2013. The nucleus accumbens as a nexus between values and goals in goal-directed behavior: a review and a new hypothesis. Front. Behav. Neurosci. 7, 135.

Manning, C.A., Parsons, M.W., Gold, P.E., 1992. Anterograde and retrograde enhancement of 24-h memory by glucose in elderly humans. Behav. Neural Biol. 58 (2), 125–130.

Manning, C.A., Stone, W.S., Korol, D.L., Gold, P.E., 1998. Glucose enhancement of 24-h memory retrieval in healthy elderly humans. Behav. Brain Res. 93 (1–2), 71–76.

Manohar, S.G., Chong, T.T., Apps, M.A., Batla, A., Stamelou, M., Jarman, P.R., Bhatia, K.P., Husain, M., 2015. Reward pays the cost of noise reduction in motor and cognitive control. Curr. Biol. 25 (13), 1707–1716.

Mathar, D., Horstmann, A., Pleger, B., Villringer, A., Neumann, J., 2015. Is it worth the effort? Novel insights into obesity-associated alterations in cost-benefit decision-making. Front. Behav. Neurosci. 9, 360.

Matthews, C.E., Chen, K.Y., Freedson, P.S., Buchowski, M.S., Beech, B.M., Pate, R.R., Troiano, R.P., 2008. Amount of time spent in sedentary behaviors in the United States, 2003-2004. Am. J. Epidemiol. 167 (7), 875–881.

Mazzoni, P., Hristova, A., Krakauer, J.W., 2007. Why don't we move faster? Parkinson's disease, movement vigor, and implicit motivation. J. Neurosci. 27 (27), 7105–7116.

McArdle, W.D., Katch, F.I., Katch, V.L., 2010. Exercise Physiology: Nutrition, Energy, and Human Performance. Lippincott Williams & Wilkins, Baltimore.

McGuire, J.T., Botvinick, M.M., 2010. Prefrontal cortex, cognitive control, and the registration of decision costs. Proc. Natl. Acad. Sci. U. S. A. 107 (17), 7922–7926.

Meikle, A., Riby, L.M., Stollery, B., 2004. The impact of glucose ingestion and glucoregulatory control on cognitive performance: a comparison of younger and middle aged adults. Hum. Psychopharmacol. 19 (8), 523–535.

Meyniel, F., Sergent, C., Rigoux, L., Daunizeau, J., Pessiglione, M., 2013. Neurocomputational account of how the human brain decides when to have a break. Proc. Natl. Acad. Sci. U. S. A. 110 (7), 2641–2646.

Meyniel, F., Safra, L., Pessiglione, M., 2014. How the brain decides when to work and when to rest: dissociation of implicit-reactive from explicit-predictive computational processes. PLoS Comput. Biol. 10 (4), e1003584.

Mingote, S., Font, L., Farrar, A.M., Vontell, R., Worden, L.T., Stopper, C.M., Port, R.G., Sink, K.S., Bunce, J.G., Chrobak, J.J., Salamone, J.D., 2008. Nucleus accumbens adenosine A2A receptors regulate exertion of effort by acting on the ventral striatopallidal pathway. J. Neurosci. 28 (36), 9037–9046.

Mitchell, J.A., Pate, R.R., Beets, M.W., Nader, P.R., 2013. Time spent in sedentary behavior and changes in childhood BMI: a longitudinal study from ages 9 to 15 years. Int. J. Obes. (Lond.) 37 (1), 54–60.

Mokdad, A.H., Ford, E.S., Bowman, B.A., Dietz, W.H., Vinicor, F., Bales, V.S., Marks, J.S., 2003. Prevalence of obesity, diabetes, and obesity-related health risk factors, 2001. JAMA 289 (1), 76–79.

Molden, D.C., Hui, C.M., Scholer, A.A., Meier, B.P., Noreen, E.E., D'Agostino, P.R., Martin, V., 2012. Motivational versus metabolic effects of carbohydrates on self-control. Psychol. Sci. 23 (10), 1137–1144.

Müller, C., Luehrs, M., Baecke, S., Adolf, D., Luetzkendorf, R., Luchtmann, M., Bernarding, J., 2012. Building virtual reality fMRI paradigms: a framework for presenting immersive virtual environments. J. Neurosci. Methods 209, 290–298.

Niv, Y., Daw, N.D., Joel, D., Dayan, P., 2007. Tonic dopamine: opportunity costs and the control of response vigor. Psychopharmacology (Berl.) 191, 507–520.

O'Dwyer, N.J., Neilson, P.D., 2000. Metabolic energy expenditure and accuracy in movement: relation to levels of muscle and cardiorespiratory activation and the sense of effort. In: Sparrow, W.A. (Ed.), Energetics of Human Activity. Human Kinetics, Champaign, IL, pp. 1–42.

O'Doherty, J., Dayan, P., Schultz, J., Deichmann, R., Friston, K., Dolan, R.J., 2004. Dissociable roles of ventral and dorsal striatum in instrumental conditioning. Science 304 (5669), 452–454.

Okon-Singer, H., Mehnert, J., Hoyer, J., Hellrung, L., Schaare, H.L., Dukart, J., Villringer, A., 2014. Neural control of vascular reactions: impact of emotion and attention. J. Neurosci. 34 (12), 4251–4259.

Orquin, J.L., Kurzban, R., 2015. A meta-analysis of blood glucose effects on human decision making. Psychol. Bull.

Ostrander, S., Cazares, V.A., Kim, C., Cheung, S., Gonzalez, I., Izquierdo, A., 2011. Orbitofrontal cortex and basolateral amygdala lesions result in suboptimal and dissociable reward choices on cue-guided effort in rats. Behav. Neurosci. 125 (3), 350–359.

Ousdal, O.T., Reckless, G.E., Server, A., Andreassen, O.A., Jensen, J., 2012. Effect of relevance on amygdala activation and association with the ventral striatum. Neuroimage 62 (1), 95–101.

Palmiter, R.D., 2007. Is dopamine a physiologically relevant mediator of feeding behavior? Trends Neurosci. 30 (8), 375–381.

Palmiter, R.D., 2008. Dopamine signaling in the dorsal striatum is essential for motivated behaviors: lessons from dopamine-deficient mice. Ann. N. Y. Acad. Sci. 1129, 35–46.

Paret, C., Kluetsch, R., Ruf, M., Demirakca, T., Hoesterey, S., Ende, G., Schmahl, C., 2014. Down-regulation of amygdala activation with real-time fMRI neurofeedback in a healthy female sample. Front. Behav. Neurosci. 8, 299.

Pasquereau, B., Turner, R.S., 2013. Limited encoding of effort by dopamine neurons in a cost-benefit trade-off task. J. Neurosci. 33 (19), 8288–8300.

Pessiglione, M., Schmidt, L., Draganski, B., Kalisch, R., Lau, H., Dolan, R.J., Frith, C.D., 2007. How the brain translates money into force: a neuroimaging study of subliminal motivation. Science 316 (5826), 904–906.

Peters, J., Büchel, C., 2011. The neural mechanisms of inter-temporal decision-making: understanding variability. Trends Cogn. Sci. 15 (5), 227–239.

Phillips, P.E., Walton, M.E., Jhou, T.C., 2007. Calculating utility: preclinical evidence for cost-benefit analysis by mesolimbic dopamine. Psychopharmacology (Berl.) 191 (3), 483–495.

Pietilainen, K.H., Kaprio, J., Borg, P., Plasqui, G., Yki-Jarvinen, H., Kujala, U.M., Rose, R.J., Westerterp, K.R., Rissanen, A., 2008. Physical inactivity and obesity: a vicious circle. Obesity (Silver Spring) 16 (2), 409–414.

Prevost, C., Pessiglione, M., Metereau, E., Clery-Melin, M.L., Dreher, J.C., 2010. Separate valuation subsystems for delay and effort decision costs. J. Neurosci. 30 (42), 14080–14090.

Proffitt, D.R., 2006. Embodied perception and the economy of action. Perspect. Psychol. Sci. 1 (2), 110–122.

Rhodes, R.E., Mark, R.S., Temmel, C.P., 2012. Adult sedentary behavior: a systematic review. Am. J. Prev. Med. 42 (3), e3–e28.

Riby, L.M., Meikle, A., Glover, C., 2004. The effects of age, glucose ingestion and glucoregulatory control on episodic memory. Age Ageing 33 (5), 483–487.

Rigoli, F., Chew, B., Dayan, P., Dolan, R.J., 2016. The dopaminergic midbrain mediates an effect of average reward on pavlovian vigor. J. Cogn. Neurosci. 1–15.

Rigoux, L., Guigon, E., 2012. A model of reward- and effort-based optimal decision making and motor control. PLoS Comput. Biol. 8 (10), e1002716.

Rudebeck, P.H., Walton, M.E., Smyth, A.N., Bannerman, D.M., Rushworth, M.F., 2006. Separate neural pathways process different decision costs. Nat. Neurosci. 9 (9), 1161–1168.

Salamone, J.D., Correa, M., Farrar, A., Mingote, S.M., 2007. Effort-related functions of nucleus accumbens dopamine and associated forebrain circuits. Psychopharmacology (Berl.) 191 (3), 461–482.

Scharnowski, F., Weiskopf, N., 2015. Cognitive enhancement through real-time fMRI neurofeedback. Curr. Opin. Behav. Sci. 4, 122–127.

Scholey, A.B., Harper, S., Kennedy, D.O., 2001. Cognitive demand and blood glucose. Physiol. Behav. 73 (4), 585–592.

Schott, B.H., Minuzzi, L., Krebs, R.M., Elmenhorst, D., Lang, M., Winz, O.H., Seidenbecher, C.I., Coenen, H.H., Heinze, H.J., Zilles, K., Düzel, E., Bauer, A., 2008. Mesolimbic functional magnetic resonance imaging activations during reward anticipation correlate with reward-related ventral striatal dopamine release. J. Neurosci. 28 (52), 14311–14319.

Schouppe, N., Demanet, J., Boehler, C.N., Ridderinkhof, K.R., Notebaert, W., 2014. The role of the striatum in effort-based decision-making in the absence of reward. J. Neurosci. 34 (6), 2148–2154.

Selinger, J.C., O'Connor, S.M., Wong, J.D., Donelan, J.M., 2015. Humans can continuously optimize energetic cost during walking. Curr. Biol. 25 (18), 2452–2456.

Sevgi, M., Rigoux, L., Kuhn, A.B., Mauer, J., Schilbach, L., Hess, M.E., Gruendler, T.O., Ullsperger, M., Stephan, K.E., Bruning, J.C., Tittgemeyer, M., 2015. An obesity-predisposing variant of the FTO gene regulates D2R-dependent reward learning. J. Neurosci. 35 (36), 12584–12592.

Shen, J., Zhang, G., Yao, L., Zhao, X., 2015. Real-time fMRI training-induced changes in regional connectivity mediating verbal working memory behavioral performance. Neuroscience 289, 144–152.

Shenhav, A., Botvinick, M.M., Cohen, J.D., 2013. The expected value of control: an integrative theory of anterior cingulate cortex function. Neuron 79 (2), 217–240.

Smith, M.A., Riby, L.M., Eekelen, J.A., Foster, J.K., 2011. Glucose enhancement of human memory: a comprehensive research review of the glucose memory facilitation effect. Neurosci. Biobehav. Rev. 35 (3), 770–783.

Stauffer, W.R., Lak, A., Schultz, W., 2014. Dopamine reward prediction error responses reflect marginal utility. Curr. Biol. 24 (21), 2491–2500.

Stice, E., Spoor, S., Bohon, C., Small, D.M., 2008. Relation between obesity and blunted striatal response to food is moderated by TaqIA A1 allele. Science 322 (5900), 449–452.

Stice, E., Burger, K.S., Yokum, S., 2015. Reward region responsivity predicts future weight gain and moderating effects of the TaqIA allele. J. Neurosci. 35 (28), 10316–10324.

Sulzer, J., Haller, S., Scharnowski, F., Weiskopf, N., Birbaumer, N., Blefari, M.L., Bruehl, A.B., Cohen, L.G., deCharms, R.C., Gassert, R., Goebel, R., Herwig, U., LaConte, S., Linden, D., Luft, A., Seifritz, E., Sitaram, R., 2013a. Real-time fMRI neurofeedback: progress and challenges. Neuroimage 76, 386–399.

Sulzer, J., Sitaram, R., Blefari, M.L., Kollias, S., Birbaumer, N., Stephan, K.E., Luft, A., Gassert, R., 2013b. Neurofeedback-mediated self-regulation of the dopaminergic midbrain. Neuroimage 83, 817–825.

Sun, X., Veldhuizen, M.G., Wray, A.E., de Araujo, I.E., Sherwin, R.S., Sinha, R., Small, D.M., 2014. The neural signature of satiation is associated with ghrelin response and triglyceride metabolism. Physiol. Behav. 136, 63–73.

Sun, X., Kroemer, N.B., Veldhuizen, M.G., Babbs, A.E., de Araujo, I.E., Gitelman, D.R., Sherwin, R.S., Sinha, R., Small, D.M., 2015. Basolateral amygdala response to food cues in the absence of hunger is associated with weight gain susceptibility. J. Neurosci. 35 (20), 7964–7976.

Syed, E.C., Grima, L.L., Magill, P.J., Bogacz, R., Brown, P., Walton, M.E., 2016. Action initiation shapes mesolimbic dopamine encoding of future rewards. Nat. Neurosci. 19 (1), 34–36.

Tellez, L.A., Han, W., Zhang, X., Ferreira, T.L., Perez, I.O., Shammah-Lagnado, S.J., van den Pol, A.N., de Araujo, I.E., 2016. Separate circuitries encode the hedonic and nutritional values of sugar. Nat. Neurosci. 19 (3), 465–470.

Thibault, R.T., Lifshitz, M., Birbaumer, N., Raz, A., 2015. Neurofeedback, self-regulation, and brain imaging: clinical science and fad in the service of mental disorders. Psychother. Psychosom. 84 (4), 193–207.

Treadway, M.T., Buckholtz, J.W., Cowan, R.L., Woodward, N.D., Li, R., Ansari, M.S., Baldwin, R.M., Schwartzman, A.N., Kessler, R.M., Zald, D.H., 2012. Dopaminergic mechanisms of individual differences in human effort-based decision-making. J. Neurosci. 32 (18), 6170–6176.

Varazzani, C., San-Galli, A., Gilardeau, S., Bouret, S., 2015. Noradrenaline and dopamine neurons in the reward/effort trade-off: a direct electrophysiological comparison in behaving monkeys. J. Neurosci. 35 (20), 7866–7877.

Verguts, T., Vassena, E., Silvetti, M., 2015. Adaptive effort investment in cognitive and physical tasks: a neurocomputational model. Front. Behav. Neurosci. 9, 57.

Vernon, D.J., 2005. Can neurofeedback training enhance performance? An evaluation of the evidence with implications for future research. Appl. Psychophysiol. Biofeedback 30 (4), 347–364.

Volkow, N.D., Wang, G.J., Fowler, J.S., Tomasi, D., Telang, F., 2011. Addiction: beyond dopamine reward circuitry. Proc. Natl. Acad. Sci. U. S. A. 108 (37), 15037–15042.

Walton, M.E., Bannerman, D.M., Alterescu, K., Rushworth, M.F., 2003. Functional specilization within medial frontal cortex of the anterior cingulate for evaluating effort-related decisions. J. Neurosci. 23 (16), 6475–6479.

Walton, M.E., Kennerley, S.W., Bannerman, D.M., Phillips, P.E., Rushworth, M.F., 2006. Weighing up the benefits of work: behavioral and neural analyses of effort-related decision making. Neural Netw. 19 (8), 1302–1314.

Wanat, M.J., Kuhnen, C.M., Phillips, P.E., 2010. Delays conferred by escalating costs modulate dopamine release to rewards but not their predictors. J. Neurosci. 30 (36), 12020–12027.

Wang, G.J., Volkow, N.D., Logan, J., Pappas, N.R., Wong, C.T., Zhu, W., Netusil, N., Fowler, J.S., 2001. Brain dopamine and obesity. Lancet 357 (9253), 354–357.

Wang, A.Y., Miura, K., Uchida, N., 2013. The dorsomedial striatum encodes net expected return, critical for energizing performance vigor. Nat. Neurosci. 16 (5), 639–647.

Weiskopf, N., 2012. Real-time fMRI and its application to neurofeedback. Neuroimage 62, 682–692.

Westbrook, A., Braver, T.S., 2016. Dopamine does double duty in motivating cognitive effort. Neuron 89 (4), 695–710.

White, O., Davare, M., Andres, M., Olivier, E., 2013. The role of left supplementary motor area in grip force scaling. PLoS One 8 (2), e83812.

Worden, L.T., Shahriari, M., Farrar, A.M., Sink, K.S., Hockemeyer, J., Muller, C.E., Salamone, J.D., 2009. The adenosine A2A antagonist MSX-3 reverses the effort-related effects of dopamine blockade: differential interaction with D1 and D2 family antagonists. Psychopharmacology (Berl.) 203 (3), 489–499.

Zadra, J.R., Weltman, A.L., Proffitt, D.R., 2016. Walkable distances are bioenergetically scaled. J. Exp. Psychol. Hum. Percept. Perform. 42 (1), 39–51.

Zenon, A., Sidibe, M., Olivier, E., 2015. Disrupting the supplementary motor area makes physical effort appear less effortful. J. Neurosci. 35 (23), 8737–8744.

Zotev, V., Krueger, F., Phillips, R., Alvarez, R.P., Simmons, W.K., Bellgowan, P., Drevets, W.C., Bodurka, J., 2011. Self-regulation of amygdala activation using real-time FMRI neurofeedback. PLoS One 6 (9), e24522.

CHAPTER

Involvement of opioid signaling in food preference and motivation: Studies in laboratory animals

7

I. Morales*, L. Font†, P.J. Currie*, R. Pastor*,†,1

*Reed College, Portland, OR, United States

†Area de Psicobiología, Universitat Jaume I, Castellón, Spain

[1]Corresponding author: Tel.: +34-964-729-844; Fax: +34-964-729-267, e-mail address: raul.pastor@uji.es

Abstract

Motivation is a complex neurobiological process that initiates, directs, and maintains goal-oriented behavior. Although distinct components of motivated behavior are difficult to investigate, appetitive and consummatory phases of motivation are experimentally separable. Different neurotransmitter systems, particularly the mesolimbic dopaminergic system, have been associated with food motivation. Over the last two decades, however, research focusing on the role of opioid signaling has been particularly growing in this area. Opioid receptors seem to be involved, via neuroanatomically distinct mechanisms, in both appetitive and consummatory aspects of food reward. In the present chapter, we review the pharmacology and functional neuroanatomy of opioid receptors and their endogenous ligands, in the context of food reinforcement. We examine literature aimed at the development of laboratory animal techniques to better understand different components of motivated behavior. We present recent data investigating the effect of opioid receptor antagonists on food preference and effort-related decision making in rats, which indicate that opioid signaling blockade selectively affects intake of relatively preferred foods, resulting in reduced willingness to exert effort to obtain them. Finally, we elaborate on the potential role of opioid system manipulations in disorders associated with excessive eating and obesity.

Keywords

Motivation, Effort, Decision making, Food preference, Opioid system, Eating disorders

Progress in Brain Research, Volume 229, ISSN 0079-6123, http://dx.doi.org/10.1016/bs.pbr.2016.06.002

1 INTRODUCTION

The understanding of the central nervous system's regulation of eating behavior has become an increasingly studied topic in behavioral neuroscience. Research aimed at elucidating the neurobiological determinants of food pleasure, palatability, appetite, food salience, feeding microstructure, and instrumental responding for food reinforcement has yielded noteworthy knowledge regarding key psychological processes (ie, motivation, emotion, learning, and memory). It has also fueled an interest in understanding the neuropathology of eating disorders associated with dysregulation of motivational circuits, decision-making processes, cognitive biases, and compulsivity (for reviews, see Baldo and Kelley, 2007; Castro and Berridge, 2014b; Kessler et al., 2016; Salamone and Correa, 2013; Voon, 2015). The present chapter focuses on the biological basis of motivational aspects of food intake regulation, with a special emphasis on animal research methodology and the role of opioid signaling in food preference and effort-related decision making.

2 STUDYING FOOD INTAKE: THEORETICAL CONSIDERATIONS

The study of the neural processes by which organisms identify, seek, and learn about biologically relevant stimuli represents a major area in behavioral neuroscience research. Broadly, this field of study is often referred to as the neurobiology of reward[1] or reinforcement. Reinforcement is a complex process that supports activation, direction, and maintenance of goal-oriented behavior. This process involves emotion, motivation, and learning and memory mechanisms (Berridge et al., 2009). From a

[1]The term "reward" is present in a vast body of literature in psychology and behavioral neuroscience. However, it is not always clear what is meant when this term is used, as it is often not properly defined. Reward has been used interchangeably with positive reinforcer, reinforcement, primary motivation, and hedonic responses; thus, reward has been used to refer to a stimulus, a process, or an emotion. The broad application of this term within the scientific literature makes it challenging to integrate comprehensive information. For these reasons, this chapter will maintain a distinction between reinforcement and reward. Reinforcement will refer to the adaptive process that allows organisms to identify, seek, obtain, and learn about biologically important stimuli and experiences; a process that describes how an organism's behavior changes. Objects or stimuli that modify behavioral output will be described as positive or negative reinforcers. To avoid confusion, when referring to positive affect or hedonic, we will simply describe the dependent variable measured in a particular study; for example, taste-dependent affective facial reactions. We understand that this is especially important when discussing data obtained with animal research. It is important to minimize interpretation based on the assumption that positive reinforcers always regulate behavior because of their intrinsic emotionally positive properties. Decades of research have shown that behavior (ie, in well-learned responses, habits, or pathologies such as addiction) can be largely mediated by mechanisms that are not necessarily dependent on the hedonic properties of positive reinforcers per se. Although these terms will be here explored mostly in the context of eating, it should be noted that these psychological constructs could be applied to a wide range of behaviors. In addition, while motivation and emotion are mostly described in terms of positive reinforcement, they are also involved in processes mediating aversive consequences.

traditional behavioral perspective, reinforcement refers to the process by which stimuli or events can act to strengthen behavior (Shahan, 2010; Skinner, 1938, 1953; White and Milner, 1992). Reinforcing stimuli can be described as positive or negative. A positive reinforcer, such as palatable food, increases response frequency with its addition, while a negative reinforcer, such as a painful stimulus, increases response frequency through its removal (Dinsmoor, 2004; Slocum and Vollmer, 2015). Reinforcers can be unconditioned (primary), innately biologically relevant stimuli such as food, water, and sex, or conditioned (secondary; originally neutral stimuli paired with a primary reinforcer), such as a particular environment that, once paired with a fearful response, elicits fear by itself. Importantly, reinforcers promote acquisition and storage of information surrounding the events in which they are encountered (Everitt et al., 2001; Hyman et al., 2006; Packard and Knowlton, 2002; White and Milner, 1992). Interactions with relevant stimuli not only generate reinforcement-related learning but also produce emotional responses, which can be positive (ie, pleasure) or negative (ie, displeasure; Berridge, 2000; Cardinal et al., 2002; Salamone and Correa, 2012). In general, foods (and other stimuli or behaviors) that produce positive affect are more likely to be consumed relative to those that are not preferred, indicating that emotional responses play a key role in reinforcement processes. Reinforcement is not a unitary phenomenon and likely cannot be explained in reference to a single feature of a particular stimulus or process. Constellations of smaller neural processes, emotional and motivational, interact and contribute to this larger mechanism, so it has become increasingly important to gain a better understanding of the role of each individual component and the behavioral processes they give rise to (Bickel et al., 2000; Colwill and Rescorla, 1986; Dickinson and Balleine, 1994; Everitt et al., 2001; Salamone and Correa, 2002).

2.1 MOTIVATION AND EMOTION

Reinforcers are often said to be motivators (Salamone et al., 2007; White and Milner, 1992; for a detailed review of classical psychology literature, see Salamone and Correa, 2002). One property of reinforcers is their ability to promote behavior; they induce activation and maintenance of goal-directed actions (Berridge and Kringelbach, 2013; Dickinson and Balleine, 1994; Everitt et al., 2001; Salamone and Correa, 2002, 2012). Motivation is often defined as a process that enables organisms to regulate their internal and external environments (Nader et al., 1997; Salamone, 2010; Salamone and Correa, 2002). As organisms seek biologically relevant stimuli, their behavior occurs in distinct, experimentally separable phases. The initial phase, often called appetitive or instrumental, involves identifying and changing the proximity of goal objects. Appetitive behaviors have also been described as anticipatory, preparatory, or seeking actions (Blackburn, 2002; Blackburn et al., 1989; Czachowski et al., 2002; Foltin, 2001; Ikemoto and Panksepp, 1996). The consummatory or concluding end of motivated behavior describes the direct interactions that take place between organisms and their target stimuli. Consummatory behaviors tend to be stereotypical species-specific movements, such as chewing, swallowing, drinking, licking, and tongue protrusions (Berridge, 2004). As Salamone and Correa (2012) note, motivated behavior can be further organized into qualitatively different, directional components that describe how organisms avoid or actively seek out certain stimuli. They also highlight the activational properties of reinforcers due to their capacity to stimulate arousal and maintain activity (Cofer and Appley, 1964; Parkinson et al., 2002; Robbins and Koob, 1980; Salamone, 1988; Salamone and Correa, 2012; White and Milner, 1992). This facet of motivated behavior is particularly important as significant stimuli are not always readily available. Both in the laboratory and the natural world, animals must exert significant effort to obtain their target goals. The ability to energize in this way, either by speed (in wheel running), vigor (when lever pressing), or persistence (when climbing a barrier), is highly adaptive as it allows organisms to overcome obstacles necessary for survival (Salamone and Correa, 2012). In summary, motivation is a complex process involving a wide range of behaviors that allow organisms to bring their goals closer in proximity, interact with their environments, and avoid or delay particular circumstances. Motivation should not be thought of as a single entity as it can be further organized into temporal, activational, and directional components. A number of behaviors can be considered an expression of motivation and it is important to specify what type of behavior is being referenced as different neural mechanisms might be responsible for producing them. Although motivation is a key component, it is not the only constituent of the process of reinforcement.

Emotions are powerful physiological responses; subjective, internal states that can guide reinforced behavior. They may initially regulate the direction of behavior (approach vs. avoidance) and the degree of resources (ie, energy) required in the execution of such behavior. Although emotions are difficult to objectively define, the experience of emotions is at the core of the mechanisms that regulate an organism's interaction with motivational stimuli. Generally speaking, one can suggest that all interactions with biologically relevant objects involve some level of emotional processing. For example, consuming preferred foods gives rise to pleasure, which can affect our likelihood of eating that food again in the future. However, pleasure is not just a sensory property of a given stimulus, as it involves the coordination of mechanisms that add hedonic value to its experience (Berridge and Kringelbach, 2008, 2013; Craig, 1918; Finlayson et al., 2007; Kringelbach, 2004; Robinson and Berridge, 1993, 2003; Sherrington, 1906). Pleasure is a complex affective emotion that can manifest in two different ways as hedonic responses have subjective and objective properties (Berridge et al., 2009). Pleasure can arise through conscious experience, allowing people to self-report on it. While in certain contexts this can be a useful tool, the conscious experience of pleasure also appears to involve the activity of other cognitive mechanisms (Berridge and Kringelbach, 2013; Kringelbach, 2015; Shin et al., 2009), which makes isolating its neural signatures rather difficult. In addition, experiments with animals cannot make use of these measures, forcing researchers to use other methods of investigation. It has been suggested that emotions likely evolved from simple brain mechanisms that conferred animals some adaptive advantage. This, together with the fact that pleasure can also occur in the absence of conscious experience, suggests that it can be objectively measured given the right set of tools (Berridge and Kringelbach, 2013; Cardinal et al., 2002). Using a test of taste reactivity, researchers have found highly conserved reactions to presentations of sweet and bitter solutions in adults, babies, nonhuman primates, and rodents (Berridge, 1996; Berridge and Robinson, 1998; Cabanac and Lafrance, 1991; Ekman, 2006; Steiner, 1973, 1974). Positive hedonic responses include lip smacking and tongue protrusions to presentations of sweet, sucrose solutions. Bitter quinine solutions elicit aversive gapes, lip retractions, and arm and hand flailing (Berridge, 2000). The fact that animals share certain emotional responses with humans suggests that we can use neuroscientific tools to better understand the brain circuits and mechanisms responsible for producing these responses. Measuring observable, objective, hedonic responses to natural reinforcers has important implications as it may help researchers understand their relation to more cognitive forms of pleasure. It might also help dissociate between neural processes that underlie emotional and motivational aspects of reinforcement.

2.2 INCENTIVE SALIENCE MODEL—LIKING VS WANTING

The previous section briefly outlined some of the terminology used in the study of reinforcement and motivated behavior that is relevant for the present chapter. This section will more closely describe a theoretical view of reinforcement and related concepts that are important for the understanding of results and experiments reviewed here. The incentive salience hypothesis, which built on earlier theories of incentive motivation (Cofer and Appley, 1964), was developed by Berridge and colleagues in the late 1990s (Berridge and Robinson, 1998, see also Robinson and Berridge, 2003, 2008). For these authors, reinforcement is a multifaceted psychological process that involves learning (Pavlovian mechanisms), motivation (or incentive value), and emotional (hedonic) aspects. Sometimes the motivational and hedonic mechanisms are here referred to as *wanting* and *liking*, respectively. *Wanting* and *liking* are presented as different components under the control of different neural systems. *Wanting*, or incentive salience, is described as the psychological salience that becomes attributed to a given stimulus, turning it from something neutral to something *wanted*, and influencing the energy an animal will exert to obtain it. *Liking*, on the other hand, is pure hedonic affect. Although these processes are distinct they normally occur together; thus, stimuli that tend to be more liked are also more wanted (Berridge et al., 2009; Tibboel et al., 2015).

This model of incentive salience posits that animals will often encounter unconditioned stimuli (US), such as sweet foods, that produce positive affective responses (eg, pleasure), making these stimuli *liked*. These primary reinforcers also inherently carry some incentive or motivational value that causes animals to seek them (producing *wanting*) when they are available. Through Pavlovian learning systems, US and their consequences become associated with normally neutral cues (eg, light) that come to predict them. Through these associative mechanisms, the incentive value of the primary reinforcer is transferred, becoming a property of the conditioned stimulus (CS). Originally a CS has no control over an organism, but through learning mechanisms, it gains the ability to recruit *wanting* and *liking* processes. When the organism encounters these stimuli in the future, attribution of incentive salience to the CS will trigger *wanting* and direct behavior. Although interactions with the CS can also produce *liking* responses, the main behavior-directing component of such a model is incentive salience attribution. In addition, this model is also used to explain how certain physiological states can influence behavior. During a state of energy depletion, regulatory mechanisms interact with external motivational stimuli to enhance or attenuate their incentive value; for example, food palatability is amplified by hunger (Berridge, 2004, 2012; Robinson and Berridge, 1993; Toates, 1986).

The incentive salience hypothesis is similar to other theoretical frameworks in that it also posits that emotion, motivation, and learning are critically involved. However, important differences across individual approaches exist. Salamone and others highlight the importance of dissecting different aspects of motivation and focus on a microanalysis of different elements and types of motivated behaviors, while for Berridge and colleagues, motivation is not necessarily defined by a given behavior per se. Rather, it is seen as the attribution of incentive salience to a given stimulus. While this stamping in of incentive salience can give rise to a number of different behaviors, they all fall under the umbrella term *wanting* (Berridge and Robinson, 1998). The differences in the two approaches described earlier can be reconciled. As Salamone and Correa (2002) point out, the incentive salience model capitalizes on the dissociable nature of reinforcement phenomena, namely *liking* and *wanting*. Just as these two processes can be separated, *wanting* may also be separated into a number of subcomponents (ie, temporal, activational, and directional), with distinct neurobiological signatures. New data from our laboratory, described in further detail later in this chapter, show how opioid receptor antagonism decreases the incentive value of a preferred reinforcer (sucrose pellets) when measured in an effort-free preference intake test. This, we propose, ultimately resulted in decreased responding for that preferred food type when animals where tested in an effort-dependent operant test. These two tests might be measuring substantially different expressions of motivated behavior, and perhaps different subcomponents of *wanting*. Progress in experimental psychology and behavioral neuroscience has allowed researchers to learn about reinforcement and motivated behavior, and a broader theoretical integration across different perspectives, such as those presented here, can only help to understand the implications of this knowledge for applied research.

3 LABORATORY ANIMAL RESEARCH IN MOTIVATED BEHAVIOR

As mentioned previously, distinct functional aspects of motivated behavior can be described. Appetitive behaviors regulate the proximity of motivational stimuli while consummatory behaviors allow organisms to interact with their goals. Thus, the types of behaviors that can arise in response to feeding can vary (for a review, see Benoit and Tracy, 2008). Because of this, a number of different behavioral tests have been developed that allow researchers to study certain aspects of motivation. When combined with neuropharmacology, these procedures can help identify brain mechanisms that contribute to very specific aspects of motivation. Some of the most commonly employed behavioral paradigms, with relevance for the data discussed here, will be described in this section.

3.1 INTAKE TESTS

In general, intake tests are conducted to measure consummatory behaviors, or direct interactions with food reinforcers. Animals can either be given a single food option over a number of sessions or be offered concurrent options freely available in order to assess preference (Altizer and Davidson, 1999; Benoit et al., 2000; Davidson et al., 1997; Johnson and Bickel, 2006). Preference tests can be helpful when they are used in parallel with operant tasks, as they can serve to explain why an animal might have ceased to engage in instrumental behavior. For example, in choice situations, an animal might be more willing to work for one reinforcer over the other, which might be explained by its preference for that food type. It is important to note that preference does not necessarily indicate hedonia. An organism may prefer one of two options, but still not find either particularly pleasurable. These tests are helpful as they measure aspects of consummatory behavior, which is closer in time to the experience of emotion than instrumental behaviors (Benoit and Tracy, 2008; Berridge and Robinson, 1998). Although some researchers will often use food preference or intake procedures to assess the "reward value" of a reinforcer, it is not a commonly accepted way of doing so. Consumption tests only indirectly assess whether a given reinforcer produces pleasure. In other words, *liking* is assumed from observed *wanting*. Although these mechanisms often work together, they can be experimentally dissociated, meaning *wanting* measures are not perfect predictors of *liking*.

3.2 OPERANT PROCEDURES

Operant procedures have been used for many decades to study the behavior of animals (primarily rodents and pigeons) under various schedules of reinforcement. Although they were not originally developed with the intention to assess motivation per se, they are certainly useful for these purposes. Animals are generally trained to peck a key or press a lever for food reinforcement. After this behavior has been established, the lever pressing or time requirements can be modified to better suit the researchers needs. Operant schedules of reinforcement can be set up on fixed-ratio schedules, where the number of lever presses required for reinforcement is held constant (Bickel et al., 2000; Ferster and Skinner, 1957). These procedures can give researchers valuable indices of motor function and motivation in general. Also, progressive ratio (PR) schedules have been extensively used and sometimes favored by many scientists (Arnold and Roberts, 1997; Brown et al., 1998; Ferguson and Paule, 1997; Hodos, 1961; Richardson and Roberts, 1996). During a PR schedule, the response requirements are gradually increased every time an animal is reinforced. For example, on a PR2 schedule, an animal may first have to press a lever once for food, followed by 3 the next time, 5 the third time, and so on until the session is programmed to end. The highest ratio achieved is sometimes termed the break point, a commonly used measure of reinforcement efficacy, or the ability of a given reinforcer to maintain goal-directed behavior (Arnold and Roberts, 1997; Bickel et al., 2000; Bradshaw and Killeen, 2012; Hodos, 1961; Hodos and Kalman, 1963). Because of the changing work requirements, PR schedules are well suited to directly assess motor function and, particularly, work expenditure for a given reinforcer. However, it is important to note that while PR schedules are commonly used indices of motivation, no single schedule is ideal. Studies have found that changing a number of unrelated external variables such as lever height and distance can affect response outcomes (Bradshaw and Killeen, 2012; Hamill et al., 1999; Richardson and Roberts, 1996). A more comprehensive approach incorporating various schedules and measures might be better suited given the multidimensional nature of motivated behavior.

3.3 CONCURRENT FEEDING LEVER-PRESSING/CHOW INTAKE TASK

Originally developed by Salamone et al. (1991), the concurrent feeding task was designed to dissociate between disruptions in primary motivational and activational components more closely related to effort expenditure. It helped showcase how dopamine signaling can selectively alter some aspects of food motivation. In this task, animals can either complete a lever schedule of reinforcement for a preferred palatable food option or approach and consume chow that is concurrently available within the chamber (Farrar et al., 2010; Koch et al., 2000; Nowend et al., 2001; Salamone et al., 1991). Within a given session, animals have to make a series of economic decisions between alternative options with competing requirements (Hursh et al., 1988). These procedures were originally administered using FR5 schedules of reinforcement (Cousins et al., 1994; Salamone et al., 1991), but have been recently extended to use PR schedules of reinforcement (Randall et al., 2012, 2014). Given that it has been said that PR break points can be thought of as good indices of the amount of effort an animal is willing to exert for food reinforcement (Salamone et al., 2009; Stewart, 1975), the use of PR schedules in the context of choice serve as a good model of effort-based decision-making processes. A T-maze procedure has also been developed to study effort-related choice in rats and mice (Correa et al., 2016; Denk et al., 2005; Mai et al., 2012; Pardo et al., 2012; Salamone et al., 1994; Yohn et al., 2015), which serves as a validation of the aforementioned lever-pressing task (for a detailed description, see Salamone and Correa, 2012).

An advantage of the concurrent feeding tasks is their ability to dissociate distinct motivational components. In addition, they carry a naturalistic advantage as organisms must often decide between competing resources and not single options. The development of these paradigms also fits well with the literature aimed at using economic concepts in the analysis of behavior (Hursh, 1984, 1993). These studies often stress the importance that response costs, like lever-pressing requirements, help determine behavioral output (Collier and Jennings, 1969; Johnson and Collier, 1987). In economic terms, animals in these procedures are making cost/benefit decisions related to the price of food in terms of the effort necessary. Finally, apart from the abovementioned procedures, delay-discounting tasks and tandem schedules of reinforcement that have ratio requirements attached to time interval requirements have also been used to evaluate aspects of primary motivation and reinforcement (Floresco et al., 2008; Koffarnus et al., 2011; Mingote et al., 2005, 2008; Wade et al., 2000; Winstanley et al., 2005).

4 NEUROBIOLOGY OF FOOD INTAKE: MOTIVATION, DOPAMINE, AND OPIOID SIGNALING

The previous sections showed that certain aspects of motivated behavior can be experimentally dissociated into distinct components. Pioneering work by neuroscientists has come to show that pharmacology and brain manipulations offer a great tool for researchers to determine the dissociable contributions of particular neurotransmitter systems in mediating motivational and emotional components of food reinforcement. Numerous central and peripheral neuroendocrine signals are involved in the control of eating behavior and energy homeostasis. Comprehensive reviews of the neurobiology that regulates food intake can be found elsewhere (Alonso-Alonso et al., 2015; Currie, 2003; Kelley et al., 2005). In the following section, we will focus on the contribution of opioid signaling on food intake, with a special emphasis on hedonic processing and motivated behavior. As the role of the opioid system in these processes is often suggested to be mediated by its actions on dopamine neurotransmission, we will first briefly summarize key proposals of the role of brain dopamine systems (in particular, mesolimbic dopamine) in the neurobiology of reinforcement.

4.1 DOPAMINE

The study of the role of DA in reinforcement, as a central topic of research in behavioral neuroscience, started to take prominence in the 1970s. The use of intracranial self-stimulation during this decade was really common as researchers hoped this technique could shed some light on the nature of reinforcement (Crow, 1972). Scientists found that animals would stop administering intracranial self-stimulation if they were treated with dopamine receptor antagonists or had lesions to DA-rich areas (reviewed in Wise, 2008). The same DA manipulations were also found to block self-administration of drugs like amphetamine and cocaine (Wise, 2008). It was also shown that DA receptor antagonists would produce reductions in lever pressing or running for food reinforcement (Wise et al., 1978). The wealth of the literature was interpreted to mean that DA was responsible for mediating the "rewarding" effects produced by natural reinforcers and drugs, so administration of DA antagonists would produce anhedonia in animals (Wise, 1982). Although very popular in the literature, the notion that DA mediates hedonic responses has been at the core of continuous stimulating debate and reformulation (Berridge and Kringelbach, 2015; Salamone and Correa, 2012). In the 1990s, Berridge and colleagues, with their incentive salience model, argued that DA was responsible for the stamping in of incentive salience of a given reinforcer. DA would mediate and provide motivational importance, in turn making the animal more likely to engage in actions to interact with a reinforcer. In other words, DA signaling played a role in the *wanting* of motivational stimuli, but had no effect on the *liking*. Using facial taste reactivity as a measure of hedonic response, Berridge and colleagues have consistently shown that DA depletions or antagonists fail to alter hedonic responses to sweet solutions in animals, although they can in some circumstances reduce intake of those foods (Berridge and Kringelbach, 2015; Castro et al., 2015). Salamone and colleagues proposed that DA was involved in other aspects of motivation; they have argued that DA is responsible for mediating the energizing and effort-dependent aspects of motivational stimuli (Salamone and Correa, 2012). DA antagonism has been shown to alter highly active instrumental behaviors, such as demanding lever pressing, leaving primary motivational aspects of motivation like appetite unchanged (Randall et al., 2012; Salamone and Correa, 2012). Thus based on these results, Salamone and colleagues argued that DA was responsible for mediating the activational and directional aspects of food motivation but played very little role in hedonia or consummatory behaviors. Some of the current proposed hypotheses of brain DA function (and in particular of mesocorticolimbic circuits) vary in their language and implications, but are not necessarily mutually exclusive.

4.2 THE ENDOGENOUS OPIOID SYSTEM

The endogenous opioid system (EOS) consists of various endogenously produced opioid peptides and the receptors they bind to, which are distributed throughout peripheral tissues and the central nervous system (CNS). The widespread localization of the EOS throughout the body is likely related to this system's involvement in a number of proposed biological functions including analgesia, respiration, hormone regulation, fluid balance, motor function, motivation, learning and memory, and hedonic processing (Berridge, 1996; Ghelardini et al., 2015; Kelley et al., 2005; Kieffer and Evans, 2009; Mansour et al., 1988). An exhaustive review of opioid involvement in these processes is beyond the scope of this chapter's goal (for reviews, see Bodnar, 2004, 2016), as our focus will remain on opioid system involvement in food intake, motivation, and hedonic processing.

Mammalian binding sites for opiates in the brain were first discovered in the early 1970s (Pert and Snyder, 1973). Subsequent pharmacological characterizations revealed that these receptors were not homogenous. To date, four main receptor types have been cloned, mu, delta, kappa, and the nociceptin receptor (Bunzow et al., 1994; Chen et al., 1993; Evans et al., 1992; Li et al., 1993; Meng et al., 1993; Mollereau et al., 1994; Thompson et al., 1993; Wang et al., 1993; Zastawny et al., 1994). Opioid receptors belong to a larger class of G-protein coupled receptors with inhibitory postsynaptic actions. They are activated by endogenously produced peptides, but also by exogenous compounds such as the opiates morphine and heroin. Four main opioid precursors, proopiomelanocortin, proenkephalin, prodynorphin, and prepronociceptin, contain the genetic specificity needed to produce a number of opioid peptides that are then released at the synaptic terminals of various opioidergic neurons. Opioid precursors give rise to beta-endorphins, enkephalins, dynorphins, and nociceptin, respectively (for reviews, see Dores et al., 2002; Larhammar et al., 2015). Although there are no ligands exclusively associated with one receptor type, they do have different binding affinities for each receptor. Mu-opioid receptors have high affinity for beta-endorphin and enkephalins, but a low affinity for dynorphins. Delta receptors show high affinity for enkephalins, whereas dynorphins bind to kappa receptors (Lutz et al., 1985; Mansour et al., 1994; Pert and Snyder, 1973; Simon et al., 1973; Terenius, 1973; also see Dietis et al., 2011; Pasternak, 2014). The study of the pharmacology of opioid receptors and ligands continues to be a very active area of research. For instance, mu-opioid receptor subtypes, based on the complexity of the mu-opioid receptor gene and its different splice variants (Pasternak, 2014), have been proposed. EOS components are found throughout the periphery and the CNS, including areas such as the pituitary, arcuate nucleus of the hypothalamus, nucleus of the solitary tract, the adrenal medulla, the gut, and gastrointestinal tract, where they help regulate a number of biological functions (Dietis et al., 2011; Khachaturian et al., 1985; Mollereau and Mouledous, 2000; Sauriyal et al., 2011). The opioid system has also been found to play a key role in regulating food intake and reinforcement processes. Opioid receptors and peptides are densely localized in brain areas that control several aspects of reinforcement, including the ventral tegmental area (VTA), nucleus accumbens (NAc), prefrontal cortex (PFC), hypothalamus, and amygdala (Mansour et al., 1994, 1995; Sauriyal et al., 2011; Zhang et al., 2015). In the next sections, we will review current knowledge about the opioid system's contribution to food intake and food reinforcement mechanisms, with a special focus on research conducted in laboratory animals.

4.3 OPIOID SIGNALING AND FOOD-MOTIVATED BEHAVIOR

Studies suggesting that the EOS was involved in food intake regulation date back to the 1970s (Holtzman, 1975, 1979). It was initially shown that administration of opioid receptor agonists caused robust increases in food intake in animals and, by contrast, opioid receptor antagonists had inhibitory effects on energy intake (Brown and Holtzman, 1979; Cooper, 1980; Frenk and Rogers, 1979; Holtzman, 1975; Levine et al., 1990; MacDonald et al., 2003, 2004; Taber et al., 1998). Later studies confirmed a number of hypothalamic and limbic brain areas where opioid peptides as well as mu, delta, and kappa receptors are found. When administered into the lateral hypothalamus, VTA, NAc, or amygdala, mu-opioid receptor agonists were shown to have prophagic outcomes (Bakshi and Kelley, 1993; Katsuura et al., 2011; Mucha and Iversen, 1986; Nathan and Bullmore, 2009; Stanley et al., 1988; Zhang and

Kelley, 2000). Similar effects have been shown using delta receptor agonist microinjections in the ventromedial hypothalamus, PVN, NAc, VTA, and amygdala (Ardianto et al., 2016; Burdick et al., 1998; Gosnell et al., 1986; Jenck et al., 1987; Majeed et al., 1986; McLean and Hoebel, 1983; Ruegg et al., 1997). The effect of kappa receptor manipulations appears to be more complex and site specific. Systemic administration of a kappa-opioid receptor agonist did not change food intake. However, antagonism of these receptors in the LH and VTA, but not in the NAc, decreased food intake (Ikeda et al., 2015).

Mu-opioid receptor agonists like morphine have also been seen to increase consumption of highly palatable high-fat and carbohydrate-rich foods (Katsuura et al., 2011; Marks-Kaufman, 1982; Ottaviani and Riley, 1984). Also, mu-opioid receptor antagonists appear to be most potent in reducing intake of highly palatable sweet solutions or foods high in fat content, prompting researchers to question whether the EOS was responsible for regulating intake of specific macronutrients (Apfelbaum and Mandenoff, 1981; Calcagnetti et al., 1990; Cooper et al., 1985; Levine et al., 1982, 1995; Marks-Kaufman et al., 1984). Interestingly, it has been found that baseline preference, not macronutrients per se, might be the determining factor (Glass et al., 2000; Gosnell et al., 1990; Olszewski et al., 2002; Taha, 2010; Welch et al., 1994); animals that prefer high-fat foods will alter their eating of fat in response to opioid receptor stimulation or antagonism, while animals that prefer carbohydrates will be most affected in their consumption of this macronutrient. Areas involved in mediating these processes include the NAc (Kelley et al., 2002; Le Merrer et al., 2009; Zhang and Kelley, 2000). This baseline preference is relevant as it is also correlated with opioid agonists and antagonists' ability to alter taste reactivity (Doyle et al., 1993; Parker et al., 1992; Peciña and Berridge, 1994, 2005; Rideout and Parker, 1996; Smith et al., 2011). It is important to suggest that the role of the EOS in regulating hedonic aspects of eating might take place outside of caloric needs. Antagonism of mu-opioid receptors has been seen to reduce intake of sweet solutions without caloric content such as saccharin (Beczkowska et al., 1993). Classic food intake and preference tests, however, are not commonly accepted measures of positive affect. As mentioned before, taste-dependent hedonic responses can be studied investigating affective facial reactions. Findings from studies employing taste reactivity tests suggest that the EOS is involved in mediating hedonic or *liking* responses to food. When administered at very specific sites (hedonic hotspots; reviewed in Castro and Berridge, 2014b; Castro et al., 2015; Richard et al., 2013) of the ventral striatum and ventral pallidum, administration of a number of opioid receptor agonists increases hedonic responses to palatable foods and sweet solutions (Castro and Berridge, 2014a; Peciña and Berridge, 1994, 2005; Smith and Berridge, 2005).

In addition to regulating food intake and hedonic responses to palatable food, the EOS also affects an animal's willingness to exert effort to obtain food. Solinas and Goldberg (2005) tested the effects of the primarily mu-opioid receptor antagonist naloxone (systemic, 1.0 mg/kg) on PR responding in food-restricted Sprague Dawley rats and found significant suppression effects at this dose. Similarly, Barbano et al. (2009) found that systemic naloxone (1 mg/kg) reduced break points on a PR3 schedule in both food-sated and -restricted Wistar rats, although the effects were more pronounced in satiated animals. In addition, Levine et al. (1995) showed that naloxone (3 mg/kg) attenuated food intake in 24-h-deprived animals, but the magnitude of the effect varied by food type. Here, we present novel data (Fig. 1) using a FR5/chow procedure where rats can choose between completing an FR5 lever-pressing task for a preferred food (banana-flavored sucrose pellets) or consuming freely available standard rodent chow.[2] Our data indicate that, when given systemic injections of naloxone (3 mg/kg), rats reduced lever pressing for the more palatable reinforcer (therefore earning less sucrose pellets), while chow intake is unaffected. These data show that opioid signal inhibition does not reduce overall, unspecific appetite, but rather reduced the amount of effort devoted to obtain a more preferred food. We also present data (Fig. 2) showing that the same dose of naloxone used in our first study reduced sucrose pellet intake (without altering chow intake) when tested on an effort-free food preference test. In our experiment, rats might have experienced a reduced hedonic response associated with eating sucrose pellets, thereby showing reduced willingness to work for this preferred food. As suggested before, altered palatability might in turn translate into impaired motivation to obtain the reinforcer (Barbano and Cador, 2007; Kelley et al., 2002). It is not entirely clear what neural circuits translate decreased palatability to reduced motivation, although evidence suggests that interactions between opioid and DAergic systems are involved (Barbano et al., 2009; Berridge, 1996). A study conducted by Wassum et al. (2009) has suggested that although palatability and motivational aspects of reinforcement depend on opioid receptor activation, they are both functionally and neuroanatomically dissociable. The authors showed opioid receptors in the NAc shell and ventral pallidum affected palatability, whereas basolateral amygdala opioid signals were important for encoding the motivational value.

DA agonists and antagonists have been shown to affect instrumental responding for food in a similar manner to opioid manipulations, suggesting that opioid systems might recruit mesolimbic DA circuitry (Le Merrer et al., 2009; Ting-A-Kee and Van

[2]We used 19 adult male Long Evans rats purchased from Envigo (Indianapolis, IN). The colony was kept on a 12:12 light/dark cycle, with the lights on at 0700, and temperature controlled at 22 ± 2°C. Rats were housed in pairs and handled daily throughout the experiment. Prior to experiment initiation, animals were given food and water ad libitum. Once testing began, they were given free access to water in their home cages but were food restricted for the duration of the experiment. On experimental days, animals were allowed to consume all of the food obtained during behavioral tests and were given 1 h access to laboratory chow (Lab Diet 5012, St. Louis, MO) after each session. Following procedures described in Farrar et al. (2010), rats were trained to lever press for palatable pellets under an FR5/chow schedule. Upon achievement of a stable baseline, pharmacological testing was conducted. Pharmacology was administered on two consecutive Fridays, with doses (saline and 3 mg/kg of naloxone) counterbalanced across individuals. Rats continued baseline training from Monday through Thursday, with weekends off. Two weeks after completion of the FR5/chow study, rats ($n = 9$, randomly selected) were used to evaluate the effects of naloxone on an effort-free food preference test; animals had both, sucrose pellets and chow available. All procedures were conducted in accordance with the Institutional Animal Care and Use Guidelines of Reed College and the National Institute of Health (NIH) guidelines for the Care and Use of Laboratory Animals.

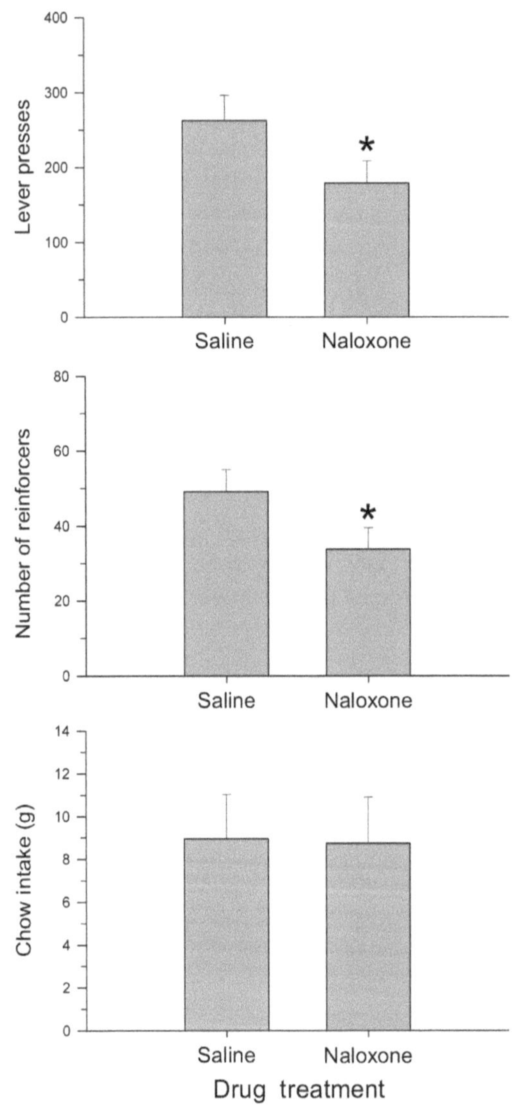

FIG. 1

Effects of the opioid receptor antagonist naloxone on FR5/chow performance. Animals ($n=19$) received intraperitoneal (IP) injections of saline or naloxone (3 mg/kg) 30 min before FR5/chow testing (sessions were 30-min long). Data are represented as means $\pm$ standard error of means (SEM) for number of lever presses (to obtain banana-flavored sucrose pellets; *top panel*), number of reinforcers earned (number of sucrose pellets obtained following a FR5 schedule; *middle panel*), and chow intake (concurrently and freely available standard rat laboratory food; *lower panel*). Statistical analysis (dependent *t*-test) indicated that naloxone significantly decreased lever presses [$t_{(18)}=3.2$, $p<0.01$], and the number of reinforcers earned [$t_{(18)}=3.3$, $p<0.01$], but had no effect on chow consumption (*$p<0.01$, compared to saline).

FIG. 2

Effects of systemic naloxone administration on free food intake and preference. Animals ($n=9$) received IP injections of saline or naloxone (3 mg/kg), 30 min before testing (sessions were 30-min long). Data are represented as means ± SEM for grams of food consumed (banana pellets or chow). A repeated measures, two-way analysis of variance (ANOVA) indicated a main effect of naloxone treatment [$F_{(1,24)}=15.6$, $p<0.01$], food type [$F_{(1,24)}=5.1$, $p<0.05$], as well as an significant interaction between factors [$F_{(1,24)}=22.9$, $p<0.01$]. Tukey's HSD post hoc test showed that animals, when treated with saline, significantly preferred banana-flavored sucrose pellets over chow ($^{\#}p<0.01$). However, this preference was not seen in animals treated with naloxone ($^{*}p<0.01$; saline vs naloxone effects on banana pellet consumption).

der Kooy, 2012). As mentioned before, DA systems in these brain areas are known to regulate behavioral processes like incentive salience and exertion of effort (Robinson and Berridge, 2008; Salamone and Correa, 2012). It is well documented that opioid receptors regulate activity of VTA DA neurons (Margolis et al., 2014). Mu-opioid receptor activation in the NAc increases Fos expression within the VTA, the origin of mesolimbic DA neurons (Bontempi and Sharp, 1997; Zhang and Kelley, 2000). In addition, central administration of mu-opioid agonists into the ventricles increases DA activity within the NAc (Shippenberg et al., 1993; Spanagel et al., 1990, 1992; Yoshida et al., 1999). Administration of exogenous opioid compounds such as morphine or heroin stimulates DA release through activation of mu- and delta-opioid receptors (Hirose et al., 2005; Murakawa et al., 2004; Okutsu et al., 2006; Yoshida et al., 1999). Mu-opioid receptor activity in the VTA decreases inhibition of GABAergic interneurons, which in turn affects DA release in the NAc (Bonci and Williams, 1997; Fields and Margolis, 2015; Johnson and North, 1992; Ting-A-Kee and Van der Kooy, 2012). By contrast, activation of kappa receptors appears to have the opposite effect (Di Chiara and Imperato, 1988; Spanagel et al., 1994; Zhang et al., 2004).

Growing evidence clearly indicates that opioid receptors, and in particular mu-opioid receptors, play an important role in regulating food palatability, eating behavior and, according to new data presented here, effort-related decision making. Opioid signaling appears to play a role in mediating palatability of preferred food, which in turn might translate into altered motivation to obtain that reinforcer. The mechanisms by which decreased palatability translates into decreased motivation, however, remain to be fully understood. As suggested before, it is possible that (palatability and effort expenditure) might be mediated by independent opioid signaling pathways, or that opioids only directly act on primary hedonic processing and indirectly affect effort-related functions downstream (either by opioid receptor modulation of DA neurons or through some other mechanism). Further research, however, will need to better identify specific brain systems involved in those processes and to what extend they can be dissociated at an experimental level. In this regard, direct comparisons of opioid and DA manipulations using PR/chow tasks might be effective and advantageous.

5 CLINICAL APPLICATIONS AND FUTURE DIRECTIONS

The EOS has been implicated in a number of disorders including drug and alcohol addiction, pathological alterations of mood, and eating disorders (Fattore et al., 2015; Giuliano and Cottone, 2015; Kulkarni and Dhir, 2009; Kurbanov et al., 2012; Tejeda et al., 2012). Compulsive overeating is a maladaptive behavior associated with a number of eating disorders such as obesity, bulimia, and binge-eating disorder (Alonso-Alonso et al., 2015; Nathan and Bullmore, 2009). At the heart of this feeding behavior lies increased responsiveness to food and food-associated environmental cues. A number of human studies have found relationships between binge-eating behavior, obesity, and certain polymorphisms of the DRD2 (dopamine D_2 receptor) gene, DAT1 (dopamine transporter) gene, and the OPRM1 (mu-opioid receptor) gene (Davis et al., 1983, 2008, 2009, 2011; Epstein et al., 2007; Haghighi et al., 2014; Shinohara et al., 2004). In addition to the work done in humans, researchers using animal models have also begun to more precisely explain the neural mechanisms behind binge-like behavior. Over the years, a number of animal models of binge-like eating have been developed. Most of these have focused on dietary restraint access by disrupting food intake through restrictions of caloric availability, limiting duration of food access, combining access to food with environmental stress, and intermittingly offering sugar and chow, and each has shown these to be crucial in instigating binge-like behavior (Corwin and Buda-Levin, 2004; Geary, 2003; Giraudo et al., 1993; Hagan et al., 2002, 2003; Howard and Porzelius, 1999; Inoue et al., 2004). Studies using a variety of binge-eating-like behavior models have consistently shown that opioid receptor antagonists, particularly those acting on mu-opioid receptors, reduce or attenuate the expression of binge-like food consumption in animals (Barbano and Cador, 2006; Bodnar et al., 1995; Cooper, 1980; Davis et al., 1983; Giraudo et al., 1993; Glass et al., 2001; Hadjimarkou et al., 2004; Hagan et al., 1997; Kelley et al., 1996; Levine and Billington, 1997). It is still debated whether eating disorders can be thought of as food addictions (Salamone and Correa, 2013) as contention still exists about how the neural signatures of eating disorders are similar to the neuroadaptations that take place in the development of drug addiction (Ifland et al., 2009; Pelchat, 2009; Rogers and Smit, 2000). Regardless, future research concerning the role of opioids in both appetitive and consummatory aspects of food-motivated behavior can help bring to light how these processes might be similar or different from those involved in addiction. Additionally, better understanding of the connection or dissociation between the more hedonic aspects of food, or *liking*, and the more motivational, or *wanting* (ie, does decreased *liking* translate into attenuated *wanting*?), might help explain how compulsive food-taking patterns characteristic of binging behavior emerge.

Of special interest, and particularly relevant to western societies, is the overconsumption of sugary foods. In pathological cases, patterns of sugar ingestion can be so severe that they could mimic those observed in drug and alcohol addiction. Obsessive cravings and compulsive intake habits, often in the face of severe personal and medical consequences, are characteristic of both drug abuse and binge eating. In recent years, scientists have placed increasing emphasis on understanding the neural mechanisms that mediate the transition from manageable to the unmanageable patterns of food consumption seen in some eating disorders. Animal models such as the ones highlighted in this chapter can give key insights into the role that the EOS and other systems play in specific aspects of the processes that support eating disorders. Understanding the brain processes by which vulnerable individuals lose control is key to developing better treatment and prevention methods.

ACKNOWLEDGMENTS

This research was funded in part by a grant from the M.J. Murdock Charitable Trust (Life Sciences) to P.J.C., and a Reed College Initiative grant to I.M. The authors gratefully acknowledge the technical assistance provided by Emma Brockway, Joaquín A. Selva, Hannah Baumgartner, and Lia Zallar, and the animal colony care provided by Greg Wilkinson. Dr. Timothy D. Hackenberg critically revised earlier versions of this manuscript.

REFERENCES

Alonso-Alonso, M., Woods, S.C., Pelchat, M., Grigson, P.S., Stice, E., Farooqi, S., Khoo, C.S., Mattes, R.D., Beauchamp, G.K., 2015. Food reward system: current perspectives and future research needs. Nutr. Rev. 73, 296–307.

Altizer, A.M., Davidson, T.L., 1999. The effects of NPY and 5-TG on responding to cues for fats and carbohydrates. Physiol. Behav. 65, 685–690.

Apfelbaum, M., Mandenoff, A., 1981. Naltrexone suppresses hyperphagia induced in the rat by a highly palatable diet. Pharmacol. Biochem. Behav. 15, 89–91.

Ardianto, C., Yonemochi, N., Yamamoto, S., Yang, L., Takenoya, F., Shioda, S., Nagase, H., Ikeda, H., Kamei, J., 2016. Opioid systems in the lateral hypothalamus regulate feeding behavior through orexin and GABA neurons. Neuroscience 320, 183–293.

Arnold, J.M., Roberts, D.C.S., 1997. A critique of fixed and progressive ratio schedules used to examine the neural substrates of drug reinforcement. Pharmacol. Biochem. Behav. 57, 441–447.

Bakshi, V.P., Kelley, A.E., 1993. Feeding induced by opioid stimulation of the ventral striatum: role of opiate receptor subtypes. J. Pharmacol. Exp. Ther. 265, 1253–1260.

Baldo, B.A., Kelley, A.E., 2007. Discrete neurochemical coding of distinguishable motivational processes: insights from nucleus accumbens control of feeding. Psychopharmacology 191, 439–459.

Barbano, M.F., Cador, M., 2006. Differential regulation of the consummatory, motivational and anticipatory aspects of feeding behavior by dopaminergic and opioidergic drugs. Neuropsychopharmacology 31, 1371–1381.

Barbano, M.F., Cador, M., 2007. Opioids for hedonic experience and dopamine to get ready for it. Psychopharmacology 191, 497–506.

Barbano, M.F., Le Saux, M., Cador, M., 2009. Involvement of dopamine and opioids in the motivation to eat: influence of palatability, homeostatic state, and behavioral paradigms. Psychopharmacology 203, 475–487.

Beczkowska, I.W., Koch, J.E., Bostock, M.E., Leibowitz, S.F., Bodnar, R.J., 1993. Central opioid receptor subtype antagonists differentially reduce intake of saccharin and maltose dextrin solutions in rats. Brain Res. 618, 261–270.

Benoit, S.C., Tracy, A.L., 2008. Behavioral controls of food intake. Peptides 29, 139–147.

Benoit, S.C., Morell, J.R., Davidson, T.L., 2000. Lesions of the amygdala central nucleus interfere with blockade of satiation for peanut oil by Na-2-mercaptoacetate. Psychobiology 28, 387–393.

Berridge, K.C., 1996. Food reward: brain substrates of wanting and liking. Neurosci. Biobehav. Rev. 20, 1–25.

Berridge, K.C., 2000. Reward learning: reinforcement, incentives, and expectations. In: Medin, D.L. (Ed.), Psychology of Learning and Motivation, vol. 40. Academic Press, Cambridge, MA, pp. 223–278.

Berridge, K.C., 2004. Motivation concepts in behavioral neuroscience. Physiol. Behav. 81, 179–209.

Berridge, K.C., 2012. From prediction error to incentive salience: mesolimbic computation of reward motivation. Eur. J. Neurosci. 35, 1124–1143.

Berridge, K.C., Kringelbach, M.L., 2008. Affective neuroscience of pleasure: reward in humans and animals. Psychopharmacology 199, 457–480.

Berridge, K.C., Kringelbach, M.L., 2013. Neuroscience of affect: brain mechanisms of pleasure and displeasure. Curr. Opin. Neurobiol. 23, 294–303.

Berridge, K.C., Kringelbach, M.L., 2015. Pleasure systems in the brain. Neuron 86 (3), 646–664. http://dx.doi.org/10.1016/j.neuron.2015.02.018.

Berridge, K.C., Robinson, T.E., 1998. What is the role of dopamine in reward: hedonic impact, reward learning, or incentive salience? Brain Res. Brain Res. Rev. 28, 309–369.

Berridge, K.C., Robinson, T.E., Aldridge, J.W., 2009. Dissecting components of reward: 'liking', 'wanting', and learning. Curr. Opin. Pharmacol. 9, 65–73.

Bickel, W.K., Marsch, L.A., Carroll, M.E., 2000. Deconstructing relative reinforcement efficacy and situating the measures of pharmacological reinforcement with behavioral economics: a theoretical proposal. Psychopharmacology 153, 44–56.

Blackburn, K., 2002. A new animal model of binge eating: key synergistic role of past caloric restriction and stress. Physiol. Behav. 77, 45–54.

Blackburn, J.R., Phillips, A.G., Fibiger, H.C., 1989. Dopamine and preparatory behavior: III. Effects of metoclopramide and thioridazine. Behav. Neurosci. 103, 903–906.

Bodnar, R.J., 2004. Endogenous opioids and feeding behavior: a 30-year historical perspective. Peptides 25, 697–725.

Bodnar, R.J., 2016. Endogenous opiates and behavior: 2014. Peptides 75, 18–70.

Bodnar, R.J., Glass, M.J., Ragnauth, A., Cooper, M.L., 1995. General, μ and k opioid antagonists in the nucleus accumbens alter food intake under deprivation, glucoprivic and palatable conditions. Brain Res. 700, 205–212.

Bonci, A., Williams, J.T., 1997. Increased probability of GABA release during withdrawal from morphine. J. Neurosci. 17, 796–803.

Bontempi, B., Sharp, F.R., 1997. Systemic morphine-induced Fos protein in the rat striatum and nucleus accumbens is regulated by mu opioid receptors in the substantia nigra and ventral tegmental area. J. Neurosci. 17, 8596–8612.

Bradshaw, C.M., Killeen, P.R., 2012. A theory of behaviour on progressive ratio schedules, with applications in behavioural pharmacology. Psychopharmacology 222, 549–564.

Brown, D.R., Holtzman, S.G., 1979. Suppression of deprivation-induced food and water intake in rats and mice by naloxone. Pharmacol. Biochem. Behav. 11, 567–573.

Brown, C., Fletcher, P., Coscina, D., 1998. Neuropeptide Y-induced operant responding for sucrose is not mediated by dopamine. Peptides 19, 1667–1673.

Bunzow, J.R., Saez, C., Mortrud, M., Bouvier, C., Williams, J.T., Low, M., Grandy, D.K., 1994. Molecular cloning and tissue distribution of a putative member of the rat opioid receptor gene family that is not a mu, delta or kappa opioid receptor type. FEBS Lett. 347, 284–288.

Burdick, K., Yu, W.Z., Ragnauth, A., Moroz, M., Pan, Y.X., Rossi, G.C., Pasternak, G.W., Bodnar, R.J., 1998. Antisense mapping of opioid receptor clones: effects upon 2-deoxy-D-glucose-induced hyperphagia. Brain Res. 794, 359–363.

Cabanac, M., Lafrance, L., 1991. Facial consummatory responses in rats support the ponderostat hypothesis. Physiol. Behav. 50, 179–183.

Calcagnetti, D.J., Calcagnetti, R.L., Fanselow, M.S., 1990. Centrally administered opioid antagonists, nor-binaltorphimine, 16-methyl cyprenorphine and MR2266, suppress intake of a sweet solution. Pharmacol. Biochem. Behav. 35, 69–73.

Cardinal, R.N., Parkinson, J.A., Hall, J., Everitt, B.J., 2002. Emotion and motivation: the role of the amygdala, ventral striatum, and prefrontal cortex. Neurosci. Biobehav. Rev. 26, 321–352.

Castro, D.C., Berridge, K.C., 2014a. Opioid hedonic hotspot in nucleus accumbens shell: mu, delta, and kappa maps for enhancement of sweetness "liking" and "wanting". J. Neurosci. 34, 4239–4250.

Castro, D.C., Berridge, K.C., 2014b. Advances in the neurobiological bases for food 'liking' versus 'wanting'. Physiol. Behav. 136, 22–30.

Castro, D.C., Cole, S.L., Berridge, K.C., 2015. Lateral hypothalamus, nucleus accumbens, and ventral pallidum roles in eating and hunger: interactions between homeostatic and reward circuitry. Front. Syst. Neurosci. 15, 9–90.

Chen, Y., Mestek, A., Liu, J., Hurley, J.A., Yu, L., 1993. Molecular cloning and functional expression of a m-opioid receptor from rat brain. Mol. Pharmacol. 44, 8–12.

Cofer, C., Appley, M., 1964. Motivation: Theory and Research. John Wiley, Oxford, England.

Collier, G., Jennings, W., 1969. Work as a determinant of instrumental performance. J. Comp. Physiol. Psychol. 68, 659–662.
Colwill, R.M., Rescorla, R.A., 1986. Associative structures in instrumental learning. In: Bower, G.H. (Ed.), The Psychology of Learning and Motivation. Academic Press, New York, pp. 55–104.
Cooper, S.J., 1980. Naloxone: effects on food and water consumption in the non-deprived and deprived rat. Psychopharmacology 71, 1–6.
Cooper, S.J., Barber, D.J., Barbour-McMullen, J., 1985. Selective attenuation of sweetened milk consumption by opiate receptor antagonists in male and female rats of the Roman strains. Neuropeptides 5, 349–352.
Correa, M., Pardo, M., Bayarri, P., López-Cruz, L., San Miguel, N., Valverde, O., Ledent, C., Salamone, J.D., 2016. Choosing voluntary exercise over sucrose consumption depends upon dopamine transmission; effects of haloperidol in wild type and adenosine A_{2a} KO mice. Psychopharmacology 233, 393–404.
Corwin, R.L., Buda-Levin, A., 2004. Behavioral models of binge-type eating. Physiol. Behav. 82, 123–130.
Cousins, M.S., Wei, W., Salamone, J.D., 1994. Pharmacological characterization of performance on a concurrent lever pressing/feeding choice procedure: effects of dopamine antagonist, cholinomimetic, sedative and stimulant drugs. Psychopharmacology 116, 529–537.
Craig, W., 1918. Appetites and aversions as constituents of instincts. Biol. Bull. 34, 91–107.
Crow, T.J., 1972. Catecholamine-containing neurones and electrical self-stimulation: 1. A review of some data. Psychol. Med. 2, 414–421.
Currie, P.J., 2003. Integration of hypothalamic feeding and metabolic signals: focus on neuropeptide Y. Appetite 41, 335–337.
Czachowski, C.L., Santini, L.A., Legg, B.H., Samson, H.H., 2002. Separate measures of ethanol seeking and drinking in the rat: effects of remoxipride. Alcohol 28, 39–46.
Davidson, T.L., Altizer, A.M., Benoit, S.C., Walls, E.K., Powley, T.L., 1997. Encoding and selective activation of "metabolic memories" in the rat. Behav. Neurosci. 111, 1014–1130.
Davis, J.M., Lowy, M.T., Yim, G.K.W., Lamb, D.R., Malven, P.V., 1983. Relationship between plasma concentrations of immunoreactive beta-endorphin and food intake in rats. Peptides 4, 79–83.
Davis, C., Levitan, R.D., Kaplan, A.S., Carter, J., Reid, C., Curtis, C., Patte, K., Hwang, R., Kennedy, J.L., 2008. Reward sensitivity and the D2 dopamine receptor gene: a case-control study of binge eating disorder. Prog. Neuropsychopharmacol. Biol. Psychiatry 32, 620–628.
Davis, C.A., Levitan, R.D., Reid, C., Carter, J.C., Kaplan, A.S., Patte, K.A., King, N., Curtis, C., Kennedy, J.L., 2009. Dopamine for "wanting" and opioids for "liking": a comparison of obese adults with and without binge eating. Obesity 17, 1220–1225.
Davis, C., Zai, C., Levitan, R.D., Kaplan, A.S., Carter, J.C., Reid-Westoby, C., Curtis, C., Wight, K., Kennedy, J.L., 2011. Opiates, overeating and obesity: a psychogenetic analysis. Int. J. Obes. 35, 1347–1354.
Denk, F., Walton, M.E., Jennings, K.A., Sharp, T., Rushworth, M.F.S., Bannerman, D.M., 2005. Differential involvement of serotonin and dopamine systems in cost-benefit decisions about delay or effort. Psychopharmacology 179, 587–596.
Di Chiara, G., Imperato, A., 1988. Opposite effects of mu and kappa opiate agonists on dopamine release in the nucleus accumbens and in the dorsal caudate of freely moving rats. J. Pharmacol. Exp. Ther. 244, 1067–1080.

Dickinson, A., Balleine, B., 1994. Motivational control of goal-directed action. Anim. Learn. Behav. 22, 1–18.

Dietis, N., Rowbotham, D.J., Lambert, D.G., 2011. Opioid receptor subtypes: fact or artifact? Br. J. Anaesth. 107, 8–18.

Dinsmoor, J.A., 2004. The etymology of basic concepts in the experimental analysis of behavior. J. Exp. Anal. Behav. 82, 311–316.

Dores, R.M., Lecaude, S., Bauer, D., Danielson, P.B., 2002. Analyzing the evolution of the opioid/orphanin gene family. Mass Spectrom. Rev. 21, 220–243.

Doyle, T.G., Berridge, K.C., Gosnell, B.A., 1993. Morphine enhances hedonic taste palatability in rats. Pharmacol. Biochem. Behav. 46, 745–749.

Ekman, P., 2006. Darwin and Facial Expression: A Century of Research in Review. Malor Books, Los Altos, CA.

Epstein, L.H., Temple, J.L., Neaderhiser, B.J., Salis, R.J., Erbe, R.W., Leddy, J.J., 2007. Food reinforcement, the dopamine D2 receptor genotype, and energy intake in obese and non-obese humans. Behav. Neurosci. 121, 877–886.

Evans, C.J., Keith Jr., D.E., Morrison, H., Magendzo, K., Edwards, R.H., 1992. Cloning of a delta opioid receptor by functional expression. Science 258, 1952–1955.

Everitt, B.J., Dickinson, A., Robbins, T.W., 2001. The neuropsychological basis of addictive behaviour. Brain Res. Rev. 36, 129–138.

Farrar, A.M., Segovia, K.N., Randall, P.A., Nunes, E.J., Collins, L.E., Stopper, C.M., Port, R.G., Hockenmeyer, J., Muller, C.E., Correa, M., Salamone, J.D., 2010. Nucleus accumbens and effort-related functions: behavioral and neural markers of the interactions between adenosine A2A and dopamine D2 receptors. Neuroscience 16, 1056–1067.

Fattore, L., Fadda, P., Antinori, S., Fratta, W., 2015. Role of opioid receptors in the reinstatement of opioid-seeking behavior: an overview. Methods Mol. Biol. 1230, 281–293.

Ferguson, S.A., Paule, M.G., 1997. Progressive ratio performance varies with body weight in rats. Behav. Process. 40, 177–182.

Ferster, C.B., Skinner, B.F., 1957. Schedules of Reinforcement. Appleton-Century-Crofts, New York.

Fields, H.L., Margolis, E.B., 2015. Understanding opioid reward. Trends Neurosci. 38, 217–225.

Finlayson, G., King, N., Blundell, J.E., 2007. Liking vs. wanting food: importance for human appetite control and weight regulation. Neurosci. Biobehav. Rev. 31, 987–1002.

Floresco, S.B., Tse, M.T.L., Ghods-Sharifi, S., 2008. Dopaminergic and glutamatergic regulation of effort- and delay-based decision making. Neuropsychopharmacology 33, 1966–1979.

Foltin, R.W., 2001. Effects of amphetamine, dexfenfluramine, diazepam, and other pharmacological and dietary manipulations on food "seeking" and "taking" behavior in non-human primates. Psychopharmacology 158, 28–38.

Frenk, H., Rogers, G.H., 1979. The suppressant effects of naloxone on food and water intake in the rat. Behav. Neural Biol. 26, 23–40.

Geary, N., 2003. A new animal model of binge eating. Int. J. Eat. Disord. 34, 198–199.

Ghelardini, C., Di Cesare Mannelli, L., Bianchi, E., 2015. The pharmacological basis of opioids. Clin. Cases Miner. Bone Metab. 12, 219–221.

Giraudo, S.Q., Grace, M.K., Welch, C.C., Billington, C.J., Levine, A.S., 1993. Naloxone's anorectic effect is dependent upon the relative palatability of food. Pharmacol. Biochem. Behav. 46, 917–921.

Giuliano, C., Cottone, P., 2015. The role of the opioid system in binge eating disorder. CNS Spectr. 20, 537–545.
Glass, M.J., Billington, C.J., Levine, A.S., 2000. Naltrexone administered to central nucleus of amygdala or PVN: neural dissociation of diet and energy. Am. J. Physiol. Regul. Integr. Comp. Physiol. 279, R86–R92.
Glass, M.J., Grace, M.K., Cleary, J.P., Billington, C.J., Levine, A.S., 2001. Naloxone's effect on meal microstructure of sucrose and cornstarch diets. Am. J. Physiol. Regul. Integr. Comp. Physiol. 281, R1605–R1612.
Gosnell, B.A., Morley, J.E., Levine, A.S., 1986. Opioid-induced feeding: localization of sensitive brain sites. Brain Res. 369, 177–184.
Gosnell, B.A., Krahn, D.D., Majchrzak, M.J., 1990. The effects of morphine on diet selection are dependent upon baseline diet preferences. Pharmacol. Biochem. Behav. 37, 207–212.
Hadjimarkou, M.M., Singh, A., Kandov, Y., Israel, Y., Pan, Y.X., Rossi, G.C., Pasternak, G.W., Bodnar, R.J., 2004. Opioid receptor involvement in food deprivation-induced feeding: evaluation of selective antagonist and antisense oligodeoxynucleotide probe effects in mice and rats. J. Pharmacol. Exp. Ther. 311, 1188–1202.
Hagan, M.M., Holguin, F.D., Cabello, C.E., Hanscom, D.R., Moss, D.E., 1997. Combined naloxone and fluoxetine on deprivation-induced binge eating of palatable foods in rats. Pharmacol. Biochem. Behav. 58, 1103–1107.
Hagan, M.M., Wauford, P.K., Chandler, P.C., Jarrett, L.A., Rybak, R.J., Blackburn, K., 2002. A new animal model of binge eating: key synergistic role of past caloric restriction and stress. Physiol. Behav. 77, 45–54.
Hagan, M.M., Chandler, P.C., Wauford, P.K., Rybak, R.J., Oswald, K.D., 2003. The role of palatable food and hunger as trigger factors in an animal model of stress induced binge eating. Int. J. Eat. Disord. 34, 183–197.
Haghighi, A., Melka, M.G., Bernard, M., Abrahamowicz, M., Leonard, G.T., Richer, L., Perron, M., Veillette, S., Xu, C.J., Greenwood, C.M., Dias, A., El-Sohemy, A., Gaudet, D., Paus, T., Pausova, Z., 2014. Opioid receptor mu 1 gene, fat intake and obesity in adolescence. Mol. Psychiatry 19, 63–68.
Hamill, S., Trevitt, J.T., Nowend, K.L., Carlson, B.B., Salamone, J.D., 1999. Nucleus accumbens dopamine depletions and time-constrained progressive ratio performance: effects of different ratio requirements. Pharmacol. Biochem. Behav. 64, 21–27.
Hirose, N., Murakawa, K., Takada, K., Oi, Y., Suzuki, T., Nagase, H., Cools, A.R., Koshikawa, N., 2005. Interactions among mu- and delta-opioid receptors, especially putative delta1- and delta2-opioid receptors, promote dopamine release in the nucleus accumbens. Neuroscience 135, 213–225.
Hodos, W., 1961. Progressive ratio as a measure of reward strength. Science 134, 943–944.
Hodos, W., Kalman, G., 1963. Effects of increment size and reinforcer volume on progressive ratio performance. J. Exp. Anal. Behav. 6, 387.
Holtzman, S.G., 1975. Effects of narcotic antagonists on fluid intake in the rat. Life Sci. 16, 1465–1470.
Holtzman, S.G., 1979. Suppression of appetitive behaviour in the rat by naloxone: lack of prior morphine dependence. Life Sci. 24, 219–226.
Howard, C.E., Porzelius, L.K., 1999. The role of dieting in binge eating disorder: etiology and treatment implications. Clin. Psychol. Rev. 19, 25–44.
Hursh, S.R., 1984. Behavioral economics. J. Exp. Anal. Behav. 42, 435–452.

Hursh, S.R., 1993. Behavioral economics of drug self-administration: an introduction. Drug Alcohol Depend. 33, 165–172.

Hursh, S.R., Raslear, T.G., Shurtleff, D., Bauman, R., Simmons, L., 1988. A cost-benefit analysis of demand for food. J. Exp. Anal. Behav. 50, 419–440.

Hyman, S.E., Malenka, R.C., Nestler, E.J., 2006. Neural mechanisms of addiction: the role of reward-related learning and memory. Annu. Rev. Neurosci. 29, 565–598.

Ifland, J.R., Preuss, H.G., Marcus, M.T., Rourke, K.M., Taylor, W.C., Burau, K., Jacobs, W.S., Kadish, W., Manso, G., 2009. Refined food addiction: a classic substance use disorder. Med. Hypotheses 72, 518–526.

Ikeda, H., Ardianto, C., Yonemochi, N., Yang, L., Ohashi, T., Ikegami, M., Nagase, H., Kamei, J., 2015. Inhibition of opioid systems in the hypothalamus as well as the mesolimbic area suppresses feeding behavior of mice. Neuroscience 311, 9–21.

Ikemoto, S., Panksepp, J., 1996. Dissociations between appetitive and consummatory responses by pharmacological manipulations of reward-relevant brain regions. Behav. Neurosci. 110, 331–345.

Inoue, K., Zorrilla, E.P., Tabarin, A., Valdez, G.R., Iwasaki, S., Kiriike, N., Koob, G.F., 2004. Reduction of anxiety after restricted feeding in the rat: implication for eating disorders. Biol. Psychiatry 55, 1075–1081.

Jenck, F., Gratton, A., Wise, R.A., 1987. Opioid receptor subtypes associated with ventral tegmental facilitation of lateral hypothalamic brain stimulation reward. Brain Res. 423, 34–38.

Johnson, M.W., Bickel, W.K., 2006. Replacing relative reinforcing efficacy with behavioral economic demand curves. J. Exp. Anal. Behav. 85, 73–93.

Johnson, D.F., Collier, G.H., 1987. Caloric regulation and patterns of food choice in a patchy environment: the value and cost of alternative foods. Physiol. Behav. 39, 351–359.

Johnson, S.W., North, R.A., 1992. Opioids excite dopamine neurons by hyperpolarization of local interneurons. J. Neurosci. 12, 483–488.

Katsuura, Y., Heckmann, J.A., Taha, S.A., 2011. Mu opioid receptor stimulation in the nucleus accumbens elevates fatty tastant intake by increasing palatability and suppressing satiety signals. Am. J. Physiol. Regul. Integr. Comp. Physiol. 301, R244–R254.

Kelley, A.E., Bless, E.P., Swanson, C.J., 1996. Investigation of the effects of opiate antagonists infused into the nucleus accumbens on feeding and sucrose drinking in rats. J. Pharmacol. Exp. Ther. 278, 1499–1507.

Kelley, A.E., Bakshi, V.P., Haber, S.N., Steininger, T.L., Will, M.J., Zhang, M., 2002. Opioid modulation of taste hedonics within the ventral striatum. Physiol. Behav. 76, 365–377.

Kelley, A.E., Baldo, B.A., Pratt, W.E., Will, M.J., 2005. Corticostriatal-hypothalamic circuitry and food motivation: integration of energy, action and reward. Physiol. Behav. 86, 773–795.

Kessler, R.M., Hutson, P.H., Herman, B.K., Herman, B.K., Potenza, M.N., 2016. The neurobiological basis of binge-eating disorder. Neurosci. Biobehav. Rev. 63, 223–238. pii: S0149-7634(15)30254-2.

Khachaturian, H., Lewis, E.J., Schafer, M.K., Watson, S.J., 1985. Anatomy of the CNS opioid systems. Trends Neurosci. 3, 111–119.

Kieffer, B.L., Evans, C.J., 2009. Opioid receptors: from binding sites to visible molecules in vivo. Neuropharmacology 56 (Suppl. 1), 205–212.

Koch, M., Schmid, A., Schnitzler, H.U., 2000. Role of nucleus accumbens dopamine D1 and D2 receptors in instrumental and Pavlovian paradigms of conditioned reward. Psychopharmacology 152, 67–73.

Koffarnus, M.N., Newman, A.H., Grundt, P., Rice, K.C., Woods, J.H., 2011. Effects of selective dopaminergic compounds on a delay-discounting task. Behav. Pharmacol. 22, 300–311.

Kringelbach, M.L., 2004. Food for thought: hedonic experience beyond homeostasis in the human brain. Neuroscience 126, 807–819.

Kringelbach, M.L., 2015. The pleasure of food: underlying brain mechanisms of eating and other pleasures. Flavour 4, 20.

Kulkarni, S.K., Dhir, A., 2009. Sigma-1 receptors in major depression and anxiety. Expert. Rev. Neurother. 9, 1021–1034.

Kurbanov, D.B., Currie, P.J., Simonson, D.C., Borsook, D., Elman, I., 2012. Effects of naltrexone on food intake and weight gain in olanzapine-treated rats. J. Psychopharmacol. 26, 1244–1251.

Larhammar, D., Bergqvist, C., Sundström, G., 2015. Ancestral vertebrate complexity of the opioid system. Vitam. Horm. 97, 95–122.

Le Merrer, J., Becker, J.A., Befort, K., Kieffer, B.L., 2009. Reward processing by the opioid system in the brain. Physiol. Rev. 89, 1379–1412.

Levine, A.S., Billington, C.J., 1997. Why do we eat? A neural systems approach. Annu. Rev. Nutr. 7, 597–619.

Levine, A.S., Murray, S.S., Kneip, J., Grace, M., Morley, J.E., 1982. Flavor enhances the antidipsogenic effect of naloxone. Physiol. Behav. 28, 23–25.

Levine, A.S., Grace, M., Billington, C.J., 1990. The effect of centrally administered naloxone on deprivation and drug-induced feeding. Pharmacol. Biochem. Behav. 36, 409–412.

Levine, A.S., Weldon, D.T., Grace, M., Cleary, J.P., Billington, C.J., 1995. Naloxone blocks that portion of feeding driven by sweet taste in food-restricted rats. Am. J. Physiol. 268, R248–R252.

Li, L.Y., Su, Y.F., Zhang, Z.M., Wong, C.S., Chang, K.J., 1993. Purification and cloning of opioid receptors. NIDA Res. Monogr. 134, 146–164.

Lutz, R.A., Cruciani, R.A., Munson, P.J., Rodbard, D., 1985. Mu1: a very high affinity subtype of enkephalin binding sites in rat brain. Life Sci. 36, 2233–2338.

MacDonald, A.F., Billington, C.J., Levine, A.S., 2003. Effects of the opioid antagonist naltrexone on feeding induced by DAMGO in the ventral tegmental area and in the nucleus accumbens shell region in the rat. Am. J. Physiol. Regul. Integr. Comp. Physiol. 285, R999–R1004.

MacDonald, A.F., Billington, C.J., Levine, A.S., 2004. Alterations in food intake by opioid and dopamine signaling pathways between the ventral tegmental area and the shell of the nucleus accumbens. Brain Res. 1018, 78–85.

Mai, B., Sommer, S., Hauber, W., 2012. Motivational states influence effort-based decision making in rats: the role of dopamine in the nucleus accumbens. Cogn. Affect. Behav. Neurosci. 12, 74–84.

Majeed, N.H., Przewklocka, B., Wędzony, K., Przewklocki, R., 1986. Stimulation of food intake following opioid microinjection into the nucleus accumbens septi in rats. Peptides 7, 711–716.

Mansour, A., Khachaturian, H., Lewis, M.E., Akil, H., Watson, S.J., 1988. Anatomy of CNS opioid receptors. Trends Neurosci. 11, 308–314.

Mansour, A., Fox, C.A., Burke, S., Meng, F., Thompson, R.C., Akil, H., Watson, S.J., 1994. Mu, delta, and kappa opioid receptor mRNA expression in the rat CNS: an in situ hybridization study. J. Comp. Neurol. 350, 412–438.

Mansour, A., Fox, C.A., Akil, H., Watson, S.J., 1995. Opioid-receptor mRNA expression in the rat CNS: anatomical and functional implications. Trends Neurosci. 18, 22–29.

Margolis, E.B., Hjelmstad, G.O., Fujita, W., Fields, H.L., 2014. Direct bidirectional μ-opioid control of midbrain dopamine neurons. J. Neurosci. 34, 14707–14716.

Marks-Kaufman, R., 1982. Increased fat consumption induced by morphine administration in rats. Pharmacol. Biochem. Behav. 16, 949–955.

Marks-Kaufman, R., Balmagiya, T., Gross, E., 1984. Modifications in food intake and energy metabolism in rats as a function of chronic naltrexone infusions. Pharmacol. Biochem. Behav. 20, 911–916.

McLean, S., Hoebel, B.G., 1983. Feeding induced by opiates injected into the paraventricular hypothalamus. Peptides 4, 287–292.

Meng, F., Xie, G.X., Thompson, R.C., Mansour, A., Goldstein, A., Watson, S.J., Akil, H., 1993. Cloning and pharmacological characterization of a rat kappa opioid receptor. Proc. Natl. Acad. Sci. U.S.A. 90, 9954–9958.

Mingote, S., Weber, S.M., Ishiwari, K., Correa, M., Salamone, J.D., 2005. Ratio and time requirements on operant schedules: effort-related effects of nucleus accumbens dopamine depletions. Eur. J. Neurosci. 21, 1749–1757.

Mingote, S., Font, L., Farrar, A.M., Vontell, R., Worden, L.T., Stopper, C.M., Port, R.G., Sink, K.S., Bunce, J.G., Chrobak, J.J., Salamone, J.D., 2008. Nucleus accumbens adenosine A2A receptors regulate exertion of effort by acting on the ventral striatopallidal pathway. J. Neurosci. 28, 9037–9046.

Mollereau, C., Mouledous, L., 2000. Tissue distribution of the opioid receptor-like (ORL1) receptor. Peptides 21, 907–917.

Mollereau, C., Parmentier, M., Mailleux, P., Butour, J.L., Moisand, C., Chalon, P., Caput, D., Vassart, G., Meunier, J.C., 1994. ORL1, a novel member of the opioid receptor family. Cloning, functional expression and localization. FEBS Lett. 341, 33–38.

Mucha, R.F., Iversen, S.D., 1986. Increased food intake after opioid microinjections into nucleus accumbens and ventral tegmental area of rat. Brain Res. 397, 214–224.

Murakawa, K., Hirose, N., Takada, K., Suzuki, T., Nagase, H., Cools, A.R., Koshikawa, N., 2004. Deltorphin II enhances extracellular levels of dopamine in the nucleus accumbens via opioid receptor-independent mechanisms. Eur. J. Pharmacol. 491, 31–36.

Nader, K., Bechara, A., Van der Kooy, D., 1997. Neurobiological constraints on behavioral models of motivation. Annu. Rev. Psychol. 48, 85–114.

Nathan, P.J., Bullmore, E.T., 2009. From taste hedonics to motivational drive: central μ-opioid receptors and binge-eating behaviour. Int. J. Neuropsychopharmacol. 12, 995–1008.

Nowend, K.L., Arizzi, M., Carlson, B.B., Salamone, J.D., 2001. D1 or D2 antagonism in nucleus accumbens core or dorsomedial shell suppresses lever pressing for food but leads to compensatory increases in chow consumption. Pharmacol. Biochem. Behav. 69, 373–382.

Okutsu, H., Watanabe, S., Takahashi, I., Aono, Y., Saigusa, T., Koshikawa, N., Cools, A.R., 2006. Endomorphin-2 and endomorphin-1 promote the extracellular amount of accumbal dopamine via nonopioid and mu-opioid receptors, respectively. Neuropsychopharmacology 31, 375–383.

Olszewski, P.K., Grace, M.K., Sanders, J.B., Billington, C.J., Levine, A.S., 2002. Effect of nociceptin/orphanin FQ on food intake in rats that differ in diet preference. Pharmacol. Biochem. Behav. 73, 529–535.

Ottaviani, R., Riley, A.L., 1984. Effect of chronic morphine administration on the self-selection of macronutrients in the rat. Nutr. Behav. 2, 27–36.

Packard, M.G., Knowlton, B.J., 2002. Learning and memory functions of the basal ganglia. Annu. Rev. Neurosci. 25, 563–593.

Pardo, M., Lopez-Cruz, L., Valverde, O., Ledent, C., Baqi, Y., Müller, C.E., Salamone, J.D., Correa, M., 2012. Adenosine A2A receptor antagonism and genetic deletion attenuate the effects of dopamine D2 antagonism on effort-based decision making in mice. Neuropharmacology 62, 2068–2077.

Parker, L.A., Maier, S., Rennie, M., Crebolder, J., 1992. Morphine- and naltrexone-induced modification of palatability: analysis by the taste reactivity test. Behav. Neurosci. 106, 999–1010.

Parkinson, J.A., Dalley, J.W., Cardinal, R.N., Bamford, A., Fehnert, B., Lachenal, G., Rudarakanchana, N., Halkerston, K.M., Robbins, T.W., Everitt, B.J., 2002. Nucleus accumbens dopamine depletion impairs both acquisition and performance of appetitive Pavlovian approach behaviour: implications for mesoaccumbens dopamine function. Behav. Brain Res. 137, 149–163.

Pasternak, G.W., 2014. Opioids and their receptors: are we there yet? Neuropharmacology 76, 198–203.

Peciña, S., Berridge, K.C., 1994. Central enhancement of taste pleasure by intraventricular morphine. Neurobiology 3, 269–280.

Peciña, S., Berridge, K.C., 2005. Hedonic hot spot in nucleus accumbens shell: where do μ-opioids cause increased hedonic impact of sweetness? J. Neurosci. 25, 11777–11786.

Pelchat, M.L., 2009. Food addiction in humans. J. Nutr. 139, 620–622.

Pert, C.B., Snyder, S.H., 1973. Properties of opiate-receptor binding in rat brain. Proc. Natl. Acad. Sci. U.S.A. 70, 2243–2247.

Randall, P.A., Pardo, M., Nunes, E.J., López Cruz, L., Vemuri, V.K., Makriyannis, A., Baqi, Y., Müller, C.E., Correa, M., Salamone, J.D., 2012. Dopaminergic modulation of effort-related choice behavior as assessed by a progressive ratio chow feeding choice task: pharmacological studies and the role of individual differences. PLoS One 7, e47934.

Randall, P.A., Lee, C.A., Nunes, E.J., Yohn, S.E., Nowak, V., Khan, B., Shah, P., Pandit, S., Vemuri, V.K., Makriyannis, A., Baqi, Y., Müller, C.E., Correa, M., Salamone, J.D., 2014. The VMAT-2 inhibitor tetrabenazine affects effort-related decision making in a progressive ratio/chow feeding choice task: reversal with antidepressant drugs. PLoS One 9, e99320.

Richard, J.M., Castro, D.C., Difeliceantonio, A.G., Robinson, M.J., Berridge, K.C., 2013. Mapping brain circuits of reward and motivation: in the footsteps of Ann Kelley. Neurosci. Biobehav. Rev. 37, 1919–1931.

Richardson, N.R., Roberts, D.C.S., 1996. Progressive ratio schedules in drug self-administration studies in rats: a method to evaluate reinforcing efficacy. J. Neurosci. Methods 66, 1–11.

Rideout, H.J., Parker, L.A., 1996. Morphine enhancement of sucrose palatability: analysis by the taste reactivity test. Pharmacol. Biochem. Behav. 53, 731–734.

Robbins, T.W., Koob, G.F., 1980. Selective disruption of displacement behaviour by lesions of the mesolimbic dopamine system. Nature 285, 409–412.

Robinson, T.E., Berridge, K.C., 1993. The neural basis of drug craving: an incentive-sensitization theory of addiction. Brain Res. Brain Res. Rev. 18, 247–291.

Robinson, T.E., Berridge, K.C., 2003. Addiction. Annu. Rev. Psychol. 54, 25–53.

Robinson, T.E., Berridge, K.C., 2008. The incentive sensitization theory of addiction: some current issues. Philos. Trans. R Soc. Lond. B Biol. Sci. 363, 3137–3146.

Rogers, P.J., Smit, H.J., 2000. Food craving and food "addiction": a critical review of the evidence from a biopsychosocial perspective. Pharmacol. Biochem. Behav. 66, 3–14.

Ruegg, H., Yu, W.Z., Bodnar, R.J., 1997. Opioid-receptor subtype agonist-induced enhancements of sucrose intake are dependent upon sucrose concentration. Physiol. Behav. 62, 121–128.

Salamone, J.D., 1988. Dopaminergic involvement in activational aspects of motivation: effects of haloperidol on schedule-induced activity, feeding, and foraging in rats. Psychobiology 16, 196–206.

Salamone, J.D., 2010. Motor function and motivation. In: In: Koob, G., Le Moal, M., Thompson, R.F. (Eds.), Encyclopedia of Behavioral Neuroscience, vol. 3. Academic Press, Oxford, pp. 267–276.

Salamone, J.D., Correa, M., 2002. Motivational views of reinforcement: implications for understanding the behavioral functions of nucleus accumbens dopamine. Behav. Brain Res. 137, 3–25.

Salamone, J.D., Correa, M., 2012. The mysterious motivational functions of mesolimbic dopamine. Neuron 76, 470–485.

Salamone, J.D., Correa, M., 2013. Dopamine and food addiction: lexicon badly needed. Biol. Psychiatry 73, e15–e24.

Salamone, J.D., Steinpreis, R.E., McCullough, L.D., Smith, P., Grebel, D., Mahan, K., 1991. Haloperidol and nucleus accumbens dopamine depletion suppress lever pressing for food but increase free food consumption in a novel food choice procedure. Psychopharmacology 104, 515–521.

Salamone, J.D., Cousins, M.S., Bucher, S., 1994. Anhedonia or anergia? Effects of haloperidol and nucleus accumbens dopamine depletion on instrumental response selection in a T-maze cost/benefit procedure. Behav. Brain Res. 65 (2), 221–229.

Salamone, J.D., Correa, M., Farrar, A.M., Mingote, S.M., 2007. Effort-related functions of nucleus accumbens dopamine and associated forebrain circuits. Psychopharmacology 191, 461–482.

Salamone, J.D., Correa, M., Farrar, A.M., Nunes, E.J., Pardo, M., 2009. Dopamine, behavioral economics, and effort. Front. Behav. Neurosci. 3, 13.

Sauriyal, D.S., Jaggi, A.S., Singh, N., 2011. Extending pharmacological spectrum of opioids beyond analgesia: multifunctional aspects in different pathophysiological states. Neuropeptides 45, 175–188.

Shahan, T.A., 2010. Conditioned reinforcement and response strength. J. Exp. Anal. Behav. 93, 269–289.

Sherrington, C.S., 1906. The Integrative Action of the Nervous System. C. Scribner's Sons, New York.

Shin, A.C., Zheng, H., Berthoud, H.R., 2009. An expanded view of energy homeostasis: neural integration of metabolic, cognitive, and emotional drives to eat. Physiol. Behav. 97, 572–580.

Shinohara, M., Mizushima, H., Hirano, M., Shioe, K., Nakazawa, M., Hiejima, Y., Ono, Y., Kanba, S., 2004. Eating disorders with binge-eating behaviour are associated with the s allele of the 3′-UTR VNTR polymorphism of the dopamine transporter gene. J. Psychiatry Neurosci. 29, 134–137.

Shippenberg, T.S., Bals-Kubik, R., Herz, A., 1993. Examination of the neurochemical substrates mediating the motivational effects of opioids: role of the mesolimbic dopamine system and D-1 vs. D-2 dopamine receptors. J. Pharmacol. Exp. Ther. 265, 53–59.

Simon, E.J., Hiller, J.M., Edelman, I., 1973. Stereospecific binding of the potent narcotic analgesic (3H) etorphine to rat-brain homogenate. Proc. Natl. Acad. Sci. U.S.A. 70, 1947–1949.

Skinner, B.F., 1938. The Behavior of Organisms: An Experimental Analysis. Appleton-Century, Oxford, England.

Skinner, B.F., 1953. Science and Human Behavior. Simon and Schuster, New York.

Slocum, S.K., Vollmer, T.R., 2015. A comparison of positive and negative reinforcement for compliance to treat problem behavior maintained by escape. J. Appl. Behav. Anal. 48, 563–574.

Smith, K.S., Berridge, K.C., 2005. The ventral pallidum and hedonic reward: neurochemical maps of sucrose "liking" and food intake. J. Neurosci. 25, 8637–8649.

Smith, K.S., Berridge, K.C., Aldridge, J.W., 2011. Disentangling pleasure from incentive salience and learning signals in brain reward circuitry. Proc. Natl. Acad. Sci. U.S.A. 108, E255–E264.

Solinas, M., Goldberg, S.R., 2005. Motivational effects of cannabinoids and opioids on food reinforcement depend on simultaneous activation of cannabinoid and opioid systems. Neuropsychopharmacology 30, 2035–2045.

Spanagel, R., Herz, A., Shippenberg, T.S., 1990. The effects of opioid peptides on dopamine release in the nucleus accumbens: an in vivo microdialysis study. J. Neurochem. 55, 1734–1740.

Spanagel, R., Herz, A., Shippenberg, T.S., 1992. Opposing tonically active endogenous opioid systems modulate the mesolimbic dopaminergic pathway. Proc. Natl. Acad. Sci. U.S.A. 89, 2046–2050.

Spanagel, R., Almeida, O.F., Bartl, C., Shippenberg, T.S., 1994. Endogenous kappa-opioid systems in opiate withdrawal: role in aversion and accompanying changes in mesolimbic dopamine release. Psychopharmacology 115, 121–127.

Stanley, B., Lanthier, D., Leibowitz, S.F., 1988. Multiple brain sites sensitive to feeding stimulation by opioid agonists: a cannula-mapping study. Pharmacol. Biochem. Behav. 31, 825–832.

Steiner, J.E., 1973. The gustofacial response: observation on normal and anencephalic newborn infants. In: Oral Sensation and Perception: Development in the Fetus and Infant: Fourth Symposium. US Government Printing Office, Dhew, Oxford, England. pp. xix, 419.

Steiner, J.E., 1974. Discussion paper: innate, discriminative human facial expressions to taste and smell stimulation. Ann. N. Y. Acad. Sci. 237, 229–233.

Stewart, W.J., 1975. Progressive reinforcement schedules: a review and evaluation. Aust. J. Psychol. 27, 9–22.

Taber, M.T., Zernig, G., Fibiger, H.C., 1998. Opioid receptor modulation of feeding-evoked dopamine release in the rat nucleus accumbens. Brain Res. 785, 24–30.

Taha, S.A., 2010. Preference or fat? Revisiting opioid effects on food intake. Physiol. Behav. 100, 429–437.

Tejeda, H.A., Shippenberg, T.S., Henriksson, R., 2012. The dynorphin/k-opioid receptor system and its role in psychiatric disorders. Cell. Mol. Life Sci. 69, 857–896.

Terenius, L., 1973. Characteristics of the 'receptor' for narcotic analgesics in synaptic plasma membrane fraction from rat brain. Acta Pharmacol. Toxicol. 33, 377–384.

Thompson, R.C., Mansour, A., Akil, H., Watson, S.J., 1993. Cloning and pharmacological characterization of a rat mu opioid receptor. Neuron 11, 903–913.

Tibboel, H., De Houwer, J., Van Bockstaele, B., 2015. Implicit measures of "wanting" and "liking" in humans. Neurosci. Biobehav. Rev. 57, 350–364.
Ting-A-Kee, R., Van der Kooy, D., 2012. The neurobiology of opiate motivation. Cold Spring Harb. Perspect. Med. 2, a012096.
Toates, F., 1986. Motivational Systems. Cambridge University Press, Cambridge.
Voon, V., 2015. Cognitive biases in binge eating disorder: the hijacking of decision making. CNS Spectr. 20, 566–573.
Wade, T.R., De Wit, H., Richards, J.B., 2000. Effects of dopaminergic drugs on delayed reward as a measure of impulsive behavior in rats. Psychopharmacology 150, 90–101.
Wang, J.B., Imai, Y., Eppler, C.M., Gregor, P., Spivak, C.E., Uhl, G.R., 1993. Mu opiate receptor: cDNA cloning and expression. Proc. Natl. Acad. Sci. U.S.A. 90, 10230–10234.
Wassum, K.M., Ostlund, S.B., Maidment, N.T., Balleine, B.W., 2009. Distinct opioid circuits determine the palatability and the desirability of rewarding events. Proc. Natl. Acad. Sci. U.S.A. 106, 12512–12517.
Welch, C.C., Grace, M.K., Billington, C.J., Levine, A.S., 1994. Preference and diet type affect macronutrient selection after morphine, NPY, norepinephrine, and deprivation. Am. J. Physiol. 266, R426–R433.
White, N.M., Milner, P.M., 1992. The psychobiology of reinforcers. Annu. Rev. Psychol. 43, 443–471.
Winstanley, C.A., Theobald, D.E.H., Dalley, J.W., Robbins, T.W., 2005. Interactions between serotonin and dopamine in the control of impulsive choice in rats: therapeutic implications for impulse control disorders. Neuropsychopharmacology 30, 669–682.
Wise, R.A., 1982. Neuroleptics and operant behavior: the anhedonia hypothesis. Behav. Brain Sci. 5, 39–53.
Wise, R.A., 2008. Dopamine and reward: the anhedonia hypothesis 30 years on. Neurotox. Res. 14, 169–183.
Wise, R.A., Spindler, J., DeWit, H., Gerberg, G.J., 1978. Neuroleptic-induced "anhedonia" in rats: pimozide blocks reward quality of food. Science 201, 262–264.
Yohn, S.E., Thompson, C., Randall, P.A., Lee, C.A., Müller, C.E., Baqi, Y., Correa, M., Salamone, J.D., 2015. The VMAT-2 inhibitor tetrabenazine alters effort-related decision making as measured by the T-maze barrier choice task: reversal with the adenosine A2A antagonist MSX-3 and the catecholamine uptake blocker bupropion. Psychopharmacology 232, 1313–1323.
Yoshida, Y., Koide, S., Hirose, N., Takada, K., Tomiyama, K., Koshikawa, N., Cools, A.R., 1999. Fentanyl increases dopamine release in rat nucleus accumbens: involvement of mesolimbic mu- and delta-2-opioid receptors. Neuroscience 92, 1357–1365.
Zastawny, R.L., George, S.R., Nguyen, T., Cheng, R., Tsatsos, J., Briones-Urbina, R., O'Dowd, B.F., 1994. Cloning, characterization, and distribution of a mu-opioid receptor in rat brain. J. Neurochem. 62, 2099–2105.
Zhang, M., Kelley, A.E., 2000. Enhanced intake of high-fat food following striatal mu-opioid stimulation: microinjection mapping and fos expression. Neuroscience 99, 267–277.
Zhang, Y., Butelman, E.R., Schlussman, S.D., Ho, A., Kreek, M.J., 2004. Effect of the endogenous kappa opioid agonist dynorphin A(1-17) on cocaine-evoked increases in striatal dopamine levels and cocaine-induced place preference in C57BL/6J mice. Psychopharmacology 172, 422–429.
Zhang, J., Muller, J.F., McDonald, A.J., 2015. Mu opioid receptor localization in the basolateral amygdala: an ultrastructural analysis. Neuroscience 303, 352–363.

CHAPTER

Exploring individual differences in task switching: Persistence and other personality traits related to anterior cingulate cortex function

8

A. Umemoto*[,1], C.B. Holroyd[†]

*Institute of Biomedical and Health Sciences, Hiroshima University, Hiroshima, Japan

[†]University of Victoria, Victoria, BC, Canada

[1]Corresponding author: Tel.: +81-82-257-1722; Fax: 81-82-257-1723, e-mail address: akumemoto@gmail.com

Abstract

Anterior cingulate cortex (ACC) is involved in cognitive control and decision-making but its precise function is still highly debated. Based on evidence from lesion, neurophysiological, and neuroimaging studies, we have recently proposed a critical role for ACC in motivating extended behaviors according to learned task values (Holroyd and Yeung, 2012). Computational simulations based on this theory suggest a hierarchical mechanism in which a caudal division of ACC selects and applies control over task execution, and a rostral division of ACC facilitates switches between tasks according to a higher task strategy (Holroyd and McClure, 2015). This theoretical framework suggests that ACC may contribute to personality traits related to persistence and reward sensitivity (Holroyd and Umemoto, 2016). To explore this possibility, we carried out a voluntary task switching experiment in which on each trial participants freely chose one of two tasks to perform, under the condition that they try to select the tasks "at random" and equally often. The participants also completed several questionnaires that assessed personality trait related to persistence, apathy, anhedonia, and rumination, in addition to the Big 5 personality inventory. Among other findings, we observed greater compliance with task instructions by persistent individuals, as manifested by a greater facility with switching between tasks, which is suggestive of increased engagement of rostral ACC.

Keywords

Individual differences, Anterior cingulate cortex function, Personality, Persistence, Voluntary task switching, Task selection

Progress in Brain Research, Volume 229, ISSN 0079-6123, http://dx.doi.org/10.1016/bs.pbr.2016.06.003

Anterior cingulate cortex (ACC) constitutes a broad swath of neural territory along the frontal midline of the brain that is widely believed to contribute to cognitive control. Cognitive control is said to facilitate the execution of nonautomatic or effortful behaviors, especially when these are associated with response conflict or occur in novel environments (Norman and Shallice, 1986). Despite decades of research on this subject (Cohen et al., 1990; Miller and Cohen, 2001), the exact function of ACC is still highly debated. Prominent theories propose a role for ACC in performance or conflict monitoring (Botvinick et al., 2001; Carter et al., 1998; Ridderinkhof et al., 2004) and in reinforcement learning (RL) (Holroyd and Coles, 2002; Rushworth et al., 2007; see Holroyd and Yeung, 2011 for review). Yet, although these theories have received substantial empirical support from the human neuroimaging literature, they have been challenged by observations that ACC damage typically spares these functions (Holroyd and Yeung, 2011, 2012). The fact that ACC damage does not manifestly disrupt the behavioral concomitants of these control processes indicates that these functions are not uniquely implemented by ACC.

To address this issue, we recently proposed a novel theory of ACC function (Holroyd and McClure, 2015; Holroyd and Yeung, 2011, 2012) based on recent advances in RL theory related to hierarchical reinforcement learning (HRL) (Botvinick, 2012; Botvinick et al., 2009). By our account, ACC is responsible for motivating the execution of extended, goal-directed behaviors. This theory holds that, rather than learning the reward value of individual actions according to standard principles of RL, the ACC learns the reward value of entire *tasks*. For example, on this view the ACC would learn that dining out has a high reward value by way of reinforcing the task set (a value associated with the entire action policy of going out to a restaurant) rather than by the exhaustive process of reinforcing each individual action that comprises the policy (such as opening the front door, walking to the car, opening the car door, and so on). The ACC would then decide to eat at a restaurant instead of cooking at home by comparing the relative values of these tasks, rather than by acting on the values of the individual actions that comprise the tasks. In this way, HRL affords increased computational efficiency for complex problems characterized by hierarchical structure.

Recent computational simulations illustrate how the ACC could implement this function (Holroyd and McClure, 2015). The model proposes a multilevel hierarchy for action selection and regulation. At the lowest level, the striatum, in conjunction with other brain areas, carries out behaviors that directly act on the external environment. This low-level system is assumed to be effort-averse such that it eschews the production of effortful behaviors, especially when these are associated with low immediate reward value. One level higher, caudal ACC (cACC) is said to select tasks for execution based on their learned average reward values, in the presence of a cost that penalizes switches between tasks, which are assumed to be effortful. Further, the cACC applies a control signal that attenuates the effortful costs incurred by the low-level action selection mechanism. In so doing cACC ensures that the lower-level system produces behaviors that comply with the selected task. Thus, if the cACC selected a task to run up a steep mountain but the striatum resisted the effort in doing so, the control signal produced by cACC would attenuate that cost, thereby motivating the individual to run to the top.

Further, the model proposes that rostral ACC (rACC) implements an even higher level of the hierarchy responsible for regulating the function of cACC. On this view, rACC selects so-called meta-tasks, each of which affords different task sets. For example, the decision to go to work (a meta-task in this framework) would afford various ways of traveling there (by bus, car, taxi, bicycle, waking, and so on). In this example, whereas the rACC would decide *to* travel to work (rather than to do something else, such as spend the day at the park), the cACC would decide on *how* to travel to work (ie, the mode of transport), and the low-level system would implement the series of actions that fulfill these goals. Finally, in parallel to the control mechanism by which cACC attenuates effortful costs incurred by action selection, the rACC is hypothesized to apply a control signal that attenuates effortful costs incurred when switching between tasks. Thus, rACC helps cACC switch from one task to a different task that is more appropriate for the current context, consistent with empirical evidence from both human and nonhuman animal studies (Holroyd and McClure, 2015).

Using the HRL-ACC theory as an organizing framework, we have proposed that individual differences in ACC function contribute to differences in personality (Holroyd and Umemoto, 2016). In particular, the theory suggests that individual differences in ACC function should express as personality traits that relate to the motivation of extended behaviors. In fact, a growing body of evidence suggests that ACC contributes to the personality traits of persistence, apathy, reward sensitivity, and rumination—a repetitive, maladaptive style of thinking about oneself (Nolen-Hoeksema, 1991; see Holroyd and Umemoto, 2016 for a comprehensive review on the subject of ACC and personality). For example, a variety of findings suggest that ACC activity is associated with persevering through challenges (Blanchard et al., 2015; Gusnard et al., 2003; Kurniawan et al., 2010; Parvizi et al., 2013). In a functional magnetic resonance imaging (fMRI) experiment, the cACC of persistent individuals was relatively more activated compared to that of other individuals when the participants rejected low-effort choices with low payoffs in favor of high-effort choices with high payoffs (Kurniawan et al., 2010). Relatedly, apathy—which is associated with a reduction of voluntary, goal-directed behaviors—is a common consequence of ACC damage (Eslinger and Damasio, 1985; Levy and Dubois, 2006; van Reekum et al., 2005). Electrophysiological and functional neuroimaging studies also suggest that cACC function also contributes to reward sensitivity (Bress and Hajcak, 2013; Keedwell et al., 2005; Liu et al., 2014; Proudfit, 2015), and that rACC contributes to rumination (Pizzagalli, 2011 for review). Consistent with the proposed function for rACC, rumination also impedes task switching (Altamirano et al., 2010; Davis and Nolen-Hoeksema, 2000; Whitmer and Banich, 2007). These observations align with the position that ACC serves as a computational hub that links motivation and control processes (Gläscher et al., 2012; Holroyd and Yeung, 2012; Holroyd and Umemoto, 2016; Shenhav et al., 2013; see Botvinick and Braver, 2015 for review). They also dovetail with the idea that the motivation to perform a given task is determined by a comparison between the subjective value of completing the task and the costs incurred in doing so, a balance that should reflect individual differences in motivation (Westbrook et al., 2013) and reward sensitivity (eg, Braem et al., 2012; Engelmann et al., 2009; Locke and Braver, 2008).

Here we utilized a voluntary task switching paradigm to investigate whether particular personality traits related to ACC function influence task selection and execution as predicted by the HRL-ACC theory (Holroyd and McClure, 2015; Holroyd and Yeung, 2012). The task switching paradigm requires participants to switch back and forth between executing two simple tasks, which induces switch cost (SC): slower responses and more errors when switching between tasks as compared to repeating the same task (Allport et al., 1994; Jersild, 1927; Meiran, 1996; Rogers and Monsell, 1995; Spector and Bierderman, 1976). Although the underlying mechanisms of SCs are still highly debated, a commonly accepted theory relates the phenomenon to task sets, which have been defined as "a configuration of cognitive processes that is actively maintained for subsequent task performance" (Sakai, 2008). According to this view, SCs result from reconfiguring the task set when switching to the new task, which does not occur when the same task is repeated (Monsell, 2003). Seemingly paradoxically, furthermore, numerous studies have reported larger SCs when participants switch to relatively easy, more automatic tasks (such as reading the word in the Stroop task; Stroop, 1935) compared to when they switch to relatively difficult, more effortful tasks (such as naming the color in the Stroop task). It has been proposed that this "paradoxical" asymmetrical SC results from the need to release control over the harder task when switching to the easier task, whereas no such release of control is necessary when switching from the easier task to the harder task (Gilbert and Shallice, 2002; Monsell, 2003; Yeung and Monsell, 2003; but see also Kiesel et al., 2010).

In a voluntary task switching paradigm we used in this study, participants were instructed to freely choose which of two tasks to perform on each trial while selecting both tasks about equally often and at random, as if they were flipping a coin on each trial (Arrington and Logan, 2004, 2005; Yeung, 2010). In this version of the paradigm, participants actually choose to perform the harder task more often than the easier task, evidently because the cost of switching from the hard task to the easy task is prohibitive (Masson and Carruthers, 2014; Millington et al., 2013; Yeung, 2010). Although this finding may appear contrary to the Law of Least Effort (Hull, 1943), the inference is that switching away from the harder task is actually harder than *doing* the harder task. Note that participants tend to select the less demanding task when the instructions permit them to freely choose which task to perform (eg, Kool et al., 2010), whereas they tend to select the more demanding task when the instructions indicate that they should perform each task about equally often, as in our study. Functional neuroimaging studies indicate greater engagement by ACC in voluntary task selection, suggesting that this paradigm may be optimal for testing ACC function (eg, Deiber et al., 1999; Forstmann et al., 2006; Vassena et al., 2014).

The HRL-ACC theory makes specific predictions about individual differences in task switching behavior, depending on whether the differences relate to cACC or rACC function (Holroyd and McClure, 2015). First, enhanced cACC activity would increase top-down control over task execution, which in turn would increase SCs and impede task switching. Second, reduced cACC activity would decrease control over task execution, which in turn would reduce SCs and facilitate task switching (but at the cost of slower responses and a higher error rate). Third, enhanced rACC activity would increase top-down control over task switching, which in turn would facilitate task switching by cACC and attenuate SCs. Fourth, reduced rACC activity would decrease control over task switching by cACC, leading to larger SCs. Note that these predictions indicate that individual differences in the expression of the two brain regions could produce identical behavioral effects; in particular, increased SCs could result either from enhanced cACC activity (because increased control is applied over the given task, rendering it difficult to reconfigure the task set when switching) or to decreased rACC activity (removing the control signal that would otherwise attenuate the SC).

For this reason, our predictions are based on existing literature about which areas of ACC are most associated with the personality traits of interest: rumination, apathy, anhedonia, and persistence. First, rumination has been associated with rACC function (Pizzagalli, 2011 for review) and is said to reflect perseveration of task-inappropriate processes (Altamirano et al., 2010; Davis and Nolen-Hoeksema, 2000; Whitmer and Banich, 2007), which can result from rACC damage (Holroyd and McClure, 2015). For these reasons, we reasoned that impaired rACC function associated with rumination would be revealed in larger SCs, including larger paradoxical asymmetrical SCs. Further, we predicted that the increased asymmetry of the SCs would impede the high ruminators from switching to the easier task, with the result that they would choose to execute the harder task relatively more often (see also Altamirano et al., 2010).

Second, apathy and anhedonia have been associated with reduced cACC function. For instance, apathetic individuals who are otherwise healthy exhibit significantly less cACC activation for actions that demand higher levels of effort (Bonnelle et al., 2016). Likewise, individuals high in anhedonia as it relates to depression exhibit reduced cACC activity as revealed by electrophysiological studies (Proudfit, 2015 for review). For these reasons, we predicted that high levels of apathy and anhedonia would be associated with decreased application of top-down control over task performance. This should be revealed in smaller SCs, due to the reduced control over task execution, together with slower responses and an increased error rate.

Third, persistence has been associated with both cACC (Blanchard et al., 2015; Kurniawan et al., 2010; Parvizi et al., 2013) and rACC (Gusnard et al., 2003) activity, rendering specific predictions about this trait more difficult to make. Complicating matters further, participants in our experiment were in fact given two sets of instructions, either of which they could comply with to a larger or lesser degree: first, to execute each task quickly and accurately, and second, to switch between the tasks at random with an equal probability. Whereas the former process aligns with cACC function (which is concerned about control over individual tasks), the latter process would seem to align with rACC function (which is concerned about the meta-task, which here concerns choosing each task equally often). To foreshadow our results, we found that persistent individuals were more concerned with the higher-level aspects of the meta-task (concerned with switching between the specific tasks) than about performance on the tasks per se.

We also examined how these personality traits aligned with the traits assessed by the Big 5 personality inventory (John et al., 2008), several of which are also related to motivational factors and reward sensitivity. For instance, extroverted individuals report higher levels of positive affect and exhibit enhanced activity in cortical areas concerned with reward processing (eg, orbitofrontal cortex and cACC), as observed in fMRI (DeYoung et al., 2010) and ERP (Cooper et al., 2014) experiments. By contrast, neurotic individuals report relatively more negative affect (Watson and Clark, 1992) and, in one fMRI study involving an affect-neutral oddball task, exhibited reduced rACC activation and increased cACC activation (Eisenberger et al., 2005; see also Bishop et al., 2004; DeYoung et al., 2010; Gray and Braver, 2002). This observation suggests that these individuals may exhibit increased SCs, particularly for the easier task—ie, increased paradoxical asymmetrical SCs—similar to the predicted effect of rumination.

Finally, conscientiousness is closely related to persistence (Cloninger et al., 1993), both of which have been associated with increased rACC activity (Gusnard et al., 2003). The latter finding is compatible with the common notion that conscientious individuals should particularly be concerned with carrying out a given task correctly. This in turn predicts that SCs should be attenuated in these individuals and that they may be concerned about the meta-task similar to persistent individuals. Therefore, the Big 5 personality traits were expected to complement the relation between the ACC-related traits and task performance.

1 MATERIALS AND METHODS

1.1 PARTICIPANTS

One hundred and thirty-two undergraduate students participated in either of two versions of the task, which was slightly altered mid-way as described below in order to speed data collection[1]: Fifty-seven of them (15 male) participated in version 1 and 75 (17 male) undergraduate students participated in version 2. Participants were recruited from the University of Victoria Department of Psychology subject pool

[1]The two versions yielded similar results (including the task bias and the proportion of switch trials, $p=0.2$ and $p=0.48$, respectively) except that the average RT for the first version was statistically significantly slower than the second version by 68 ms ($p=0.02$).

to fulfill a course requirement. All subjects (32 males, age range = 18–33 years, mean age = 21 ± 3 years) had normal or corrected-to-normal vision. All subjects provided informed consent as approved by the local research ethics committee. The experiment was conducted in accordance with the ethical standards prescribed in the 1964 Declaration of Helsinki.

1.2 TASK DESIGN

Participants performed a voluntary task switching task (Yeung, 2010) in which they freely chose to respond to a given stimulus based either on its location (location task) or on its shape (shape task). On each trial, one of three shapes (a circle, a square, or a triangle) appeared in one of three locations inside a grid composed of three adjacent boxes (5.5 × 15 cm^2) (Fig. 1). The stimulus and location were pseudorandomly selected such that each shape was equally likely to appear in each of the three grids; stimulus repetitions were allowed (ie, the same stimulus could appear in the same location consecutively across trials). Half of the participants used their right (left) hand to respond to the shape of the stimulus and their left (right) hand to respond to the location of the stimulus. Participants used the three middle fingers (ie, the index, middle, and ring fingers) of each hand to respond to the stimulus by pressing either of the "Q," "W," and "E" keys with their left hand or the "P," "[," and "]" keys with their right hand on a standard keyboard. Stimulus–response mappings were compatible for the location task, such that participants used their leftmost finger for the stimulus appearing in the left box, their middle finger for the stimulus appearing in the middle box, and their rightmost finger for the stimulus appearing in the right box. For the shape task, the leftmost finger was always used for the circle, the middle finger for the square, and the rightmost finger for the triangle. Each block of trials started with presentation of the grid, which remained on the screen throughout the block. On each trial, the shape stimulus appeared in one of the grid locations and remained on the screen until the participant made a response. Two hundred milliseconds following the response, the next trial began with the presentation of the next stimulus.

1.2.1 Procedure

Participants first practiced each task separately (27 trials each). They then practiced switching between the two tasks within the same block of trials (two blocks of 45 trials each). Task instructions were identical to that used in Yeung (2010, p. 351). After each block of practice trials participants received feedback regarding their average reaction times (RTs) and accuracy. When switching between tasks during the practice blocks, they were further informed about the number of trials in which they chose the shape and the location tasks, as well as how often they switched between tasks. They were also reminded to perform the task quickly and accurately and that the two tasks should be performed about equally often by switching back and forth between them. The feedback on RT and accuracy were provided in order to ensure that the participants remained engaged in the task while adhering to the task instructions.

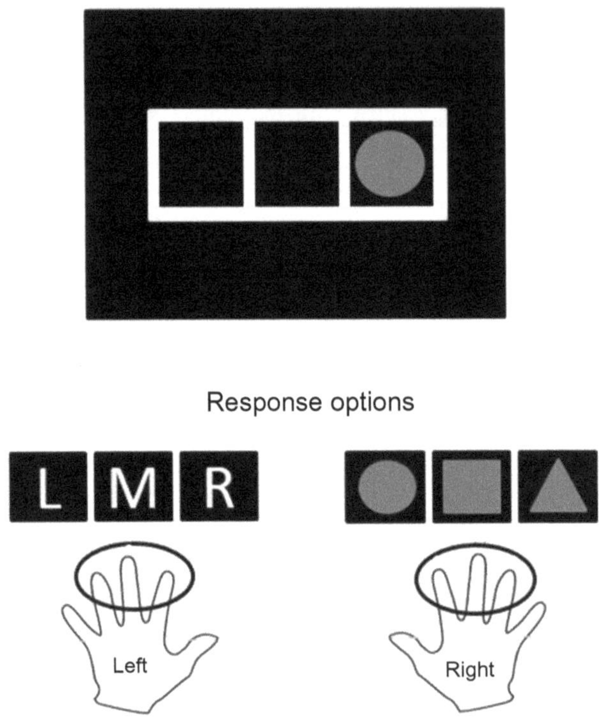

FIG. 1

An example trial of the voluntary task-selection experiment. The *top panel* illustrates an example trial as presented to the participant on a computer screen. The *bottom panel* depicts the response options, which were not presented to the participants, for the purpose of illustration. In this example, key presses with the three middle fingers of the right hand are individually mapped to three stimuli that differed in shape (*circle*, *square*, and *triangle* from the leftmost finger to the rightmost finger) and key presses with the three middle fingers of the left hand are mapped to the corresponding grid location (*left*, "L," *middle*, "M," and *right*, "R," from the leftmost finger to the rightmost finger). Task-hand mappings were counterbalanced across participants (see text). Here, if a participant were to decide to respond to the shape, then the correct response would entail pressing with the leftmost finger of the right hand (corresponding to the *circle*). By contrast, if the participant were to decide to respond to the location of the stimulus, then the correct response would entail pressing with the rightmost finger of the left hand (corresponding to the right location). Location responses are typically faster and more accurate than shape responses in this task, indicating that the location task is easier than the shape task.

For instance, switching tasks half-way through the experiment would result in performing the two tasks equally often but would go against the instruction to perform the tasks in a random order. Likewise, a strategy of systematically alternating between the two tasks would also fail to comply with the instructions.

The experiment proper, which was comprised of 8 blocks of 90 trials each, began following the practice period. Two groups of participants performed slightly different versions of the task. Fifty-seven participants performed the task in a single room in our laboratory (version 1) and 75 participants performed the experiment in groups of up to 10 participants in a computer laboratory at the University of Victoria (version 2). For both groups, performance feedback was provided after each block of trials as in the practice block, except the group performing version 2 did not receive feedback on the number of trials selected for each task.

1.3 QUESTIONNAIRES

Following task completion, participants answered five personality questionnaires administered via LimeSurvey (https://www.limesurvey.org/) on the same computer where the task was performed. These included the 20-item Persistence Scale (PS; Cloninger et al., 1993), which assesses the tendency to overcome daily challenges; the 22-item Ruminative Responses Scale (RRS; Treynor et al., 2003), which measures the propensity to ruminate in response to depressed mood; the 14-item Apathy Scale, which assesses the level of goal-directed behavior as it relates to cognitive activities (eg, "Are you interested in learning new things?"), to emotion (eg, "Are you indifferent to things?"), and to behavior (eg, "Does someone have to tell you what to do each day?") (Starkstein et al., 1992); the 14-item Snaith–Hamilton Pleasure Scale (SHAPS; Snaith et al., 1995), which assesses the extent to which individuals experience pleasure (ie, the level of anhedonia); and the 44-item Big 5 Personality Inventory, which assesses five core personality factors (openness, conscientiousness, extroversion, agreeableness, and neuroticism) (John et al., 2008). Each questionnaire was answered on a Likert scale ranging from 1 (definitely false) to 5 (definitely true) for the PS, from 1 (almost never) to 4 (almost always) for the RRS, from 1 (strongly/definitely agree) to 4 (strongly disagree) for the SHAPS, from 0 to 3 for the Apathy Scale (from 0 = a lot to 3 = not at all for the question 1–8, and from 0 = not at all to 3 = a lot for the question 9–14), and from 1 (disagree strongly) to 5 (agree strongly) for the Big 5 Personality Inventory. Higher scores indicate higher expression of these traits (ie, high in persistence, rumination, anhedonia, apathy, and the Big 5 personality factors).

1.4 STATISTICAL ANALYSES

The first trial of each block, error trials, trials following errors (for the RT analyses only), and trials with response repetitions (18% of the total trials) were excluded from statistical analysis. Response repetitions have been commonly excluded from statistical analyses in task switching studies because they can differentially affect switch and repeat trials (ie, the SCs), particularly when using two tasks that differ in task difficulty (eg, Bryck and Mayr, 2008; Yeung, 2010; Yeung and Monsell, 2003). Trials with RT ± 2 standard deviations (SDs) of the RTs for each subject were also eliminated from analysis to eliminate the effect of outliers on average RTs.

SCs were calculated for each measure as switch trials minus repeat trials, separately for the two tasks (ie, *SC-shape*, the location-to-shape switch trials minus the shape-to-shape repeat trials, and *SC-location*, the shape-to-location switch trials minus the location-to-location repeat trials), separately for RTs and error rates. SCs for the two tasks were also averaged together to create *average SCs*, separately for RTs and error rates. Additionally, SCs for the shape task was subtracted from SCs for the location task to generate a difference in SC (ie, asymmetrical SCs), separately for the RTs and error rates. The task bias was examined as in the previous studies (Millington et al., 2013; Yeung, 2010), measured by subtracting the number of trials participants selected the location task over all the trials from the number of trials participants selected the shape task over all the trials; positive values indicate that participants chose the harder shape task more often than the easier location task. Data were combined across the two versions of the task to increase statistical power.

In order to address possible speed–accuracy trade-offs between these measures, we also created measures that collapsed across RTs and error rates. First, to generate an *overall performance* measure, the average RTs and error rates across the two tasks for each participant were separately z-scored across participants. Then, the standardized values were added together for each participant, such that higher values indicate worse performance (ie, longer RTs and increased error rates). Second, to generate an *overall SC*, the average RT-SCs to the shape and to the location task for each participant were pooled into a single distribution across participants. These values were then z-scored across participants, and subsequently sorted back into separate distributions for shape and location. This procedure was then repeated on the error rate-SCs to the shape and location task. The standardized RT-SCs and error rate-SCs were then summed together for each participant, separately for the shape and location tasks, thereby generating *overall SC-shape* and *overall SC-location* measures. Finally, the difference in the overall SCs (ie, asymmetry in the overall SCs between the two tasks) was calculated by subtracting the overall SC-shape from the overall SC-location. Larger values indicate larger asymmetry in SCs between the two tasks (ie, larger overall SCs-location than the overall SCs-shape).

Multiple linear regression analyses were conducted on the overall performance measure, the overall SCs (ie, the standardized performance measures), and the proportion of switch trials, with the personality traits (including the PS, RRS, Apathy Scale, SHAPS, and the five factors from the Big 5 personality inventory as indicated above) as predictors. The regressions utilized the backward method in which all of the predictors were entered into the model, and noncontributing predictors were stepwise eliminated (removal criteria set at $F=0.1$). To account for the potential influence of outliers, we adopted the following jackknife approach. For each dependent variable, the same multiple regression analysis was performed multiple times by a method of leave-one-out (ie, by excluding the data for a different participant at each iteration) (Hewig et al., 2011). Based on the result of each iteration, if any single participant was found to contribute uniquely to the final regression model—in that removing their data resulted in an inclusion or exclusion of one or more personality predictors from the model, and the same result was not obtained by the other iterations within the same analysis—then the data of this participant were excluded from the given analysis. This procedure was applied to each multiple regression analysis. The degrees of freedom indicate the number of participants included in each analysis. This method is free of experimenter bias by providing objective criteria for the systematic removal of outliers and ensures that the results are robust against the contribution of any single participant. Across all of the tests reported below, this method excluded the data of between zero and three participants, with an average of 1.4 participants.

2 RESULTS

The data of participants who reported multiple major concussions or acquired brain injury (two participants), who exhibited difficulty understanding the task instructions in English (four participants), or who performed with less than 70% accuracy (one participant) were excluded from analysis. Additionally, we inspected the data visually to determine whether participants had performed the task in a systematic order (eg, alternating tasks every few trials, switching tasks at the beginning of each block). This excluded data from one participant for repeating the same task continuously for the first two blocks. Therefore, the data of 124 participants total were included in the analyses. In addition, for the error-related analyses, the data of 10 more participants were excluded due to a technical error, leaving the data of 114 participants total.

2.1 QUESTIONNAIRES

A summary of the personality questionnaire scores is provided in Table 1, and a summary of zero-order correlations among these questionnaires is provided in Table 2.

Table 1 Summary Statistics for the Personality Questionnaire Scores

	Mean	SD	Range
Persistence	71	11.6	37–99
RRS	42	10.5	24–82
SHAPS	21	7.4	14–55
Apathy	12	4.5	3–23
Big 5			
Extraversion	3.3	0.75	1.4–4.9
Agreeableness	3.5	0.53	2.2–4.8
Conscientiousness	3.5	0.53	2.1–4.8
Neuroticism	3.1	0.62	1.3–4.8
Openness	3.4	0.48	1.8–4.5

RRS, *Ruminative Responses Scale;* SD, *standard deviation;* SHAPS, *Snaith–Hamilton pleasure scale.*

Table 2 Zero-Order Correlations Between the Personality Questionnaire Scores

	PS	RRS	SHAPS	AS	Ext	Agr	Con	Neu	Ope
PS									
RRS	−0.32**								
SHAPS	−0.27**	0.17							
AS	−0.53**	0.44**	0.27**						
Big 5									
Ext	0.24**	−0.16	−0.24**	−0.31**					
Agr	0.13	−0.12	0	−0.13	0.08				
Con	0.5**	−0.38**	−0.16	−0.47**	0.08	0.31**			
Neu	−0.12	0.42**	0.12	0.18*	−0.08	−0.26**	−0.29**		
Ope	0.24**	−0.11	−0.12	−0.33**	0.17	0	0.06	−0.1	

AS, *Apathy Scale;* PS, *Persistence Scale;* RRS, *Ruminative Responses Scale;* SHAPS, *Snaith–Hamilton pleasure scale. From the Big 5 personality inventory:* Agr, *agreeableness;* Con, *conscientiousness;* Ext, *extroversion;* Neu, *neuroticism;* Ope, *openness.*
*$p < 0.05$.
**$p < 0.01$.

Table 3 Means and Standard Deviations for the Shape and the Location Task, Separately for the Switch and Repeat Trials in the Reaction Times (RTs), and Error Rates

	Switch	Repeat	*p* Value	SC
RT-shape	924 (212)	806 (172)	<0.01	118 (126)
RT-location	829 (234)	542 (123)	<0.01	287 (186)
Error-shape	7 (4.8)	5.6 (4.2)	<0.01	1.4 (3.8)
Error-location	5.1 (3.6)	1.9 (2.5)	<0.01	3.2 (3)
	Shape	*Location*	p *Value*	
Task choice	0.51 (0.04)	0.49 (0.04)	<0.01	

Switch costs (SCs) are calculated by subtracting the repeat trials from the switch trials in the RTs or the errors. Task choice represents the proportion of trials each task was performed.
RTs are shown in ms, errors in %, and task choice in proportion.
Standard deviations are shown in parenthesis.

2.2 BEHAVIORS

Table 3 provides the means and SDs for the shape and the location task. As commonly observed, RTs were slower following switch trials than following repeat trials for both the shape task, $t(123) = -10.4$, $p < 0.01$, and the location task, $t(123) = -17.1$, $p < 0.01$ (Fig. 2A). Likewise, error rates were larger following switch trials than following repeat trials for both the shape task, $t(113) = -4$, $p < 0.01$, and the location task, $t(113) = -11.1$, $p < 0.01$ (Fig. 2B). As expected, the location task was performed faster and with fewer errors compared to the shape task (RTs: $t(123) = 18.9$, $p < 0.01$, and error rates: $t(113) = 9.9$, $p < 0.01$), indicating that the location task was easier than the shape task (Fig. 2A and B). Also as expected, SCs were asymmetrical between the two tasks so that the SCs to the location task were larger than the SCs to the shape task for both RTs, $t(123) = -11.9$, $p < 0.01$, and error rates, $t(113) = -4.2$, $p < 0.01$ (Fig. 2C and D). Consistent with these observations, there was a significant overall SCs-location (combined across the RT and error rates data), $t(113) = 5.7$, $p < 0.01$, and overall SC-shape (combined across the RT and error rates data), $t(113) = -5.8$, $p < 0.01$, indicating that the findings do not result from a speed–accuracy trade-off. Furthermore, the overall SCs-location was significantly larger than the overall SCs-shape, $t(113) = -9.2$, $p < 0.01$, confirming that the asymmetry in SCs was not due to a speed–accuracy trade-off. Finally, we observed a small but significant task-selection bias, indicating that participants chose the shape task more often than the location task (Table 3), $t(123) = 3.5$, $p < 0.01$. Thus, consistent with previous studies (Millington et al., 2013; Yeung, 2010), we replicated the finding that participants voluntarily selected the harder (shape) task more often than the easier (location) task.

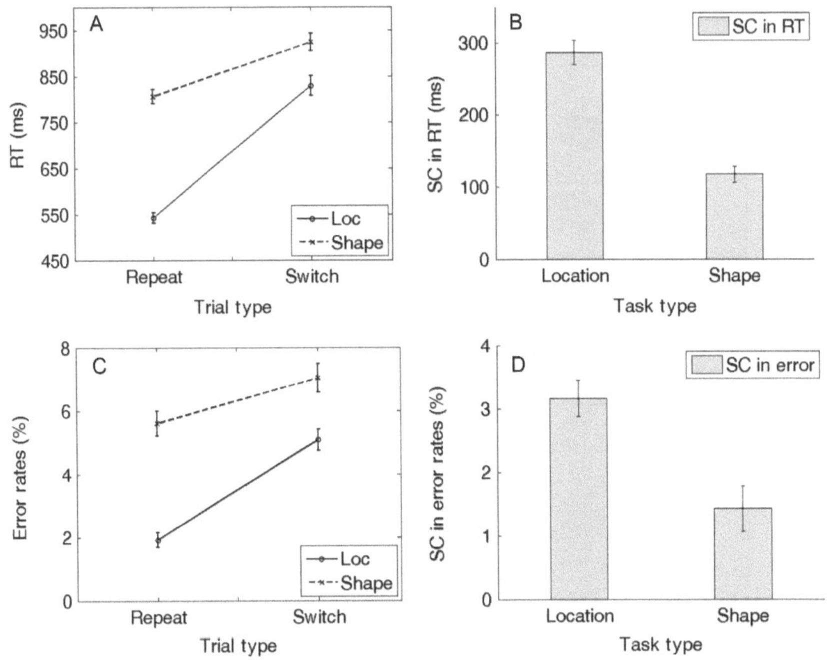

FIG. 2

Task performance. (A) Reaction times (RTs) in milliseconds (ms) for the repeat and switch trials (*x*-axis) for the location (Loc) and the shape task. (B) Error rates (%) for the repeat and switch trials (*x*-axis) for the location (Loc) and the shape task. (C) Switch costs (SCs) in RT to the location (*left*) and the shape (*right*) task. (D) Switch costs (SCs) in error rates to the location (*left*) and the shape (*right*) task. Error bars indicate within-subject standard errors of the mean.

2.3 RELATIONS BETWEEN BEHAVIORS AND QUESTIONNAIRE SCORES

Table 4 provides the results of the multiple regression analyses performed on the overall performance measure and the overall SCs; note that these measures incorporate both RTs and error rate data, so individual differences in performance do not reflect simple trade-offs between speed and accuracy. A multiple regression analysis on the overall performance measure indicated that participants high in extroversion, neuroticism, and persistence performed the task relatively worse than others, $F(3107)=4.3$, $p=0.01$, accounting for 11% of the variance. Comparable analyses indicated that overall SCs were larger for participants high in extroversion and agreeableness, $F(2111)=9.4$, $p<0.01$, explaining 15% of the variance, and overall asymmetrical SCs were larger for participants high in extroversion and neuroticism and low in persistence, $F(3108)=3.8$, $p=0.01$, explaining 10% of the variance. Further,

Table 4 Results of Multiple Regression Analyses on the Overall Performance Measure, Overall Switch Costs (SCs), Overall Asymmetrical SCs (Asym-SCs), Proportion of Trials Participants Switched Tasks (Proportion Switch), and Trait Persistence

Multiple Linear Regression on Task Performance

	Predictors	Beta	t	p Value	Final Model	R^2
Overall performance	Extroversion	0.21	2.3	0.03	$F(3107)=4.3$, $p=0.01$	0.11
	Neuroticism	0.17	1.9	0.06		
	Persistence	0.17	1.8	0.08		
Overall SCs	Extroversion	0.21	2.4	0.02	$F(2111)=9.4$, $p<0.01$	0.15
	Agreeableness	0.3	3.4	<0.01		
Overall asym-SCs	Extroversion	0.17	1.8	0.07	$F(3108)=3.8$, $p=0.01$	0.1
	Neuroticism	0.2	2.2	0.03		
	Persistence	−0.18	−2	0.05		
Proportion switch	Extroversion	−0.33	−3.7	<0.01	$F(2118)=7.4$, $p<0.01$	0.11
	Anhedonia	−0.19	−2.1	0.04		
Persistence	Conscientiousness	0.35	4.2	<0.01	$F(5108)=15$, $p<0.01$	0.41
	Anhedonia	−0.14	−1.8	0.07		
	Apathy	−0.32	−3.7	<0.01		
	Overall asym-SCs	−0.13	−1.7	0.09		
	Task bias	−0.14	−1.9	0.07		

Note that the overall performance measure and overall SCs incorporated both RTs and error rates (see Section 1).

a multiple regression analysis on the proportion of trials participants switched between tasks (ie, the number of switch trials over all the trials) revealed that participants high in extroversion and anhedonia were significantly less likely to switch tasks compared to their counterparts, $F(2118)=7.4$, $p<0.01$, explaining 11% of the variance (Table 4). Although we found no significant correlation between the overall performance measure and the proportion of switch trials ($r=-0.07$, $p=0.5$), there was a significant negative correlation between the overall SCs (standardized) and the proportion of switch trials ($r=-0.30$, $p<0.01$), such that participants who produced larger overall SCs were less likely to switch tasks. No personality traits predicted the proportion of trials one task was chosen over another task (ie, the task bias). Finally, as an exploratory analysis, a multiple regression analysis on the persistence scores with the remaining personality trait scores, the difference in the overall SCs (ie, the asymmetrical SCs), and the task bias as predictors indicated that participants high in persistence were characterized by high conscientiousness, low anhedonia, low apathy, smaller overall differences in SCs, and reduced task bias, $F(5108)=15$, $p<0.01$, explaining 41% of the variance (Fig. 3).

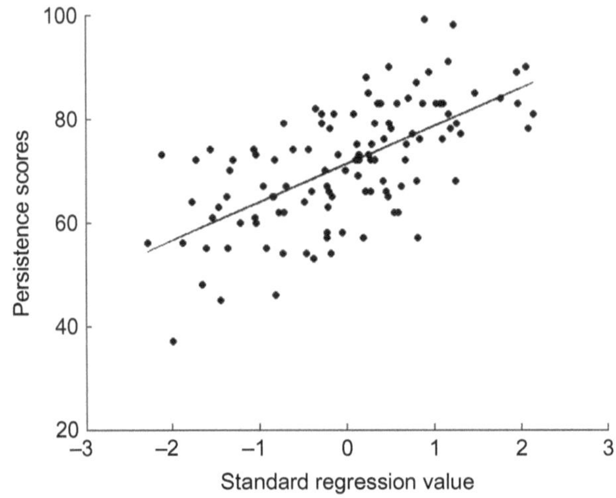

FIG. 3

The result of a multiple linear regression analysis on the trait persistence (*y*-axis) with conscientiousness, anhedonia, apathy, the overall difference in SCs (ie, asymmetrical SCs), and the task bias, together (ie, the standard regression value on the *x*-axis) explaining 41% of the variance.

3 DISCUSSION

The HRL-ACC theory holds that two subdivisions of ACC implement a hierarchical mechanism for task selection and execution (Holroyd and McClure, 2015; Holroyd and Yeung, 2012). On this account, the cACC is said to select tasks for execution and to apply a control signal that motivates and sustains performance until their successful completion; as a consequence, the application of control impedes dynamic shifts between different tasks, which require a reconfiguration of the task set. This proposal is consistent with the SC phenomenon in task switching paradigms, as observed when switching between the shape and location task in the present study, which are said to arise from difficulty releasing control when switching between tasks (Monsell, 2003). A computational model based on the HRL-ACC theory incorporates these observations by imposing a penalty that biases cACC toward repeating the same task rather than switching between them, even when an alternative task is associated with a higher reward value than the current task under execution (Holroyd and McClure, 2015).

Further, on this view the rACC is said to select and implement the higher task strategy (ie, the meta-task) according to comparable principles. In the present context, the choice is between following the instructions of the experimenter or doing something else (such as daydreaming, pressing the keys at random, abandoning the experiment prematurely, and so on); the behavioral data indicate that most participants indeed followed the task instructions, which was to execute the two tasks while switching between them at random but about equally often. Further, the model holds that the rACC applies a control signal that attenuates the SCs experienced by the cACC (Holroyd and McClure, 2015; see also Gläscher et al., 2012; Pollmann et al., 2000; Wager et al., 2005), thereby facilitating switches by cACC from one task to another. These considerations suggest that task switching performance should relate to particular personality traits associated with ACC function, such as persistence, apathy, reward sensitivity, and rumination (Holroyd and Umemoto, 2016). We explored this question in a voluntary task switching paradigm, which allowed for analyses of task selection and cognitive control as revealed by the patterns of SCs and other performance measures. The Big 5 personality inventory was included in this analysis to compare the putative ACC-related traits with a more normative set of personality measures.

We successfully replicated the standard task switching paradigm phenomena. First, the switch trials were slower and more error-prone compared to the repeat trials, indicating SCs. Second, the location task was performed faster and with fewer errors as compared to the shape task, indicating that the former task was easier than the latter task. Third, the SCs to the location task were larger than the SCs to the shape task, indicating paradoxical asymmetrical SCs. And fourth, participants were more likely to choose the harder shape task than the easier location task, indicating a bias for the harder task. These findings were replicated even when collapsing the speed and accuracy measures into a standardized measure, which indicates that they did not result from speed–accuracy trade-offs. We therefore utilized these standardized measures when examining the relationships between task performance and personality.

Several personality traits related to task performance. First, persistence scores were predicted by high conscientiousness, low anhedonia, and low apathy (Fig. 3). Given the instructions to switch between the tasks about equally often, persistent, and conscientious people might be expected to especially activate rACC, which according to the HRL-ACC model implements a high-level strategy that reduces SCs incurred by cACC (Holroyd and McClure, 2015). Consistent with this possibility, persistence was associated with smaller differences in the overall SCs (ie, reduced overall asymmetrical SCs), indicating that the overall SCs for shape and location were more comparable in persistent participants compared to other individuals. Further, persistence was associated with reduced task bias, indicating that the persistent participants selected both tasks about equally often, unlike other participants who tended to select the harder task more often (Fig. 3 and Table 4). These results suggest that persistent participants attended relatively more to the task instructions. Yet seemingly at odds with this inference is the fact that these individuals also performed the tasks relatively poorly, as revealed by larger RTs and higher error rates. Given that the task instructions also involved being fast and accurate, the poorer performance might indicate relative noncompliance with the instructions. We suggest that strong performance at the meta-task level impairs performance at the task level, and vice versa, such that these individuals performed poorly simply because it is difficult to perform both tasks well when switching often between them.

Our supposition—in which the persistent individuals were concerned more with complying with the task instruction of switching between tasks equally often than with responding quickly and accurately for a given task—is consistent with a previous fMRI study wherein increased rACC activation was associated with higher trait persistence (Gusnard et al., 2003). To be clear, our replication of the standard paradoxical asymmetrical SCs and the task bias indicates that most participants tended to stick with the harder shape task rather than switch to the easier location task, evidently because of the increased SCs to the easier task. However, this was not true for participants who self-reported as persistent: for these individuals both the paradoxical asymmetrical SCs and the task bias were reduced.

Alternatively, persistent participants may simply have performed the tasks relatively poorly, not because they were complying with the instructions to switch between tasks, but simply because they were not especially engaged in the task (consistent with reduced cACC activation). As a consequence, the relative decrease in control levels would yield smaller SCs, which in turn would promote switching between tasks and a reduced task bias—independently of greater rACC activation. Several observations argue against this interpretation. First, persistence and conscientiousness exhibited a strong, positive correlation in this population (Fig. 3 and Table 2); given that conscientiousness is especially sensitive to the propensity to comply with task instructions, this group would be expected to try to execute the task successfully. Second, in three other unpublished experiments we have found that high persistence is associated with better performance and increased self-reports of engagement across a variety of tasks (Umemoto, 2016). Thus, under most circumstance the trait persistence appears to predict better task performance. For these reasons, we believe it unlikely that the persistent individuals exhibited a smaller task bias and reduced asymmetrical SCs simply because they were unengaged in the task. Rather, it may be that worse performance overall is an inevitable consequence when participants are required to shift frequently between tasks.

The emergence of the two personality traits from the Big 5 personality inventory—extroversion and neuroticism—is also interesting given that both traits have been associated with ACC function in the previous literature (eg, Eisenberger et al., 2005; DeYoung et al., 2010; see also Bishop et al., 2004; Gray and Braver, 2002, using similar personality traits). We found that extroversion was associated with high persistence and low apathy and anhedonia. Although we might expect to see similar patterns of task performance between extroversion and persistence given their positive association (Table 2), and both traits were associated with worse performance (ie, slow RTs with increased error rates), extroversion, and persistence were associated with larger and smaller paradoxical asymmetrical SCs, respectively. Further, extroversion but not persistence was associated with larger overall SCs and a smaller proportion of task switches, consistent with our finding that participants who produced larger overall SCs were less likely to switch tasks (ie, a significant negative correlation between the overall SCs (standardized) and the proportion of switch trials). We speculate that extroverted individuals were concerned more with task execution than with the meta-strategy of switching between tasks, which would predict hyperactive cACC (yielding an increase in SCs due to heightened control over the task at hand) and/or reduced rACC activation (resulting in an inability to attenuate these SCs).

Contrary to our predictions, there was no effect of rumination on task performance. Nevertheless, rumination was strongly positively correlated with neuroticism (Table 2), which was associated with worse overall task performance and increased paradoxical asymmetrical SCs. This is consistent with our prediction that rumination and neuroticism would impair rACC function, as suggested by previous neuroimaging studies (Eisenberger et al., 2005; see also Bishop et al., 2004; Pizzagalli, 2011). On this view, task switches to the easier location task, which normally impose a larger SC penalty for most participants, may have been especially difficult for the more neurotic participants due to decreased rACC activation. Relatedly, it may be the case that rumination scores did not correlate with any of the task performance measures because rumination commonly occurs in a depressive state (Nolen-Hoeksema, 1991). A future study could investigate whether rumination affects task switching ability when participants are naturally in such state, or when a negative mood is induced experimentally.

An important direction for future research would be to assess individual differences in task switching as they relate to subcomponents of these personality traits. For example, recent evidence suggests that agentic extroversion (as measured, for instance, by the Eysenck personality questionnaire; Eysenck and Eysenck, 1991) is related to dopamine-dependent behaviors associated with motivation and reward-seeking, whereas affiliative extroversion is related to enjoyment of close social bonds (Smillie, 2013 for review). Likewise, neuroticism is strongly correlated with anxiety and depression, but these two disorders are also functionally dissociable (see Proudfit, 2015; Weinberg et al., 2012). Future studies could examine how these subcomponents relate to task selection and cognitive control. Further, even though apathy is closely associated with ACC function (Holroyd and Umemoto, 2016), we did not observe any relationship between apathy and task performance. One possibility is that different apathy subtypes may predict task performance. For instance, a recent fMRI study indicated a strong association between behavioral apathy, which is characterized by a lack of self-initiated actions, and cACC activity (Bonnelle et al., 2016). Another possibility is that the Apathy Scale (Starkstein et al., 1992) was developed primarily for patients with Parkinson's disease, and thus this scale may be more sensitive to apathy in clinical populations.

Consistent with previous suggestions that have linked aspects of personality to neurocognitive processes responsible for behavioral control (Carver et al., 2000; see also Gray and Braver, 2002), our study provides an initial, exploratory step on the role of ACC in personality. Of course, other brain areas also contribute to personality and to task switching (Holroyd and Umemoto, 2016), and several other theories make contrasting predictions about ACC function (eg, Alexander and Brown, 2011; Shenhav et al., 2013, 2014; Silvetti et al., 2014). Therefore it remains possible that this pattern of results could be explained by alternative accounts. Nevertheless, our predictions are based on existing literature about which areas of ACC are most associated with the personality traits of interest, and that ACC function should be expressed along a gradient from weaker to stronger activation levels (Holroyd and Umemoto, 2016); whether and how other theories can account for these findings remain to be determined. Our results also suggest follow-up fMRI experiments that would elucidate the functional divisions between the two ACC regions as they relate to the personality. Of particular interest is the role of trait persistence in high-level, hierarchical tasks; further examinations of persistence and its involvement in the proposed rACC–cACC functional divisions would constitute a fruitful avenue for investigation.

REFERENCES

Alexander, W.H., Brown, J.W., 2011. Medial prefrontal cortex as an action-outcome predictor. Nat. Neurosci. 14 (10), 1338–1344.

Allport, D.A., Styles, E.A., Hsieh, S., 1994. Shifting international set: exploring the dynamic control of tasks. In: Umilta, C., Moscovitch, M. (Eds.), Attention and Performance XV: Conscious and Nonconscious Information Processing. MIT Press, Cambridge, pp. 421–452.

Altamirano, L.J., Miyake, A., Whitmer, A.J., 2010. When mental inflexibility facilitates executive control beneficial side effects of ruminative tendencies on goal maintenance. Psychol. Sci. 21, 1377–1382.

Arrington, C.M., Logan, G.D., 2004. The cost of a voluntary task switch. Psychol. Sci. 15, 610–615.

Arrington, C.M., Logan, G.D., 2005. Voluntary task switching: chasing the elusive homunculus. J. Exp. Psychol. Learn. Mem. Cogn. 31, 683–702.

Bishop, S., Duncan, J., Brett, M., Lawrence, A.D., 2004. Prefrontal cortical function and anxiety: controlling attention to threat-related stimuli. Nat. Neurosci. 7 (2), 184–188.

Blanchard, T.C., Strait, C.E., Hayden, B.Y., 2015. Ramping ensemble activity in dorsal anterior cingulate neurons during persistent commitment to a decision. J. Neurophysiol. 114 (4), 2439–2449.

Bonnelle, V., Manohar, S., Behrens, T., Husain, M., 2016. Individual differences in premotor brain systems underlie behavioral apathy. Cereb. Cortex 26, 807–819.

Botvinick, M.M., 2012. Hierarchical reinforcement learning and decision making. Curr. Opin. Neurobiol. 22 (6), 956–962.

Botvinick, M., Braver, T., 2015. Motivation and cognitive control: from behavior to neural mechanism. Annu. Rev. Psychol. 66 (1), 83.

Botvinick, M.M., Braver, T.S., Barch, D.M., Carter, C.S., Cohen, J.D., 2001. Conflict monitoring and cognitive control. Psychol. Rev. 108 (3), 624.

Botvinick, M.M., Niv, Y., Barto, A.C., 2009. Hierarchically organized behavior and its neural foundations: a reinforcement learning perspective. Cognition 113 (3), 262–280.

Braem, S., Verguts, T., Roggeman, C., Notebaert, W., 2012. Reward modulates adaptations to conflict. Cognition 125 (2), 324–332.

Bress, J.N., Hajcak, G., 2013. Self-report and behavioral measures of reward sensitivity predict the feedback negativity. Psychophysiology 50 (7), 610–616.

Bryck, R.L., Mayr, U., 2008. Task selection cost asymmetry without task switching. Psychon. Bull. Rev. 15, 128–134.

Carter, C.S., Braver, T.S., Barch, D.M., Botvinick, M.M., Noll, D., Cohen, J.D., 1998. Anterior cingulate cortex, error detection, and the online monitoring of performance. Science 280 (5364), 747–749.

Carver, C.S., Sutton, S.K., Scheier, M.F., 2000. Action, emotion, and personality: emerging conceptual integration. Pers. Soc. Psychol. Bull. 26 (6), 741–751.

Cloninger, C.R., Svrakic, D.M., Przybeck, T.R., 1993. A psychobiological model of temperament and character. Arch. Gen. Psychiatry 50 (12), 975–990.

Cohen, J.D., Dunbar, K., McClelland, J.L., 1990. On the control of automatic processes: a parallel distributed processing account of the Stroop effect. Psychol. Rev. 97 (3), 332.

Cooper, A.J., Duke, É., Pickering, A.D., Smillie, L.D., 2014. Individual differences in reward prediction error: contrasting relations between feedback-related negativity and trait measures of reward sensitivity, impulsivity and extraversion. Front. Hum. Neurosci. 8, 248.

Davis, R.N., Nolen-Hoeksema, S., 2000. Cognitive inflexibility among ruminators and nonruminators. Cogn. Ther. Res. 24 (6), 699–711.

Deiber, M.P., Honda, M., Ibañez, V., Sadato, N., Hallett, M., 1999. Mesial motor areas in self-initiated versus externally triggered movements examined with fMRI: effect of movement type and rate. J. Neurophysiol. 81 (6), 3065–3077.

DeYoung, C.G., Hirsh, J.B., Shane, M.S., Papademetris, X., Rajeevan, N., Gray, J.R., 2010. Testing predictions from personality neuroscience brain structure and the big five. Psychol. Sci. 21, 820–828.

Eisenberger, N.I., Lieberman, M.D., Satpute, A.B., 2005. Personality from a controlled processing perspective: an fMRI study of neuroticism, extraversion, and self-consciousness. Cogn. Affect. Behav. Neurosci. 5 (2), 169–181.

Engelmann, J.B., Damaraju, E., Padmala, S., Pessoa, L., 2009. Combined effects of attention and motivation on visual task performance: transient and sustained motivational effects. Front. Hum. Neurosci. 3, 4.

Eslinger, P.J., Damasio, A.R., 1985. Severe disturbance of higher cognition after bilateral frontal lobe ablation patient EVR. Neurology 35 (12), 1731–1741.

Eysenck, H.J., Eysenck, S.B.G., 1991. Manual of Eysenck Personality Acales (EPS Adult). Hodder & Stoughton.

Forstmann, B.U., Brass, M., Koch, I., von Cramon, D.Y., 2006. Voluntary selection of task sets revealed by functional magnetic resonance imaging. J. Cogn. Neurosci. 18, 388–398.

Gilbert, S.J., Shallice, T., 2002. Task switching: a PDP model. Cogn. Psychol. 44, 297–337.

Gläscher, J., Adolphs, R., Damasio, H., Bechara, A., Rudrauf, D., Calamia, M., et al., 2012. Lesion mapping of cognitive control and value-based decision making in the prefrontal cortex. Proc. Natl. Acad. Sci. U.S.A. 109 (36), 14681–14686.

Gray, J.R., Braver, T.S., 2002. Personality predicts working-memory—related activation in the caudal anterior cingulate cortex. Cogn. Affect. Behav. Neurosci. 2 (1), 64–75.

Gusnard, D.A., Ollinger, J.M., Shulman, G.L., Cloninger, C.R., Price, J.L., Van Essen, D.C., Raichle, M.E., 2003. Persistence and brain circuitry. Proc. Natl. Acad. Sci. U.S.A. 100 (6), 3479–3484.

Hewig, J., Kretschmer, N., Trippe, R.H., Hecht, H., Coles, M.G., Holroyd, C.B., Miltner, W.H., 2011. Why humans deviate from rational choice. Psychophysiology 48 (4), 507–514.

Holroyd, C.B., Coles, M.G., 2002. The neural basis of human error processing: reinforcement learning, dopamine, and the error-related negativity. Psychol. Rev. 109 (4), 679.

Holroyd, C.B., McClure, S.M., 2015. Hierarchical control over effortful behavior by rodent medial frontal cortex: a computational model. Psychol. Rev. 122 (1), 54–83.

Holroyd, C.B., Umemoto, A., 2016. The Research Domain Criteria framework: the case for anterior cingulate cortex. manuscript under revision.
Holroyd, C.B., Yeung, N., 2011. An integrative theory of anterior cingulate cortex function: option selection in hierarchical reinforcement learning. In: Mars, R.B., Sallet, J., Rushworth, M.F.S., Yeung, N. (Eds.), Neural Basis of Motivational and Cognitive Control. MIT Press, Cambridge, pp. 333–349.
Holroyd, C.B., Yeung, N., 2012. Motivation of extended behaviors by anterior cingulate cortex. Trends Cogn. Sci. 16 (2), 122–128.
Hull, C., 1943. Principles of Behavior: An Introduction to Behavior Theory. Appleton-Century, Oxford, England.
Jersild, A.T., 1927. Mental set and shift. Arch. Psychol. 89, 5–89.
John, O.P., Naumann, L.P., Soto, C.J., 2008. Paradigm shift to the integrative big five trait taxonomy: history, measurement, and conceptual issues. In: John, O.P., Robins, R.W., Pervin, L.A. (Eds.), Handbook of Personality: Theory and Research, third ed. Guilford Press, New York, NY, pp. 114–158.
Keedwell, P.A., Andrew, C., Williams, S.C., Brammer, M.J., Phillips, M.L., 2005. The neural correlates of anhedonia in major depressive disorder. Biol. Psychiatry 58 (11), 843–853.
Kiesel, A., Steinhauser, M., Wendt, M., Falkenstein, M., Jost, K., Philipp, A.M., et al., 2010. Control and interference in task switching—a review. Psychol. Bull. 136 (5), 849.
Kool, W., McGuire, J.T., Rosen, Z.B., Botvinick, M.M., 2010. Decision making and the avoidance of cognitive demand. J. Exp. Psychol. Gen. 139 (4), 665–682.
Kurniawan, I.T., Seymour, B., Talmi, D., Yoshida, W., Chater, N., Dolan, R.J., 2010. Choosing to make an effort: the role of striatum in signaling physical effort of a chosen action. J. Neurophysiol. 104 (1), 313–321.
Levy, R., Dubois, B., 2006. Apathy and the functional anatomy of the prefrontal cortex–basal ganglia circuits. Cereb. Cortex 16 (7), 916–928.
Liu, W.H., Wang, L.Z., Shang, H.R., Shen, Y., Li, Z., Cheung, E.F., Chan, R.C., 2014. The influence of anhedonia on feedback negativity in major depressive disorder. Neuropsychologia 53, 213–220.
Locke, H.S., Braver, T.S., 2008. Motivational influences on cognitive control: behavior, brain activation, and individual differences. Cogn. Affect. Behav. Neurosci. 8 (1), 99–112.
Masson, M.E., Carruthers, S., 2014. Control processes in voluntary and explicitly cued task switching. Q. J. Exp. Psychol. 67 (10), 1944–1958.
Meiran, N., 1996. Reconfiguration of processing mode prior to task performance. J. Exp. Psychol. Learn. Mem. Cogn. 22 (6), 1423.
Miller, E.K., Cohen, J.D., 2001. An integrative theory of prefrontal cortex function. Annu. Rev. Neurosci. 24 (1), 167–202.
Millington, R.S., Poljac, E., Yeung, N., 2013. Between-task competition for intentions and actions. Q. J. Exp. Psychol. 66 (8), 1504–1516.
Monsell, S., 2003. Task switching. Trends Cogn. Sci. 7 (3), 134–140.
Nolen-Hoeksema, S., 1991. Responses to depression and their effects on the duration of depressive episodes. J. Abnorm. Psychol. 100 (4), 569.
Norman, D.A., Shallice, T., 1986. Attention to action. In: Schwartz, G.E., Shapiro, D., Davidson, R.J. (Eds.), Consciousness and Self-Regulation. Springer, USA, pp. 1–18.
Parvizi, J., Rangarajan, V., Shirer, W.R., Desai, N., Greicius, M.D., 2013. The will to persevere induced by electrical stimulation of the human cingulated gyrus. Neuron 80 (6), 1359–1367.

Pizzagalli, D.A., 2011. Frontocingulate dysfunction in depression: toward biomarkers of treatment response. Neuropsychopharmacology 36 (1), 183–206.

Pollmann, S., Weidner, R., Müller, H.J., Cramon, D.Y., 2000. A fronto-posterior network involved in visual dimension changes. J. Cogn. Neurosci. 12 (3), 480–494.

Proudfit, G.H., 2015. The reward positivity: from basic research on reward to a biomarker for depression. Psychophysiology 52 (4), 449–459.

Ridderinkhof, K.R., Ullsperger, M., Crone, E.A., Nieuwenhuis, S., 2004. The role of the medial frontal cortex in cognitive control. Science 306 (5695), 443–447.

Rogers, R.D., Monsell, S., 1995. The cost of a predictable switch between simple cognitive tasks. J. Exp. Psychol. Gen. 124, 207–231.

Rushworth, M.F.S., Behrens, T.E.J., Rudebeck, P.H., Walton, M.E., 2007. Contrasting roles for cingulate and orbitofrontal cortex in decisions and social behaviour. Trends Cogn. Sci. 11 (4), 168–176.

Sakai, K., 2008. Task set and prefrontal cortex. Annu. Rev. Neurosci. 31, 219–245.

Shenhav, A., Botvinick, M.M., Cohen, J.D., 2013. The expected value of control: an integrative theory of anterior cingulate cortex function. Neuron 79 (2), 217–240.

Shenhav, A., Straccia, M.A., Cohen, J.D., Botvinick, M.M., 2014. Anterior cingulate engagement in a foraging context reflects choice difficulty, not foraging value. Nat. Neurosci. 17 (9), 1249–1254.

Silvetti, M., Alexander, W., Verguts, T., Brown, J.W., 2014. From conflict management to reward-based decision making: actors and critics in primate medial frontal cortex. Neurosci. Biobehav. Rev. 46, 44–57.

Smillie, L.D., 2013. Extraversion and reward processing. Curr. Dir. Psychol. Sci. 22 (3), 167–172.

Snaith, R.P., Hamilton, M., Morley, S., Humayan, A., Hargreaves, D., Trigwell, P., 1995. A scale for the assessment of hedonic tone the Snaith-Hamilton Pleasure Scale. Br. J. Psychiatry 167 (1), 99–103.

Spector, A., Biederman, I., 1976. Mental set and mental shift revisited. Am. J. Psychol. 89, 669–679.

Starkstein, S.E., Mayberg, H.S., Preziosi, T., Andrezejewski, P., Leiguarda, R., Robinson, R.G., 1992. Reliability, validity, and clinical correlates of apathy in Parkinson's disease. J. Neuropsychiatry Clin. Neurosci. 4 (2), 134–139.

Stroop, J.R., 1935. Studies of interference in serial verbal reaction. J. Exp. Psychol. 18, 643–662.

Treynor, W., Gonzalez, R., Nolen-Hoeksema, S., 2003. Rumination reconsidered: a psychometric analysis. Cogn. Ther. Res. 27 (3), 247–259.

Umemoto, A., 2016. Individual differences in personality associated with anterior cingulate cortex function: implication for understanding depression. unpublished doctoral dissertation, University of Victoria, Victoria, British Columbia.

van Reekum, R., Stuss, D.T., Ostrander, L., 2005. Apathy: why care? J. Neuropsychiatry Clin. Neurosci. 17 (1), 7–19.

Vassena, E., Krebs, R.M., Silvetti, M., Fias, W., Verguts, T., 2014. Dissociating contributions of ACC and vmPFC in reward prediction, outcome, and choice. Neuropsychologia 59, 112–123.

Wager, T.D., Jonides, J., Smith, E.E., Nichols, T.E., 2005. Toward a taxonomy of attention shifting: individual differences in fMRI during multiple shift types. Cogn. Affect. Behav. Neurosci. 5 (2), 127–143.

Watson, D., Clark, L.A., 1992. On traits and temperament: general and specific factors of emotional experience and their relation to the five-factor model. J. Pers. 60 (2), 441–476.

Weinberg, A., Klein, D.N., Hajcak, G., 2012. Increased error-related brain activity distinguishes generalized anxiety disorder with and without comorbid major depressive disorder. J. Abnorm. Psychol. 121 (4), 885.

Westbrook, A., Kester, D., Braver, T.S., 2013. What is the subjective cost of cognitive effort? Load, trait, and aging effects revealed by economic preference. PLoS One 8 (7), e68210.

Whitmer, A.J., Banich, M.T., 2007. Inhibition versus switching deficits in different forms of rumination. Psychol. Sci. 18 (6), 546–553.

Yeung, N., 2010. Bottom-up influences on voluntary task switching: the elusive homunculus escapes. J. Exp. Psychol. Learn. Mem. Cogn. 36 (2), 348.

Yeung, N., Monsell, S., 2003. The effects of recent practice on task switching. J. Exp. Psychol. Hum. Percept. Perform. 29, 919–936.

CHAPTER

Competition, testosterone, and adult neurobehavioral plasticity

9

A.B. Losecaat Vermeer*[,1], I. Riečanský[†,‡], C. Eisenegger*[,1]

**Neuropsychopharmacology and Biopsychology Unit, Faculty of Psychology, University of Vienna, Vienna, Austria*

[†]Social, Cognitive and Affective Neuroscience Unit, Faculty of Psychology, University of Vienna, Vienna, Austria

[‡]Laboratory of Cognitive Neuroscience, Institute of Normal and Pathological Physiology, Slovak Academy of Sciences, Bratislava, Slovakia

[1]Corresponding authors: Tel.: +43-1-4277-47186; Fax: +43-1-4277-847186 (A.B.L.V.); Tel.: +43-1-4277-47139; Fax: +43-1-4277-847139 (C.E.), e-mail address: annabel.losecaat.vermeer@univie.ac.at; christoph.eisenegger@univie.ac.at

Abstract

Motivation in performance is often measured via competitions. Winning a competition has been found to increase the motivation to perform in subsequent competitions. One potential neurobiological mechanism that regulates the motivation to compete involves sex hormones, such as the steroids testosterone and estradiol. A wealth of studies in both nonhuman animals and humans have shown that a rise in testosterone levels before and after winning a competition enhances the motivation to compete. There is strong evidence for acute behavioral effects in response to steroid hormones. Intriguingly, a substantial testosterone surge following a win also appears to improve an individual's performance in later contests resulting in a higher probability of winning again. These effects may occur via androgen and estrogen pathways modulating dopaminergic regions, thereby behavior on longer timescales. Hormones thus not only regulate and control social behavior but are also key to adult neurobehavioral plasticity. Here, we present literature showing hormone-driven behavioral effects that persist for extended periods of time beyond acute effects of the hormone, highlighting a fundamental role of sex steroid hormones in adult neuroplasticity. We provide an overview of the relationship between testosterone, motivation measured from objective effort, and their influence in enhancing subsequent effort in competitions. Implications for an important role of testosterone in enabling neuroplasticity to improve performance will be discussed.

Keywords

Competition, Motivation, Testosterone, Neuroplasticity, Winner effect

Progress in Brain Research, Volume 229, ISSN 0079-6123, http://dx.doi.org/10.1016/bs.pbr.2016.05.004

The focus of this review is on how competitions can be used to measure motivation, enhance motivation, and improve performance. We will describe the neurobiological mechanisms underlying motivation in competitions and we will provide insight into how sex hormones can enhance performance via their effects on neuroplasticity.

1 COMPETITION AND MOTIVATION

Competition is essential for survival in virtually all-living organisms. Organisms compete to gain access to limited resources such as food, water, territory, protection, and sexual mates. In humans and nonhuman animals, competition is often a means to achieve and maintain a higher social status in a hierarchy that allows access to such valued resources. While nonhuman animals often compete by displaying aggression, in humans competitions are not only expressed in aggressive ways (ie, by causing physical harm to others through violence, for example; Archer, 2006) but often occur in nonaggressive forms, for instance, via personal achievements, performance or in negotiations. In humans, competition can be a powerful incentivizing tool. It enhances motivation and performance output across many domains such as in business, market economies, law, politics, education, and sports (Deci et al., 1981; Hirshleifer, 1978).

Competition can be regarded as both an intrinsic and extrinsic incentive to perform. Sports tournaments are a good example to illustrate the distinction between the two incentives. Athletes need to be highly motivated to train long hours and perform to their maximum ability during competitions in order to outperform others and, for example, to achieve a higher position in the rankings, or win a prize (ie, external reward). On the other hand, athletes' motivation to work harder can also be prompted by the internal drive to challenge themselves to improve performance and thereby increase their competence and skills in the task. This is referred to as intrinsic motivation, which is the motivation to perform an action due to the enjoyment and self-determination resulting from the activity (Deci et al., 1981). The importance of intrinsic motivation has also been shown in anonymous competitions in the lab. Related to this is a study (Kuhnen and Tymula, 2012) where individuals had to compete against others in a math task (eg, solving as many equations as possible within an allocated time). It was found that the opportunity to privately compare own performance with the performance of the others significantly increased their motivation to exert effort in the subsequent rounds. Motivation was assessed from performance, that is, the number of correctly solved equations. The motivation to be able to compare own performance levels relative to the performance levels of other players is assumed to be driven by a desire to gain or maintain high self-esteem or feelings of competence (Kuhnen and Tymula, 2012). The increased self-esteem or competence from performing the task presumably intrinsically motivates them to exert more effort in the task. This suggests that the intrinsic value one can gain from the competitive activity is an important factor that incentivizes performance in competition. In sum, in addition to the extrinsic incentives competition offers, it can also increase intrinsic motivation, which is relevant concerning the strong effects intrinsic motivation can have on behavior (Reeve and Deci, 1996). Both types of motivation influence the extent to which competitions can incentivize an individual's performance.

2 EXPERIMENTAL APPROACHES TO MEASURE THE MOTIVATION TO COMPETE IN THE LABORATORY

Studies have assessed individuals' motivation to compete mostly in two ways. One approach is using choice preference as a measure of the motivation to compete (ie, competitiveness), by asking individuals to make a decision, for instance, between engaging into competition or rather performing an alternative task with no competitive element (eg, McGee and McGee, 2013; Mehta et al., 2015; Niederle and Vesterlund, 2007), or bidding at auctions where people have to decide between how much, if any, they want to bid on the target item (van den Bos et al., 2013). The other approach that is used by many studies is to assess individuals' motivation to compete directly from performance in real effort-based tasks. These real effort-based tasks can either require mostly cognitive effort, for example, solving mazes and puzzles (eg, Gneezy et al., 2003; Niederle and Vesterlund, 2007; Reeve et al., 1985), anagrams (Charness and Villeval, 2009), mathematical problems (Rutström and Williams, 2000), performing in trivia challenges (Hoffman et al., 1994), or playing videogames such as Tetris (Zilioli and Watson, 2014; Zilioli et al., 2014). In addition, there are numerous studies using real effort-based competitions where physical effort (ie, physical/motor tasks) is used as an index of competitiveness. These include, for example, cracking walnuts (Fahr and Irlenbusch, 2000), moving as many sliders to a fixed target as fast as possible (Gill and Prowse, 2012), performing a handgrip force endurance task (Cooke et al., 2013; Le Bouc and Pessiglione, 2013), or cycling in a head-to-head competition (Corbett et al., 2012). A meta-analysis by Stanne et al. (1999) using physical effort competitions found an overall enhancing effect of competition on performance.

Some studies have used both choice and real effort in the same design to assess individuals' motivation to compete. An influential study by Niederle and Vesterlund (2007) used a multistage competition design involving an incentivized cognitive effort task with monetary incentives (ie, payment was contingent on correctly adding up as many sets of five two-digit random numbers, within 5 min). In the first stage all players performed the effort task and received the same monetary reward for every correctly solved equation (noncompetitive piece rate payment). In the second stage participant performed a forced competition in groups of four and were only paid if they were the winner (tournament, involving fourfold higher payment for the winner). In the third stage of the experiment, participants were asked to decide according to which of the two payment schemes they wanted to perform the task. In men, an increased performance was found in the tournament condition relative to piece rate condition, as well as in those who chose to compete again in the tournament condition in comparison to those who decided to perform in the piece rate condition (Niederle and Vesterlund, 2007). Together, these studies demonstrate that competitions can have motivation-enhancing effects, which can be measured in at least two fundamental ways, either by choice or by real effort. In addition to using dichotomous decisions (ie, to compete or not to compete), using a continuous measure such as real effort as an index of competitiveness is expected to receive increasing attention in competition research, as it provides a powerful measure of motivation and performance that might be more sensitive to pharmacological and context manipulations.

While much social and applied psychological as well as behavioral economics research has been devoted to further our understanding of the motivation-enhancing effects of competition, research on the underlying neurobiological mechanisms in humans has only recently begun. In the following part, we will describe the neurobiological basis of motivation in competition by relying on existing models from both animal and human research and discuss behavioral, psychopharmacological, and neuroimaging evidence in humans. We will show that testosterone is a key hormone involved in competition, and we will shed light on the mechanisms that could promote motivation in competition on longer timescales. Further, we will describe the neurobiological mechanisms underlying the so called "winner effect" in detail, which represents a particularly strong case of the motivation-enhancing effects of competition.

3 NEUROENDOCRINOLOGICAL FACTORS THAT INFLUENCE COMPETITIVENESS

In addition to the modulating effects of psychological variables and social contexts on social behavior, neurotransmitters (eg, dopamine, serotonin, norepinephrine) and hormones (eg, testosterone, estradiol, oxytocin) have been found to play a crucial role in regulating behavior. These neuroactive hormones and neurotransmitters can regulate and adapt behavior by modifying neuronal dynamics, excitability, and synaptic function (Crockett and Fehr, 2014).

Research investigating human competition has shown an increasing interest in the role of neurotransmitters and hormones, of which primarily testosterone. The steroid androgen testosterone, a product of the hypothalamic–pituitary–gonadal axis, is produced in both men and women with approximately 95% of circulating testosterone in men produced by the Leydig cells of the testes, while in women approximately 50% of circulating levels are produced by the ovaries and placenta. The adrenal cortex also secretes testosterone; however, in men it contributes only 5% of circulating testosterone, while in women it accounts for roughly 50% of testosterone (Burger, 2002). Testosterone has an important role in the development of secondary sexual attributes, such as increased muscle tissue, bone mass, and body hair in males. Besides the developmental characteristics, testosterone also plays an important role in socioemotional and decision-making behavior (Bos et al., 2012; Eisenegger et al., 2011). The substantial influence of testosterone on the brain in archetypical situations, such as fight, flight, mating, and the search and struggle for status, makes it an important variable in studying competition (Mazur and Booth, 1998).

In a broad number of animal species, including humans, testosterone secretion is modulated in the context of competitive interactions. So far, several models have described the relevance and function of androgen modulation associated with competition. The challenge hypothesis, which was originally postulated for birds (Wingfield et al., 1990), states that testosterone levels rise in males in response to challenges (eg, during the mating season), while they are low during periods of social stability. These fluctuations occur presumably to avoid the costs for keeping testosterone levels chronically high (Folstad and Karter, 1992). It also predicts that animal species were more likely to evolve the ability to increase androgens following a social dispute if such a response helped facilitate reproduction such as mate guarding, territory defense, and fighting ability during male–male competition. Whether the challenge hypothesis also applies to humans has been reviewed by Archer (2006). In humans, for instance, basal testosterone levels correlate positively with psychometric measures such as the self-reported willingness to win in competition (Suay et al., 1999; Williams et al., 1982; Zumoff et al., 1984), and betting strategies in auctions (van den Bos et al., 2013).

4 TESTOSTERONE LIKELY INFLUENCES COMPETITIVENESS VIA MODULATION OF DOPAMINERGIC FUNCTION

Accumulating evidence from animal research suggests that testosterone modulates the motivation to compete via actions on the mesocorticolimbic dopaminergic system (see Box 1). Work in animals has so far supported this assumption by identifying

BOX 1 THE LINK BETWEEN DOPAMINE, MOTIVATION, AND EFFORT

Integration of motivation and effort is assumed to be mediated primarily by the striatal dopamine system (Hosp et al., 2011; Salamone and Correa, 2002; Westbrook and Braver, 2016). Effort is often referred to as strenuous physical or mental exertion typically with the aim of achieving a desired outcome or goal. Thus, effort is generally considered costly, and if given a choice most animals will choose actions that are less effortful (Salamone et al., 2007). In rats, dopamine depletion decreases tolerance for effort (ie, increases effort costs), whereas drugs enhancing dopamine have the reverse effect (Salamone and Correa, 2002; Salamone et al., 2007). Evidence in humans also suggests that striatal dopamine is required to overcome costs when high levels of effort are necessary to obtain a desired goal (Botvinick et al., 2009; Kurniawan et al., 2010). Activity in the ventral striatum and midbrain correlates with the expected amount of reward, discounted by the amount of effort to be invested (Croxson et al., 2009). In addition, the anterior cingulate cortex might play a prominent role in the value of exerting effort for a potential reward. That is, dopamine in anterior cingulate cortex is posited to promote persistence of effort via its function in integrating action–outcome associations (see Kurniawan et al., 2011; Westbrook and Braver, 2016 for review).

the neurobiological pathways linking testosterone, and its primary metabolites 5α-dihydrotestosterone (DHT) and estradiol to several brain regions of the dopaminergic system (for more details on testosterone metabolization, see Section 7.2). For instance, fundamental neurobiological studies have demonstrated that dopamine neurons contain androgen receptors (ARs) (Creutz and Kritzer, 2004) and estrogen receptors alpha (ER-α) and beta (ER-β) (for a recent review see, eg, Almey et al., 2015). In the substantia nigra of adolescent male rats, testosterone is also able to change—via direct action at ARs—the levels of dopamine receptors, and levels of the dopamine transporter protein, which regulates dopamine availability in the synaptic cleft (Purves-Tyson et al., 2012, 2014). Testosterone also appears to affect dopaminergic neurotransmission in prefrontal brain areas. Recent rodent research has shown that less than one quarter of the dopamine cells of the ventral tegmental area (VTA) that project to prefrontal cortex contain ARs (Aubele et al., 2008). However, all of the major afferent projections to the VTA—those arising from pyramidal cells of the prefrontal cortex itself, are by far the most AR enriched (Aubele and Kritzer, 2011).

Systemic manipulations of testosterone levels, for instance via castration, reduce the concentration of dopamine in the striatum in rodents, an effect that can be prevented by supplementation with testosterone, DHT, but also estradiol (Alderson and Baum, 1981; Mitchell and Stewart, 1989). In rats, activity of tyrosine hydroxylase, which is the rate-limiting enzyme in dopamine biosynthesis, is reduced in the striatum following orchidectomy, and this reduction can be prevented by testosterone supplementation (Abreu et al., 1988). Moreover, administration of testosterone in gonadally intact adult male rats increases dopamine concentration (de Souza Silva et al., 2009) and dopamine turnover in the striatum (Thiblin et al., 1999). Furthermore, in rhesus macaques, circulating testosterone levels were found to correlate positively with concentration of striatal tyrosine hydroxylase (Morris et al., 2010). In sum, these studies highlight a close relationship between testosterone, including its metabolites, and the mesostriatal dopaminergic system.

Behaviorally, rodents can be conditioned with acute peripheral and intranucleus accumbens administration of testosterone and DHT, such that they show a place preference for where they received the hormone (for review see, eg, Wood, 2008). This effect has been localized to the nucleus accumbens shell (Frye et al., 2002), an important reward region in rodents (Robbins and Everitt, 1996) that corresponds to the ventral striatum in humans. Place preference for testosterone can be blocked by both D1 or D2 dopamine receptor antagonists (SCH23390 or sulpiride, respectively; Packard et al., 1998), which suggests that some of the rewarding effects of testosterone are mediated via the dopaminergic system. Not surprisingly, intracerebral testosterone self-administration protocols in rodents have shown that some animals overdose to lethal doses (Wood et al., 2004). Intriguingly, some of the general reinforcing effects of testosterone have been observed within time periods as short as 30 min poststimulus, suggesting that testosterone may have rapid effects on dopaminergic function (Nyby, 2008).

In humans, evidence is generally consistent with animal research. Clinically low testosterone levels as observed in hypogonadal men appear to be associated with apathy and lack of motivation (Bhasin et al., 2006). Single-dose testosterone administration in healthy female subjects increases motivation to engage in cued behaviors (Aarts and van Honk, 2009) and increases BOLD activation in the ventral striatum during reward anticipation, which is most pronounced in women with low appetitive motivation (Hermans et al., 2010). Together, this shows the tight neurobiological coupling between the androgen and the dopamine system and suggests that testosterone increases motivation to compete via activating selective dopaminergic pathways (see Box 1).

5 HOW DO COMPETITION OUTCOMES MODULATE SUBSEQUENT COMPETITIVENESS?

Not surprisingly, competitiveness varies substantially as a function of prior experiences in competitions such as a previous victory or defeat. What is intriguing, however, is the observation that testosterone secretion varies with the outcome of competition. Specifically, in many species testosterone levels fluctuate as a function of whether the competition was won or lost, such that the winner experiences a surge of testosterone levels, while the loser experiences a drop of testosterone levels (biosocial model of status; Mazur, 1985). Work by Monaghan and Glickman (2001) illustrated this in rhesus monkeys, where they found that in a competition to establish rank the winning male emerged with a 10-fold increase in testosterone, while the loser experienced a drop to 10% of baseline levels within 24 h postcompetition, which persisted for several weeks. Similar results were observed in humans, for instance, in sports competitions testosterone levels increased before the contest started, and further increased after a win. These findings were found across a range of tournaments such as tennis (Booth et al., 1989), wrestling (Elias, 1981), and also nonphysical contests such as chess (Mazur et al., 1992). Again, effects can be large; for example, athlete hockey players merely watching themselves win a match on video produced a 40% testosterone surge from baseline (Carré and Putnam, 2010). These changes occur relatively quickly, observed within approximately 15 min postoutcome in humans in most studies (for relevant reviews see, eg, Carré and Olmstead, 2015; Oliveira and Oliveira, 2014). However, not all studies in humans have consistently observed a testosterone surge following a win, and a drop of testosterone following a loss, as losers also often show an increase in testosterone levels (eg, Mehta and Josephs, 2006; Van Anders and Watson, 2007). The overall evidence in support of testosterone increases exclusively after wins in males appears to be small, albeit significant (Archer, 2006; Carré and Olmstead, 2015; Oliveira and Oliveira, 2014). In humans, this appears to be due to a large extent to moderating variables, such as the cognitive appraisal of the competition (eg, whether it is experienced as a threat or as a challenge, mood, personality), but also variables like the physical location (eg, home versus away), which is evident in both humans and animals (Carré, 2009; Carré et al., 2006; Fuxjager and Marler, 2010; Oyegbile and Marler, 2005). If the extent of such moderating variables is being considered, the relationship appears to be more pronounced (reviewed in, eg, Carré and Olmstead, 2015; Salvador and Costa, 2009).

It is noteworthy that competition outcomes are often not clear cut—when there is no clear difference in performance between the winner and loser. The behavioral effects of such close outcomes have been well established for nonsocial contexts, where positive effects on motivation have been observed for outcomes that are experienced as "small" losses, or "near-misses" (ie, close but objective losses; Reid, 1986). Indeed, participants who nearly won in a gambling task, for instance, were more motivated to continue to gamble as compared to those who clearly lost (Berger and Pope, 2011; Clark et al., 2009). This phenomenon also extends to competitive contexts. A typical example is when an individual just ends up at second place with a small difference in performance compared to the winner, or in hierarchies that are not yet established or unstable. Recently, such situations have also been modeled in the laboratory (Zilioli et al., 2014). In this study, Zilioli and colleagues (2014) examined how victories and defeats in unstable hierarchies (ie, wherein participants experienced close victories and defeats, and where participants were uncertain of the competition outcome until the very end of the competition) can differently affect testosterone response in women (Zilioli et al., 2014). They found that participants who experienced a defeat in unstable hierarchies had larger increases in testosterone levels relative to participants who experienced a victory in unstable hierarchies. This testosterone surge has been interpreted as to boost motivation to increase performance on next encounters. Thus, close or unstable losses could increase individuals' motivation to improve performance, also in competitive contexts. In short, competitive outcomes are an important moderator of individuals' motivation to compete and are observed to have a bidirectional relationship with circulating testosterone levels.

What could be the function of these testosterone dynamics? Mazur and Booth (1998) have pointed toward a role of testosterone in guiding further status-seeking behavior. Specifically, that sustained testosterone increases motivation for subsequent status battles in winners, whereas when testosterone decreases it discourages such battles in losers. It has been shown that short-term fluctuations of testosterone correlate with a host of behavioral measures. For example, the extent of the testosterone surge has been shown to predict reactive aggression (Carré et al., 2013; Geniole et al., 2013). Importantly, the surge has also been shown to predict motivation in performance in subsequent competitions (Carré and McCormick, 2008; Mehta and Josephs, 2006). The precise role of these fluctuations remains to be further investigated. Interestingly, these behaviors were generally assessed shortly after the change in testosterone was detected (usually within 10–20 min) (Fig. 1A). In the following, we will focus on research showing how competition outcome, and acute fluctuations in testosterone can influence behavior long term.

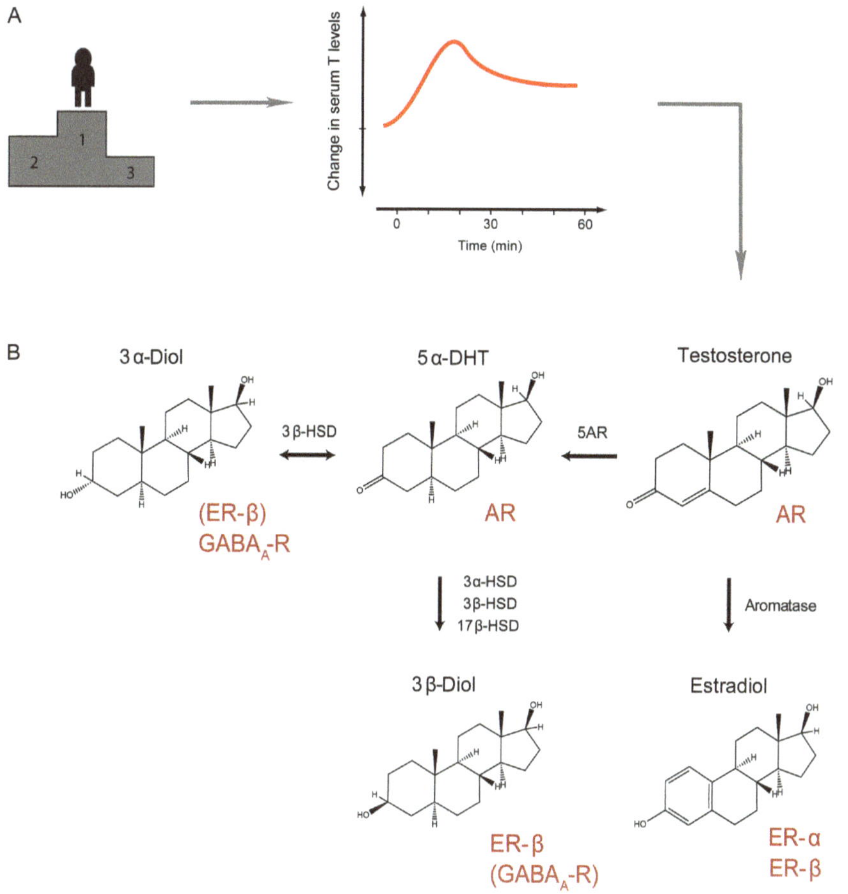

FIG. 1

Diagram depicting testosterone and its metabolites that may contribute to the winner effect in humans. (A) Illustration of a testosterone surge following victory. (B) The metabolization of testosterone that may take place in the central nervous system. Major receptor types for these metabolites are shown in *red*. Enzymes are shown in *italics*. *Single-sided arrow* depicts unidirectional catalysis, and *double-sided arrow* illustrates bidirectional catalysis. *3α-diol*, 5α-androstane-3α,17β-diol; *3β-diol*, 5α-androstane-3β,17β-diol; *3α-HSD*, 3α-hydroxysteroid-dehydrogenase; *3β-HSD*, 3β-hydroxysteroid-dehydrogenase; *17β-HSD*, 17β-hydroxysteroid-dehydrogenase; *5AR*, 5α-reductase; *AR*, androgen receptor; *ER-α*, estrogen receptor α; *ER-β*, estrogen receptor β; *GABA$_A$-R*, gamma-aminobutyric acid receptor type A.

Adapted from Handa, R.J., Pak, T.R., Kudwa, A.E., Lund, T.D., Hinds, L., 2008. An alternate pathway for androgen regulation of brain function: activation of estrogen receptor beta by the metabolite of dihydrotestosterone, 5alpha-androstane-3beta,17beta-diol. Horm. Behav. 53, 741–752. doi:10.1016/j.yhbeh.2007.09.012.

6 LONG-TERM EFFECTS OF COMPETITION OUTCOMES ON COMPETITIVENESS

An interesting observation is that winning not only promotes further competitiveness, but that it enhances the probability of winning the next contest. This is referred to as the winner effect. In contrast, the loser effect refers to the observation that losing enhances the probability of losing a subsequent contest (Chase et al., 1994; Dugatkin, 1997). The influence of prior contest outcomes on winning or losing subsequent contests has been observed in many animal species, for instance in fish, rodents (Dugatkin, 1997; Fuxjager et al., 2011a; Gleason et al., 2009; Hsu et al., 2006; Oyegbile and Marler, 2005), and also humans (see Oliveira and Oliveira, 2014 for an overview).

Here, we will first focus on the psychological aspects that may underlie the winner effect. For instance, recent research indicates that human subjects who won in a competition provided more effort as measured by the number of mathematical equations solved in a laboratory task than subjects who lost in a prior competition. This effect was specific to the link between actual performance and outcome, as subjects who randomly won in a separate experimental condition did not invest more effort subsequently (McGee and McGee, 2013). Supporting this, on a cognitive effort task testosterone responded only to actual ability-determined competition outcomes, not to competition outcomes that were based on chance (Van Anders and Watson, 2007). This suggests that the experience of an actual achievement may be important for the motivation to compete and thus may be essential for the winner effect to emerge. In addition, in motor and cognitive tasks without a competitive element providing positive feedback about performance when participants chose to receive this feedback led to an increased performance compared to providing feedback at random times (Chiviacowsky, 2007; Chiviacowsky and Wulf, 2002). Together, these findings suggest that when performance feedback is perceived as real and can be attributed to the self (ie, is self-determined), it can intrinsically motivate behavior and positively affect learning (ie, a skill in subsequent competitions). This is in line with existing theories on motivation (Ryan and Deci, 2000), describing that not only the experience of perceived competence is important for motivating individuals to act, but specifically that their performance should also be perceived as self-determined.

While the above studies suggest that psychological variables such as perceived competence and personal achievement are important moderating factors of the winner effect, animal research has also shown changes in the neurobiology underlying the winner effect. It is thus important to scrutinize the nature of these neurobiological changes as they would allow building models on how to best harvest the beneficial effects of winning on subsequent motivation in humans.

So far, research in rodents has shown that testosterone surges observed after winning a competition increase the probability of winning a future competition (Oyegbile and Marler, 2005; Trainor et al., 2004). In those studies mice were castrated and implanted with testosterone, which maintains circulating testosterone at levels typical of adult males but, in effect, prevents testosterone changes in response to social or environmental cues (Trainor et al., 2004). This procedure showed that a robust winner effect was evident if animals accumulate three separate victories in their home territory and receive additional testosterone injections after each of these contests. Mice form an intermediate winner effect when they accumulated the same number and type of victories but received postencounter saline injections (Fuxjager et al., 2011b). It has thus been proposed that postcompetition testosterone fluctuations represent a neuroendocrine substrate of the robust winner effect (Fuxjager and Marler, 2010; Oyegbile and Marler, 2005). This was also found in other animal species. For instance, in male tilapia, winners that were treated with an antiandrogen drug (ie, cyproterone acetate) were less likely to win a subsequent aggressive interaction (relative to controls) (Oliveira et al., 2009). However, the usual relationship of testosterone fluctuations and the winner effect has not always been observed (Hirschenhauser et al., 2008, 2013).

In humans, a recent study (Zilioli and Watson, 2014) found that a rise in testosterone during a laboratory competition predicted better performance 24 h later on the same competition. Concurring with the somewhat weak evidence for the existence of a "competition effect" (ie, where winners show an increase in testosterone following the competition, and losers show a decrease in testosterone) the positive relationship between reactivity of testosterone to the first competition and performance 24 h later was found in both winners and losers (Zilioli and Watson, 2014). Albeit that evidence in humans is still correlative, these findings suggest that testosterone may, in certain contexts, induce long-lasting changes in performance in competition.

7 MECHANISMS MEDIATING LONG-TERM BEHAVIORAL EFFECTS FOLLOWING A TESTOSTERONE SURGE

It is well established that both short- and long-term modifications of behavior rely on changes of synaptic connections between neurons (Sweatt, 2016). Therefore, a likely mechanism by which testosterone surges might potentiate the winner effect is that the hormone affects synaptic and (more general) neuronal plasticity. Neuroplasticity refers to functional or structural changes that occur in the brain to adjust to changes in the external environment or internal milieu (May, 2011).

7.1 ROLE OF AR IN NEURONAL PLASTICITY

ARs are expressed in many different neuronal populations in the nervous system in both males and females (Choate et al., 1998; Simerly et al., 1990). Testosterone can either directly bind to ARs, or activate ARs after conversion to DHT, which is a more potent AR agonist than testosterone. The conversion of testosterone to DHT is catalyzed by the enzyme 5α-reductase, which is expressed in the brain (Celotti et al., 1997) (see Fig. 1B).

There is accumulating evidence from animal research supporting a role of ARs in neuronal plasticity (Fester and Rune, 2015). In adult rats, castration results in massive reduction of spine synapses in the hippocampus and the prefrontal cortex that can be reversed by supplementation of both testosterone or DHT (see Hajszan et al., 2008 for review). Furthermore, changes in synaptic morphology in the hippocampus following castration in mice are associated with a decrease in levels of brain-derived neurotrophic factor (BDNF), a protein that is important for normal synaptic physiology. Such a reduction in BDNF levels can be both prevented and recovered by testosterone replacement (Li et al., 2012; for review see, eg, Pluchino et al., 2013). This is in line with in vitro studies of hippocampal preparations, showing that administration of androgens increases the number of dendritic spines (Hatanaka et al., 2009). Furthermore, there is evidence that ARs play a role in adult hippocampal neurogenesis promoted by physical exercise (Okamoto et al., 2012).

Importantly, it has been shown that genetically modified mice lacking ARs specifically in the nervous system show a deficit in long-term potentiation in the hippocampus as well as an impairment of memory consolidation (Picot et al., 2016). In rats, posttraining systemic injections of testosterone and DHT improve memory on tests performed 24 h later (Edinger et al., 2004; Frye and Lacey, 2001). Moreover, Edinger and Frye (2007a) demonstrated that intrahippocampal administration of the AR antagonist flutamide immediately after training impaired memory consolidation.

7.2 ROLE OF ER IN NEURONAL PLASTICITY

An alternative pathway for the action of testosterone involves its conversion to estradiol, which binds to ER-α and ER-β (Fig. 1B) (Almey et al., 2015). Although estradiol is primarily known as a female sex hormone, the hormone controls many physiological and behavioral responses in both females and males (for review Cornil et al., 2012). The conversion of testosterone to estradiol is mediated by the enzyme aromatase, which is widely expressed in the human brain in both sexes (see Biegon, 2016 for review). The activity of the enzyme can be rapidly regulated via phosphorylation, resulting in fast changes in local estradiol concentrations. In addition, estradiol is assumed to fulfill all criteria for being classified as a neurotransmitter (for review, see Balthazart, 2010), which suggests that the indirect testosterone effects, via conversion to estradiol, may be tightly regulated by neuronal activity (Farinetti et al., 2015).

There is extensive literature on the role of ERs in neuronal plasticity. ERs seem to play a fundamental role in regulating neurogenesis, synaptogenesis, dendritic, and axonal growth (for recent reviews see, e.g., Fester and Rune, 2015; Frick et al., 2015). Similar to testosterone, posttraining systemic but also intrahippocampal administration of estradiol results in a "off-line gain" (ie, improvement) in memory performance on cognitive tests up to 24 h later (for review see, e.g., Packard, 1998). This is in line with the finding that posttraining administration of estradiol rapidly increases (ie, within minutes) dendritic spines in the hippocampus and the prefrontal cortex (Inagaki et al., 2010). In addition, administration of an estradiol antagonist (bisphenol A) impairs memory consolidation, blocks the off-line enhancing effects of estradiol, and reduces dendritic spines in the hippocampus and the prefrontal cortex (for review see Luine and Frankfurt, 2012).

An alternative pathway by which ERs can be activated to affect off-line gains is via metabolites of DHT, which cannot be aromatized to estradiol (Fig. 1B). The DHT metabolites 3β-diol and 3α-diol act on ER-β and γ-aminobutyric acid type A ($GABA_A$) receptors (Frye et al., 2008; Handa et al., 2008). Similar to ER-β, $GABA_A$ receptors have been associated with neuronal plasticity (for review, eg, Pallotto and Deprez, 2014). 3α-diol injected into the hippocampus following training in a memory task significantly increased task performance 1 day later (Edinger and Frye, 2007b). Furthermore, inhibiting the expression of ER-β, but not ER-α, abolished the positive effect of 3α-diol (Edinger and Frye, 2007b).

Taken together, these studies demonstrate that testosterone may affect neuronal plasticity in various ways, at short timescales but with long-term consequences. Testosterone surges following a victory in competition may induce neuroplasticity via both ARs and ERs in several brain areas such as the hippocampus, prefrontal cortex, or striatum. These effects seem to occur fast enough to enhance memory consolidation and boost the behavior that led to victory in the first place. The winner effect can be modulated via the direct effects of testosterone onto ARs but also indirectly via its metabolites DHT and estradiol (Fig. 1). The reviewed literature supports the idea that estradiol might be involved in the neuroplasticity associated with the winner effect, and this might imply a potential role of aromatase in the winner effect in humans (Fuxjager et al., 2011b). An alternative pathway involves nonestrogen agonists of ER-β (3α-diol and 3β-diol) in the winner effect. As of yet, the relative importance of these alternative pathways is not clear, and in humans the potential role of aromatase in the winner effect still needs to be investigated. The work reviewed in this section suggests that postcompetition activation of both ARs and ERs might in humans improve future winning ability by increasing motivation and enhancing skills via effects on neuroplasticity. Pharmacological models involving selective blockade of these receptor systems (eg, using flutamide to block ARs, and raloxifene to block ERs) may help to shed light on the relative weight these systems might have in shaping the winner effect in humans.

8 DISCUSSION

Motivation to compete is a complex phenomenon that is influenced by psychological, neurobiological, as well as social contextual factors.

8.1 A CONCEPTUAL FRAMEWORK

In the following framework we summarize how competition influences motivation and performance. The framework is speculative and builds around existing models that describe the relationship between testosterone and the motivation to compete, as well as describing how competition outcome, and psychological and cognitive variables interact with testosterone secretion (biosocial model of status: Archer, 2006; Mazur and Booth, 1998; Salvador and Costa, 2009; challenge hypothesis: Wingfield et al., 1990). To these models we added recent insights mostly stemming from animal research into how testosterone and estradiol might affect performance in competitive contexts in the laboratory on a longer-term basis via effects on neuroplasticity.

One way to illustrate competitiveness within this framework is as the *decision to compete or not to compete*. The decision to compete is assumingly based on an evaluation of the subjective benefit weighted against the subjective cost of competing (Croxson et al., 2009; Studer and Knecht, 2016). The subjective benefits associated with competition can be determined by an individual's *expectations* of winning (ie, probability) and the *subjective value* of winning a competition. The expectations of winning has also been described as the person's resource holding potential (Hurd, 2006). That is, an individual's physical or cognitive ability or skills that determine the ability to win a competition. The subjective value of engaging into a competition can thus be conceptualized as the expected subjective benefit of, for example, a gain in status plus the intrinsic value of competing. This could be the subjective benefit from winning a prize and the subjective benefit from the feeling of competence. A low subjective value of winning has, for instance, been suggested to explain why women relative to men are less motivated to compete in videogames, because these types of competitions may have low subjective value to women compared to other types of competitions (Niederle and Vesterlund, 2011). Furthermore, the expected benefit of winning needs to account for any potential expected disutility of losing a competition such as a loss in status, or perception of reduced competence. Finally, the effort (cognitive or physical) that has to be invested in the competition is conceptualized as the subjective costs to compete.

The other way, as reviewed here, indicates that the motivation to compete can also be measured in forced competition paradigms via real effort, and that the outcome of competition can map onto the subsequent motivation to compete. In reference to the above framework, this implies that for individuals who are engaged in a forced competition, the dependent variable of their motivation to compete is reflected in the effort they exert into the task (cf. Kuhnen and Tymula, 2012). For example, in a forced real effort competition, an individual with a strong motivation to achieve or maintain high status will exert more effort because of the high subjective utility of winning. At the same time, effort can possibly also be motivated or enhanced by the high subjective disutility of losing. Here, testosterone might increase the subjective utility of winning *and* the disutility of losing, by its proposed effects on the motivation to seek and maintain social status (Eisenegger et al., 2011; Mazur and Booth, 1998). The hormone may also, by virtue of its acute effects on the mesostriatal and mesolimbic dopaminergic system, promote effort by reducing effort costs (see Box 1). In situations where effort can be directly inferred from performance, which is not limited by ability, higher effort will then increase the probability of winning (Wallin et al., 2015).

In humans it has been shown that a short-lived postcompetition testosterone surge positively correlates with performance more than 24 h postcompetition (Zilioli and Watson, 2014). The function of testosterone may in this context also be understood via its well-established role in promoting neuroplasticity within the dopaminergic reward system. Studies in rodents showed that repeated winning increased expression of ARs in the nucleus accumbens (Fuxjager et al., 2010) and potentiated the synthesis of catecholamines (Schwartzer et al., 2013). This suggests that the winner effect involves an enhancement of dopaminergic neurotransmission and also a sensitization to androgens. This is significant also considering the established role of the dopaminergic system in memory consolidation of rewarding and reward predicting events (for review see, Miendlarzewska et al., 2015; Shohamy and Adcock, 2010). For instance, it has been demonstrated that rewards significantly increase off-line gains in long-term memory retention (Abe et al., 2011; Sugawara et al., 2012). In sum, these findings indicate that testosterone and dopamine may act in concert in inducing neuroplasticity that enhances both the consolidation of successful strategies and motivation to reach the desired goals.

Similarly, a reinforcement learning mechanism might be involved in the winner effect (see also Box 1). It is plausible, for instance, that winning a competition yields a pronounced positive reward prediction error (RPE), when the outcome is uncertain (Schultz, 1997). Recent evidence also showed that serum testosterone levels are positively related to RPEs in the ventral striatum of individuals performing a reinforcement learning paradigm, which suggests a role of the hormone in shaping RPEs in humans (Morris et al., 2015). However, in this study, testosterone was only related to positive RPEs, but not negative RPEs. Together, this suggests that a testosterone surge following a win might enhance the associated positive RPE that will increase the incentive motivation to perform in a subsequent competition. Incentive motivation entails a set of processes that translate higher expected rewards into higher effort exertion (Berridge, 2004). Although the relationship between expected reward and effort exertion is complex, recent findings in humans (Schmidt et al., 2012) provide insight for generating new hypotheses of how motivation in competition and competition outcome may promote effort in subsequent competitions. In this neuroimaging study different amounts of rewards were associated with effort in two domains, a physical and a cognitive domain. Schmidt and colleagues found that the ventral striatum reflected expected reward during both cognitive and physical effort exertion. Specifically they showed that the ventral striatum mediated these incentive effects through connections of the basal ganglia and midbrain dopamine neurons, boosting task-relevant brain regions and performance (ie, cognitive circuits for cognitive real effort and motor circuits for physical real effort tasks). Based on this research we can predict that a testosterone-enhanced RPE associated with winning would increase the expected reward of winning a subsequent competition. The higher expected reward then boosts effort invested via activation of dopaminergic pathways to the ventral striatum that map onto the task-specific circuits. Future studies could apply a reinforcement learning framework using a repeated competition design to address the complex relationship of motivation, real effort, competition, and testosterone levels.

There are motivational aspects of competition that are intriguing and powerful. For instance, what are the motivational incentives when humans compete with themselves? A motive that drives people to compete with themselves is the goal to improve skills in an activity, which can be classified as an extrinsic, though self-set and integrated, motive. However, intrinsic motivation might also play an important role in this, since a self-challenge might be enjoyable. Thus, self-competition is of particular interest because it has been shown that intrinsic motivation usually has a strong and longer-lasting influence on performance relative to extrinsic motivation (Reeve and Deci, 1996; Reeve et al., 1985). Prior research in the laboratory has shown that receiving performance feedback is a clear motivational incentive (Chiviacowsky and Wulf, 2002, 2005; Kuhnen and Tymula, 2012; Widmer et al., 2016), an effect that is likely driven by the feeling of competence and self-esteem. Furthermore, successful achievements of effortful challenges enhance motivation and increase the value of the achievement as reflected in the ventral striatum (Lutz et al., 2012).

An interesting and open question is the role of testosterone in such a self-competition. Evidence supports a role of the hormone in this by showing that individuals' level of self-efficacy, effort, and motivation are positively related to testosterone levels (Costa et al., 2016; van der Meij et al., 2010). The role of testosterone in individual challenges is elusive; however, some evidence showed that testosterone concentrations only rose in social competitions among individuals who self-reported to have shown good individual performance (Trumble et al., 2012). This suggests that in a real effort self-competition, there might be an increase in testosterone secretion following "wins," that is, when performance increases across several stages of self-competitions.

8.2 SUMMARY

We highlighted research showing that competition is a powerful incentivizing tool. These motivational effects can be segregated into extrinsic and intrinsic motivations. We have argued that real effort-based competitions have the advantage of providing an assessment of the motivation to compete that allows for higher variance in behavior, as opposed to measuring motivation to compete using dichotomous decisions. The reviewed work highlights testosterone as an important neuroendocrinological variable that promotes the motivation to compete. It further emphasizes the role of testosterone in the winner effect as representing a performance increasing effect that seems to persist for extended periods of time. Such effects critically require neuroplasticity, for which testosterone has been shown to play an important role. Furthermore, animal literature suggests that testosterone might enable neuroplasticity not only via direct action on ARs but also via indirect action on ERs following aromatization of testosterone to estradiol and the DHT metabolites 3β-diol and 3α-diol. Testosterone or its metabolites may also induce neuroplasticity within the dopaminergic system and thus may have lasting effects on motivation to compete. The precise role of the different pathways of testosterone signaling in humans is still elusive; however, psychopharmacological models could provide a better understanding of this. One approach would involve a blockade of the enzyme aromatase. Alternatively, there are also selective antagonists of both ARs and ERs available and approved for use in humans that may help to further specify effects. The use of increasingly sophisticated psychopharmacological approaches and behavioral paradigms will provide more insight into the neurobiological mechanisms that link testosterone, motivation, and competition in humans.

ACKNOWLEDGMENTS

A.L.V. and C.E. were supported by the Vienna Science and Technology Fund (WWTF VRG13-007). I.R. was supported by the Slovak Research and Development Agency (Grant No. APVV-14-0840).

The authors declare no conflict of interest.

REFERENCES

Aarts, H., van Honk, J., 2009. Testosterone and unconscious positive priming increase human motivation separately. Neuroreport 20, 1300–1303. http://dx.doi.org/10.1097/WNR.0b013e3283308cdd.

Abe, M., Schambra, H., Wassermann, E.M., Luckenbaugh, D., Schweighofer, N., Cohen, L.G., 2011. Reward improves long-term retention of a motor memory through induction of offline memory gains. Curr. Biol. 21, 557–562. http://dx.doi.org/10.1016/j.cub.2011.02.030.

Abreu, P., Hernandez, G., Calzadilla, C.H., Alonso, R., 1988. Reproductive hormones control striatal tyrosine hydroxylase activity in the male rat. Neurosci. Lett. 95, 213–217. http://dx.doi.org/10.1016/0304-3940(88)90659-3.

Alderson, L.M., Baum, M.J., 1981. Differential effects of gonadal steroids on dopamine metabolism in mesolimbic and nigro-striatal pathways of male rat brain. Brain Res. 218, 189–206. http://dx.doi.org/10.1016/0006-8993(81)91300-7.

Almey, A., Milner, T.A., Brake, W.G., 2015. Estrogen receptors in the central nervous system and their implication for dopamine-dependent cognition in females. Horm. Behav. 74, 125–138. http://dx.doi.org/10.1016/j.yhbeh.2015.06.010.

Archer, J., 2006. Testosterone and human aggression: an evaluation of the challenge hypothesis. Neurosci. Biobehav. Rev. 30, 319–345. http://dx.doi.org/10.1016/j.neubiorev.2004.12.007.

Aubele, T., Kritzer, M.F., 2011. Gonadectomy and hormone replacement affects in vivo basal extracellular dopamine levels in the prefrontal cortex but not motor cortex of adult male rats. Cereb. Cortex 21, 222–232. http://dx.doi.org/10.1093/cercor/bhq083.

Aubele, T., Kaufman, R., Montalmant, F., Kritzer, M.F., 2008. Effects of gonadectomy and hormone replacement on a spontaneous novel object recognition task in adult male rats. Horm. Behav. 54, 244–252. http://dx.doi.org/10.1016/j.yhbeh.2008.04.001.

Balthazart, J., 2010. Behavioral implications of rapid changes in steroid production action in the brain [Commentary on Pradhan D.S., Newman A.E.M., Wacker D.W., Wingfield J.C., Schlinger B.A. and Soma K.K.: Aggressive interactions rapidly increase androgen

synthesis in the brain during the nonbreeding season. Hormones and Behavior, 2010]. Horm. Behav. 57, 375–378. http://dx.doi.org/10.1016/j.yhbeh.2010.02.003.
Berger, J., Pope, D., 2011. Can losing lead to winning? Manage. Sci. 57, 817–827. http://dx.doi.org/10.1287/mnsc.1110.1328.
Berridge, K.C., 2004. Motivation concepts in behavioral neuroscience. Physiol. Behav. 81, 179–209. http://dx.doi.org/10.1016/j.physbeh.2004.02.004.
Bhasin, S., Cunningham, G.R., Hayes, F.J., Matsumoto, A.M., Snyder, P.J., Swerdloff, R.S., Montori, V.M., 2006. Testosterone therapy in adult men with androgen deficiency syndromes: an endocrine society clinical practice guideline. J. Clin. Endocrinol. Metab. 91, 1995–2010. http://dx.doi.org/10.1210/jc.2005-2847.
Biegon, A., 2016. In vivo visualization of aromatase in animals and humans. Front. Neuroendocrinol. 40, 42–51. http://dx.doi.org/10.1016/j.yfrne.2015.10.001.
Booth, A., Shelley, G., Mazur, A., Tharp, G., Kittok, R., 1989. Testosterone, and winning and losing in human competition. Horm. Behav. 23, 556–571. http://dx.doi.org/10.1016/0018-506x(89)90042-1.
Bos, P.A., Panksepp, J., Bluthé, R.M., Honk, J. Van, 2012. Acute effects of steroid hormones and neuropeptides on human social-emotional behavior: a review of single administration studies. Front. Neuroendocrinol. 33, 17–35. http://dx.doi.org/10.1016/j.yfrne.2011.01.002.
Botvinick, M.M., Huffstetler, S., McGuire, J.T., 2009. Effort discounting in human nucleus accumbens. Cogn. Affect. Behav. Neurosci. 9, 16–27. http://dx.doi.org/10.3758/CABN.9.1.16.
Burger, H.G., 2002. Androgen production in women. Fertil. Steril. 77, 3–5. http://dx.doi.org/10.1016/S0015-0282(02)02985-0.
Carré, J.M., 2009. No place like home: testosterone responses to victory depend on game location. Am. J. Hum. Biol. 21, 392–394. http://dx.doi.org/10.1002/ajhb.20867.
Carré, J.M., McCormick, C.M., 2008. Aggressive behavior and change in salivary testosterone concentrations predict willingness to engage in a competitive task. Horm. Behav. 54, 403–409. http://dx.doi.org/10.1016/j.yhbeh.2008.04.008.
Carré, J.M., Olmstead, N.A., 2015. Social neuroendocrinology of human aggression: examining the role of competition-induced testosterone dynamics. Neuroscience 286, 171–186. http://dx.doi.org/10.1016/j.neuroscience.2014.11.029.
Carré, J.M., Putnam, S.K., 2010. Watching a previous victory produces an increase in testosterone among elite hockey players. Psychoneuroendocrinology 35, 475–479. http://dx.doi.org/10.1016/j.psyneuen.2009.09.011.
Carré, J.M., Muir, C., Belanger, J., Putnam, S.K., 2006. Pre-competition hormonal and psychological levels of elite hockey players: relationship to the "home advantage" Physiol. Behav. 89, 392–398. http://dx.doi.org/10.1016/j.physbeh.2006.07.011.
Carré, J.M., Hyde, L.W., Neumann, C.S., Viding, E., Hariri, A.R., 2013. The neural signatures of distinct psychopathic traits. Soc. Neurosci. 8, 122–135. http://dx.doi.org/10.1080/17470919.2012.703623.
Celotti, F., Negri-Cesi, P., Poletti, A., 1997. Steroid metabolism in the mammalian brain: 5alpha-reduction and aromatization. Brain Res. Bull. 44, 365–375. http://dx.doi.org/10.1016/s0361-9230(97)00216-5.
Charness, G., Villeval, M.-C., 2009. Cooperation and competition in intergenerational experiments in the field and the laboratory. Am. Econ. Rev. 99, 956–978. http://dx.doi.org/10.1257/aer.99.3.956.

Chase, I.D., Bartolomeo, C., Dugatkin, L.A., 1994. Aggressive interactions and inter-contest interval: how long do winners keep winning? Anim. Behav. 48, 393–400. http://dx.doi.org/10.1006/anbe.1994.1253.

Chiviacowsky, S., 2007. Feedback after good trials enhances learning. Res. Q. Exerc. Sport 78, 40–47. http://dx.doi.org/10.5641/193250307X13082490460346.

Chiviacowsky, S., Wulf, G., 2002. Self-controlled feedback: does it enhance learning because performers get feedback when they need it? Res. Q. Exerc. Sport 73, 408–415. http://dx.doi.org/10.1080/02701367.2002.10609040.

Chiviacowsky, S., Wulf, G., 2005. Self-controlled feedback is effective if it is based on the learner's performance. Res. Q. Exerc. Sport 76, 42–48. http://dx.doi.org/10.1080/02701367.2005.10599260.

Choate, J.V., Slayden, O.D., Resko, J.A., 1998. Immunocytochemical localization of androgen receptors in brains of developing and adult male rhesus monkeys. Endocrine 8, 51–60. http://dx.doi.org/10.1385/ENDO:8:1:51.

Clark, L., Lawrence, A.J., Astley-Jones, F., Gray, N., 2009. Gambling near-misses enhance motivation to gamble and recruit win-related brain circuitry. Neuron 61, 481–490. http://dx.doi.org/10.1016/j.neuron.2008.12.031.

Cooke, A., Kavussanu, M., McIntyre, D., Ring, C., 2013. The effects of individual and team competitions on performance, emotions, and effort. J. Sport Exerc. Psychol. 35, 132–143.

Corbett, J., Barwood, M.J., Ouzounoglou, A., Thelwell, R., Dicks, M., 2012. Influence of competition on performance and pacing during cycling exercise. Med. Sci. Sports Exerc. 44, 509–515. http://dx.doi.org/10.1249/MSS.0b013e31823378b1.

Cornil, C.A., Ball, G.F., Balthazart, J., 2012. Rapid control of male typical behaviors by brain-derived estrogens. Front. Neuroendocrinol. 33, 425–446. http://dx.doi.org/10.1016/j.yfrne.2012.08.003.

Costa, R., Serrano, M.A., Salvador, A., 2016. Importance of self-efficacy in psychoendocrine responses to competition and performance in women. Psicothema 28, 66–70. http://dx.doi.org/10.7334/psicothema2015.166.

Creutz, L.M., Kritzer, M.F., 2004. Mesostriatal and mesolimbic projections of midbrain neurons immunoreactive for estrogen receptor beta or androgen receptors in rats. J. Comp. Neurol. 476, 348–362. http://dx.doi.org/10.1002/cne.20229.

Crockett, M.J., Fehr, E., 2014. Social brains on drugs: tools for neuromodulation in social neuroscience. Soc. Cogn. Affect. Neurosci. 9, 250–254. http://dx.doi.org/10.1093/scan/nst113.

Croxson, P.L., Walton, M.E., O'Reilly, J.X., Behrens, T.E.J., Rushworth, M.F.S., 2009. Effort-based cost-benefit valuation and the human brain. J. Neurosci. 29, 4531–4541. http://dx.doi.org/10.1523/JNEUROSCI.4515-08.2009.

De Souza Silva, M.A., Mattern, C., Topic, B., Buddenberg, T.E., Huston, J.P., 2009. Dopaminergic and serotonergic activity in neostriatum and nucleus accumbens enhanced by intranasal administration of testosterone. Eur. Neuropsychopharmacol. 19, 53–63. http://dx.doi.org/10.1016/j.euroneuro.2008.08.003.

Deci, E.L., Betley, G., Kahle, J., Abrams, L., Porac, J., 1981. When trying to win: competition and intrinsic motivation. Personal. Soc. Psychol. Bull. 7, 79–83. http://dx.doi.org/10.1177/014616728171012.

Dugatkin, L.A., 1997. Winner and loser effects and the structure of dominance hierarchies. Behav. Ecol. 8, 583–587. http://dx.doi.org/10.1093/beheco/8.6.583.

Edinger, K.L., Frye, C.A., 2007a. Androgens' performance-enhancing effects in the inhibitory avoidance and water maze tasks may involve actions at intracellular androgen receptors in the dorsal hippocampus. Neurobiol. Learn. Mem. 87, 201–208. http://dx.doi.org/10.1016/j.nlm.2006.08.008.

Edinger, K.L., Frye, C.A., 2007b. Androgens' effects to enhance learning may be mediated in part through actions at estrogen receptor-beta in the hippocampus. Neurobiol. Learn. Mem. 87, 78–85. http://dx.doi.org/10.1016/j.nlm.2006.07.001.

Edinger, K.L., Lee, B., Frye, C.A., 2004. Mnemonic effects of testosterone and its 5alpha-reduced metabolites in the conditioned fear and inhibitory avoidance tasks. Pharmacol. Biochem. Behav. 78, 559–568. http://dx.doi.org/10.1016/j.pbb.2004.04.024.

Eisenegger, C., Haushofer, J., Fehr, E., 2011. The role of testosterone in social interaction. Trends Cogn. Sci. 15, 263–271. http://dx.doi.org/10.1016/j.tics.2011.04.008.

Elias, M., 1981. Serum cortisol, testosterone, and testosterone-binding globulin responses to competitive fighting in human males. Aggress. Behav. 7, 215–224. http://dx.doi.org/10.1002/1098-2337(1981)7:3<215::AID-AB2480070305>3.0.CO;2-M.

Fahr, R., Irlenbusch, B., 2000. Fairness as a constraint on trust in reciprocity: earned property rights in a reciprocal exchange experiment. Econ. Lett. 66, 275–282. http://dx.doi.org/10.1016/S0165-1765(99)00236-0.

Farinetti, A., Tomasi, S., Foglio, B., Ferraris, A., Ponti, G., Gotti, S., Peretto, P., Panzica, G.C., 2015. Testosterone and estradiol differentially affect cell proliferation in the subventricular zone of young adult gonadectomized male and female rats. Neuroscience 286, 162–170. http://dx.doi.org/10.1016/j.neuroscience.2014.11.050.

Fester, L., Rune, G.M., 2015. Sexual neurosteroids and synaptic plasticity in the hippocampus. Brain Res. 1621, 162–169. http://dx.doi.org/10.1016/j.brainres.2014.10.033.

Folstad, I., Karter, A.J., 1992. Parasites, bright males, and the immunocompetence handicap. Am. Nat. 139, 603–622. http://dx.doi.org/10.1086/285346.

Frick, K.M., Kim, J., Tuscher, J.J., Fortress, A.M., 2015. Sex steroid hormones matter for learning and memory: estrogenic regulation of hippocampal function in male and female rodents. Learn. Mem. 22, 472–493. http://dx.doi.org/10.1101/lm.037267.114.

Frye, C.A., Lacey, E.H., 2001. Posttraining androgens' enhancement of cognitive performance is temporally distinct from androgens' increases in affective behavior. Cogn. Affect. Behav. Neurosci. 1, 172–182. http://dx.doi.org/10.3758/CABN.1.2.172.

Frye, C.A., Rhodes, M.E., Rosellini, R., Svare, B., 2002. The nucleus accumbens as a site of action for rewarding properties of testosterone and its 5alpha-reduced metabolites. Pharmacol. Biochem. Behav. 74, 119–127. http://dx.doi.org/10.1016/s0091-3057(02)00968-1.

Frye, C.A., Koonce, C.J., Edinger, K.L., Osborne, D.M., Walf, A.A., 2008. Androgens with activity at estrogen receptor beta have anxiolytic and cognitive-enhancing effects in male rats and mice. Horm. Behav. 54, 726–734. http://dx.doi.org/10.1016/j.yhbeh.2008.07.013.

Fuxjager, M.J., Marler, C.A., 2010. How and why the winner effect forms: influences of contest environment and species differences. Behav. Ecol. 21, 37–45. http://dx.doi.org/10.1093/beheco/arp148.

Fuxjager, M.J., Forbes-Lorman, R.M., Coss, D.J., Auger, C.J., Auger, A.P., Marler, C.A., 2010. Winning territorial disputes selectively enhances androgen sensitivity in neural pathways related to motivation and social aggression. Proc. Natl. Acad. Sci. U.S.A. 107, 12393–12398. http://dx.doi.org/10.1073/pnas.1001394107.

Fuxjager, M.J., Montgomery, J.L., Marler, C.A., 2011a. Species differences in the winner effect disappear in response to post-victory testosterone manipulations. Proc. R. Soc. B Biol. Sci. 278, 3497–3503. http://dx.doi.org/10.1098/rspb.2011.0301.

Fuxjager, M.J., Oyegbile, T.O., Marler, C.A., 2011b. Independent and additive contributions of postvictory testosterone and social experience to the development of the winner effect. Endocrinology 152, 3422–3429. http://dx.doi.org/10.1210/en.2011-1099.

Geniole, S.N., Busseri, M.A., McCormick, C.M., 2013. Testosterone dynamics and psychopathic personality traits independently predict antagonistic behavior towards the perceived loser of a competitive interaction. Horm. Behav. 64, 790–798. http://dx.doi.org/10.1016/j.yhbeh.2013.09.005.

Gill, D., Prowse, V., 2012. A structural analysis of disappointment aversion in a real effort competition. Am. Econ. Rev. 102, 469–503. http://dx.doi.org/10.1257/aer.102.1.469.

Gleason, E.D., Fuxjager, M.J., Oyegbile, T.O., Marler, C.A., 2009. Testosterone release and social context: when it occurs and why. Front. Neuroendocrinol. 30, 460–469. http://dx.doi.org/10.1016/j.yfrne.2009.04.009.

Gneezy, U., Niederle, M., Rustichini, A., Brodkey, D., Vigna, S. Della, Orosel, G., Piankov, N., Roth, A., Vesterlund, L., 2003. Performance in competitive environments: gender differences*. Q. J. Econ. 118, 1049–1074. http://dx.doi.org/10.1162/00335530360698496.

Hajszan, T., MacLusky, N.J., Leranth, C., 2008. Role of androgens and the androgen receptor in remodeling of spine synapses in limbic brain areas. Horm. Behav. 53, 638–646. http://dx.doi.org/10.1016/j.yhbeh.2007.12.007.

Handa, R.J., Pak, T.R., Kudwa, A.E., Lund, T.D., Hinds, L., 2008. An alternate pathway for androgen regulation of brain function: activation of estrogen receptor beta by the metabolite of dihydrotestosterone, 5alpha-androstane-3beta,17beta-diol. Horm. Behav. 53, 741–752. http://dx.doi.org/10.1016/j.yhbeh.2007.09.012.

Hatanaka, Y., Mukai, H., Mitsuhashi, K., Hojo, Y., Murakami, G., Komatsuzaki, Y., Sato, R., Kawato, S., 2009. Androgen rapidly increases dendritic thorns of CA3 neurons in male rat hippocampus. Biochem. Biophys. Res. Commun. 381, 728–732. http://dx.doi.org/10.1016/j.bbrc.2009.02.130.

Hermans, E.J., Bos, P.A., Ossewaarde, L., Ramsey, N.F., Fernandez, G., van Honk, J., 2010. Effects of exogenous testosterone on the ventral striatal BOLD response during reward anticipation in healthy women. Neuroimage 52, 277–283. http://dx.doi.org/10.1016/j.neuroimage.2010.04.019.

Hirschenhauser, K., Wittek, M., Johnston, P., Möstl, E., 2008. Social context rather than behavioral output or winning modulates post-conflict testosterone responses in Japanese quail (Coturnix japonica). Physiol. Behav. 95, 457–463. http://dx.doi.org/10.1016/j.physbeh.2008.07.013.

Hirschenhauser, K., Gahr, M., Goymann, W., 2013. Winning and losing in public: audiences direct future success in Japanese quail. Horm. Behav. 63, 625–633. http://dx.doi.org/10.1016/j.yhbeh.2013.02.010.

Hirshleifer, J., 1978. Competition, cooperation, and conflict in economics and biology. Annu. Meet. Am. Econ. Assoc. 68, 238–243.

Hoffman, E., McCabe, K., Shachat, K., Smith, V., 1994. Preferences, property rights, and anonymity in bargaining games. Games Econ. Behav. 7, 346–380. http://dx.doi.org/10.1006/game.1994.1056.

Hosp, J.A., Pekanovic, A., Rioult-Pedotti, M.S., Luft, A.R., 2011. Dopaminergic projections from midbrain to primary motor cortex mediate motor skill learning. J. Neurosci. 31, 2481–2487. http://dx.doi.org/10.1523/JNEUROSCI.5411-10.2011.

Hsu, Y., Earley, R.L., Wolf, L.L., 2006. Modulation of aggressive behaviour by fighting experience: mechanisms and contest outcomes. Biol. Rev. Camb. Philos. Soc. 81, 33–74. http://dx.doi.org/10.1017/S146479310500686X.

Hurd, P.L., 2006. Resource holding potential, subjective resource value, and game theoretical models of aggressiveness signalling. J. Theor. Biol. 241, 639–648. http://dx.doi.org/10.1016/j.jtbi.2006.01.001.

Inagaki, T., Gautreaux, C., Luine, V., 2010. Acute estrogen treatment facilitates recognition memory consolidation and alters monoamine levels in memory-related brain areas. Horm. Behav. 58, 415–426. http://dx.doi.org/10.1016/j.yhbeh.2010.05.013.

Kuhnen, C.M., Tymula, A., 2012. Feedback, self-esteem, and performance in organizations. Manag. Sci. 58, 94–113. http://dx.doi.org/10.1287/mnsc.1110.1379.

Kurniawan, I.T., Seymour, B., Talmi, D., Yoshida, W., Chater, N., Dolan, R.J., 2010. Choosing to make an effort: the role of striatum in signaling physical effort of a chosen action. J. Neurophysiol. 104, 313–321. http://dx.doi.org/10.1152/jn.00027.2010.

Kurniawan, I.T., Guitart-Masip, M., Dolan, R.J., 2011. Dopamine and effort-based decision making. Front. Neurosci. 5, 81. http://dx.doi.org/10.3389/fnins.2011.00081.

Le Bouc, R., Pessiglione, M., 2013. Imaging social motivation: distinct brain mechanisms drive effort production during collaboration versus competition. J. Neurosci. 33, 15894–15902. http://dx.doi.org/10.1523/JNEUROSCI.0143-13.2013.

Li, M., Masugi-Tokita, M., Takanami, K., Yamada, S., Kawata, M., 2012. Testosterone has sublayer-specific effects on dendritic spine maturation mediated by BDNF and PSD-95 in pyramidal neurons in the hippocampus CA1 area. Brain Res. 1484, 76–84. http://dx.doi.org/10.1016/j.brainres.2012.09.028.

Luine, V.N., Frankfurt, M., 2012. Estrogens facilitate memory processing through membrane mediated mechanisms and alterations in spine density. Front. Neuroendocrinol. 33, 388–402. http://dx.doi.org/10.1016/j.yfrne.2012.07.004.

Lutz, K., Pedroni, A., Nadig, K., Luechinger, R., Jäncke, L., 2012. The rewarding value of good motor performance in the context of monetary incentives. Neuropsychologia 50, 1739–1747. http://dx.doi.org/10.1016/j.neuropsychologia.2012.03.030.

May, A., 2011. Experience-dependent structural plasticity in the adult human brain. Trends Cogn. Sci. 15, 475–482. http://dx.doi.org/10.1016/j.tics.2011.08.002.

Mazur, A., 1985. A biosocial model of status in face-to-face primate groups. Soc. Forces 64, 377–402. http://dx.doi.org/10.1093/sf/64.2.377.

Mazur, A., Booth, A., 1998. Testosterone and dominance in men. Behav. Brain Sci. 21, 353–363. http://dx.doi.org/10.1017/s0140525x98001228. discussion 363–397.

Mazur, A., Booth, A., Dabbs Jr., J.M., 1992. Testosterone and chess competition. Soc. Psychol. Q. 55, 70–77. http://dx.doi.org/10.2307/2786687.

McGee, A., McGee, P., 2013. After the Tournament: Outcomes and Effort Provision. Working Paper. http://ftp.iza.org/dp7759.pdf.

Mehta, P.H., Josephs, R.a., 2006. Testosterone change after losing predicts the decision to compete again. Horm. Behav. 50, 684–692. http://dx.doi.org/10.1016/j.yhbeh.2006.07.001.

Mehta, P.H., Son, V. Van, Welker, K.M., Prasad, S., Sanfey, A.G., Smidts, A., Roelofs, K., 2015. Exogenous testosterone in women enhances and inhibits competitive decision-making depending on victory–defeat experience and trait dominance. Psychoneuroendocrinology 60, 224–236. http://dx.doi.org/10.1016/j.psyneuen.2015. 07.004.

Miendlarzewska, E.A., Bavelier, D., Schwartz, S., 2015. Influence of reward motivation on human declarative memory. Neurosci. Biobehav. Rev. 61, 156–176. http://dx.doi.org/10.1016/j.neubiorev.2015.11.015.

Mitchell, J.B., Stewart, J., 1989. Effects of castration, steroid replacement, and sexual experience on mesolimbic dopamine and sexual behaviors in the male rat. Brain Res. 491, 116–127. http://dx.doi.org/10.1016/0006-8993(89)90093-0.

Monaghan, E.P., Glickman, S.E., 2001. Hormones and aggressive behavior. In: Becker, J.B., Breedlove, S.M., Crews, D. (Eds.), Behavioural Endocrinology. MIT Press, Cambridge, MA, pp. 261–287.

Morris, R.W., Fung, S.J., Rothmond, D.A., Richards, B., Ward, S., Noble, P.L., Woodward, R.A., Weickert, C.S., Winslow, J.T., 2010. The effect of gonadectomy on prepulse inhibition and fear-potentiated startle in adolescent rhesus macaques. Psychoneuroendocrinology 35, 896–905. http://dx.doi.org/10.1016/j.psyneuen.2009. 12.002.

Morris, R.W., Purves-Tyson, T.D., Weickert, C.S., Rothmond, D., Lenroot, R., Weickert, T.W., 2015. Testosterone and reward prediction-errors in healthy men and men with schizophrenia. Schizophr. Res. 168, 649–660. http://dx.doi.org/10.1016/j.schres. 2015.06.030.

Niederle, M., Vesterlund, L., 2007. Do women shy away from competition? Do men compete too much? Q. J. Econ. 122, 1067–1101. http://dx.doi.org/10.1162/qjec.122.3.1067.

Niederle, M., Vesterlund, L., 2011. Gender and competition. Annu. Rev. Econ. 3, 601–630. http://dx.doi.org/10.1146/annurev-economics-111809-125122.

Nyby, J.G., 2008. Reflexive testosterone release: a model system for studying the nongenomic effects of testosterone upon male behavior. Front. Neuroendocrinol. 29, 199–210. http://dx.doi.org/10.1016/j.yfrne.2007.09.001.

Okamoto, M., Hojo, Y., Inoue, K., Matsui, T., Kawato, S., McEwen, B.S., Soya, H., 2012. Mild exercise increases dihydrotestosterone in hippocampus providing evidence for androgenic mediation of neurogenesis. Proc. Natl. Acad. Sci. U.S.A. 109, 13100–13105. http://dx.doi.org/10.1073/pnas.1210023109.

Oliveira, G.A., Oliveira, R.F., 2014. Androgen response to competition and cognitive variables. Neurosci. Neuroecon. 3, 19–32. http://dx.doi.org/10.2147/nan.s55721.

Oliveira, R.F., Silva, A., Canário, A.V.M., 2009. Why do winners keep winning? Androgen mediation of winner but not loser effects in cichlid fish. Proc. Biol. Sci. 276, 2249–2256. http://dx.doi.org/10.1098/rspb.2009.0132.

Oyegbile, T.O., Marler, C.A., 2005. Winning fights elevates testosterone levels in California mice and enhances future ability to win fights. Horm. Behav. 48, 259–267. http://dx.doi.org/10.1016/j.yhbeh.2005.04.007.

Packard, M.G., 1998. Posttraining estrogen and memory modulation. Horm. Behav. 34, 126–139. http://dx.doi.org/10.1006/hbeh.1998.1464.

Packard, M.G., Schroeder, J.P., Alexander, G.M., 1998. Expression of testosterone conditioned place preference is blocked by peripheral or intra-accumbens injection of alpha-flupenthixol. Horm. Behav. 34, 39–47. http://dx.doi.org/10.1006/hbeh.1998.1461.

Pallotto, M., Deprez, F., 2014. Regulation of adult neurogenesis by GABAergic transmission: signaling beyond GABAA-receptors. Front. Cell. Neurosci. 8, 166. http://dx.doi.org/10.3389/fncel.2014.00166.

Picot, M., Billard, J.-M., Dombret, C., Albac, C., Karameh, N., Daumas, S., Hardin-Pouzet, H., Mhaouty-Kodja, S., 2016. Neural androgen receptor deletion impairs the temporal processing of objects and hippocampal CA1-dependent mechanisms. PLoS One 11, e0148328. http://dx.doi.org/10.1371/journal.pone.0148328.

Pluchino, N., Russo, M., Santoro, A.N., Litta, P., Cela, V., Genazzani, A.R., 2013. Steroid hormones and BDNF. Neuroscience 239, 271–279. http://dx.doi.org/10.1016/j.neuroscience.2013.01.025.

Purves-Tyson, T.D., Handelsman, D.J., Double, K.L., Owens, S.J., Bustamante, S., Weickert, C.S., 2012. Testosterone regulation of sex steroid-related mRNAs and dopamine-related mRNAs in adolescent male rat substantia nigra. BMC Neurosci. 13, 1–12. http://dx.doi.org/10.1186/1471-2202-13-95.

Purves-Tyson, T.D., Owens, S.J., Double, K.L., Desai, R., Handelsman, D.J., Weickert, C.S., 2014. Testosterone induces molecular changes in dopamine signaling pathway molecules in the adolescent male rat nigrostriatal pathway. PLoS One 9, e91151. http://dx.doi.org/10.1371/journal.pone.0091151.

Reeve, J., Deci, E.L., 1996. Elements of the competitive situation that affect intrinsic motivation. Personal. Soc. Psychol. Bull. 22, 24–33. http://dx.doi.org/10.1177/0146167296221003.

Reeve, J., Olson, B.C., Cole, S.G., 1985. Motivation and performance: two consequences of winning and losing in competition. Motiv. Emot. 9, 291–298. http://dx.doi.org/10.1007/BF00991833.

Reid, R.L., 1986. The psychology of the near miss. J. Gambl. Behav. 2, 32–39. http://dx.doi.org/10.1007/BF01019932.

Robbins, T.W., Everitt, B.J., 1996. Neurobehavioural mechanisms of reward and motivation. Curr. Opin. Neurobiol. 6, 228–236. doi:10.1016/S0959-4388(96)80077-8.

Rutström, E.E., Williams, M.B., 2000. Entitlements and fairness. J. Econ. Behav. Organ. 43, 75–89. http://dx.doi.org/10.1016/S0167-2681(00)00109-8.

Ryan, R., Deci, E., 2000. Intrinsic and extrinsic motivations: classic definitions and new directions. Contemp. Educ. Psychol. 25, 54–67. http://dx.doi.org/10.1006/ceps. 1999.1020.

Salamone, J.D., Correa, M., 2002. Motivational views of reinforcement: implications for understanding the behavioral functions of nucleus accumbens dopamine. Behav. Brain Res. 137, 3–25. http://dx.doi.org/10.1016/S0166-4328(02)00282-6.

Salamone, J.D., Correa, M., Farrar, A., Mingote, S.M., 2007. Effort-related functions of nucleus accumbens dopamine and associated forebrain circuits. Psychopharmacology (Berl) 191, 461–482. http://dx.doi.org/10.1007/s00213-006-0668-9.

Salvador, A., Costa, R., 2009. Coping with competition: neuroendocrine responses and cognitive variables. Neurosci. Biobehav. Rev. 33, 160–170. http://dx.doi.org/10.1016/j.neubiorev.2008.09.005.

Schmidt, L., Lebreton, M., Cléry-Melin, M.-L., Daunizeau, J., Pessiglione, M., 2012. Neural mechanisms underlying motivation of mental versus physical effort. PLoS Biol. 10, e1001266. http://dx.doi.org/10.1371/journal.pbio.1001266.

Schultz, W., 1997. A neural substrate of prediction and reward. Science 275, 1593–1599. http://dx.doi.org/10.1126/science.275.5306.1593.

Schwartzer, J.J., Ricci, L.A., Melloni, R.H., 2013. Prior fighting experience increases aggression in Syrian hamsters: implications for a role of dopamine in the winner effect. Aggress. Behav. 39, 290–300. http://dx.doi.org/10.1002/ab.21476.

Shohamy, D., Adcock, R.A., 2010. Dopamine and adaptive memory. Trends Cogn. Sci. 14, 464–472. http://dx.doi.org/10.1016/j.tics.2010.08.002.

Simerly, R.B., Chang, C., Muramatsu, M., Swanson, L.W., 1990. Distribution of androgen and estrogen receptor mRNA-containing cells in the rat brain: an in situ hybridization study. J. Comp. Neurol. 294, 76–95. http://dx.doi.org/10.1002/cne.902940107.

Stanne, M.B., Johnson, D.W., Johnson, R.T., 1999. Does competition enhance or inhibit motor performance: a meta-analysis. Psychol. Bull. 125, 133–154. http://dx.doi.org/10.1037/0033-2909.125.1.133.

Studer, B., Knecht, S., 2016. Chapter 2—A Benefit–cost framework of motivation for a specific activity. In: Studer, B., Knecht, S. (Eds.), Progress in Brain Research, vol. 229. Elsevier, Amsterdam, pp. 25–47.

Suay, F., Salvador, A., Gonzalez-Bono, E., Sanchis, C., Martinez, M., Martinez-Sanchis, S., Simon, V.M., Montoro, J.B., 1999. Effects of competition and its outcome on serum testosterone, cortisol and prolactin. Psychoneuroendocrinology 24, 551–566. http://dx.doi.org/10.1016/S0306-4530(99)00011-6.

Sugawara, S.K., Tanaka, S., Okazaki, S., Watanabe, K., Sadato, N., 2012. Social rewards enhance offline improvements in motor skill. PLoS One 7, e48174. http://dx.doi.org/10.1371/journal.pone.0048174.

Sweatt, J.D., 2016. Neural plasticity & behavior—sixty years of conceptual advances. J. Neurochem. 1–21. http://dx.doi.org/10.1111/jnc.13580.

Thiblin, I., Finn, A., Ross, S.B., Stenfors, C., 1999. Increased dopaminergic and 5-hydroxytryptaminergic activities in male rat brain following long-term treatment with anabolic androgenic steroids. Br. J. Pharmacol. 126, 1301–1306. http://dx.doi.org/10.1038/sj.bjp.0702412.

Trainor, B.C., Bird, I.M., Marler, C.A., 2004. Opposing hormonal mechanisms of aggression revealed through short-lived testosterone manipulations and multiple winning experiences. Horm. Behav. 45, 115–121. http://dx.doi.org/10.1016/j.yhbeh. 2003.09.006.

Trumble, B.C., Cummings, D., von Rueden, C., O'Connor, K.A., Smith, E.A., Gurven, M., Kaplan, H., 2012. Physical competition increases testosterone among Amazonian forager-horticulturalists: a test of the "challenge hypothesis" Proc. Biol. Sci. 279, 2907–2912. http://dx.doi.org/10.1098/rspb.2012.0455.

Van Anders, S.M., Watson, N.V., 2007. Effects of ability- and chance-determined competition outcome on testosterone. Physiol. Behav. 90, 634–642. http://dx.doi.org/10.1016/j.physbeh.2006.11.017.

Van den Bos, W., Golka, P.J.M., Effelsberg, D., Mcclure, S.M., Isoda, M., Medical, K., Chang, L.J., 2013. Pyrrhic victories: the need for social status drives costly competitive behavior. Front. Neurosci. 7, 1–11. http://dx.doi.org/10.3389/fnins.2013.00189.

Van der Meij, L., Buunk, A.P., Almela, M., Salvador, A., 2010. Testosterone responses to competition: the opponent's psychological state makes it challenging. Biol. Psychol. 84, 330–335. http://dx.doi.org/10.1016/j.biopsycho.2010.03.017.

Wallin, K.G., Alves, J.M., Wood, R.I., 2015. Anabolic-androgenic steroids and decision making: probability and effort discounting in male rats. Psychoneuroendocrinology 57, 84–92. http://dx.doi.org/10.1016/j.psyneuen.2015.03.023.

Westbrook, A., Braver, T.S., 2016. Dopamine does double duty in motivating cognitive effort. Neuron 89, 695–710. http://dx.doi.org/10.1016/j.neuron.2015.12.029.

Widmer, M., Ziegler, N., Held, J., Luft, A., Lutz, K., 2016. Chapter 13—Rewarding feedback promotes motor skill consolidation via striatal activity. In: Studer, B., Knecht, S. (Eds.), Progress in Brain Research, vol. 229. Elsevier, Amsterdam, pp. 303–323.

Williams, R.B., Lane, J.D., Kuhn, C.M., Melosh, W., White, A.D., Schanberg, S.M., 1982. Type A behavior and elevated physiological and neuroendocrine responses to cognitive tasks. Science 218, 483–485. http://dx.doi.org/10.2307/1689484.

Wingfield, J.C., Hegner, R.E., Dufty Jr., A.M., Ball, G.F., 1990. The "challenge hypothesis": theoretical implications for patterns of testosterone secretion, mating systems, and breeding strategies. Am. Nat. 136, 829. http://dx.doi.org/10.1086/285134.

Wood, R.I., 2008. Anabolic-androgenic steroid dependence? Insights from animals and humans. Front. Neuroendocrinol. 29, 490–506. http://dx.doi.org/10.1016/j.yfrne.2007.12.002.

Wood, R.I., Johnson, L.R., Chu, L., Schad, C., Self, D.W., 2004. Testosterone reinforcement: intravenous and intracerebroventricular self-administration in male rats and hamsters. Psychopharmacology (Berl) 171, 298–305. http://dx.doi.org/10.1007/s00213-003-1587-7.

Zilioli, S., Watson, N.V., 2014. Testosterone across successive competitions: evidence for a "winner effect" in humans? Psychoneuroendocrinology 47, 1–9. http://dx.doi.org/10.1016/j.psyneuen.2014.05.001.

Zilioli, S., Mehta, P.H., Watson, N.V., 2014. Losing the battle but winning the war: uncertain outcomes reverse the usual effect of winning on testosterone. Biol. Psychol. 103, 54–62. http://dx.doi.org/10.1016/j.biopsycho.2014.07.022.

Zumoff, B., Rosenfeld, R.S., Friedman, M., Byers, S.O., Rosenman, R.H., Hellman, L., 1984. Elevated daytime urinary excretion of testosterone glucuronide in men with the type A behavior pattern. Psychosom. Med. 46, 223–225. http://dx.doi.org/10.1097/00006842-198405000-00004.

Fatigue with up- vs downregulated brain arousal should not be confused

10

U. Hegerl*,†,1, C. Ulke*

**Research Center of the German Depression Foundation, Leipzig, Germany*
†University of Leipzig, Leipzig, Germany
[1]Corresponding author: Tel.: +49-341-9724570; Fax: +49-341-9724539, e-mail address: ulrich.hegerl@medizin.uni-leipzig.de

Abstract

Fatigue is considered to be an important and frequent factor in motivation problems. However, this term lacks clinical and pathophysiological validity. Semantic precision has to be improved. Lack of drive and tiredness with increased sleepiness as observed in fatigue in the context of inflammatory and immunological processes (hypoaroused fatigue) has to be separated from inhibition of drive and tiredness with prolonged sleep onset latency as observed in major depression (hyperaroused fatigue). Subjective experiences as reported by patients, as well as clinical, behavioral, and neurobiological findings support the validity and importance of this distinction. A practical clinical procedure for how to separate hypo- from hyperaroused fatigue will be proposed.

Keywords

Fatigue, Brain arousal, Drive, Depression, Inflammatory and immunological processes, VIGALL

1 INTRODUCTION

Fatigue is associated with the difficulty to initiate or sustain voluntary activities (Chaudhuri and Behan, 2004). Its neurobiological mechanisms are not entirely understood. It is a common symptom in the context of inflammatory and immunological processes (Morris et al., 2016), where it has been linked to proinflammatory cytokines—immunological transmitters that are involved in the sleep/wake regulation (Krueger, 2008).

Fatigue is also a highly prevalent symptom in depression (Demyttenaere et al., 2005; Vaccarino et al., 2008) and a highly prevalent residual symptom of depression (Fava et al., 2014; Hybels et al., 2005). An upregulation of the central noradrenergic

Progress in Brain Research, Volume 229, ISSN 0079-6123, http://dx.doi.org/10.1016/bs.pbr.2016.06.001

neurotransmission which is implicated in the regulation of arousal (Berridge, 2008) is discussed as an underlying factor of fatigue in depression.

Research on fatigue is hampered by semantic problems, and the ways fatigue as a clinical symptom is operationalized. Fatigue can denote both an anergic state with a lack of drive and sleepiness, as often observed in the context of immunological and inflammatory processes, and also a state of exhaustion with high inner tension, inhibition of drive and difficulty to fall asleep as typically seen in major depression. The use of the term fatigue obfuscates two very different meanings; it mixes up these clearly heterogeneous states.

Arguments will be presented that brain arousal is a dimension which has to be considered when trying to partition fatigue in clinically and pathophysiologically more homogeneous subtypes (Hegerl et al., 2013). Brain arousal has been defined as the level of the activation of the central nervous system that determines an organism's responsiveness to external (eg, threats) and internal cues (eg, pain). Different arousal levels denote various states of global brain function which shape the behavioral responses to these cues (Pfaff et al., 2008). It is evident that the regulation of brain arousal and its reliable and rapid adaptation to the environment are of crucial importance for all higher organisms. Both the ability to reduce the arousal level when being at rest or when going to sleep but also the ability to increase it in the case of danger are critical for survival.

In the following, several lines of evidence will be provided supporting the distinction between hypo- and hyperaroused fatigue followed by suggestions concerning clinical practice and a clinical procedure to separate these two subtypes. Peripheral fatigue (eg, in the context of myopathies) or the chronic fatigue syndrome (for review, see Afari and Buchwald, 2003) will not be addressed in this chapter.

2 FATIGUE AND BRAIN AROUSAL

2.1 FATIGUE AS CLINICAL SYMPTOM: SEMANTIC PROBLEMS

The term fatigue is used as a defining feature of major depressive disorder and to denote symptoms occurring in the context of other disorders comprising neuroinflammatory, inflammatory, and immunological processes, for example, multiple sclerosis (MS; Malekzadeh et al., 2015), Parkinson's disease (Rocha et al., 2015), poststroke fatigue (Kuppuswamy et al., 2015), rheumatoid arthritis (Van Steenbergen et al., 2015), and cancer-related fatigue (Bower, 2007).

The same term is used for syndromes which are profoundly different concerning behavior, neurobiology, and subjective experience (Hegerl et al., 2013; Pigeon et al., 2003). Fatigue in depression is a state of exhaustion with difficulties to fall asleep and prolonged sleep latencies in the Multiple Sleep Latency Test (MSLT; Carskadon et al., 1986). Conversely, fatigue in the inflammatory context is a state of excessive daytime sleepiness with short sleep onset latencies during the MSLT. Similarly, the terms "lack of drive" and "lack of energy" mix up two states: an apathetic state with an actual lack of drive, and a state with an inhibition of drive, upregulated arousal, and high inner tension. These states are profoundly different concerning underlying pathophysiological processes (see Fig. 1).

In the next section, we will briefly introduce the concept of brain arousal and present a novel method to objectively assess the level of brain arousal and its regulation—the Vigilance Algorithm Leipzig (VIGALL; http://research.uni-leipzig.de/vigall).

2.2 BRAIN AROUSAL

Brain arousal is a construct with a long history in neurology (Plum and Posner, 1982) and behavioral neuroscience (Pfaff et al., 2008, 2012). It is thought to determine the responsiveness of an organism to sensory stimuli and it is linked to homeostatic drives (Ribeiro et al., 2007; Zitnik, 2015), cognition (Sara and Bouret, 2012), and many other behavioral and neurophysiological functions (Pfaff et al., 2008). Within the Research Domain Criteria (RDoC) project, proposed and supported by the NIMH (National Institute of Mental Health; Insel et al., 2010), arousal has been identified as one of the five fundamental dimensions which have to be considered when transdiagnostically looking for pathophysiologically and prognostically homogeneous classifications of mental disorders. During the RDoC workshop proceedings (NIMH, 2012), arousal was further specified as follows:

- arousal facilitates interaction with the environment in a context-specific manner (eg, under conditions of threat, some stimuli must be ignored while sensitivity and responses to others are enhanced, as exemplified in the startle reflex);
- arousal can be evoked by either external/environmental stimuli or internal stimuli (eg, emotions and cognition) and modulated by the physical characteristics and motivational significance of stimuli;

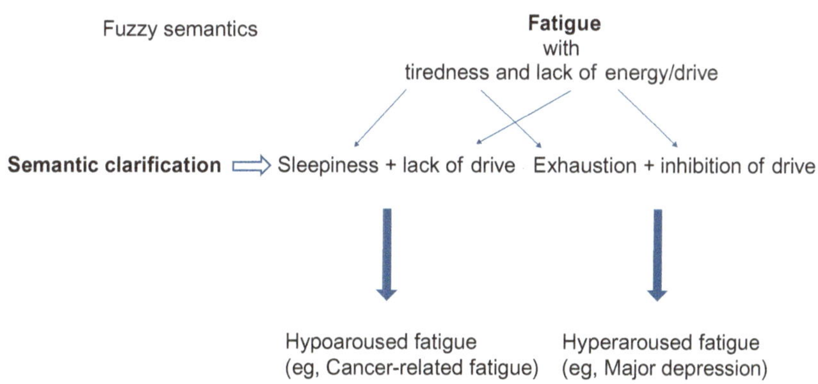

FIG. 1

Semantic clarification leads to the distinction between hypo- and hyperaroused fatigue.

- arousal varies along a continuum that can be quantified in any behavioral state, including wakefulness and low-arousal states such as sleep, anesthesia, and coma;
- arousal is distinct from motivation and valence, but can covary with intensity of motivation and valence; may be associated with increased or decreased locomotor activity; and can be regulated by homeostatic drives (eg, hunger, sleep, thirst, sex).

Many different transmitter systems (eg, noradrenergic, serotonergic, histaminergic, cholinergic, glutamatergic, orexinergic/hypocretinergic) are implicated in arousal and sleep/wake regulation (Brown et al., 2012). During the waking state, fluctuations can be observed between lower arousal levels with a decreasing sensitivity to sensory stimuli and higher arousal levels with a hypersensitivity to sensory stimuli (Berridge et al., 2012).

It is of interest in this context that the relationship between the level of brain arousal and behavioral patterns can have an inverted U-shape (Yerkes and Dodson, 1908). A lack of overt behavior can be observed with low (eg, drowsiness or sleep) but also very high brain arousal levels (eg, freezing), see Fig. 2. In line with this, impairment of motivation and executive functions is observed at low as well as high levels of arousal (Berridge and Arnsten, 2013). As a consequence, certain behavioral patterns can appear similar while having profoundly different underlying neurophysiological mechanisms. This also holds true for fatigue. The inverted U-shaped relationship between the level of arousal and the overt behavior can be observed at the neuronal level involving different transmitter systems—both, high and low levels of the catecholamines, norepinephrine, and dopamine lead to decrease of reward-seeking behavior in monkeys in a spatial working memory task (for review, see Berridge and Arnsten, 2013). Therefore, on the level of overt behavior alone (eg, social withdrawal), it is not possible to identify the underlying pathophysiological mechanism.

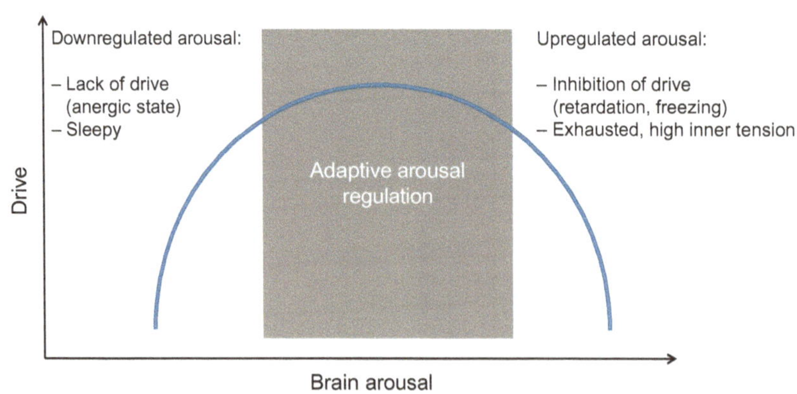

FIG. 2

Inverted U-shaped relationship between brain arousal and drive.

By far the best method to assess brain arousal in humans is the electroencephalography (EEG). EEG recordings from the scalp provide information about the temporal–spatial patterns of cortical neuronal mass activity and their changes along the sleep/wake dimension (Berridge et al., 2012). Recently, an EEG-based tool has been developed which automatically classifies 1-s EEG segments into different arousal levels and allows the study of the regulation of arousal during a 15- to 20-min EEG under resting conditions with eyes closed (Vigilance Algorithm Leipzig; VIGALL). VIGALL separates states associated with different levels of arousal ranging from active wakefulness with high alertness (EEG-vigilance stage 0) to relaxed wakefulness (stages A1, A2, A3), drowsiness (stages B1, B2/3), and sleep onset (stage C); specific details of the classification algorithm are described elsewhere (Sander et al., 2015). VIGALL has been validated in simultaneous EEG-fMRI studies (Olbrich et al., 2009), in simultaneous EEG-PET studies (Guenther et al., 2011), in relation to autonomic parameters (Olbrich et al., 2011), concerning different behavioral parameters (Bekhtereva et al., 2014; Minkwitz et al., 2011), and with regard to the MSLT (Olbrich et al., 2015).

The ability to downregulate the brain arousal level or to keep it up under circumscribed conditions is a state-modulated trait with considerable interindividual differences (Huang et al., 2015): while some individuals remain in a state of high arousal over the 15-min EEG recording (stable arousal regulation), others show a rapid decline to lower arousal states associated with drowsiness or sleep onset (unstable arousal regulation). The implication of brain arousal regulation in psychiatric research has been described elsewhere (Hegerl and Hensch, 2014; Hegerl et al., 2016).

Following, it will be argued that fatigue with an unstable arousal regulation should be separated from fatigue with a stable or hyperstable arousal regulation.

3 HYPO- VS HYPERAROUSED FATIGUE

3.1 HYPOAROUSED FATIGUE IN THE CONTEXT OF IMMUNOLOGICAL AND INFLAMMATORY PROCESSES

Fatigue is a common symptom; in the general population, the prevalence is 20% (Kroenke and Price, 1993) and in the context of disorders with inflammatory and immunological processes, the prevalence of fatigue increases dramatically (up to 80%) in specific disorders (Kroenke et al., 1999). The definition of fatigue varies with the underlying condition; eg, cancer-related fatigue is defined as "perception of unusual tiredness that varies in pattern or severity and that has a negative impact on the ability to function in people who have or have had cancer" (Barsevick et al., 2010). In MS, fatigue has been defined as "a subjective lack of physical and/or mental energy that is perceived by the individual or caregiver to interfere with usual and desired activities" (Multiple Sclerosis Council for Clinical Practice, 1998). A lack of drive (eg, loss of appetite, social withdrawal, psychomotor slowing) is often found in this type of fatigue. This lack of drive has been interpreted as an autoregulatory attempt to allow for recovery in conditions of chronic disease (sickness behavior), since it prevents the organism from overexpenditure of resources and encourages healing (Hart, 1988). Questions remain as to how these behavioral changes are mediated and as to what is the pathophysiology of fatigue in these conditions. Several models have been put forward (among others):

- a brainstem fatigue generator model from viral damage to dopaminergic pathways and the ascending reticular activating system and the brainstem (Bruno et al., 1998);
- a neural model of central fatigue on the basis of an integration failure within the basal ganglia affecting the striatal–thalamic–frontal cortical system (Chaudhuri and Behan, 2000);
- a chronic stress model (hypocortisolism from an overactivity of the hypothalamic–pituitary–adrenal axis and subsequent downregulation of the corticotropin-releasing factor; Chaudhuri and Behan, 2004; Fries et al., 2005);
- an inflammatory response model with a suggested upregulation of proinflammatory cytokines (eg, interleukin (IL)-1, IL-6, tumor necrosis factor alpha (TNF-α); for review, see Dantzer et al., 2014; Harrington, 2012).

There is converging evidence that inflammation plays a key role in the pathophysiology of chronic fatigue (Dantzer et al., 2014; Morris et al., 2016). In several medical conditions, associations between proinflammatory cytokines IL-1, IL-6, TNF-α, and fatigue (Bower and Lamkin, 2013; Heesen et al., 2006; Miller et al., 2008) and TNF-α and sleepiness were found (Krueger et al., 2011). Importantly, proinflammatory cytokines have an influence on the arousal and sleep/wake regulation and have shown to be sleep inducing (Imeri and Opp, 2009; Inui, 2001; Krueger et al., 1990, 2011). Proinflammatory cytokines have also been associated with reduced motivation and locomotor activity in animal studies (Bonsall et al., 2015; Harrington, 2012; Mccusker and Kelley, 2013).

3.1.1 Clinical examples that support the model of hypoaroused fatigue in the context of immunological and inflammatory processes

Cancer-related fatigue has been associated with excessive daytime sleepiness; in a study by Forsythe et al. (2012), long-term cancer survivors ($n = 1171$) were more likely to report excessive daytime sleepiness than healthy controls ($n = 250$). Cancer-related fatigue has also been associated with sleep-inducing proinflammatory cytokines; for example, in a prospective study involving 46 cancer patients, significant associations between IL-6 and fatigue scores (assessed by the Multidimensional Fatigue Inventory) were found (Xiao et al., 2016). In addition, several studies reported underactivation of the hypothalamo–pituitary–adrenal axis in the context of fatigue: after a standardized laboratory stressor, a blunted cortisol response was found in fatigued breast cancer survivors in a study involving 27 participants (Bower et al., 2005). In line with this, altered stress reactivity (assessed with diurnal reactive cortisol profiles) was observed in breast cancer survivors displaying a relatively flat profile following the acute stress induction in comparison to controls (Couture-Lalande et al., 2014). Further, using VIGALL, an unstable brain arousal regulation was found in 22 patients with cancer-related fatigue in comparison to healthy controls (Olbrich et al., 2012).

Poststroke fatigue has consistently been associated with excessive daytime sleepiness, assessed with various self-rating scales (eg, Epworth Sleepiness Scale) and objective measures (for review, see Ding et al., 2016). Using the MSLT to objectively measure excessive daytime sleepiness, several studies found short sleep latencies (ranging from 0.5 to 4 min) to sleep stage 1 (Bassetti et al., 1996; Khairkar and Diwan, 2012; Scammell et al., 2001). Poststroke fatigue has also been associated with sleep-inducing proinflammatory cytokines: for example, acute serum levels of sleep-inducing IL-1β were positively correlated with fatigue severity (assessed by Fatigue Severity Scale) at 6 months after the stroke, whereas acute serum levels of antiinflammatory cytokines IL-ra and IL-9 were negatively correlated with the fatigue score at 12 months after the stroke (Ormstad et al., 2011).

Fatigue in MS has repeatedly been reported to cooccur with excessive daytime sleepiness. In a study by Stanton et al. (2006) with 60 participants fatigue and excessive daytime sleepiness were both common symptoms (64% and 32%). In another study involving 32 MS patients, 47% reported hypersomnia on the Epworth Sleepiness Scale and 44% met laboratory criteria for hypersomnia with a sleep latency $\leq$8 min in the MSLT (Sater et al., 2015). Consistently, an upregulation of proinflammatory cytokines has been reported in MS-related fatigue: for example, TNF-α, but not wake-promoting IL-10 and interferon gamma (assessed by cytokine mRNA expression), has been associated with MS-related fatigue (assessed by Fatigue Severity Scale) in a study involving 37 MS patients (Flachenecker et al., 2004).

3.2 HYPERAROUSED FATIGUE IN THE CONTEXT OF DEPRESSION

Major depressive disorder is a highly prevalent recurrent or chronic illness (1-year prevalence: females 8.2%, males 4.8%; Center for Behavioral Health Statistics and Quality, 2015) which is associated with a reduction in life expectancy of about 10 years. Besides depressed mood and anhedonia, fatigue is one of the core symptoms of depression according to ICD-10 and further specified as the reduction of energy. In German translations of this core item, *erhöhte Müdigkeit* (increased fatigue, tiredness) is further specified as *verminderter Antrieb* (reduced drive). In both languages, the terms lack the semantic clarity as discussed earlier.

The neurobiological mechanisms of fatigue in depression are not fully understood (Fava et al., 2014). A dysregulation of several neurochemical systems and circuits have been implicated, including the noradrenergic, dopaminergic, acetylcholinergic, and histaminergic systems (Demyttenaere et al., 2005; Stahl et al., 2003) and more recently, immune-inflammatory pathways have also been proposed (Morris et al., 2016). A large body of work has focused on hyperactivity of the noradrenergic systems in depression (reviewed in Hegerl and Hensch, 2014). In acute depression, hyperactivity of the noradrenergic locus coeruleus (LC) has been suggested as a pathogenetic factor of behavioral inhibition, retardation, and anhedonia (Stone et al., 2011; Weinshenker and Holmes, 2015). LC neurons are major projection sites of orexin/hypocretin producing neurons which are implicated in the control of reward/feeding, wakefulness, and energy homeostasis (for review, see Sakurai, 2014). Further, the LC plays a central role in the control of arousal (Berridge, 2008; Samuels and Szabadi, 2008).

The fatigue in depression is not associated with excessive daytime sleepiness or reduced sleep onset latencies. On the contrary, typical major depression is a state with an upregulated brain arousal. This is supported by many clinical and neurophysiological data. Patients with major depressive disorder have difficulties falling asleep and maintaining sleep (Mendlewicz, 2009; Tsuno et al., 2005). The upregulation of arousal is also found during the daytime: although feeling exhausted, and despite disturbed nighttime sleep, depressive patients have prolonged sleep latencies in the MSLT in comparison to controls (Kayumov et al., 2000). Concerning subjective experiences, patients with major depression typically report high inner tension and report feelings like "as being before an exam." Many other autonomic and neurobiological parameters support the model of an upregulation of arousal in depression. Increases in heart rate, muscle tone, or skin conductance (Carney et al., 2005) and upregulations of the stress hormone system (Pariante and Lightman, 2008) are typically found in patients with major depression.

Recently, further evidence has been provided by EEG analyses: applying VIGALL, a hyperstable regulation of brain arousal has been shown in unmedicated depressed patients compared with healthy controls (Hegerl et al., 2012). As shown in Fig. 3,

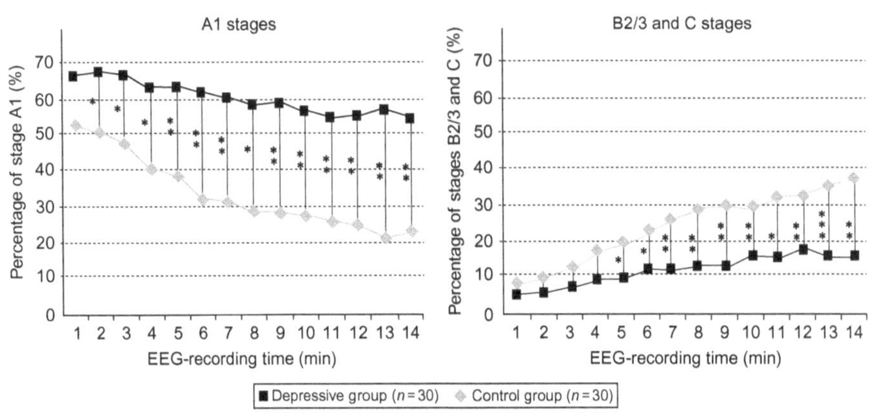

FIG. 3

Time course of EEG-vigilance stage A1 (*left side*) and EEG-vigilance stages B2/3 and C (*right side*) in depressive patients compared to healthy controls. Differences between depressives and controls were tested by Mann–Whitney *U*-test: $^{*}p<0.05$; $^{**}p<0.01$; $^{***}p<0.001$.

Figure constructed according to Hegerl, U., Wilk, K., Olbrich, S., Schoenknecht, P., Sander, C., 2012. Hyperstable regulation of vigilance in patients with major depressive disorder. World J. Biol. Psychiatry 13, 436–446.

patients remain in a more stable manner in stage A1 (indicating high arousal) and show less transitions to drowsiness and sleep onset (B2/3 and C stages) compared to healthy controls. Table 1 summarizes the proposed distinction between hypo- and hyperaroused fatigue and the associated features.

3.3 HYPO- VS HYPERAROUSED FATIGUE: CLINICAL RELEVANCE

With the fundamental role of arousal in all brain functions in mind, it appears unlikely that patients with hypoaroused fatigue will respond to treatment in the same manner as those with hyperaroused fatigue. Patients with hypoaroused fatigue might benefit from the arousal-stabilizing effects of psychostimulants. Indeed, in cancer patients, there is some evidence that psychostimulants may have positive effects. Specific effects on chemotherapy-related fatigue have been demonstrated in a phase III randomized placebo-controlled double-blind clinical trial, whereby patients with severe baseline fatigue ($n=458$) benefited from modafinil while patients with mild or moderate fatigue ($n=173$) did not (Jean-Pierre et al., 2010). In line with this, in a phase II placebo-controlled randomized trial assessing the feasibility of armodafinil in 54 cancer patients receiving brain radiation therapy, those with greater baseline fatigue experienced improved quality of life and reduced fatigue when using armodafinil (Page et al., 2015). Further, in a randomized controlled trial assessing the

Table 1 Proposed Features of Hypo- and Hyperaroused Fatigue

	Hypoaroused Fatigue (eg, in Disorders Involving Inflammatory and Immunological Processes)	**Hyperaroused Fatigue (eg, in Major Depressive Disorder)**
Drive	Lack of drive/anergic state (Barsevick et al., 2010)	Inhibition of drive/retardation (Stone et al., 2011)
Daytime wakefulness	Excessive daytime sleepiness (Forsythe et al., 2012), short sleep latency in the Multiple Sleep Latency Test (Ding et al., 2016)	Exhausted, long sleep latency in the Multiple Sleep Latency Test (Kayumov et al., 2000)
Brain arousal regulation measured with EEG	Unstable (Olbrich et al., 2012)	Hyperstable (Hegerl et al., 2012)
Hypothalamic–pituitary–adrenal axis activity	Blunted (Bower et al., 2005)	Increased (Pariante and Lightman, 2008)
Positive treatment response	Modafinil in cancer patients with higher baseline fatigue (Jean-Pierre et al., 2010; Page et al., 2015), dexmethylphenidate in cancer patients with chemotherapy-related fatigue (Lower et al., 2009)	Antidepressants (Shen et al., 2011)

benefits of dexmethylphenidate in 152 cancer patients with chemotherapy-related fatigue, significant differences in fatigue scores were demonstrated between treatment and placebo groups (Lower et al., 2009). In summary, there is some evidence for positive effects of psychostimulants in cancer- and cancer treatment-related fatigue; however, many studies are hampered by small sample sizes (for review, see Minton et al., 2011).

Since major depression is associated with a hyperstable arousal regulation (hyperaroused fatigue), psychostimulants are unlikely to be helpful. Indeed, there is no evidence for specific antidepressant effects of psychostimulants as monotherapy or add-on in typical uni- and bipolar depression. In fact several recent clinical trials and research programs with stimulants in depression have been stopped due to lack of efficacy (for review, see Hegerl and Hensch, 2014; Hegerl et al., 2016). For major depression, first-line treatments are therefore antidepressants. It is of interest that preclinical studies show consistently that antidepressants reduce the activity and firing rate of neurons in the noradrenergic LC and counteract the upregulated arousal regulation in typical depression (West et al., 2009). The same mechanisms may explain the antidepressant effect of therapeutic sleep deprivation which is unlikely to work in hypoaroused fatigue.

4 PRACTICAL CLINICAL PROCEDURE

For clinicians, it is important to clarify whether motivation problems with fatigue occur in the context of hypo- or hyperaroused fatigue. In hypoaroused fatigue, a lack of motivation results from apathy, sleepiness, and lack of drive. In hyperaroused fatigue that typically occurs in the context of depression, a lack of motivation results from exhaustion, inhibition of drive and ambivalence, making it difficult for the patient to develop goal-directed behavior.

There is a high prevalence of major depression in the context of inflammatory or immunological processes. In 10% or more of these patients, fatigue is likely to occur. In this case, the treatment would be different from that in hypoaroused fatigue.

How can clinicians separate hypoaroused fatigue from major depression? The following criteria, based mainly on clinical judgment, point to the presence of major depression and are not typically found in patients with hypoaroused fatigue:

- feelings of guilt
- profound anhedonia
- emotional numbness
- high inner tension (as being before an exam)
- difficulties relaxing and falling asleep
- previous depressive episodes
- suicidal tendencies
- delusional depression
- a family history of affective disorders

In bipolar affective disorders, previous manic or hypomanic episodes are helpful in classifying the present state of fatigue as a symptom of a depressive episode. In the case of depression, care has to be taken not to overlook an acute risk of suicide. An active exploration is necessary for every patient with a relevant depressive syndrome.

5 SUMMARY

Several lines of evidence support the distinction between hypo- and hyperaroused fatigue, the latter in most cases corresponding to a depressive disorder. To disentangle patients with these different forms of fatigue when recruiting patients for treatment studies will increase the pathophysiological homogeneity of included patients and is likely to increase the chance for relevant findings. For clinicians, it is obligatory to check for the presence of a depressive disorder in patients with fatigue and to treat this severe and often life-threatening disorder according to guidelines, if present.

ACKNOWLEDGMENT

This publication was supported within the framework of the cooperation between the German Depression Foundation and the Deutsche Bahn Stiftung gGmbH.

REFERENCES

Afari, N., Buchwald, D., 2003. Chronic fatigue syndrome: a review. Am. J. Psychiatry 160, 221–236.

Barsevick, A., Frost, M., Zwinderman, A., Hall, P., Halyard, M., G. Consortium, 2010. I'm so tired: biological and genetic mechanisms of cancer-related fatigue. Qual. Life Res. 19, 1419–1427.

Bassetti, C., Mathis, J., Gugger, M., Lovblad, K.O., Hess, C.W., 1996. Hypersomnia following paramedian thalamic stroke: a report of 12 patients. Ann. Neurol. 39, 471–480.

Bekhtereva, V., Sander, C., Forschack, N., Olbrich, S., Hegerl, U., Muller, M.M., 2014. Effects of EEG-vigilance regulation patterns on early perceptual processes in human visual cortex. Clin. Neurophysiol. 125, 98–107.

Berridge, C.W., 2008. Noradrenergic modulation of arousal. Brain Res. Rev. 58, 1–17.

Berridge, C.W., Arnsten, A.F., 2013. Psychostimulants and motivated behavior: arousal and cognition. Neurosci. Biobehav. Rev. 37, 1976–1984.

Berridge, C.W., Schmeichel, B.E., España, R.A., 2012. Noradrenergic modulation of wakefulness/arousal. Sleep Med. Rev. 16, 187–197.

Bonsall, D.R., Kim, H., Tocci, C., Ndiaye, A., Petronzio, A., Mckay-Corkum, G., Molyneux, P.C., Scammell, T.E., Harrington, M.E., 2015. Suppression of locomotor activity in female C57Bl/6J mice treated with interleukin-1β: investigating a method for the study of fatigue in laboratory animals. PLoS One 10, e0140678.

Bower, J.E., 2007. Cancer-related fatigue: links with inflammation in cancer patients and survivors. Brain Behav. Immun. 21, 863–871.

Bower, J.E., Lamkin, D.M., 2013. Inflammation and cancer-related fatigue: mechanisms, contributing factors, and treatment implications. Brain Behav. Immun. 30, S48–S57.

Bower, J.E., Ganz, P.A., Aziz, N., 2005. Altered cortisol response to psychologic stress in breast cancer survivors with persistent fatigue. Psychosom. Med. 67, 277–280.

Brown, R.E., Basheer, R., Mckenna, J.T., Strecker, R.E., Mccarley, R.W., 2012. Control of sleep and wakefulness. Physiol. Rev. 92, 1087–1187.

Bruno, R.L., Creange, S.J., Frick, N.M., 1998. Parallels between post-polio fatigue and chronic fatigue syndrome: a common pathophysiology? Am. J. Med. 105, 66S–73S.

Carney, R.M., Freedland, K.E., Veith, R.C., 2005. Depression, the autonomic nervous system, and coronary heart disease. Psychosom. Med. 67, S29–S33.

Carskadon, M.A., Dement, W.C., Mitler, M.M., Roth, T., Westbrook, P.R., Keenan, S., 1986. Guidelines for the multiple sleep latency test (MSLT): a standard measure of sleepiness. Sleep 9, 519–524.

Center for Behavioral Health Statistics and Quality, 2015. Behavioral Health Trends in the United States: Results from the 2014 National Survey on Drug Use and Health [Online]. SAMHSA, Rockville, MD. Available, http://www.samhsa.gov/data/. accessed 16.2.2016.

Chaudhuri, A., Behan, P.O., 2000. Fatigue and basal ganglia. J. Neurol. Sci. 179, 34–42.

Chaudhuri, A., Behan, P.O., 2004. Fatigue in neurological disorders. Lancet 363, 978–988.

Couture-Lalande, M.-È., Lebel, S., Bielajew, C., 2014. Analysis of the cortisol diurnal rhythmicity and cortisol reactivity in long-term breast cancer survivors. Breast Cancer Manag. 3, 465–476.

Dantzer, R., Heijnen, C.J., Kavelaars, A., Laye, S., Capuron, L., 2014. The neuroimmune basis of fatigue. Trends Neurosci. 37, 39–46.

Demyttenaere, K., De Fruyt, J., Stahl, S.M., 2005. The many faces of fatigue in major depressive disorder. Int. J. Neuropsychopharmacol. 8, 93–105.

Ding, Q., Whittemore, R., Redeker, N., 2016. Excessive daytime sleepiness in stroke survivors: an integrative review. Biol. Res. Nurs. 18, 420–431.

Fava, M., Ball, S., Nelson, J.C., Sparks, J., Konechnik, T., Classi, P., Dube, S., Thase, M.E., 2014. Clinical relevance of fatigue as a residual symptom in major depressive disorder. Depress. Anxiety 31, 250–257.

Flachenecker, P., Bihler, I., Weber, F., Gottschalk, M., Toyka, K.V., Rieckmann, P., 2004. Cytokine mRNA expression in patients with multiple sclerosis and fatigue. Mult. Scler. 10, 165–169.

Forsythe, L.P., Helzlsouer, K.J., Macdonald, R., Gallicchio, L., 2012. Daytime sleepiness and sleep duration in long-term cancer survivors and non-cancer controls: results from a registry-based survey study. Support. Care Cancer 20, 2425–2432.

Fries, E., Hesse, J., Hellhammer, J., Hellhammer, D.H., 2005. A new view on hypocortisolism. Psychoneuroendocrinology 30, 1010–1016.

Guenther, T., Schonknecht, P., Becker, G., Olbrich, S., Sander, C., Hesse, S., Meyer, P.M., Luthardt, J., Hegerl, U., Sabri, O., 2011. Impact of EEG-vigilance on brain glucose uptake measured with [(18)F]FDG and PET in patients with depressive episode or mild cognitive impairment. NeuroImage 56, 93–101.

Harrington, M.E., 2012. Neurobiological studies of fatigue. Prog. Neurobiol. 99, 93–105.

Hart, B.L., 1988. Biological basis of the behavior of sick animals. Neurosci. Biobehav. Rev. 12, 123–137.

Heesen, C., Nawrath, L., Reich, C., Bauer, N., Schulz, K.H., Gold, S.M., 2006. Fatigue in multiple sclerosis: an example of cytokine mediated sickness behaviour? J. Neurol. Neurosurg. Psychiatry 77, 34–39.

Hegerl, U., Hensch, T., 2014. The vigilance regulation model of affective disorders and ADHD. Neurosci. Biobehav. Rev. 44, 45–57.

Hegerl, U., Wilk, K., Olbrich, S., Schoenknecht, P., Sander, C., 2012. Hyperstable regulation of vigilance in patients with major depressive disorder. World J. Biol. Psychiatry 13, 436–446.

Hegerl, U., Lam, R.W., Malhi, G.S., Mcintyre, R.S., Demyttenaere, K., Mergl, R., Gorwood, P., 2013. Conceptualising the neurobiology of fatigue. Aust. N. Z. J. Psychiatry 47, 312–316.

Hegerl, U., Sander, C., Hensch, T., 2016. Arousal regulation in affective disorders. In: Frodl, T. (Ed.), Systems Neuroscience in Depression. Elsevier, San Diego, pp. 341–370.

Huang, J., Sander, C., Jawinski, P., Ulke, C., Spada, J., Hegerl, U., Hensch, T., 2015. Test-retest reliability of brain arousal regulation as assessed with VIGALL 2.0. Neuropsychiatr. Electrophysiol. 1, 1–13.

Hybels, C.F., Steffens, D.C., Mcquoid, D.R., Rama Krishnan, K.R., 2005. Residual symptoms in older patients treated for major depression. Int. J. Geriatr. Psychiatry 20, 1196–1202.

Imeri, L., Opp, M.R., 2009. How (and why) the immune system makes us sleep. Nat. Rev. Neurosci. 10, 199–210.

Insel, T., Cuthbert, B., Garvey, M., Heinssen, R., Pine, D.S., Quinn, K., Sanislow, C., Wang, P., 2010. Research domain criteria (RDoC): toward a new classification framework for research on mental disorders. Am. J. Psychiatr. 167, 748–751.

Inui, A., 2001. Cytokines and sickness behavior: implications from knockout animal models. Trends Immunol. 22, 469–473.

Jean-Pierre, P., Morrow, G.R., Roscoe, J.A., Heckler, C., Mohile, S., Janelsins, M., Peppone, L., Hemstad, A., Esparaz, B.T., Hopkins, J.O., 2010. A phase 3 randomized, placebo-controlled, double-blind, clinical trial of the effect of modafinil on cancer-related fatigue among 631 patients receiving chemotherapy: a University of Rochester Cancer Center Community Clinical Oncology Program Research base study. Cancer 116, 3513–3520.

Kayumov, L., Rotenberg, V., Buttoo, K., Auch, C., Pandi-Perumal, S.R., Shapiro, C.M., 2000. Interrelationships between nocturnal sleep, daytime alertness, and sleepiness: two types of alertness proposed. J. Neuropsychiatry Clin. Neurosci. 12, 86–90.

Khairkar, P., Diwan, S., 2012. Late-onset obsessive-compulsive disorder with comorbid narcolepsy after perfect blend of thalamo-striatal stroke and post-streptococcal infection. J. Neuropsychiatry Clin. Neurosci. 24, E29–E31.

Kroenke, K., Price, R.K., 1993. Symptoms in the community: prevalence, classification, and psychiatric comorbidity. Arch. Intern. Med. 153, 2474–2480.

Kroenke, K., Stump, T., Clark, D.O., Callahan, C.M., Mcdonald, C.J., 1999. Symptoms in hospitalized patients: outcome and satisfaction with care. Am. J. Med. 107, 425–431.

Krueger, J.M., 2008. The role of cytokines in sleep regulation. Curr. Pharm. Des. 14, 3408.

Krueger, J.M., Obal Jr., F., Opp, M., Toth, L., Johannsen, L., Cady, A.B., 1990. Somnogenic cytokines and models concerning their effects on sleep. Yale J. Biol. Med. 63, 157–172.

Krueger, J.M., Majde, J.A., Rector, D.M., 2011. Cytokines in immune function and sleep regulation. Handb. Clin. Neurol. 98, 229–240.

Kuppuswamy, A., Rothwell, J., Ward, N., 2015. A model of poststroke fatigue based on sensorimotor deficits. Curr. Opin. Neurol. 28, 582–586.

Lower, E.E., Fleishman, S., Cooper, A., Zeldis, J., Faleck, H., Yu, Z., Manning, D., 2009. Efficacy of dexmethylphenidate for the treatment of fatigue after cancer chemotherapy: a randomized clinical trial. J. Pain Symptom Manage. 38, 650–662.

Malekzadeh, A., Van De Geer-Peeters, W., De Groot, V., Teunissen, C.E., Beckerman, H., TREFAMS-ACE Study Group, 2015. Fatigue in patients with multiple sclerosis: is it related to pro- and anti-inflammatory cytokines? Dis. Markers 2015, 758314.

Mccusker, R.H., Kelley, K.W., 2013. Immune-neural connections: how the immune system's response to infectious agents influences behavior. J. Exp. Biol. 216, 84–98.

Mendlewicz, J., 2009. Sleep disturbances: core symptoms of major depressive disorder rather than associated or comorbid disorders. World J. Biol. Psychiatry 10, 269–275.

Miller, A.H., Ancoli-Israel, S., Bower, J.E., Capuron, L., Irwin, M.R., 2008. Neuroendocrine-immune mechanisms of behavioral comorbidities in patients with cancer. J. Clin. Oncol. 26, 971–982.

Minkwitz, J., Trenner, M.U., Sander, C., Olbrich, S., Sheldrick, A.J., Schonknecht, P., Hegerl, U., Himmerich, H., 2011. Prestimulus vigilance predicts response speed in an easy visual discrimination task. Behav. Brain Funct. 7, 31.

Minton, O., Richardson, A., Sharpe, M., Hotopf, M., Stone, P.C., 2011. Psychostimulants for the management of cancer-related fatigue: a systematic review and meta-analysis. J. Pain Symptom Manag. 41, 761–767.

Morris, G., Berk, M., Galecki, P., Walder, K., Maes, M., 2016. The neuro-immune pathophysiology of central and peripheral fatigue in systemic immune-inflammatory and neuro-immune diseases. Mol. Neurobiol. 53, 1195–1219.

Multiple Sclerosis Council for Clinical Practice, 1998. Fatigue and multiple sclerosis: evidence-based management strategies for fatigue in multiple sclerosis. Multiple Sclerosis Council for Clinical Practice Guidelines.

NIMH, 2012. Arousal and Regulatory Systems: Workshop Proceedings [Online]. National Institute of Mental Health, Bethesda, MD. Available, http://www.nimh.nih.gov/research-priorities/rdoc/arousal-and-regulatory-systems-workshop-proceedings.shtml. accessed 14.2.2016.

Olbrich, S., Mulert, C., Karch, S., Trenner, M., Leicht, G., Pogarell, O., Hegerl, U., 2009. EEG-vigilance and BOLD effect during simultaneous EEG/fMRI measurement. NeuroImage 45, 319–332.

Olbrich, S., Sander, C., Matschinger, H., Mergl, R., Trenner, M., Schönknecht, P., Hegerl, U., 2011. Brain and body. J. Psychophysiol. 25, 190–200.

Olbrich, S., Sander, C., Jahn, I., Eplinius, F., Claus, S., Mergl, R., Schonknecht, P., Hegerl, U., 2012. Unstable EEG-vigilance in patients with cancer-related fatigue (CRF) in comparison to healthy controls. World J. Biol. Psychiatry 13, 146–152.

Olbrich, S., Fischer, M.M., Sander, C., Hegerl, U., Wirtz, H., Bosse-Henck, A., 2015. Objective markers for sleep propensity: comparison between the Multiple Sleep Latency Test and the Vigilance Algorithm Leipzig. J. Sleep Res. 24, 450–457.

Ormstad, H., Aass, H.C.D., Amthor, K.-F., Lund-Sørensen, N., Sandvik, L., 2011. Serum cytokine and glucose levels as predictors of poststroke fatigue in acute ischemic stroke patients. J. Neurol. 258, 670–676.

Page, B.R., Shaw, E.G., Lu, L., Bryant, D., Grisell, D., Lesser, G.J., Monitto, D.C., Naughton, M.J., Rapp, S.R., Savona, S.R., 2015. Phase II double-blind placebo-controlled randomized study of armodafinil for brain radiation-induced fatigue. Neuro-Oncology 17, 1393–1401.

Pariante, C.M., Lightman, S.L., 2008. The HPA axis in major depression: classical theories and new developments. Trends Neurosci. 31, 464–468.

Pfaff, D., Ribeiro, A., Matthews, J., Kow, L.M., 2008. Concepts and mechanisms of generalized central nervous system arousal. Ann. N. Y. Acad. Sci. 1129, 11–25.

Pfaff, D.W., Martin, E.M., Faber, D., 2012. Origins of arousal: roles for medullary reticular neurons. Trends Neurosci. 35, 468–476.

Pigeon, W.R., Sateia, M.J., Ferguson, R.J., 2003. Distinguishing between excessive daytime sleepiness and fatigue: toward improved detection and treatment. J. Psychosom. Res. 54, 61–69.

Plum, F., Posner, J.B., 1982. The Diagnosis of Stupor and Coma. Oxford University Press, USA.

Ribeiro, A.C., Sawa, E., Carren-Lesauter, I., Lesauter, J., Silver, R., Pfaff, D.W., 2007. Two forces for arousal: pitting hunger versus circadian influences and identifying neurons responsible for changes in behavioral arousal. Proc. Natl. Acad. Sci. U.S.A. 104, 20078–20083.

Rocha, N.P., De Miranda, A.S., Teixeira, A.L., 2015. Insights into neuroinflammation in Parkinson's disease: from biomarkers to anti-inflammatory based therapies. BioMed Res. Int. 2015, 628192.

Sakurai, T., 2014. The role of orexin in motivated behaviours. Nat. Rev. Neurosci. 15, 719–731.

Samuels, E., Szabadi, E., 2008. Functional neuroanatomy of the noradrenergic locus coeruleus: its roles in the regulation of arousal and autonomic function part I: principles of functional organisation. Curr. Neuropharmacol. 6, 235–253.

Sander, C., Hensch, T., Wittekind, D.A., Bottger, D., Hegerl, U., 2015. Assessment of wakefulness and brain arousal regulation in psychiatric research. Neuropsychobiology 72, 195–205.

Sara, S.J., Bouret, S., 2012. Orienting and reorienting: the locus coeruleus mediates cognition through arousal. Neuron 76, 130–141.

Sater, R., Gudesblatt, M., Kresa-Reahl, K., Brandes, D., Sater, P., 2015. The relationship between objective parameters of sleep and measures of fatigue, depression, and cognition in multiple sclerosis. Mult. Scler. J. Exp, Transl. Clin. 1, 1–8.

Scammell, T., Nishino, S., Mignot, E., Saper, C., 2001. Narcolepsy and low CSF orexin (hypocretin) concentration after a diencephalic stroke. Neurology 56, 1751–1753.

Shen, J., Hossain, N., Streiner, D.L., Ravindran, A.V., Wang, X., Deb, P., Huang, X., Sun, F., Shapiro, C.M., 2011. Excessive daytime sleepiness and fatigue in depressed patients and therapeutic response of a sedating antidepressant. J. Affect. Disord. 134, 421–426.

Stahl, S.M., Zhang, L., Damatarca, C., Grady, M., 2003. Brain circuits determine destiny in depression: a novel approach to the psychopharmacology of wakefulness, fatigue, and executive dysfunction in major depressive disorder. J. Clin. Psychiatry 64 (Suppl. 14), 6–17.

Stanton, B., Barnes, F., Silber, E., 2006. Sleep and fatigue in multiple sclerosis. Mult. Scler. 12, 481–486.

Stone, E.A., Lin, Y., Sarfraz, Y., Quartermain, D., 2011. The role of the central noradrenergic system in behavioral inhibition. Brain Res. Rev. 67, 193–208.

Tsuno, N., Besset, A., Ritchie, K., 2005. Sleep and depression. J. Clin. Psychiatry 66, 1254–1269.

Vaccarino, A.L., Sills, T.L., Evans, K.R., Kalali, A.H., 2008. Prevalence and association of somatic symptoms in patients with major depressive disorder. J. Affect. Disord. 110, 270–276.

Van Steenbergen, H.W., Tsonaka, R., Huizinga, T.W., Boonen, A., Van Der Helm-Van Mil, A.H., 2015. Fatigue in rheumatoid arthritis; a persistent problem: a large longitudinal study. RMD Open 1, e000041.

Weinshenker, D., Holmes, P.V., 2015. Regulation of neurological and neuropsychiatric phenotypes by locus coeruleus-derived galanin. Brain Res. 1641, 320–337.

West, C.H., Ritchie, J.C., Boss-Williams, K.A., Weiss, J.M., 2009. Antidepressant drugs with differing pharmacological actions decrease activity of locus coeruleus neurons. Int. J. Neuropsychopharmacol. 12, 627–641.

Xiao, C., Beitler, J.J., Higgins, K.A., Conneely, K., Dwivedi, B., Felger, J., Wommack, E.C., Shin, D.M., Saba, N.F., Ong, L.Y., 2016. Fatigue is associated with inflammation in patients with head and neck cancer before and after intensity-modulated radiation therapy. Brain Behav. Immun. 52, 145–152.

Yerkes, R.M., Dodson, J.D., 1908. The relation of strength of stimulus to rapidity of habit-formation. J. Comp. Neurol. Psychol. 18, 459–482.

Zitnik, G.A., 2015. Control of arousal through neuropeptide afferents of the locus coeruleus. Brain Res. 1641, 338–350.

SECTION

4

Targeting Motivation: Novel Approaches to Enhancing Motivation, Activity and Learning in Health and Disease

CHAPTER

Intrinsic motivation, curiosity, and learning: Theory and applications in educational technologies

11

P.-Y. Oudeyer*,[1], J. Gottlieb[†], M. Lopes*

Inria and Ensta ParisTech, Paris, France

[†]*Kavli Institute for Brain Science, Columbia University, New York, NY, United States*

[1]*Corresponding author: Tel.: +33-5-24574030, e-mail address: pierre-yves.oudeyer@inria.fr*

Abstract

This chapter studies the bidirectional causal interactions between curiosity and learning and discusses how understanding these interactions can be leveraged in educational technology applications. First, we review recent results showing how state curiosity, and more generally the experience of novelty and surprise, can enhance learning and memory retention. Then, we discuss how psychology and neuroscience have conceptualized curiosity and intrinsic motivation, studying how the brain can be intrinsically rewarded by novelty, complexity, or other measures of information. We explain how the framework of computational reinforcement learning can be used to model such mechanisms of curiosity. Then, we discuss the learning progress (LP) hypothesis, which posits a positive feedback loop between curiosity and learning. We outline experiments with robots that show how LP-driven attention and exploration can self-organize a developmental learning curriculum scaffolding efficient acquisition of multiple skills/tasks. Finally, we discuss recent work exploiting these conceptual and computational models in educational technologies, showing in particular how intelligent tutoring systems can be designed to foster curiosity and learning.

Keywords

Curiosity, Intrinsic motivation, Learning, Education, Active learning, Active teaching, Neuroscience, Computational modeling, Artificial intelligence, Educational technology

1 CURIOSITY FOSTERS LEARNING AND MEMORY RETENTION

Curiosity is a form of intrinsic motivation that is key in fostering active learning and spontaneous exploration. For this reason, curiosity-driven learning and intrinsic motivation have been argued to be fundamental ingredients for efficient education

Progress in Brain Research, Volume 229, ISSN 0079-6123, http://dx.doi.org/10.1016/bs.pbr.2016.05.005

(Freeman et al., 2014). Thus, elaborating a fundamental understanding of the mechanisms of curiosity, and of which features of educational activities can make them "fun" and foster motivation, is of high importance with regard to the educational challenges of the 21st century.

While there is not yet a scientific consensus on how to define curiosity operationally (Gottlieb et al., 2013; Kidd and Hayden, 2015; Oudeyer and Kaplan, 2007), states of curiosity are often associated with a psychological interest for activities or stimuli that are surprising, novel, of intermediate complexity, or characterized by a knowledge gap or by errors in prediction, which are features that can themselves be quantified mathematically (Barto et al., 2013; Oudeyer and Kaplan, 2007; Schmidhuber, 1991). Such informational features that attract the brain's attention have been called "collative variables" by Berlyne (1965).

Recent experimental studies in psychology and neuroscience have shown that experiencing these features improved memory retention and learning in human children and adults, in other animals, and in a variety of tasks. In a famous series of experiments with monkeys, Waelti et al. (2001) showed that monkeys could learn the predictive association between a stimuli and a reward only in situations where prediction errors happened: if the reward was anticipated by other means, then learning was blocked. This experiment complied with formal models of reinforcement learning, and in particular TD learning (Sutton and Barto, 1981), predicting that "organisms only learn when events violate their expectations" (Rescorla and Wagner, 1972, p. 75). In a study mixing behavioral analysis and brain imaging, Kang et al. (2009) showed that human adults show greater long-term memory retention for verbal material for which they had expressed high curiosity than for low-curiosity questions. They observed that before the presentation of answers to high-curiosity questions, curiosity states were correlated with higher activity in the striatum and inferior frontal cortex. When subjects observed answers that did not match their predictions (ie, an error was experienced), then an increase in activation of putamen and left inferior frontal cortex was observed. The modulation of hippocampus-dependent learning by curiosity states was confirmed in Gruber et al. (2014). Recently, Stahl and Feigenson (2015) showed that a similar phenomenon happens in infants, observing that the infants created stronger associations between sounds/words and visual objects in a context where object movements violated the expected laws of physics.

Novelty, surprise, intermediate complexity, and other related features that characterize informational properties of stimuli have not only been shown to enhance memory retention, but they have also been argued to be intrinsically rewarding, motivating organisms to actively search for them. Three strands of research developed arguments and experimental evidence in this direction. First, psychologists proposed that forms of intrinsic motivation motivate the organism to search for information and competence gain. Second, neuroscientists have shown that reward-related dopaminergic circuits can be activated by information independently of extrinsic reward, and behavioral preference for novelty can be observed in various animals (as along with the apparently inconsistent observation of neophobia). Third, theoretical computational models and their experimental tests in robots have shown how such mechanisms could function and how they can improve learning efficiency by self-organizing developmental learning trajectories. In what follows, we discuss these advances in turn, and then study how this perspective on curiosity and learning opens new directions in educational technologies.

2 CURIOSITY AND INTRINSIC MOTIVATION IN PSYCHOLOGY[1]

In psychology, curiosity can be approached within the conceptual framework of intrinsic motivation (Berlyne, 1960; Ryan and Deci, 2000). Ryan and Deci (2000) proposed a distinction of intrinsic and extrinsic motivation based on the concept of instrumentalization (p. 56):

> *Intrinsic motivation is defined as the doing of an activity for its inherent satisfaction rather than for some separable consequence. When intrinsically motivated, a person is moved to act for the fun or challenge entailed rather than because of external products, pressures or reward.*

Intrinsic motivation is clearly visible in young infants, who consistently try to grasp, throw, bite, squash, or shout at new objects they encounter, without any clear external pressure to do it. Although the importance of intrinsic motivation declines during development, human adults are still often intrinsically motivated to engage in activities such as crossword puzzles, painting, gardening, read novels, or watch movies. Accordingly, Ryan and Deci define extrinsic motivation as:

> *Extrinsic motivation is a construct that pertains whenever an activity is done in order to attain some separable outcome. Extrinsic motivation thus contrasts with intrinsic motivation, which refers to doing an activity simply for the enjoyment of the activity itself, rather than its instrumental value.*
>
> **Ryan and Deci (2000)**

Given this broad distinction between intrinsic and extrinsic motivation, psychologists have proposed theories about which properties of activities make them intrinsically motivating, and in particular foster curiosity as one particular form of intrinsically motivated exploration (Oudeyer and Kaplan, 2007).

2.1 DRIVES TO MANIPULATE, DRIVES TO EXPLORE

In the 1950s, psychologists attempted to give an account of intrinsic motivation and exploratory activities on the basis of the theory of drives (Hull, 1943), defined as specific tissue deficits that the organisms try to reduce, like hunger or pain. Montgomery (1954) proposed a drive for exploration and Harlow (1950) proposed

[1]Parts of the text in this section are adapted with permission from Oudeyer and Kaplan (2007).

that subjects have a drive to manipulate. This drive naming approach had shortcomings which were criticized by White (1959): intrinsically motivated exploratory activities have a fundamentally different dynamics. Indeed, they are not homeostatic: the general tendency to explore is not a consummatory response to a stressful perturbation of the organism's body.

2.2 REDUCTION OF COGNITIVE DISSONANCE

An alternative conceptualization was proposed by Festinger's theory of cognitive dissonance (Festinger, 1957), which asserted that organisms are motivated to reduce dissonance, defined as an incompatibility between internal cognitive structures and the situations currently perceived. Fifteen years later, a related view was articulated by Kagan stating that a primary motivation for humans is the reduction of uncertainty in the sense of the "incompatibility between (two or more) cognitive structures, between cognitive structure and experience, or between structures and behavior" (Kagan, 1972). More recently, the related concept of "knowledge gap" was argued to be a driver for curiosity-driven exploration (Lowenstein, 1994). However, these theories do not provide an account of certain spontaneous exploration behaviors which increase uncertainty (Gottlieb et al., 2013). Also, they do not specify whether the brain values differently or similarly different degrees of knowledge gaps.

2.3 OPTIMAL INCONGRUITY

People seem to look for situations between completely uncertain and completely certain. In 1965, Hunt developed the idea that children and adults look for optimal incongruity (Hunt, 1965). He regarded children as information-processing systems and stated that interesting stimuli were those where there was a discrepancy between the perceived and standard levels of the stimuli. For Dember and Earl, the incongruity or discrepancy in intrinsically motivated behaviors was between a person's expectations and the properties of the stimulus (Dember and Earl, 1957). Berlyne developed similar notions as he observed that the most rewarding situations were those with an intermediate level of novelty, between already familiar and completely new situations (Berlyne, 1960). This perspective was recently echoed by Kidd et al. (2012) who showed an experiment where infants preferred stimuli of intermediate complexity.

2.4 MOTIVATION FOR COMPETENCE

A last group of researchers preferred the concept of challenge to the notion of optimal incongruity. These researchers stated that what was driving human behavior was a motivation for "effectance" (White, 1959), personal causation (De Charms, 1968), competence, and self-determination (Deci and Ryan, 1985). Basically, these approaches argue that what motivates people is the degree of control they can have on other people, external objects, and themselves. An analogous concept is that of optimal challenge as put forward in the theory of "Flow" (Csikszenthmihalyi, 1991).

2.5 BERLYNE'S INFORMATIONAL APPROACH TO CURIOSITY AND INTRINSIC MOTIVATION

These diverse theoretical approaches to intrinsic motivation and the properties that render certain activities intrinsically interesting/motivating have been proposed by diverse research communities within psychology, but so far there is no consensus on a unified view of intrinsic motivation. Even more, it could be argued that distinguishing intrinsic and extrinsic motivation based on instrumentalization can be circular (Oudeyer and Kaplan, 2007). Yet, a convincing integrated noncircular view has actually been proposed in the 1960s by Berlyne (1965), and has been used as a fruitful theoretical reference for developing formal mathematical models of curiosity, as described later. The central concept of this integrated approach to intrinsic motivation is that of "collative variables," as explained in the following quotations:

> *The probability and direction of specific exploratory responses can apparently be influenced by many properties of external stimulation, as well as by many intraorganism variables. They can, no doubt, be influenced by stimulus intensity, color, pitch, and association with biological gratification and punishment, ... [but] the paramount determinants of specific exploration are, however, a group of stimulus properties to which we commonly refer by such words as "novelty," "change," "surprisingness," "incongruity," "complexity," "ambiguity," and "indistinctiveness."*
>
> **Berlyne (1965, p. 245)**

> *... these properties possess close links with the concepts of information theory, and they can, in fact, all be discussed in information-theoretic terminology. In the case of "ambiguity" and "indistinctiveness," there is uncertainty due to a gap in available information. In some forms of "novelty" and "complexity," there is uncertainty about how a pattern should be categorized, that is, what labeling responses should be attached to it and what overt response is appropriate to it. When one portion of a "complex" pattern or of a sequence of "novel" stimuli is perceived, there is uncertainty about what will be perceived next. In the case of "surprisingness" and "incongruity," there is discrepancy between information embodied in expectations and information embodied in what is perceived. For these reasons, the term "collative" is proposed as an epithet to denote all these stimulus properties collectively, since they all depend on collation or comparison of information from different stimulus elements, whether they be elements belonging to the present, past or elements that are simultaneously present in different parts of one stimulus field.*

> *It should be pointed out that the uncertainty we are discussing here is "subjective uncertainty," which is a function of subjective probabilities, analogous to the "objective" uncertainty (that is, the standard information-theoretic concept of uncertainty) that is a function of objective probabilities.*
>
> **Berlyne (1965, pp. 245–246)**

As these psychological theories of curiosity and intrinsic motivation hypothesize that the brain could be intrinsically rewarded by experiencing information gain, novelty, or complexity, a natural question that follows is whether one could identify actual neural circuitry linking the detection of novelty with the brain reward system. We now review several strands of research that identified several dimensions of this connection.

3 INFORMATION AS A REWARD IN NEUROSCIENCE

3.1 DOPAMINERGIC SYSTEMS THAT PROCESS PRIMARY REWARDS ARE ACTIVATED BY CURIOSITY

To examine the motivational systems that are recruited by curiosity, Kang et al. used functional magnetic resonance imaging to monitor brain activity in human observers who pondered trivia questions (Kang et al., 2009). After reading a question subjects rated their curiosity and confidence regarding the question and, after a brief delay, were given the answer. The key analyses focused on activations during the *anticipatory* period—after the subjects had received the question but before they were given the answer.

Areas that showed activity related to curiosity ratings during this epoch included the left caudate nucleus, bilateral inferior frontal gyrus, and loci in the putamen and globus pallidus. In an additional behavioral task, the authors showed that subjects were willing to pay a higher price to obtain the answers to questions that they were more curious about—ie, could compare money and information on a common scale. They concluded that the value of the information, reported by subjects as a feeling of curiosity, is encoded in some of the same structures that evaluate material gains.

Two recent studies extend this result, and report that midbrain dopaminergic (DA) cells and cells in the orbitofrontal cortex (OFC), a prefrontal area that receives DA innervation, encode the anticipation of obtaining reliable information from visual cues (Blanchard et al., 2015; Bromberg-Martin and Hikosaka, 2009). In that study on DA cells, monkeys were trained on so-called observing paradigms, where they had to choose between observing two cues that had equal physical rewards but differed in their offers of information (Bromberg-Martin and Hikosaka, 2009). Monkeys began each trial with a 50% probability of obtaining a large or a small reward and, before receiving the reward, had to choose to observe one of two visual items. If the monkeys chose the informative target, this target changed to one of two patterns that reliably predicted whether the trial will yield a large or small reward (Info). If the monkeys chose the uninformative item, this target also changed to produce one of two patterns, but the patterns had only a random relation to the reward size (Rand).

After a relatively brief experience with the task, the monkeys developed a reliable and consistent preference for choosing the informative cue. Because the extrinsic rewards that the monkeys received were equal for the two options (both targets had a 50% probability of delivering a large or small reward), this showed that monkeys were motivated by some cognitive or emotional factor that assigned intrinsic value to the predictive/informational cue.

Dopamine neurons encoded both reward prediction errors and the anticipation of reliable information. The neurons' responses to reward prediction errors confirmed previous results and arose *after* the monkeys' choice, when the selected target delivered its reward information. At this time, the neurons gave a burst of excitation if the cue signaled a large reward (a better than the average outcome) but were transiently inhibited if the cue signaled a small reward (an outcome that was worse than expected).

Responses to anticipated information gains, by contrast, arose *before* the monkeys' choice and thus could contribute to motivating that choice. Just before viewing the cue, the neurons emitted a slightly stronger excitatory response if the monkeys expected to view an informative cue and a weaker response if they expected only the random cue (red vs blue traces). This early response was clearly independent of the final outcome and seemed to encode enhanced arousal or motivation associated with the informative option.

A subsequent study of area OFC extended the behavioral results by showing that the monkeys will choose the informative option even if its payoff is slightly lower than that of the uninformative option—that is, monkeys are willing to sacrifice juice reward to view predictive cues (Blanchard et al., 2015). In addition, the study showed that responses to anticipated information gains in the OFC are carried by a neural population that is different from those that encode the value of primary rewards, suggesting differences in the underlying neural computations.

Together, these investigations show that, in both humans and monkeys, the motivational systems that signal the value of primary rewards are also activated by the desire to obtain information. This conclusion is consistent with earlier reports that DA neurons respond to novel or surprising events that are critical for learning environmental contingencies (Bromberg-Martin et al., 2010). The convergence of responses related to rewards and information gains is highly beneficial in allowing subjects to compare different types of currencies—eg, knowledge and money—on a common value scale when selecting actions. At the same time, the separation between the neural representations of information value and biological value in OFC cells highlights the fact that these two types of values require distinct computations. While the value of a primary reward depends on its biological properties (eg, its caloric content), the value of a source of information depends on semantic and epistemic factors that establish the meaning of the information.

3.2 SEEKING INFORMATION FOR ITSELF: LIKING AND WANTING NOVELTY, SURPRISE, AND INTERMEDIATE COMPLEXITY

Many animal studies have shown phenomena of neophilia. Rats prefer novel environments and objects to familiar ones (Bardo and Bevins, 2000) and learn motor strategies that allow them to trigger the appearance of novel items (Myers and Miller, 1954). In certain contexts, rats have also been shown to prefer obtaining novel stimuli over obtaining food or drug or at the cost of crossing electrifying grids (see Hughes, 2007 for a review). Moreover, brain responses to novelty in rats have strong similarities with brain responses to drug rewards (Bevins, 2001). In human adults, studies by Itti and Baldi have shown that surprise, defined in the domain of visual features, attracts human saccades during free-viewing exploration (Itti and Baldi, 2009). Baranes et al. extended this result to the epistemic domain, by showing that curiosity about trivia questions elicits faster anticipatory eye movements to the expected location of the answer, suggesting that eye movements are influenced by expected gains in semantic information (Baranes et al., 2015). Kidd et al. (2012) showed that human infants had a preference for looking at stimuli of intermediate complexity in the visual or auditory domain (Kidd et al., 2014).

Another recent study suggests that novelty also recruits attentional resources through reward-independent effects (Foley et al., 2014; Peck et al., 2009). In this experiment, monkeys were trained on a task in which they had initial uncertainty about the trial's outcome, and were given cues that resolved this uncertainty, by signaling whether the trial will end in a reward or a lack of reward. When the reward contingencies were signaled by novel visual cues (abstract patterns that the monkeys had never seen before), these cues evoked enhanced visual and orienting responses in the parietal lobe. If a novel cue signaled "bad news" (a lack of reward) the monkeys quickly learned this contingency and extinguished their anticipatory licking in response to the cues. Strikingly, however, the newly learnt cues continued to produce enhanced visual and saccadic responses for dozens of presentations after the extinction of the licking response. This suggests that novelty attracts attention through reward-independent mechanisms, allowing the brain to prioritize and learn about novel items for an extended period even if these items signal negative outcomes.

3.3 THE PUZZLE OF NEOPHOBIA

Gershman and Niv (2015) recently discussed a puzzling observation. Alongside a large experimental corpus showing neophilia in several animal species, an equally large corpus demonstrates neophobia—the avoidance of novelty (Hughes, 2007). Neophobia has been observed in rats (Blanchard et al., 1974), in adult humans (Berlyne, 1960), in infants (Weizmann et al., 1971), and in nonhuman primates (Weiskrantz and Cowey, 1963). To explain the apparent contradiction between these results, Gershman and Niv (2015) studied the hypothesis that certain kinds of novelty (characterized by their cues) can be selectively and aversively reinforced. That is, an individual may learn and generalize that, in different families of situations, novelty may be associated with positive or with negative outcomes, and thus learn to avoid novelty when their associated outcome is negative.

However, another complementary hypothesis to explain this apparent contradiction is the "intermediate novelty" hypothesis proposed by Berlyne (1960). Following this hypothesis, approach or avoidance of novelty would depend on the degree of novelty, ie, the degree of distance/similarity between the perceived stimuli and existing internal representations in the brain.

4 THE LEARNING PROGRESS HYPOTHESIS

Berlyne's concept of intermediate novelty, as well as the related concept of intermediate challenge of Csikszenthmihalyi, has the advantage of allowing intuitive explanations of many behavioral manifestations of curiosity and intrinsic motivation. However, recent developments in theories of curiosity, and in particular its computational theories, have questioned its applicability as an operant concept capable to generate an actual mechanism for curiosity. A first reason is that the concept of "intermediate" appears difficult to define precisely, as it implies the use of a relatively arbitrary frame of reference to assess levels of novelty/complexity. A second reason is that while novelty or complexity in themselves may be the basis of useful exploration heuristics for organisms in some particular contexts, there is in general no guarantee that observing a novel or intermediate complexity stimulus provides information that can improve the organism's prediction and control in the world. Indeed, as computational theory of learning and exploration has shown, our environment is full of novel and complex stimuli of all levels, and among them only a few may convey useful or learnable patterns. As curiosity-driven spontaneous exploration may have evolved as a mean to acquire information and skills in rapidly changing environments (Barto, 2013), it appears that heuristics based on searching for novelty and complexity can be inefficient in large or nonstationary environments (Oudeyer et al., 2007; Schmidhuber, 1991).

For these reasons, computational learning theory has explored an alternative mechanism, in which *learning progress* generates intrinsic reward (Oudeyer et al., 2007; Schmidhuber, 1991), and it was hypothesized that this mechanism could be at play in humans and animals (Kaplan and Oudeyer, 2007a,b; Oudeyer and Smith, 2016). This hypothesis proposes that the brain, seen as a predictive machine constantly trying to anticipate what will happen next, is intrinsically motivated to pursue activities in which predictions are *improving*, ie, where uncertainty is decreasing and learning is actually happening. This means that the organism loses interest in activities that are too easy or too difficult to predict (ie, where uncertainty is low or where uncertainty is high but not reducible) and focuses specifically on learnable activities that are just beyond its current predictive capacities. So, for example, an infant will be more interested in exploring how its arm motor commands can allow her to predict the movement of her hand in the visual field (initially difficult but learnable) rather than predicting the movement of walls (too easy) or the color of the next car passing through the window (novel but not learnable). As shown by the computational studies we discuss later, a practical consequence of behaviors driven by the search for LP is the targeted exploration of activities and stimuli of "intermediate complexity." Yet, an explicit measure of intermediate complexity is not computed by this mechanism: it is an emergent property of selecting actions and stimuli that maximize the derivative of errors in prediction.

5 THE LP HYPOTHESIS POSITS A POSITIVE FEEDBACK LOOP BETWEEN CURIOSITY AND LEARNING

The LP hypothesis posits a new causal link between learning and curiosity. As described in Sections 1 and 2 of this chapter, previous work in neuroscience and psychology considered a unidirectional causal chain: the brain would be motivated to search for (intermediate) novelty or complexity, and then when finding it would be in a curiosity state that would foster learning and memory retention (see Fig. 1A). In this view (Kang et al., 2009; Stahl and Feigenson, 2015), learning in itself does not have consequences on state curiosity and motivation. On the contrary, the LP hypothesis proposes that experiencing learning in a given activity (rather than just intermediate novelty) triggers an intrinsic reward, and thus that learning in itself causally influences state curiosity and intrinsic motivation (see Fig. 1B). *Thus, this hypothesis argues that there is a closed self-reinforcing feedback loop between learning and curiosity-driven intrinsic motivation. Here the learner becomes*

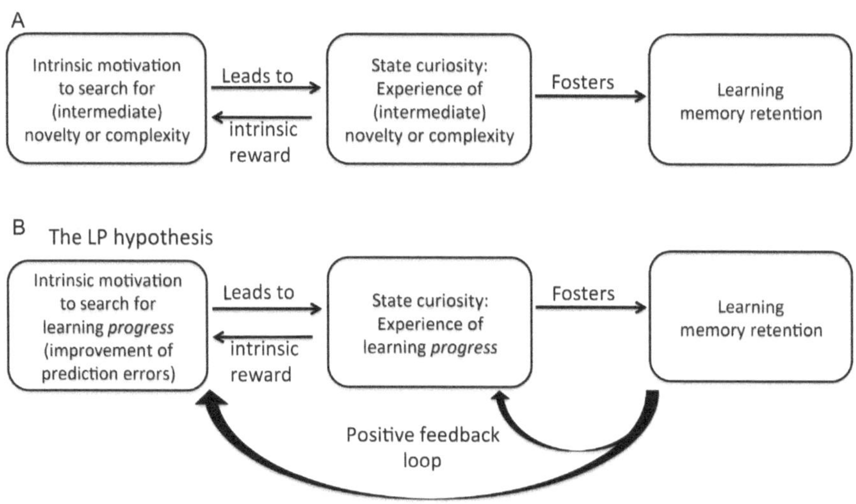

FIG. 1

Many studies of curiosity and learning have considered a one-directional causal relationship between state curiosity and learning (A). The learning progress hypothesis suggests that learning progress itself, measured as the improvement of prediction errors, can be intrinsically rewarding: this introduces a positive feedback loop between state curiosity and learning (B). This positive feedback loop in turn introduces a complex learning dynamics self-organizing learning curriculum with phases of increasing complexity, such as in the Playground Experiment (Oudeyer et al., 2007; see Figs. 2 and 3).

fundamentally active, searching for niches of learning progress, in which in turn memory retention is facilitated. As shown by computational experiments outlined later, this feedback loop has important consequences on the organization of learning experiences on the long term: as learners actively seek for situations and activities which maximize LP, they will first focus on simple learnable activities before shifting to more complex ones (see Fig. 2), and the activities they select shape their

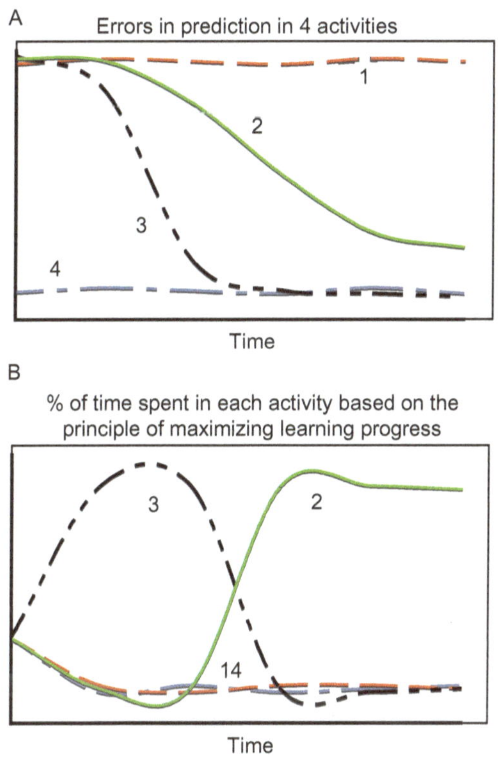

FIG. 2

The LP hypothesis proposes that active spontaneous exploration will favor exploring activities which are providing maximum improvement of prediction errors. If one imagines four activities with different learning rate profiles (A), then LP-driven exploration will avoid activities that are either too easy (4) or too difficult (1) as they do not provide learning progress, then first focus on an activity which initially provides maximal learning progress (3) (see Panel B), before reaching a learning plateau in this activity and shifting to another one (2) which at this point in the curriculum provides maximum progress (potentially thanks to skills acquired in activity (3)). As a consequence, an ordering of exploration phases forms spontaneously, generating a structured developmental trajectory.

Adapted from Kaplan, F., Oudeyer, P.-Y., 2007a. The progress-drive hypothesis: an interpretation of early imitation. In: Dautenhahn, K., Nehaniv, C. (Eds.), Models and Mechanisms of Imitation and Social Learning: Behavioural, Social and Communication Dimensions, Cambridge University Press, pp. 361–377; Kaplan, F., Oudeyer, P.-Y., 2007b. In search of the neural circuits of intrinsic motivation. Front. Neurosci. 1 (1), 225–236.

knowledge and skills, which will in turn change the potential progress in other activities and thus shape their future exploratory trajectories. As a consequence, the LP hypothesis does not only introduce a causal link between learning and curiosity but also introduces the idea that curiosity may be a key mechanism in shaping developmental organization. Later, we will outline computational experiments that have shown that such an active learning mechanisms can self-organize a progression in learning, with automatically generated developmental phases that have strong similarities with infant developmental trajectories.

6 THE LP HYPOTHESIS UNIFIES VARIOUS QUALITATIVE THEORIES OF CURIOSITY

The LP hypothesis is also associated to a mathematical formalism (outlined in Section 7) that allows to bridge several hypotheses related to curiosity and intrinsic motivation that had been so far conceptually separated (Oudeyer and Kaplan, 2007). Within the LP hypothesis, the central concept of prediction errors (and the associated measure of improvement) applies to multiple kinds of predictions. It applies to predicting the properties of external perceptual stimuli (and thus relates to the notion of perceptual curiosity Berlyne, 1960), as well as the conceptual relations among symbolic items of knowledge (and this relates to the notion of epistemic curiosity, and to the subjective notion of information gap proposed by Lowenstein, 1994). Here the maximization of LP leads to behaviors that were previously understood through Berlyne's concept of intermediate novelty/complexity, and such mechanisms correspond to a class of intrinsic motivation that has been called "knowledge-based intrinsic motivation" (Mirolli and Baldassarre, 2013; Oudeyer and Kaplan, 2007). It also applies to predicting the consequences of one's own actions in particular situations, or to predicting how well one's current skills are capable to solve a given goal/problem: here the maximization of LP, measuring a form of progress in competences related to an activity or a goal, can be used to model Csikszenthmihalyi's concept of intermediate challenge in the flow theory as well as related theories of intrinsic motivation based on self-measures of competences (Csikszenthmihalyi, 1991; White, 1959). This second form of the LP hypothesis, where LP is measured in terms of how much competences improve with experience, corresponds to a class of intrinsic motivation mechanisms that has been called "competence-based intrinsic motivation" (Mirolli and Baldassarre, 2013; Oudeyer and Kaplan, 2007).

7 COMPUTATIONAL MODELS: CURIOSITY-DRIVEN REINFORCEMENT LEARNING

Computational and robotic models have recently thrived in order to conceptualize more precisely theories of curiosity-driven learning and intrinsic motivation, as well as to study the associated learning dynamics and make experimental predictions (Baldassare and Mirolli, 2013; Gottlieb et al., 2013). A general formal framework that has been used most often to model learning and motivational systems is computational reinforcement learning (Sutton and Barto, 1998). In reinforcement learning, one considers a set of states S (characterizing the state of the world as sensed by sensors as well as the state of internal memory); a set of actions A that the organism can make; a reward function R(s,a) that provides a number r(s,a) that depends on states and actions and that should be maximized; an action policy P(a|s) which determines which actions should be made in each state so as to maximize future expected reward; and finally a learning mechanism L that allows to update the action policy in order to improve rewards in the future. Many works in computational neuroscience and psychology have focused on the details of the learning mechanism, for example, to explain differences in model-based vs model-free learning (Gershman, in press). However, the same framework can be used to model motivational mechanisms, through modeling the structure and semantics of the reward function. For example, extrinsic motivational mechanisms associated to food/energy search can be modeled through a reward function that measures the quantity of food gathered (Arkin, 2005). A motivation for mating can be modeled similarly, and as each motivational mechanism is modeled as a real number that should be maximized, such numbers can be used as a common motivational currency to make trade-offs among competing motivations (Konidaris and Barto, 2006).

Similarly, it is possible to use this framework to provide formal models of intrinsic motivation and curiosity as formulated by most theories mentioned earlier, in architecture called "intrinsically motivated reinforcement learning" (Singh et al., 2004a,b) and as reviewed in Baldassare and Mirolli (2013) and Oudeyer and Kaplan (2007). In this context, an intrinsic motivation system that pushes organisms to search for novelty can be formalized, for example, by considering a mechanism which counts how often each state of the environment has already been visited, and then using a reward function that is inversely proportional to these counts. This corresponds to the concept of exploration bonus studied by Dayan and Sejnowski (1996) and Sutton (1990). If one considers a model-based RL system that learns to predict which states will be observed upon a series of actions, as well as measures of uncertainty of these predictions, one can formalize surprise (and automatically derive an associated reward) as situations in which the subject makes an unexpected high error in predictions.

To understand how the LP hypothesis can be formally modeled in this framework, let us consider the model used in the Playground Experiment (see Fig. 3A). In this experiment, a quadruped "learning" robot (the learner) is placed on an infant play mat with a set of nearby objects and is joined by an "adult" robot (the teacher), see Fig. 3A (Kaplan and Oudeyer, 2007b; Oudeyer and Kaplan, 2006; Oudeyer et al., 2007). On the mat and near the learner are objects for discovery: an elephant (which can be bitten or "grasped" by the mouth), a hanging toy (which can be "bashed" or pushed with the leg). The teacher is preprogrammed to imitate the sounds made by the learner when the learning robot looks to the teacher while vocalizing at the same time.

The learner is equipped with a repertoire of motor primitives parameterized by several continuous numbers that control movements of its legs, head, and a simulated vocal production system. Each motor primitive is a dynamical system controlling various forms of actions: (a) turning the head in different directions; (b) opening

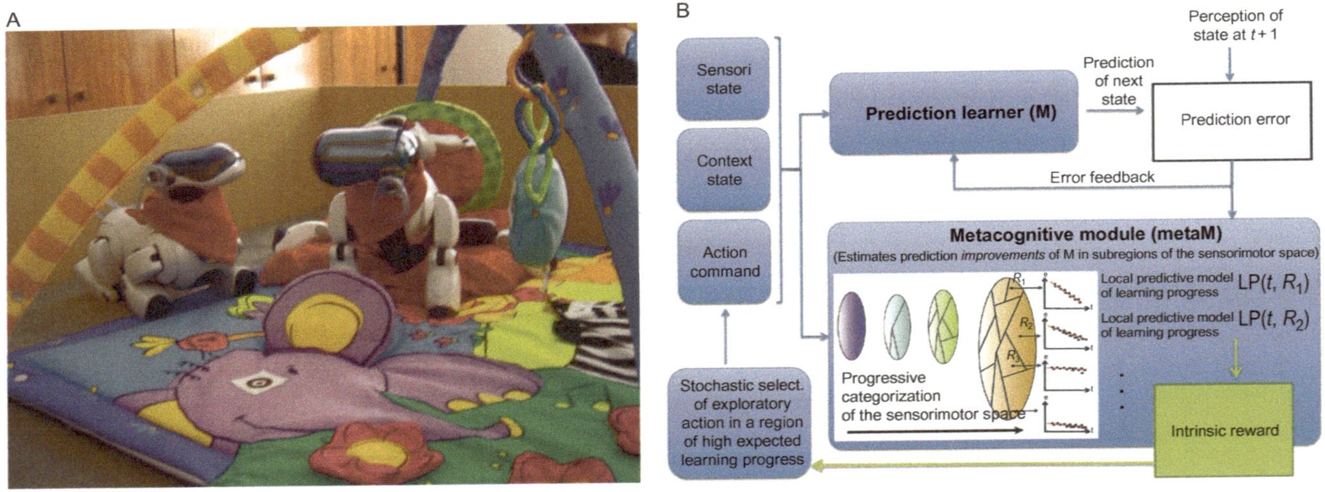

FIG. 3

(A) The Playground Experiment: a robot explores and learns the contingencies between its movement and the effect they produce on surrounding objects. To drive its exploration, it uses the active learning architecture described in (B). In this architecture, a meta-learning module tracks the evolution of errors in predictions that the robot makes using various kinds of movements in various situations. Then, an action selection module selects probabilistically actions and situations which have recently provided high improvement of predictions (learning progress), using this measure to heuristically expect further learning progress in similar situations.

Adapted from Oudeyer, P.-Y., Kaplan, F., Hafner, V., 2007. Intrinsic motivation systems for autonomous mental development. IEEE Trans. Evol. Comput. 11 (2), 265–286.

and closing the mouth while crouching with varying strengths and timing; (c) rocking the leg with varying angles and speed; (d) vocalizing with varying pitches and lengths. These primitives are parameterized by real numbers and can be combined to form a large continuous space of possible actions. Similarly, sensory primitives allow the robot to detect visual movement, salient visual properties, proprioceptive touch in the mouth, and pitch and length of perceived sounds. For the robot, these motor and sensory primitives are initially black boxes and he has no knowledge about their semantics, effects, or relations.

The robot learns how to use and tune these primitives to produce various effects on its surrounding environment, and exploration is driven by the maximization of *learning progress*, by choosing physical experiences (experiments) that improve the quality of predictions of the consequences of its actions. As data are collected though this exploration process, the robot builds a model of the world dynamics that can be reused later on for new tasks that were not known at the time of exploration (for example, using model-based reinforcement learning mechanisms).

Fig. 3B outlines a computational architecture, called R-IAC (Moulin-Frier et al., 2014; Oudeyer et al., 2007). A prediction machine (M) learns to predict the consequences of actions taken by the robot in given sensory contexts. For example, this module might learn to predict which visual movements or proprioceptive perceptions result from using a leg motor primitive with certain parameters (this model learning can be done with a neural network or any other statistical m':ùxc nklninference algorithm). Another module (metaM) estimates the evolution of errors in prediction of M in various regions of the sensorimotor space.[2] This module estimates how much errors decrease in predicting an action in certain situations, for example, in predicting the consequence of a leg movement when this action is applied toward a particular area of the environment. These estimates of error reduction are used to compute the intrinsic reward from progress in learning. This reward is an internal quantity that is proportional to the decrease of prediction errors, and the maximization of this quantity is the goal of action selection within a computational reinforcement learning architecture (Kaplan and Oudeyer, 2003; Oudeyer and Kaplan, 2007; Oudeyer et al., 2007). Importantly, the action selection system chooses most often to explore activities where the estimated reward from LP is high. However, this choice is probabilistic, which leaves the system open to learning in new areas and open to discovering other activities that may also yield progress in learning.[3] Since the sensorimotor flow

[2]In this instantiation of the LP hypothesis, an internal module metaM monitors how learning progresses to generate intrinsic rewards. However, the LP hypothesis in general does not require such an internal capacity for measuring learning progress: such information may also be provided by the environment, either directly by objects or games children play with or by adults/social peers.

[3]Here, action selection is made within a simplified form of reinforcement learning: learning progress is maximized only on the short term, and the environment is configured so that it returns to a rest position after each sensorimotor experiment. This corresponds to what is called episodic reinforcement learning, and action selection can be handled efficiently in this case using multiarmed bandit algorithms (Audibert et al., 2009). Other related computational models have considered maximizing forms of LP over the long term through RL planning techniques in environments which dynamics is state dependent (Kaplan and Oudeyer, 2003; Schmidhuber, 1991) and nonstationary (Lopes et al., 2012).

does not come presegmented into activities and tasks, a system that seeks to maximize differences in learnability is also used to progressively categorize the sensorimotor space into regions. This categorization thereby models the incremental creation and refining of cognitive categories differentiating activities/tasks.

In all of the runs of the experiment, one observes the self-organization of structured developmental trajectories, where the robot explores objects and actions in a progressively more complex stage-like manner while acquiring autonomously diverse affordances and skills that can be reused later on and that change the LP in more complicated tasks. Typically, after a phase of random body babbling, the robot focuses on performing various kinds of actions toward objects, and then focuses on some objects with particular actions that it discovers are relevant for the object. In the end, the robot is able to acquire sensorimotor skills such as how to push or grasp objects, as well as how to perform simple vocal interactions with another robot, as a side effect of its general drive to maximize LP. This typical trajectory can be explained as gradual exploration of new progress niches (zones of the sensorimotor space where it progresses in learning new skills), and those stages and their ordering can be viewed as a form of attractor in the space of developmental trajectories. Yet, one also observes diversity in the developmental trajectories observed in the experiment. With the same mechanism and same initial parameters, individual trajectories may generate qualitatively different behaviors or even invert stages. This is due to the stochasticity on the policy, to even small variability in the physical realities, and to the fact that this developmental dynamic system has several attractors with more or less extended and strong domains of attraction (characterized by amplitude of LP). This diversity can be seen as an interesting modeling outcome since individual development is not identical across different individuals but is always, for each individual, unique in its own ways. This kind of approach, then, offers a way to understand individual differences as emergent in developmental process itself and makes clear how developmental process might vary across contexts, even with an identical learning mechanism.

8 HOW LP-DRIVEN CURIOSITY GENERATES DEVELOPMENTAL TRAJECTORIES THAT REPRODUCE INFANT DEVELOPMENT SEQUENCES AND CAN ACT IN SYNERGY WITH SOCIAL LEARNING

Focusing on vocal development, Moulin-Frier et al. conducted experiments where a robot explored the control of a realistic model of the vocal tract in interaction with vocal peers through a drive to maximize LP (Moulin-Frier et al., 2014). This model relied on a physical model of the vocal tract, its motor control, and the auditory system. It also included an additional mechanism allowing the active learner to take into account social signals provided by peers. As a simulated caretaker would himself produce vocalizations organized around the systematic reuse of certain phonemes, the curiosity-driven learning system could decide whether it should try to reproduce these external speech sounds (imitation) using its current know-how, or whether it should self-explore other kinds of speech sounds. The choice was made hierarchically: first, it decided to imitate or self-explore based on how much each strategy provided LP in the past. Second, if self-exploration was selected, it decided which part of the sensorimotor space to explore based on how much LP could be expected. The experiments showed how such a mechanism generated automatically the adaptive transition from vocal self-exploration with little influence from the speech environment to a later stage where vocal exploration becomes influenced by vocalizations of peers, as typically observed in human infants (Oller, 2000). Within the initial self-exploration phase, a sequence of vocal production stages self-organizes and shares properties with infant data: the vocal learner first discovers how to control phonation, then vocal variations of unarticulated sounds, and finally articulated protosyllables. In this initial phase, imitation is rarely tried by the learner as the sounds produced by caretakers are too complicated to make any progress. But as the vocal learner becomes more proficient at producing complex sounds through self-exploration, the imitating vocalizations of the teacher begin to provide high LP, resulting in a shift from self-exploration to vocal imitation. This also illustrates how intrinsically motivated self-exploration can guide the system to efficiently and autonomously acquire basic sensorimotor skills that are instrumental to learn faster other more complicated skills.

9 INTRINSICALLY MOTIVATED EXPLORATION SCAFFOLDS EFFICIENT MULTITASK LEARNING

Computational models in the literature have shown how various forms of intrinsically motivated exploration and learning could guide efficiently the autonomous acquisition of repertoires of skills in large and difficult spaces.

A first reason is that intrinsically motivated exploration can be used as an active learning algorithm that learns efficient forward and inverse models of the world dynamics through efficient selection of experiences. Indeed, such models can be reused either directly (Baranes and Oudeyer, 2013; Oudeyer et al., 2007), or through model-based planning mechanisms (Lopes et al., 2012; Schmidhuber, 1991; Singh et al., 2004a,b), to solve repertoires of tasks that were not specified during exploration (hence without the need for long reexperiencing of the world for each new task). For example, Baranes and Oudeyer (2013) have shown how intrinsically motivated goal exploration could allow robots to sample sensorimotor spaces by actively controlling the complexity of explored sensorimotor goals, and avoiding goals which were either too easy or unreachable. This allowed the robots to learn fast repertoires of high-dimensional continuous action skills to solve distributions of sensorimotor problems such as omnidirectional legged locomotion or how to manipulate flexible objects. Lopes et al. (2012) showed how intrinsically motivated model-based reinforcement learning, driven by the maximization of empirical LP, allows efficient learning of world models when this dynamics is nonstationary, and how this accelerates the learning of a policy that targets to maximize an extrinsic reward (task predefined by experimenters).

A second reason for the efficiency of intrinsic motivation is that by fostering spontaneous exploration of novel skills, and leveraging opportunistically potential synergies among skills, it can create learning pathways toward certain skills that would have remained difficult to reach if they had been the sole target of the learning system. Indeed, in many contexts, learning a single predefined skill can be difficult as it amounts to searching (the parameters of) a solution with very rare feedback until one is very close to the solution, or with deceptive feedback due to the phenomenon of local minima. A strategy to address these issues is to direct exploration with intrinsic rewards, leading the system to explore a diversity of skills and contingencies which often result in the discovery of new subspaces/areas in the problem space, or in mutual skill improvement when exploring one goal/skill provides data that can be used to improve other goals/skills, such as in goal babbling (Benureau and Oudeyer, 2013, 2016) or off-policy reinforcement learning (see the Horde architecture, Sutton et al., 2011). For example, Lehman and Stanley (2011) showed that searching for pure novelty in the behavioral space a robot to find a reward in a maze more efficiently than if it had been searching for behavioral parameters that optimized directly the reward. In another model, Forestier and Oudeyer (2016) showed that intrinsically motivated exploration of a hierarchy of sensorimotor models allowed a simulated robot to scaffold the successive acquisition of object reaching, tool grasping, and tool use (and where direct search for tool use behaviors was vastly less efficient).

A third related reason for the efficiency of intrinsically motivated exploration is that it can drive the acquisition of macroactions, or sensorimotor primitives, which can be combinatorially reused as building block to accelerate the search for complex solutions in structured reinforcement learning problems. For example, Singh et al. (2004a,b) showed how intrinsic rewards based on measures of saliency could guide a reinforcement learner to progressively learn "options," which are temporally extended macroactions, reshaping the structure of the search space and finally learning action policies that solve an extrinsic (abstract) task that is very difficult to solve through standard RL exploration. Related uses of intrinsic motivation with a hierarchical reinforcement learning framework were demonstrated in Bakker and Schmidhuber (2004) and Kulkarni et al. (2016).

In a related line of research studying the function and origins of intrinsic motivation, Singh et al. (2010) have shown through evolutionary computational modeling that given a distribution of changing environments and an extrinsic reward that organisms need to maximize, it could be more robust for RL agents to represent and use a surrogate reward function that does not directly correspond to this extrinsic reward, but rather includes a component of intrinsic motivation that pushes the system to explore its environment beyond the direct search for the extrinsic reward.

10 APPLICATIONS IN EDUCATIONAL TECHNOLOGIES AND VIDEO GAMES

Given the strong causal interactions between curiosity-driven exploration and learning that we just reviewed, these topics have attracted the attention of theorists and experimenters on the application domain of education. Long before recent controlled experimental results showing how intrinsic motivation and curiosity could enhance learning, educational experimenters like Montessori (1948/2004) and Froebel (1885) have studied how open-ended learning environments could foster individual child development, where learners are active and where the tutor's role is to scaffold challenges of increasing complexity and provide feedback (rather than instruction). Such experimental approaches have more recently influenced the development of hands on educational practices, such as the pioneering LOGO experiments of Papert (1980), where children learn advanced concepts of mathematics, computer science, and robotics, and now disseminating at large scales in several countries (Resnick et al., 2009; Roy et al., 2015).

In parallel, philosophers and psychologists like Dewey, Vygotski, Piaget, and Bruner developed theories of constructivist learning which directly pointed toward the importance of fostering curiosity and free play and exploration in the classroom. Recently, the large body of research in educational psychology has began to study systematically how states of intrinsic motivation can be fostered, or on the contrary weakened, in the classroom, for example, when the educational context provides strong extrinsic rewards (Deci et al., 2001).

As educational technologies are now thriving, in particular with the wide spreading of Massive Open Online Courses (MOOCs) and educational applications on tablets and smartphones, it has become natural to enquire how fundamental understanding of curiosity, intrinsic motivation, and learning could be leveraged and incorporated in these educational tools to increase their efficiency.

A first line of investigation has been to embed educational training within motivating and playful video games. In a pioneer study, Malone (1980) used and refined theories of intrinsic motivation as proposed by Berlyne, White, and psychologists of the 1950–70s period, to evaluate which properties of video games could make them intrinsically motivating, and to study how such contexts could be used to distill elements of scholarly knowledge to children. In particular, he showed that video games were more intrinsically motivating when including clear goals of progressively increasing complexity, when the system provided clear feedback on the performance of users, and when outcomes were uncertain to entertain curiosity. For example, he showed how arithmetic concepts could be taught in an intrinsically motivating scenarized dart video game. As an outcome of their studies, they could generate a set of guidelines for the design of education-oriented video games.

In a similar study, studying the impact of several of the factors identified by Malone (1980) and Cordova and Lepper (1996) presented a study of a population of elementary schoolchildren using a game targeting the acquisition of arithmetic order-of-operation rules, scenarized in a "space quest" story. In this specific experimental context, they showed that embedding personalization in the math exercises (based on preferences expressed through a prequestionnaire) significantly improved intrinsic motivation, task engagement, and learning efficiency, and that this effect was heightened if in addition the software offered personalization of visual displays and a variety of exercise levels children could choose from.

Beyond explicitly including educational elements in video games, it was also shown that "pure" entertainment games such as certain types of action games can enhance attentional control, cognitive flexibility, and learning capabilities by exercising them in an intrinsically motivating playful context (Cardoso-Leite and Bavelier, 2014). Within this perspective, Merrick and Maher (2009) suggested that implementing artificial curiosity in nonplayer characters in video games could enhance the interestingness of video games. In another line of research, Law et al. (2016) showed that crowdsourcing tasks could be made more engaging by incorporating information rewards that stimulated curiosity.

A second line of investigation has considered how formal and computational models of curiosity and intrinsic motivation could be applied to intelligent tutoring systems (ITS; Clement et al., 2015; Nkambou et al., 2010), as well as MOOCs (Liyanagunawardena et al., 2013; Steels, 2015a,b). ITS, and more recently MOOCs, have targeted the design of software systems that could help students acquire new knowledge and skills, using artificial intelligence techniques to personalize teaching sequences, or the way teaching material is presented, and in particular proposing exercises that match the particular difficulties or talents of each individual learner. In this context, several approaches were designed and experimented so as to promote intrinsic motivation and learning.

Clement et al. (2015) have presented and evaluated and ITS system that directly reused computational models of curiosity-driven learning based on the LP hypothesis described earlier (Oudeyer et al., 2007). This study considered teaching arithmetic decomposition of integer and decimal numbers, in a scenarized context of money handling, to a population of 7–8 years old children (see Fig. 4). To design the ITS system, a human teacher first provided pedagogical material in the form of exercises grouped along coarsely defined levels and coarsely defined types. Then, an algorithm called ZPDES was used to automatically personalize the sequence of exercises for each student, and this personalization was made incrementally during the course of interaction with each student. This personalization was achieved by probabilistically proposing to students' exercises that maximized LP at their current level, ie, the exercises where their errors decrease fastest. In order to identify dynamically these exercises, and shift automatically to new ones when LP becomes low, the system used a multiarmed bandit algorithm that balanced exploring new exercises to assess their potential for LP, and exploiting exercises that recently lead the student to LP. During this process, the coarse structure organizing exercises that was provided by a human teacher is used to guide the algorithm toward finding fast which exercises provide maximal LP: the system starts with exercise types that are at the bottom of the difficulty hierarchy, and when some of them show a plateau in the learning

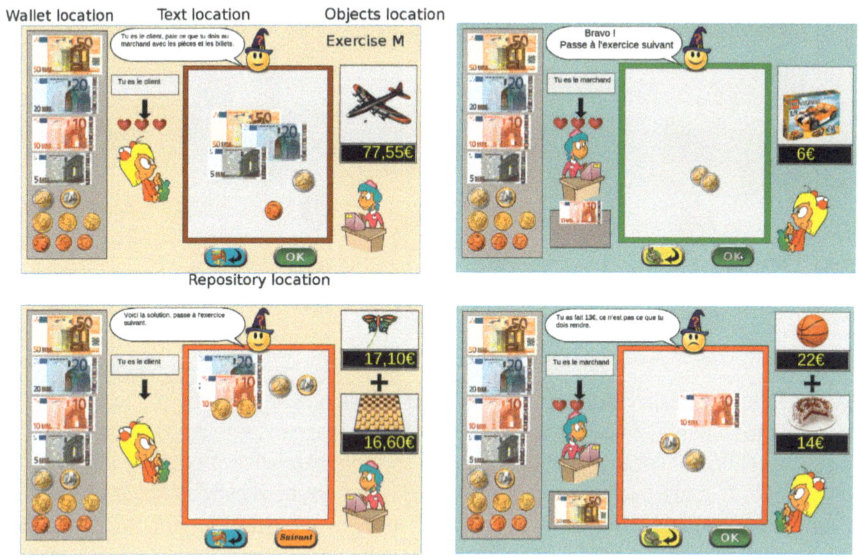

FIG. 4

Educational game used in Clement et al. (2015): a scenario where elementary schoolchildren have to learn to manipulate money is used to teach them the decomposition of integer and decimal numbers. Four principal regions are defined in the graphical interface. The first is the wallet location where users can pick and drag the money items and drop them on the repository location to compose the correct price. The object and the price are present in the object location. Four different types of exercises exist: M (customer/one object), R (merchant/one object), MM (customer/two objects), RM (merchant/two objects). The ITS system then dynamically proposes to students the exercises in which they are currently making maximal learning progress, targeting to maximize intrinsic motivation and learning efficiency.

curve, they are deactivated and new exercises upper in the hierarchy are made available to the student (see Fig. 5). The use of LP as a measure to drive the selection of exercises had two interacting purposes, relying on the bidirectional interaction described earlier. First, it targeted to propose exercises that could stimulate the intrinsic motivation of students by dynamically and continuously proposing them challenges that were neither too difficult nor too easy. Second, by doing this using LP, it targeted to generate exercise sequences that are highly efficient for maximizing the average scores over all types of exercises at the end of the training session. Indeed, Lopes and Oudeyer (2012) showed in a theoretical study that when faced with the problem of strategically choosing which topic/exercise type to work on, selecting topics/exercises that maximize LP is quasioptimal for important classes of learner models. Experiments with 400 children from 11 schools were performed, and the impact of this algorithm selecting exercises that maximize LP was compared to the impact of a sequence of exercises hand-defined by an expert teacher (that included sophisticated

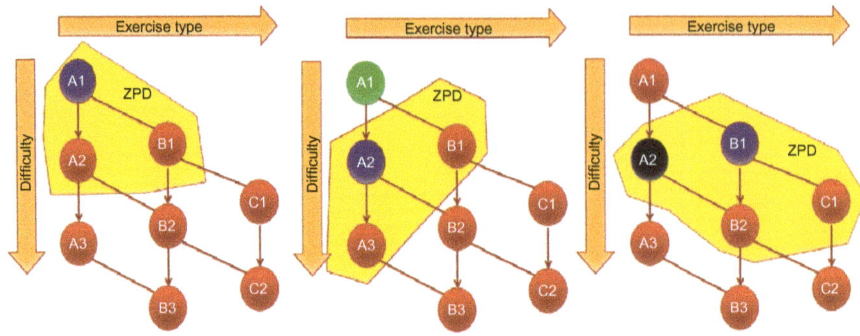

FIG. 5

Example of the evolution of the zone-of-proximal development based on the empirical results of the student. The ZPD is the set of all activities that can be selected by the algorithm. The expert defines a set of preconditions between some of the activities (A1 → A2 → A3 ...), and activities that are qualitatively equal (A==B). Upon successfully solving A1 the ZPD is increased to include A3. When A2 does not achieve any progress, the ZPD is enlarged to include another exercise type C, not necessarily of higher or lower difficulty, eg, using a different modality, and A3 is temporarily removed from the ZPD.

Adapted from Clement, B., Roy, D., Oudeyer, P.-Y., Lopes, M., 2015. Multi-armed bandits for intelligent tutoring systems. J. Educ. Data Mining 7 (2).

branching structures based on the errors-repair strategies the teacher could imagine). Results showed that the ZPDES algorithm, maximizing LP, allowed students of all levels to reach higher levels of exercises. Also, an analysis of the degree of personalization showed that ZPDES proposed a higher diversity of exercises earlier in the training sessions. Finally, a pre- and posttest comparison showed that students who were trained by ZPDES progressed better than students who used a hand-defined teaching sequence.

Several related ITS systems were developed and experimented. For example, Beuls (2013) described a system targeting the acquisition of Spanish verb conjugation, where the ITS attempts to propose exercises that are just above the current capabilities of the learner. Recently, a variation of this system was designed to foster the learning of musical counterpoint (Beuls and Loeckx, 2015). In another earlier study, Pachet (2004) presented a computer system targeting to help children discover and learn how to play musical instruments, but also capable to support creativity in experienced musicians, through fostering the experience of Flow (Csikszenthmihalyi, 1991). This system, called the Continuator (Pachet, 2004), continuously learnt the style of the player (be it a child beginner or expert) and used automatic improvization algorithm to respond to the user's musical phrases with musical phrases of the same style and complexity, but different from those actually played by users. Pachet observed that both children and expert musicians most often experience an "Eureka moment." Their interest and attention appeared to be strongly attracted by playing with the system, leading children to try and discover different modes of play and to increase the complexity of what they could do. Expert musicians also reported that the system allowed them to discover novel musical ideas and to support creation interactively.

11 DISCUSSION: CONVERGENCES, OPEN QUESTIONS, AND EDUCATIONAL DESIGN

Converging research strands in psychology, neuroscience, and computational learning theory indicate that curiosity and learning are strongly connected along several dimensions, and that these connections have wide implications for education.

As often informally observed by many education practitioners, recently developed experimental protocols showed that experiencing situations with novelty, complexity, and prediction errors fostered memory retention. Furthermore, several lines of evidence showed that the brain is equipped with neural circuits which consider information as an intrinsic reward, and thus actively searches for these situations featuring novelty and/or prediction errors.

At the same time, there are several open scientific questions. One of them is to characterize precisely which informational features are intrinsically rewarding. As mathematical formalization shows, novelty, complexity, and prediction errors can be computed in many different ways, and then used potentially in equally different ways to determine intrinsic reward in curiosity-driven exploration. For example, some hypotheses approach curiosity as a mechanism maximizing surprise (Itti and Baldi, 2009), or intermediate complexity (Kidd et al., 2012), or LP (Oudeyer et al., 2007). While on some situations these formal variations may be equivalent, they can also generate vastly different learning dynamics. For example, the LP hypothesis introduces a positive feedback loop between learning and state curiosity, and in turn this involves deep consequences on the long-term formation of learning and developmental trajectories (Oudeyer and Smith, 2016). A related question is whether the brain includes a unified mechanism for spontaneous exploration, or whether it combines several of these heuristics (and how this combination happens). Designing experimental protocols to disentangle these various hypotheses is the subject of active current research (Baranes et al., 2014; Markant et al., 2015; Meder and Nelson, 2012; Taffoni et al., 2014).

However, even if these questions are still unresolved, existing results suggest several guidelines for educational practice and the design of educational technologies. First, they highlight the importance of providing students with learning materials that are informationally engaging (surprising or with the right level of complexity/learnability) in order to foster memory retention. Second, they suggest the importance of personalization, active learning, and active teaching. Indeed, features like (intermediate) novelty or LP are fundamentally subjective in the sense that they are a measure of the relation between a particular educational material and the state of knowledge of each particular student at a given time of its learning trajectory. As a consequence, what triggers curiosity and learning will be different for different students. Human or computational teachers can address this issue by tracking the errors and behaviors of each student in order to present sequences of items that are personalized to maximize their experience of features associated to states of curiosity and motivation. Learners have also a fundamental capability that should be leveraged: as their brain is intrinsically rewarded by features like novelty or LP, they will spontaneously and actively search for these features and select adequate learning materials if the environment/teacher provides sufficient choices. While most existing studies have focused on either active learning or active teaching, the study of the dynamic interaction between active learners and teachers is still a largely open question that should be addressed to understand how this dynamics could scaffold mutual guidance toward efficient curiosity-driven learning.

REFERENCES

Arkin, R., 2005. Moving up the food chain: motivation and emotion in behavior based robots. In: Fellous, J., Arbib, M. (Eds.), Who Needs Emotions: The Brain Meets the Robot. Oxford University Press, pp. 245–270.

Audibert, J.-Y., Munos, R., Szepesvari, C., 2009. Exploration-exploitation tradeoff using variance estimates in multi-armed bandits. Theor. Comput. Sci. 410 (19), 1876–1902.

Bakker, B., Schmidhuber, J., 2004. Hierarchical reinforcement learning based on subgoal discovery and subpolicy specialization. In: Proceedings of the 8th International Conference on Intelligent Autonomous Systems, pp. 438–445.

Baldassare, G., Mirolli, M., 2013. Intrinsically Motivated Learning in Natural and Artificial Systems. Springer-Verlag, Berlin.

Baranes, A., Oudeyer, P.-Y., 2013. Active learning of inverse models with intrinsically motivated goal exploration in robots. Robot. Auton. Syst. 61 (1), 49–73.

Baranes, A.F., Oudeyer, P.Y., Gottlieb, J., 2014. The effects of task difficulty, novelty and the size of the search space on intrinsically motivated exploration. Front. Neurosci. 8, 1–9.

Baranes, A., Oudeyer, P.Y., Gottlieb, J., 2015. Eye movements reveal epistemic curiosity in human observers. Vis. Res. 117, 81–90.

Bardo, M., Bevins, R., 2000. Conditioned place preference: what does it add to our preclinical understanding of drug reward? Psychopharmacology 153 (1), 31–43.

Barto, A., 2013. Intrinsic motivation and reinforcement learning. In: Baldassarre, G., Mirolli, M. (Eds.), Intrinsically Motivated Learning in Natural and Artificial Systems. Springer, pp. 17–47.

Barto, A., Mirolli, M., Baldasarre, G., 2013. Novelty or surprise? Front. Cogn. Sci. 11. 1–15. http://dx.doi.org/10.3389/fpsyg.2013.00907.

Benureau, F.C.Y., Oudeyer, P.-Y., 2016. Behavioral diversity generation in autonomous exploration through reuse of past experience. Front. Robot. AI 3, 8. http://dx.doi.org/10.3389/frobt.2016.00008.

Berlyne, D., 1960. Conflict, Arousal and Curiosity. McGraw-Hill, New York.

Berlyne, D., 1965. Structure and Direction in Thinking. John Wiley and Sons, Inc., New York.

Beuls, K., 2013. Towards an Agent-Based Tutoring System for Spanish Verb Conjugation. PhD thesis, Vrije Universiteit Brussel.

Beuls, K., Loeckx, J., 2015. Steps towards intelligent MOOCs: a case study for learning counterpoint. In: Steels, L. (Ed.), Music Learning with Massive Open Online Courses. The Future of Learning, vol. 6. IOS Press, Amsterdam, pp. 119–144.

Bevins, R., 2001. Novelty seeking and reward: implications for the study of high-risk behaviors. Curr. Dir. Psychol. Sci. 10 (6), 189.

Blanchard, R., Kelley, M., Blanchard, D., 1974. Defensive reactions and exploratory behavior in rats. J. Comp. Physiol. Psychol. 87 (6), 1129–1133.

Blanchard, T.C., Hayden, B.Y., Bromberg-Martin, E.S., 2015. Orbitofrontal cortex uses distinct codes for different choice attributes in decisions motivated by curiosity. Neuron 85 (3), 602–614.

Bromberg-Martin, E.S., Hikosaka, O., 2009. Midbrain dopamine neurons signal preference for advance information about upcoming rewards. Neuron 63 (1), 119–126.

Bromberg-Martin, E.S., Matsumoto, M., Hikosaka, O., 2010. Dopamine in motivational control: rewarding, aversive, and alerting. Neuron 68 (5), 815–834. http://dx.doi.org/10.1016/j.neuron.2010.11.022.

Cardoso-Leite, P., Bavelier, D., 2014. Video game play, attention, and learning: how to shape the development of attention and influence learning? Curr. Opin. Neurol. 27 (2), 185–191.

Clement, B., Roy, D., Oudeyer, P.-Y., Lopes, M., 2015. Multi-armed bandits for intelligent tutoring systems. J. Educ. Data Mining 7 (2), 20–48.

Cordova, D.I., Lepper, M.R., 1996. Intrinsic motivation and the process of learning: beneficial effects of contextualization, personalization, and choice. J. Educ. Psychol. 88 (4), 715.

Csikszenthmihalyi, M., 1991. Flow—The Psychology of Optimal Experience. Harper Perennial.

Dayan, P., Sejnowski, T.J., 1996. Exploration bonuses and dual control. Mach. Learn. 25, 5–22.

De Charms, R., 1968. Personal Causation: The Internal Affective Determinants of Behavior. Academic Press, New York.

Deci, E., Ryan, R., 1985. Intrinsic Motivation and Self-Determination in Human Behavior. Plenum, New York.

Deci, E.L., Koestner, R., Ryan, R.M., 2001. Extrinsic rewards and intrinsic motivation in education: reconsidered once again. Rev. Educ. Res. 71 (1), 1–27.

Dember, W.N., Earl, R.W., 1957. Analysis of exploratory, manipulatory and curiosity behaviors. Psychol. Rev. 64, 91–96.

Festinger, L., 1957. A Theory of Cognitive Dissonance. Row, Peterson, Evanston, IL.

Foley, N.C., Jangraw, D.C., Peck, C., Gottlieb, J., 2014. Novelty enhances visual salience independently of reward in the parietal lobe. J. Neurosci. 34 (23), 7947–7957.

Forestier, S., Oudeyer, P.-Y., 2016. Curiosity-driven development of tool use precursors: a computational model. In: Proceedings of the 38th Annual Conference of the Cognitive Science Society.

Freeman, S., Eddy, S.L., McDonough, M., Smith, M.K., Okoroafor, N., Jordt, H., Wenderoth, M.P., 2014. Active learning increases student performance in science, engineering, and mathematics. PNAS 111 (23), 8410–8415.

Froebel, F., 1885. The Education of Man. A. Lovell & Company, New York.

Gershman, S.J., in press. Reinforcement learning and causal models. In: Waldmann, M. (Ed.), Oxford Handbook of Causal Reasoning. Oxford University Press.

Gershman, S.J., Niv, Y., 2015. Novelty and inductive generalization in human reinforcement learning. Top. Cogn. Sci. 7 (3), 1–25.

Gottlieb, J., Oudeyer, P.-Y., Lopes, M., Baranes, A., 2013. Information seeking, curiosity and attention: computational and neural mechanisms. Trends Cogn. Sci. 17 (11), 585–596.

Gruber, M.J., Gelman, B.D., Ranganath, C., 2014. States of curiosity modulate hippocampus-dependent learning via the dopaminergic circuit. Neuron 84, 486–496.

Harlow, H., 1950. Learning and satiation of response in intrinsically motivated complex puzzle performances by monkeys. J. Comp. Physiol. Psychol. 43, 289–294.

Hughes, R., 2007. Neotic preferences in laboratory rodents: issues, assessment and substrates. Neurosci. Biobehav. Rev. 31 (3), 441–464.

Hull, C.L., 1943. Principles of Behavior: An Introduction to Behavior Theory. Appleton-Century-Croft, New York.

Hunt, J.M., 1965. Intrinsic motivation and its role in psychological development. Neb. Symp. Motiv. 13, 189–282.

Itti, L., Baldi, P., 2009. Bayesian surprise attracts human attention. Vis. Res. 49 (10), 1295–1306.

Kagan, J., 1972. Motives and development. J. Pers. Soc. Psychol. 22, 51–66.

Kang, M.J., Hsu, M., Krajbich, I.M., Loewenstein, G., McClure, S.M., Wang, J.T., Camerer, C.F., 2009. The wick in the candle of learning: epistemic curiosity activates reward circuitry and enhances memory. Psychol. Sci. 20 (8), 963–973.

Kaplan, F., Oudeyer, P.-Y., 2003. Motivational principles for visual know-how development. In: Prince, C.G., Berthouze, L., Kozima, H., Bullock, D., Stojanov, G., Balkenius, C. (Eds.), Proceedings of the 3rd International Workshop on Epigenetic Robotics: Modeling Cognitive Development in Robotic Systems, vol. 101. Lund University Cognitive Studies, Lund, pp. 73–80.

Kaplan, F., Oudeyer, P.-Y., 2007a. The progress-drive hypothesis: an interpretation of early imitation. In: Dautenhahn, K., Nehaniv, C. (Eds.), Models and Mechanisms of Imitation and Social Learning: Behavioural, Social and Communication Dimensions. Cambridge University Press, Cambridge, pp. 361–377.

Kaplan, F., Oudeyer, P.-Y., 2007b. In search of the neural circuits of intrinsic motivation. Front. Neurosci. 1 (1), 225–236.

Kidd, C., Hayden, B.Y., 2015. The psychology and neuroscience of curiosity. Neuron 88 (3), 449–460.

Kidd, C., Piantadosi, S.T., Aslin, R.N., 2012. The Goldilocks effect: human infants allocate attention to visual sequences that are neither too simple nor too complex. PLoS One 7 (5), e36399.

Kidd, C., Piantadosi, S.T., Aslin, R.N., 2014. The Goldilocks effect in infant auditory cognition. Child Dev. 85 (5), 1795–1804.

Konidaris, G.D., Barto, A.G., 2006. An adaptive robot motivational system. In: Proceedings of the 9th International Conference on Simulation of Adaptive Behavior: From Animals to Animats 9 (SAB-06), CNR, Roma, Italy.

Kulkarni, T.D., Narasimhan, K.R., Saeedi, A., Tenenbaum, J.B., 2016. Hierarchical deep reinforcement learning: integrating temporal abstraction and intrinsic motivation. https://arxiv.org/abs/1604.06057.

Law, E., Yin, M., Joslin Goh, K.C., Terry, M., Gajos, K.Z., 2016. Curiosity killed the cat, but makes crowdwork better. In: Proceedings of CHI'16.

Lehman, J., Stanley, K.O., 2011. Abandoning objectives: evolution through the search for novelty alone. Evol. Comput. 19 (2), 189–223.

Liyanagunawardena, T.R., Adams, A.A., Williams, S.A., 2013. MOOCs: a systematic study of the published literature 2008–2012. Int. Rev. Res. Open Distrib. Learn. 14 (3), 202–227.

Lopes, M., Oudeyer, P.Y., 2012. The strategic student approach for life-long exploration and learning. In: IEEE International Conference on Development and Learning and Epigenetic Robotics (ICDL). IEEE, pp. 1–8.

Lopes, M., Lang, T., Toussaint, M., Oudeyer, P.-Y., 2012. Exploration in Model-Based Reinforcement Learning by Empirically Estimating Learning Progress. In: Proceedings of Neural Information Processing Systems (NIPS 2012). NIPS, Tahoe, USA.

Lowenstein, G., 1994. The psychology of curiosity: a review and reinterpretation. Psychol. Bull. 116 (1), 75–98.

Malone, T.W., 1980. What Makes Things Fun to Learn? A Study of Intrinsically Motivating Computer Games: Technical report. Xerox Palo Alto Research Center, Palo Alto, CA. forthcoming.

Markant, D.B., Settles, B., Gureckis, T.M., 2015. Self-directed learning favors local, rather than global, uncertainty. Cogn. Sci. 40 (1), 100–120.

Meder, B., Nelson, J.D., 2012. Information search with situation-specific reward functions. Judgm. Decis. Mak. 7, 119–148.

Merrick, K.E., Maher, M.L., 2009. Motivated Reinforcement Learning: Curious Characters for Multiuser Games. Springer Science & Business Media.

Mirolli, M., Baldassarre, G., 2013. Functions and mechanisms of intrinsic motivations. In: Mirolli, M., Baldassarre, G. (Eds.), Intrinsically Motivated Learning in Natural and Artificial Systems. Springer, Berlin, Heidelberg, pp. 49–72.

Montessori, M., 1948/2004. The Discovery of the Child. Aakar Books, Delhi.

Montgomery, K., 1954. The role of exploratory drive in learning. J. Comp. Physiol. Psychol. 47, 60–64.

Moulin-Frier, C., Nguyen, M., Oudeyer, P.-Y., 2014. Self-organization of early vocal development in infants and machines: the role of intrinsic motivation. Front. Cogn. Sci. 4, 1–20. http://dx.doi.org/10.3389/fpsyg.2013.01006.

Myers, A., Miller, N., 1954. Failure to find a learned drive based on hunger; evidence for learning motivated by exploration. J. Comp. Physiol. Psychol. 47 (6), 428.

Nkambou, R., Mizoguchi, R., Bourdeay, J., 2010. Advances in Intelligent Tutoring Systems, vol. 308. Springer, Heidelberg.

Oller, D.K., 2000. The Emergence of the Speech Capacity. Lawrence Erlbaum and Associates, Inc, Mahwah, NJ.

Oudeyer, P.-Y., Kaplan, F., 2006. Discovering communication. Connect. Sci. 18 (2), 189–206.

Oudeyer, P.-Y., Kaplan, F., 2007. What is intrinsic motivation? A typology of computational approaches. Front. Neurorobot. 1, 6. http://dx.doi.org/10.3389/neuro.12.006.2007.

Oudeyer, P.-Y., Smith, L., 2016. How evolution can work through curiosity-driven developmental process. Top. Cogn. Sci. 8 (2), 492–502.

Oudeyer, P.-Y., Kaplan, F., Hafner, V., 2007. Intrinsic motivation systems for autonomous mental development. IEEE Trans. Evol. Comput. 11 (2), 265–286.

Pachet, F., 2004. On the design of a musical flow machine. In: Tokoro, M., Steels, L. (Eds.), A Learning Zone of One's Own. IOS Press, Amsterdam.

Papert, S., 1980. Mindstorms: Children, Computers, and Powerful Ideas. Basic Books, Inc, New York.

Peck, C.J., Jangraw, D.C., Suzuki, M., Efem, R., Gottlieb, J., 2009. Reward modulates attention independently of action value in posterior parietal cortex. J. Neurosci. 29 (36), 11182–11191.

Rescorla, R.A., Wagner, A.R., 1972. A theory of Pavlovian conditioning: variations in the effectiveness of reinforcement and nonreinforcement. In: Black, A.H., Prokasy, W.F. (Eds.), Classical Conditioning II: Current Research and Theory, vol. 2, pp. Appleton Century Crofts, New York, pp. 64–99.

Resnick, M., Maloney, J., Monroy-Hernández, A., Rusk, N., Eastmond, E., Brennan, K., Kafai, Y., 2009. Scratch: programming for all. Commun. ACM 52 (11), 60–67.

Roy, D., Gerber, G., Magnenat, S., Riedo, F., Chevalier, M., Oudeyer, P.Y., Mondada, F., 2015. IniRobot: a pedagogical kit to initiate children to concepts of robotics and computer science. In: Proceedings of RIE.

Ryan, R., Deci, E., 2000. Intrinsic and extrinsic motivations: classic definitions and new directions. Contemp. Educ. Psychol. 25, 54–67.

Schmidhuber, J., 1991. Curious model-building control systems. Proceedings of the International Joint Conference on Neural Network, vol. 2, Singapore. pp. 1458–1463.

Singh, S.P., Barto, A.G., Chentanez, N., 2004a. Intrinsically motivated reinforcement learning. In: Saul, L.K., Weiss, Y., Bottou, L. (Eds.), Proceedings of Advances in Neural Information Processing Systems (NIPS 2004), pp. 1281–1288.

Singh, S., Barto, A.G., Chentanez, N., 2004b. Intrinsically motivated reinforcement learning. In: 18th Annual Conference on Neural Information Processing Systems (NIPS), Vancouver, B.C., Canada.

Singh, S.P., Lewis, R.L., Barto, A.G., Sorg, J., 2010. Intrinsically motivated reinforcement learning: an evolutionary perspective. IEEE Trans. Auton. Ment. Dev. 2 (2), 70–82.

Stahl, A.E., Feigenson, L., 2015. Observing the unexpected enhances infants' learning and exploration. Science 348 (6230), 91–94.

Steels, L., 2015a. Social flow in social MOOCs. In: Steels, L. (Ed.), Music Learning with Massive Open Online Courses. IOS Press, Amsterdam, pp. 119–144.

Steels, L., 2015b. Music Learning with Massive Open Online Courses. IOS Press, Amsterdam, pp. 119–144.

Sutton, R.S., 1990. Integrated architectures for learning, planning, and reacting based on approximating dynamic programming. In: Proceedings of the 7th International Conference on Machine Learning, ICML, pp. 216–224.

Sutton, R.S., Barto, A.G., 1981. Toward a modern theory of adaptive networks: expectation and prediction. Psychol. Rev. 88 (2), 135.

Sutton, R.S., Barto, A.G., 1998. Reinforcement Learning: An Introduction. MIT Press, Cambridge, MA.

Sutton, R.S., Modayil, J., Delp, M., Degris, T., Pilarski, P.M., White, A., Precup, D., 2011. Horde: a scalable real-time architecture for learning knowledge from unsupervised sensorimotor interaction. In: The 10th International Conference on Autonomous Agents and Multiagent Systems, vol. 2. International Conference for Autonomous Agents and Multiagent Systems, Taipei, Taiwan, pp. 761–768.

Taffoni, F., Tamilia, E., Focaroli, V., Formica, D., Ricci, L., Di Pino, G., Baldassarre, G., Mirolli, M., Guglielmelli, E., Keller, F., 2014. Development of goal-directed action selection guided by intrinsic motivations: an experiment with children. Exp. Brain Res. 232 (7), 2167–2177.

Waelti, P., Dickinson, A., Schultz, W., 2001. Dopamine responses comply with basic assumptions of formal learning theory. Nature 412 (6842), 43–48.

Weiskrantz, L., Cowey, A., 1963. The aetiology of food reward in monkeys. Anim. Behav. 11 (2–3), 225–234.

Weizmann, F., Cohen, L., Pratt, R., 1971. Novelty, familiarity, and the development of infant attention. Dev. Psychol. 4 (2), 149–154.

White, R., 1959. Motivation reconsidered: the concept of competence. Psychol. Rev. 66, 297–333.

CHAPTER

Applied economics: The use of monetary incentives to modulate behavior

12

S. Strang*[,1,2], S.Q. Park*[,1], T. Strombach[†], P. Kenning[†]

*University of Lübeck, Lübeck, Germany

[†]Heinrich-Heine-University Düsseldorf, Düsseldorf, Germany

[2]Corresponding author: Tel.: +49-451-3101-3611; Fax: +49-451-3101-3604, e-mail address: sabrina.strang@uni-luebeck.de

Abstract

According to standard economic theory higher monetary incentives will lead to higher performance and higher effort independent of task, context, or individual. In many contexts this standard economic advice is implemented. Monetary incentives are, for example, used to enhance performance at workplace or to increase health-related behavior. However, the fundamental positive impact of monetary incentives has been questioned by psychologists as well as behavioral economists during the last decade, arguing that monetary incentives can sometimes even backfire. In this chapter, studies from proponents as well as opponents of monetary incentives will be presented. Specifically, the impact of monetary incentives on performance, prosocial, and health behavior will be discussed. Furthermore, variables determining whether incentives have a positive or negative impact will be identified.

Keywords

Extrinsic motivation, Intrinsic motivation, Crowding out, Monetary incentives

1 INTRODUCTION

Does performance-based salary, as bonuses or profit sharing increase productivity of employees? Should people be rewarded for exercising regularly or quitting to smoke? Do people show more prosocial behavior when monetarily incentivized? Or generally asked, do monetary incentives always modulate human motivation and, thereby, change behavior in a desired way? According to standard economic theory the answer is clearly yes; higher incentives will automatically lead to higher effort (Baker et al., 1988). Based on this standard economic theory a variety of

[1]These authors contributed equally to this paper.

Progress in Brain Research, Volume 229, ISSN 0079-6123, http://dx.doi.org/10.1016/bs.pbr.2016.06.010

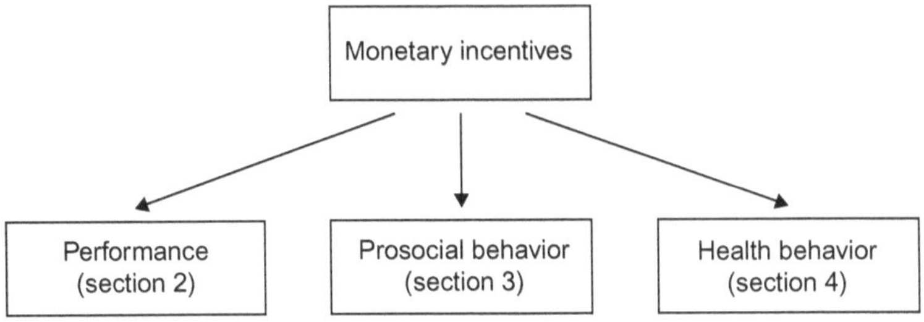

FIG. 1

Overview of chapter structure.

behavior in everyday life is incentivized. Monetary rewards are, for example, frequently used as a method for motivating employees to work better or people to live healthier (Heinrich and Marschke, 2010; Rothstein, 2008). Some companies offer their employees a bonus for good performance and health insurance companies offer a similar bonus for documented sport activities and courses. However, psychologist and behavioral economist doubt the positive effect of these monetary incentives on human motivation (Ariely et al., 2009a,b; Deci et al., 1999). They argue that monetary incentives do even backfire in some situation, meaning that they decrease motivation.

There is thus no consensus on the motivation enhancing effect of monetary incentives among different research disciplines. It is therefore still under debate whether incentive schemes used in organizations and other domains keep what standard economic theories promise. Against this background, in this chapter, scientific research on the supposedly beneficial effects of monetary incentives will be reviewed. Monetary incentives are used in a variety of context, however, for the sake of briefness, we will focus on the influence of monetary incentives on performance, prosocial behavior, and health behavior (Fig. 1). Doing so, we will show that even if monetary incentives do function as a motivator in some contexts, they appear to be counterproductive in others. Finally, we will discuss the context characteristics responsible for these ambiguous and puzzling results.

2 INCENTIVIZING PERFORMANCE: THE MORE MONEY THE BETTER?

In standard economics the principle of higher incentives leading to higher performance is believed to be a basic law of behavior (Baker et al., 1988). Incentives thus influence the degree of motivation, which results in analogous performance as outcome. Similar predictions come from behaviorists like Skinner (1963) and are known as "operant conditioning." He and other behaviorists have argued that reward and punishment modulate behavior. A behavior will increase in frequency, when it is followed by reward. Analogously, a behavior that is followed by punishment will decrease in frequency (Skinner, 1963). This theory was initially supported by animal research showing that rats' behavior (lever pressing) was strengthened by food rewards (Uhl and Young, 1967).

Human studies have also supported this direct link between incentives and behavioral increase. Toppen (1965) demonstrated that "humans responding to monetary reward are like other animals responding to other types of reward. They show greater performance in response to greater magnitude and greater ratio of reward incidence to units of effort expended" (p. 267, Toppen, 1965). In his study, students were asked to pull a handle that had certain constant tension. Half of the students received a very small reward for a specific amount of pulls (1 cent), whereas the other half received a rather large reward (25 cent). Those participants who received a larger reward showed a higher work output. These results were supported by a variety of other laboratory studies using different measures of performance (Chung and Vickery, 1976; London and Oldham, 1977; Terborg and Miller, 1978; Wimperis and Farr, 1979). The number of nut-and-bolt connections, supervisor ratings, and the number of correctly sorted cards are examples for performance measurements used in laboratory settings and were shown to increase when monetary incentives were introduced (London and Oldham, 1977; Wimperis and Farr, 1979).

Furthermore, an increase in performance due to incentives could also be demonstrated in field experiments (Latham and Dossett, 1978; Luthans et al., 1981; Pritchard et al., 1976; Yukl et al., 1976). For example, students in an Air Force technical training environment who received performance-dependent monetary rewards required less time to complete the training compared to those who did not receive monetary incentives (Pritchard et al., 1976). Trappers, whose job it is to catch mountain beavers, caught more beavers when receiving a monetary reward per beaver (Latham and Dossett, 1978) and salespersons in a department store sold more when monetarily rewarded for their sales (Luthans et al., 1981).

2.1 DOES MONEY MAKE EVERYTHING BETTER?

A metaanalysis encompassing 39 studies (mostly psychological) on the effect of monetary incentives on performance indicated that monetary incentives are positively associated with performance quantity, but not with performance quality (Jenkins et al., 1998). Thus, introducing monetary incentives increases performance quantity and, thereby, supports standard economic theory. However, the latter finding on performance quality raises some doubts on the general positive impact of monetary incentives. Another metaanalysis reported 74 studies (mainly economic) in which the level of incentives varied in different kinds of tasks (Camerer and Hogarth, 1999). Here, results were rather inconclusive. Incentives improved performance in easy tasks like problem-solving or recalling items from memory; however, incentives did not increase performance in difficult tasks, like auctions.

Moreover, recent studies have indicated that the relation between incentives and performance is not as monotonic as previous studies suggest (Ariely et al., 2009b;

Gneezy & Rustichini, 2000). For example, in a study by Gneezy and Rustichini (2000) students were allocated into four different groups receiving ascending levels of rewards (nothing, very small reward, large reward, very large reward) for correctly solved quizzes (50 in total), which were chosen to make the probability of a correct answer dependent on effort. However, the participants did not know that participants in other groups were paid differently. Students in the large and very large reward group answered most questions correctly (both 34), students in the no reward group answered 28 questions correctly, and students in the small reward group answered only 23 correctly. Thus, while large rewards increase performance, small rewards can even decrease performance compared to no reward at all.

However, very large rewards can also decrease performance. Ariely and colleagues (2009b) incentivized residents of an Indian village with either small, medium, or very large monetary rewards depending on the group they were allocated to. The participants had to execute six different tasks and their rewards were based on their performance. Maximum performance in all six tasks yielded a total reward approximately equal to half of the mean yearly consumer expenditure in the village. Interestingly, individuals' performance increased as the level of incentive increased only up to a point after which greater incentives became detrimental to performance. This decrease was observed across all six tasks (Ariely et al., 2009b).

Thus, as a body of literature indicated, higher incentives do not strictly increase performance, sometimes incentives can even have a negative effect on performance. Furthermore, although incentives might sometimes be effective in the short run, the long-term effects of incentives are not covered by most of the reported studies (Camerer and Hogarth, 1999; Gneezy and Rustichini, 2000). The question is, what happens when monetary incentives are removed after the reward-dependent performance change has taken place? Does performance stay on a higher level? Does is go back to baseline or even below? This question will be answered in the following.

2.2 WHEN MONETARY INCENTIVES BACKFIRE

Already in the 1950s some animal researcher started doubting the more or less general positive relation between rewards and performance as described in the previous paragraph. Monkeys, for example, were shown to work on a "puzzle apparatus" over an extended period without receiving any reward. When rewards were introduced (in form of raisins) in one group, performance of this group increased initially. However, when rewards were removed the group made more errors and had fewer correct solutions compared to the group who did not receive any reward (Harlow et al., 1950). This drop in performance can be explained by a decrease in intrinsic motivation. According to Deci (1971) "one is said to be intrinsically motivated to perform an activity when he receives no apparent rewards except the activity itself" (p. 105, Deci, 1971). Thus, intrinsic motivation modulates behavior, due to the rewarding aspect of the behavior per se. In contrast, extrinsic motivation changes behavior only because one receives a reward for it (Deci, 1971). This phenomen is called "crowding out" or "hidden costs of rewards" (Ariely et al., 2009b; Frey and Oberholzer-Gee,

1997; Gneezy and Rustichini, 2000). In the earlier mentioned study monkeys were first intrinsically motivated to solve the puzzles; they did not receive any reward for solving the puzzles and, therefore, are thought to have enjoyed the task itself. However, after receiving a reward for the task, their motivation shifted from intrinsic to extrinsic. After the rewards were withdrawn extrinsic motivation diminished as well and there was no motivation for solving the puzzles any more. Thus, while monetary incentives do have a positive impact on external motivation, they might undermine intrinsic motivation (Arnold, 1976; Daniel and Esser, 1980; Deci, 1971; Deci et al., 1999; Earn, 1982).

This decrease in intrinsic motivation through the introduction of external incentives was shown in a variety of human studies as well (Arnold, 1976; Daniel & Esser, 1980; Deci, 1971; Deci et al., 1999; Earn, 1982). In a pioneering study by Deci (1971) students were asked to solve interesting puzzles within a time limit of 13 min. The experiment consisted of three phases and two conditions; an experimental and a control condition. In both conditions subjects were not paid during phase one and three. In phase two, however, participants in the treatment condition were paid $1 when they solved a puzzle while subjects in the control group were not paid. In the middle of each phase the experimenter left the room for 8 min. Motivation was measured by the amount of time participants spend on solving the puzzle during these 8 min. Participants who were paid during the second phase spend less time on solving the puzzles compared to those who did not receive any reward (Deci, 1971). This crowding out effect could be replicated in a variety of studies using different tasks (Arnold, 1976; Daniel & Esser, 1980; Deci et al., 1999; Earn, 1982). Most studies measured motivation as either voluntary time spent on a task during a free-choice period or completed trials during a free-choice period.

2.3 MONEY OR FEEDBACK?

Interestingly, different types of rewards do differ in their impact on intrinsic motivation. While monetary rewards, as discussed earlier, often decrease intrinsic motivation, verbal feedback/rewards were shown to have a positive influence on intrinsic motivation (Butler, 1987; Harackiewicz, 1979; Rosenfield et al., 1980; Zinser et al., 1982). In a study similar to the one by Deci (1971) students were asked to solve puzzles in several phases (Harackiewicz, 1979). Some of the students received verbal feedback about their performance; they were told that their performance was better than the average. Compared to students who did not receive any feedback, students who received feedback showed an increase in intrinsic motivation (Harackiewicz, 1979). The informational aspect of the feedback provides a critical role in its effect on intrinsic motivation, meaning that while rewards that do not indicate level of ability led to less intrinsic motivation, rewards that reflected ability (higher rewards for greater skill) increase intrinsic motivation (Rosenfield et al., 1980).

In a meta study including 128 experiments Deci and colleagues (1999) provided summary statistics on the effect of external incentives on intrinsic motivation. Their results indicated that most extrinsic incentives schemes using monetary rewards undermine intrinsic motivation. They distinguished between task-noncontingent, task-contingent, and performance-contingent rewards. Task-noncontingent rewards are given independent of the performance or engagement in the target task; they are assigned for simple participation. Task-contingent rewards, in contrast, are conditional on completing the target task and performance-contingent rewards on performing the target task well (eg, doing better than 80% of the other participants). While monetary rewards had no impact on intrinsic motivation in the first incentive scheme, in the latter two incentive schemes, intrinsic motivation decreased. In contrast verbal feedback had either no impact or even a positive influence on intrinsic motivation independent of the incentive scheme (Deci et al., 1999). Thus, in the long run (after incentives are withdrawn) monetary incentives seem to backfire, leading to decreasing performance instead of increasing.

Monetary incentives were also shown to have a negative impact in the short run while incentives are present. In a field study by Gneezy and Rustichini (2000) students collected donations for a charity in a door-to-door charity campaign. Students were distributed into three groups, while two groups received a payment depending on their collected donations (either 1% or 10%), the third group was not monetarily incentivized. For those students who could keep 1% of the collected money, the amount collected was reduced by 36% and for those who could keep 10% of the money collected the reduction was still 8% compared to those who were not monetarily incentivized (Gneezy and Rustichini, 2000). These results show that under specific conditions monetary incentives can even have an undermining effect on performance in the short run.

2.4 EXPLAINING THE CROWDING OUT EFFECT

Cognitive evaluation theory (Deci and Ryan, 1985, 2012), psychological contract theory (Rousseau, 1998, 2001), adaptation level theory (Helson, 1948), and crowding theory (Frey and Jegen, 2001; Frey and Oberholzer-Gee, 1997) can specify under which conditions intrinsic motivation is decreased or increased by incentives. While the former three are psychological theories the latter is an economic approach.

According to cognitive evaluation theory three psychological needs underlie intrinsic motivation; the need for competence, the need for relatedness, and the need for autonomy (Deci and Ryan, 1985, 2012). The effect of incentives on intrinsic motivation depends on how rewards affect these needs. In case that incentives have a positive effect on these needs by enhancing perceived relatedness, autonomy, and/or competence, this leads to an increase of intrinsic motivation and, accordingly, performance in a task. In contrast, if the effect is negative, perceived relatedness, autonomy, and competence are reduced and intrinsic motivation will be decreased. While some incentives have a positive effect on these needs, others have a negative effect. Monetary rewards are perceived as controlling one's behavior and, thereby, reducing perceived autonomy and decreasing intrinsic motivation. Informational verbal reinforcement on the other hand is perceived as emphasizing competence and, as a consequence, increasing intrinsic motivation (Deci and Ryan, 1985, 2012).

Psychological contract theory suggests that social relations establish implicit contracts (Rousseau, 1998, 2001). These contracts involve emotional ties and loyalties that go beyond transactional exchanges but include a reciprocal appreciation of intrinsic motivation. Introducing monetary incentives influences the contract by derogating the reciprocal appreciation. This results in a transformation from an intrinsically motivated into an extrinsically motivated contract. The perception of fairness plays an important role in psychological contract theory (Rousseau, 1998, 2001). Perceived fairness strengthens emotional ties and therefore increases intrinsic motivation; incentives thus need to account for the type of contract and motives in order to increase intrinsic motivation (Rousseau, 1998, 2001).

Whereas the former two theories are based on classical psychological concepts, adaptation level theory is based on psychophysical findings (Helson, 1948) and assumes that exposure to previous stimuli serves as a reference for subsequent stimuli. In detail, the weighted log mean of previous stimuli forms the personal adaptation level, which is used to judge subsequent stimuli (Helson, 1948). Every new stimulus is integrated and therefore the adaptation level can shift as a consequence. In the context of performance payment this means, that introducing monetary incentives results in a large difference between the current adaptation level and the payment, resulting in a positive response, which can be interpreted as motivation (Bowling et al., 2005). However, with repetition the adaptation level shifts upwards resembling the new payment level. Thus, similar monetary incentives only have a positive impact on performance as long as they differ from the personal adaptation level, as soon as someone gets adapted to the new level, even higher incentives are needed to yield identical results.

In contrast to standard economic theories crowding theory accounts for intrinsic motivation (Frey and Jegen, 2001; Frey and Oberholzer-Gee, 1997). In addition to the three previous psychological theories and in line with standard economic theories, crowding theory also accounts for relative prizes. Thus, according to crowding theory monetary incentives can have a positive or negative effect on performance, depending on whether their effect dominates relative prizes or intrinsic motivation. If there is no intrinsic motivation, an increase in incentives will always improve performance. However, if there is some amount of intrinsic motivation already given, improved performance is observable even if no monetary incentives are offered. Here, the introduction of monetary incentives would initially decrease performance to a lower level. However, raising incentives would improve performance based on this lower level (Frey and Jegen, 2001; Frey and Oberholzer-Gee, 1997). Monetary incentives thus improve performance depending on the level of intrinsic motivation.

2.5 NEW INSIGHTS INTO OLD CONCEPTS ON MOTIVATION

As described earlier the impact of monetary incentives on motivation has already been investigated for decades. In particular, the reported negative impact on intrinsic motivation has fascinated and puzzled researchers in psychology and economics (Deci et al., 1999; Gneezy et al., 2011). The underlying neural mechanism of this phenomenon, however, has not been investigated until recently. With the technical advances, functional magnetic resonance imaging (fMRI) nowadays allows to get new insight into the old debate by investigating which brain processes underlie this phenomenon (Albrecht et al., 2014; Chib et al., 2012; Mobbs et al., 2009; Murayama et al., 2010; Strombach et al., 2015).

The first fMRI studies on the influence of increasing monetary incentives on motivation investigated the effect of different reward sizes on performance and brain activity (Chib et al., 2012; Mobbs et al., 2009). A motor task was used as a target task, where participants had to move a mass from a start to a target position 20 cm away. Performance was incentivized differently across trials ranging from \$0 to \$100 (Chib et al., 2012). In line with Ariely et al. (2009b) and Gneezy and Rustichini (2000; both mentioned in Section 2.1), their results indicated that participants showed performance improvement with increasing incentives up to a certain level. However, beyond this point no further performance enhancement could be observed, even though incentives further increased. They further showed that brain activity in reward areas predicts behavior. Activity during the actual task decreased with respect to the magnitude of incentives. While Mobbs and colleagues (2009) interpreted this drop in activation as "overmotivation" signal for high rewards, Chib and colleagues (2012) explained the decrease in activity with loss aversion. According to the latter explanation people are afraid of losing money when incentives are high, which in turn decreases performance as well as brain activity.

Another line of fMRI studies investigated the "crowding out effects" of monetary rewards on intrinsic motivation. Therefore, Murayama and colleagues used a set up analogous to previous behavioral paradigms (Murayama et al., 2010). Here, half of the participants were paid for performing a given task, while the others did not receive any task-related payment. The intrinsic motivation was assessed during a free-choice phase after incentives were withdrawn. In this study participants saw a stopwatch that started automatically and their task was to press a button when 50 ms were displayed. Those participants who were in the incentive group received \$2.20 for each successful button press. Participants in the other group only saw the feedback, but did not receive any money. During the free-choice phase participants who were incentivized showed less engagement in the task, replicating previous results. Furthermore, during this phase, activity in the ventral striatum, an area known to be involved in reward processing (Fig. 2; Haber and Knutson, 2010; Park et al., 2012), was decreased in participants who were incentivized (Murayama et al., 2010). In contrast, in the incentivized phase, activity in the ventral striatum showed an increase along with performance compared to the control group. The reward value of performing the task, thus, first increased when monetary incentives were introduced and then decreased when those were withdrawn, indicating that the undermining effect of monetary incentives is also reflected on the brain level.

In a similar paradigm Strombach and colleagues (2015) could replicate the increase of activity in reward-related regions due to introducing monetary incentives (Fig. 2). Interestingly, they did not find any neural changes related to the task participants executed. Instead, they could show a decrease in activity in the

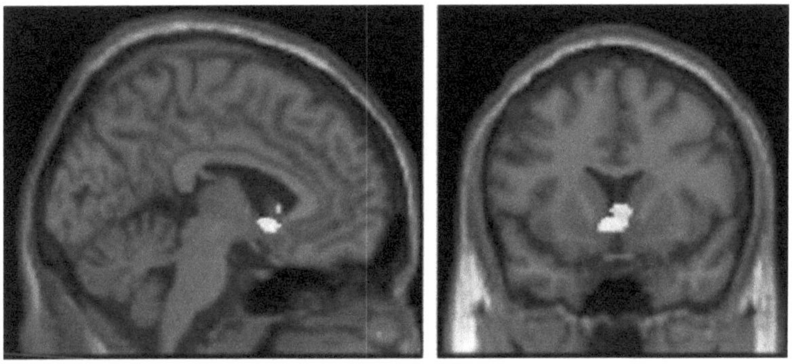

FIG. 2

Visualization of activity within the ventral striatum, no monetary incentives vs monetary incentives (Strombach et al., 2015).

ventromedial prefrontal cortex, representing subjective value. As a consequence, Strombach et al (2015) suggested that incentives do not directly affect performance by modulating neural activity in task-relevant regions, but affects the reward representation during task completion.

In a similar vein, Albrecht and colleagues (2014) investigated the "crowding out effect" and its neural basis. However, in addition they were interested in the difference between verbal and monetary rewards specifically. In their study participants in two groups received either 1€ or verbal feedback (Very well done!) for every correctly solved task. In a control group participants received neither money nor verbal feedback. In both groups the task was to find differences between two pictures. The authors could replicate the increase in activity in the ventral striatum when monetary incentives were introduced. However, they could not detect a decrease in activation after monetary incentives were withdrawn as shown by Murayama et al. (2010). Thus, although participants showed a decrease in performance, they could not show this pattern in the brain data. Verbal feedback in contrast was shown to increase performance as well as activity in the ventral striatum beyond the feedback phase, thereby reflecting the positive effect of verbal feedback on intrinsic motivation (Albrecht et al., 2014).

2.6 SUMMARY

The previously mentioned literature shows that introducing monetary incentivizes does not always increase performance as predicted by standard economic theory. Very large or very low rewards were, for example, shown not to increase performance (Ariely et al., 2009b; Gneezy and Rustichini, 2000). Furthermore, the withdrawal of previously introduced incentives can even decrease performance by decreasing intrinsic motivation (Deci, 1971). This effect is called "crowding out" or "hidden costs of rewards." Cognitive evaluation theory (Deci and Ryan, 1985, 2012), psychological contract theory (Rousseau, 1998, 2001), crowding theory

(Frey and Jegen, 2001), and adaptation level theory (Helson, 1948) can explain this decrease in intrinsic motivation. The decrease in intrinsic motivation due to withdrawing incentives is specific to monetary rewards; in contrast, verbal rewards do not have a negative effect on intrinsic motivation (Deci et al., 1999). In addition, neuroimaging studies provide evidence supporting the behavioral results contradicting standard economic theory. It could, for example, be demonstrated that large rewards (Chib et al., 2012; Mobbs et al., 2009) or the withdrawal of rewards (Murayama et al., 2010; Strombach et al., 2015) resulted in a decrease in activity in reward-related brain areas resembling the behavioral results.

Thus, in summary, monetary rewards can have a positive effect on performance in the short run; however, in the long run they might backfire and even decrease performance.

3 INCENTIVIZING PROSOCIAL BEHAVIOR

The standard economic theory often assumes that people are purely selfish. However, plenty of research has shown that people take into account the welfare of others (Andreoni, 1990; Fehr and Rockenbach, 2004). Many people spend own resources, as, for example, time or money, in order to increase the utility of others without directly benefitting themselves, this behavior can be regarded as prosocial (Gintis et al., 2003). Increasing prosocial behavior could enhance total welfare. For this purpose, incentives schemes could potentially be applied. Does paying people for helping others increase their prosocial behavior? For instance, do people donate more money or blood when they are incentivized to do so? There are a variety of factors driving people to behave prosocially (Strang and Park, 2016), among other things, reputation and a positive feeling from doing good (known as "warm glow"). Some people donate, for example, because they want to signal others that they are "nice" thereby improving their reputation (Fehr, 2004). Others might donate money because they derive a good feeling from doing good for others (Dunn et al., 2008). Both motives might be dramatically influenced by monetary incentives. When observing prosocial behavior, the introduction of monetary incentives makes it difficult to know whether the person shows increased performance due to doing good (high performance because really being prosocial) or to doing well (high performance in order to receive money). This problem holds for others observing the behavior but also for the person him or herself. When the costs of being prosocial are compensated, the act loses its prosocial character. Thus, the signal to oneself and others of a prosocial act becomes unclear when incentives are introduced.

3.1 MONEY DESTROYS PROSOCIALITY

The negative impact of monetary incentives on prosocial behavior was first mentioned in the context of blood donations. Monetary compensation for blood donation was explained to decrease people's willingness to donate blood (Titmus, 1971).

This effect could be empirically demonstrated especially in woman (Mellström and Johannesson, 2008). Participants were either offered nothing or $7 for their blood donation. In general, the number of blood donors decreased from 43% to 33% when a payment was offered. For women it even decreased from 52% to 30%. Interestingly, when participants had the possibility to donate the monetary compensation to a charity organization the number of blood donors was comparable to when no incentives were given (Mellström and Johannesson, 2008).

Crowding out of prosocial behavior could also be shown in an important real life setting; the so-called Not in my backyard (NIMBY) projects (Frey and Oberholzer-Gee, 1997). These are projects that increase overall welfare but are often unwanted by local communities. Typical examples are airports, prisons, or repositories for nuclear waste. Most people demand these facilities but refuse to have them in their home region. Since the projects increase overall welfare and include some personal costs, it can be regarded as prosocial to approve such a project in the home region. In 1993 the Swiss Government planned to build two repositories for nuclear waste. In this context, 305 personal interviews were conducted to get an impression of citizens' opinions on the project. When asked whether they were willing to permit repositories for nuclear waste in their neighborhood, surprisingly, a large fraction of participants (50%) agreed (Frey and Oberholzer-Gee, 1997). However, when offered a compensation for their permission by the Swiss Government, agreements declined. Compensation ranged from $2.175 to $6.525 per individual and year. Acceptance levels of the repository dropped to 24.6% when compensation was introduced. This decline was independent of the amount offered as compensation. Even when offered $2.000 in addition the participants did not change their original opinion (Frey and Oberholzer-Gee, 1997).

A contrasting result was shown by Ariely and colleagues (2009a). In a study, they investigated the differential effect of monetary incentives on reputation and "warm glow" or other prosocial motives. They called their paradigm "Click for Charity." Here, participants were asked to click two keys on a computer keyboard in order to earn money for a charity organization. In another condition participants were additionally incentivized with money for clicking. The total amount donated per subject was either kept private or made public. Without monetary incentives, participants showed greater effort (more clicks) in the public compared to the private condition. While monetary incentives declined effort in the public condition they increased effort in the private condition (Ariely et al., 2009a). People, thus, seem to be concerned about their reputation. The authors suggested that participants showed decreased performance in the public condition when monetary incentives were introduced because they want others to think that they perform driven by altruistic and not egoistic reasons.

3.2 MATCHING DONATIONS

Prosocial behavior was shown to decline when receiving money for it, probably because money takes away the prosocial character of this act. However, since people are concerned about being prosocial out of altruistic and not egoistic reasons, spending the money on the charity organization instead of for the participant might increase prosocial behavior in the charity context. One example for this procedure is called "matching," meaning in case of donation, a certain percentage of their donation will be additionally provided, thereby increasing the total amount donated. Therefore, matching does not alter the prosocial character of donations.

Laboratory experiments indicate that matching the donations of participants indeed increases donations (Eckel and Grossman, 2003). Participants received an endowment and could decide how much to keep for themselves and how much to donate to a charity organization. Participants were informed that their donations were matched by a certain percentage of their own donations (25%, 33%, or 100%). Matching the donations led to a higher amount of charitable giving than other incentive mechanisms (Eckel and Grossman, 2003). This effect could be replicated in a field experiment. At the University of Zurich in Switzerland each semester students are asked anonymously whether they want to contribute to one or two social funds or not. In 1 year, donations of 600 randomly selected students were matched (by 25% or 50%) under the condition that they contribute to both funds. Compared to a control group, donations increased when they were matched. However, in the period after the matching procedure their donation behavior decreases even below the prematching period behavior (Meier, 2007).

4 INCENTIVIZING HEALTH BEHAVIOR

The increasing prevalence of overweight and obesity has dramatic economic consequences for society but is of course also harmful for affected individuals (Wang et al., 2011). Addressing the causes of overweight, namely a lack of physical activity and high calorie intake, is therefore beneficial on individual as well as on societal level. Consequently, the question arises, whether incentive schemes can either increase physical activity or reduce calorie intake.

The effect of incentives on physical activity was investigated by Charness and Gneezy (2009). They conducted a field experiment in which students were offered monetary incentives when attending the university's gym. Participants were divided into three groups. In one group participants did not receive any incentives, in the other two group participants received $25 for visiting the gym at least once during the following week. After this week, one of the two groups received additionally $100 for visiting the gym at least eight times in the following 4 weeks. For those who did not visit the gym before the experiment, ie, those who were not intrinsically motivated to do sports, the incentive scheme of the third group increased attendance rates. Paying participants to visit the gym at least eight times increased the number of visits during and even after the intervention time compared to the control group. In addition they could show that participants in the "eight times gym" group derived health benefits from the intervention; they showed changes in several health-related biometric indices (as, for example, body fat and weight) compared to the other groups (Charness and Gneezy, 2009).

Analogously, it has also been shown that negative incentives can be introduced in order to decrease calorie intake. Taxes on high caloric or sugared food, similar to taxes on alcohol or tobacco, could, for example, be used as negative incentives in order to decrease the intake of these products. Introducing a tax of one-penny-per-ounce on sugar beverages is proposed to decrease the consumption 13% (Blum et al., 2009). In an experimental study Epstein and colleagues (2010) could confirm this positive effect of taxes. They set up a supermarket in their laboratory and gave two groups of participants $15 to buy products. Prizes for high and low caloric food items differed between groups. They could demonstrate that an increase in prizes of high caloric food by 10% decreased the total amount of calories purchased by 6.5% (fat calories by 12.8% and carbohydrate calories by 6.2%).

Thus, an increase in physical activity as well as a decrease in caloric intake could be achieved by introducing positive or negative incentives, respectively, at least for those who displayed unhealthy behavior before. However, whether incentives have a positive impact on health behavior in the long run is not known so far.

5 CONCLUSION

Monetary incentives do have a positive impact on behavior in specific situations. However, this positive impact is influenced by a variety of moderators and mediators. The initial type of motivation (intrinsic or extrinsic), the type of incentive scheme (task-noncontingent, task-contingent, or performance-contingent), and time (short-term or long-term effects), the type of incentive (monetary or verbal feedback), the type of task (easy or difficult), and the type of context (working, social, or health context) are variables influencing the impact of monetary incentives on behavior.

However, while we have reason to believe, that in the short run monetary incentives do have a positive impact on performance in case people are not intrinsically motivated in advance and incentives are based on the task or performance (Latham and Dossett, 1978; Luthans et al., 1981; Pritchard et al., 1976; Toppen, 1965). Nevertheless, when people are intrinsically motivated or incentives are not task or performance related, introducing monetary incentives does not have a positive effect and, as in case of intrinsic motivation, can even backfire (Deci et al., 1999). Furthermore, when withdrawing monetary incentives, the impact is reversed and performance drops to a level lower than before incentives where introduced (Deci et al., 1999). This negative effect on intrinsic motivation is specific to monetary incentives, verbal feedback, addressing competence, in contrast has a positive effect on intrinsic motivation (Harackiewicz, 1979; Rosenfield et al., 1980). Since people are often intrinsically motivated to solve difficult tasks, monetary incentives better work for easy tasks (Camerer and Hogarth, 1999). Furthermore, the context plays a critical role, while monetary incentives can have a positive impact in the working and health context; they mostly have a negative impact in the social context (Frey and Oberholzer-Gee, 1997; Mellström and Johannesson, 2008; Toppen, 1965).

Thus, when aiming to increase behavior by introducing monetary incentives, all these variables need to be considered. In the working context this may mean, that monetary incentives should only be provided to workers who are not intrinsically motivated. Relating this to the differential effect on easy vs difficult tasks, probably only workers executing very easy tasks are purely extrinsically motivated, indicating that only those will be positively influenced by monetary incentives. However, since most workers are probably intrinsically motivated to some extent, introducing monetary incentives entails the risk to decrease performance in the long run. The increase in prosocial behavior by introducing monetary incentives does not seem to be effective. Here, the intrinsic nature of prosocial behavior plays a crucial role. Introducing monetary incentives takes away the prosocial character. However, when spending the money on the charity organization (via matching procedures) instead of on the individual person, monetary incentives were shown to increase prosocial behavior in the short run. However, in the long run, as for performance, a drop below baseline could be observed. Interventions in the health context mainly focus on improving behavior of those who are not intrinsically motivated, thus here monetary incentives can increase health-related behavior. The long-term effect should, however, be investigated in more detail.

To sum up, monetary incentives should be used with caution, while effective when certain situational and personal factors are given, they backfire in most situations. However, since these factors were mostly studied in isolation, a unifying model including both situational as well as personal variables is needed in order to allow for concrete predictions about when monetary incentives are effective. The initial type of motivation is, for example, a personal variable lacking in most models on motivation, presumably explaining a large fraction of the variance. Furthermore, incentive scheme, task type, and incentive type should be investigated jointly and integrated into one model. Finally, models should be more explicit about the time window of the predicted effect. Thus, although research from the field of economics, psychology, and neuroscience has already disentangled parts of the discussion, the debate about the effect of monetary incentives on motivation continues and further research is needed to receive a global, unified picture.

ACKNOWLEDGMENTS

This work was supported by Deutsche Forschungsgemeinschaft (DFG) Grants INST 392/125-1 and PA 2682/1-1 (to S.Q.P).

REFERENCES

Albrecht, K., Abeler, J., Weber, B., Falk, A., 2014. The brain correlates of the effects of monetary and verbal rewards on intrinsic motivation. Front. Neurosci. 8, 1–10.

Andreoni, J., 1990. Impure altruism and donations to public good: a theory of warm-glow giving. Econ. J. 100, 464–477.

Ariely, D., Bracha, A., Meier, S., 2009a. Doing good or doing well? Image motivation and monetary incentives in behaving prosocially. Am. Econ. Rev. 99, 544–555.

Ariely, D., Gneezy, U., Loewenstein, G., Mazar, N., 2009b. Large stakes and big mistakes. Rev. Econ. Stud. 76, 451–469.

Arnold, H.J., 1976. Effects of performance feedback and extrinsic reward upon high intrinsic motivation. Organ. Behav. Hum. Perform. 17, 275–288.

Baker, G.P., Jensen, M.C., Murphy, K.J., 1988. Compensation and incentives: practice vs. theory. J. Financ. 43, 593–616.

Blum, J.D., Conway, P.H., Sharfstein, J.M., 2009. Ounce of prevention—the public policy case for taxes on sugared beverages. N. Engl. J. Med. 30, 1805–1808.

Bowling, N.A., Beehr, T.A., Wagner, S.H., Libkuman, T.M., 2005. Adaptation-level theory, opponent process theory, and dispositions: an integrated approach to the stability of job satisfaction. J. Appl. Psychol. 90, 1044–1053.

Butler, R., 1987. Task-involving and ego-involving properties of evaluation: effects of different feedback conditions on motivational perceptions, interest, and performance. J. Educ. Psychol. 79, 474–482.

Camerer, C.F., Hogarth, R.M., 1999. The effects of financial incentives in experiments: a review and capital labor production framework. J. Risk Uncertain. 19, 7–42.

Charness, G.B., Gneezy, U., 2009. Incentives to exercise. Econometrica 77, 909–931.

Chib, V.S., De Martino, B., Shimojo, S., O'Doherty, J.P., 2012. Neural mechanisms underlying paradoxical performance for monetary incentives are driven by loss aversion. Neuron 74, 582–594.

Chung, K.H., Vickery, W.D., 1976. Relative effectiveness and joint effects of three selected reinforcements in a repetitive task situation. Organ. Behav. Hum. Perform. 16, 114–142.

Daniel, T.L., Esser, J.K., 1980. Intrinsic motivation as influenced by rewards, task interest, and task structure. J. Appl. Psychol. 65, 566–573.

Deci, E.L., 1971. Effects of externally mediated rewards on intrinsic motivation. J. Pers. Soc. Psychol. 18, 105–115.

Deci, E.L., Ryan, R.M., 1985. Intrinsic Motivation and Self-Determination in Human Behavior. New York: Pantheon; Berlin, Heidelberg.

Deci, E.L., Ryan, R.M., 2012. Motivation, personality, and development within embedded social contexts: an overview of self-determination theory. In: Ryan, R.M. (Ed.), Oxford Handbook of Human Motivation. Oxford University Press, Oxford, UK, pp. 85–107.

Deci, E.L., Koestner, R., Ryan, R.M., 1999. A meta-analytic review of experiments examining the effects of extrinsic rewards on intrinsic motivation. Psychol. Bull. 125, 627–668.

Dunn, E.W., Aknin, L.B., Norton, M.I., 2008. Spending money on others promotes happiness. Science 319, 1687–1688.

Earn, B.M., 1982. Intrinsic motivation as a function of extrinsic financial rewards and subjective locus of control. J. Pers. 50, 360–373.

Eckel, C.C., Grossman, P.J., 2003. Rebate versus matching: does how we subsidize charitable contributions matter? J. Public Econ. 87, 681–701.

Epstein, L.H., Dearing, K.K., Roba, L.G., Finkelstein, E., 2010. The influence of taxes and subsidies on energy purchased in an experimental purchasing study. Psychol. Sci. 21, 406–414.

Fehr, E., 2004. Don't lose your reputation. Nature 432, 449–450.

Fehr, E., Rockenbach, B., 2004. Human altruism: economic, neural, and evolutionary perspectives. Curr. Opin. Neurobiol. 14, 784–790.

Frey, B.S., Jegen, R., 2001. Motivation crowding theory. J. Econ. Surv. 15, 589–611.
Frey, B.S., Oberholzer-Gee, F., 1997. The cost of price incentives: an empirical analysis of motivation crowding-out. Am. Econ. Rev. 87, 746–755.
Gintis, H., Bowles, S., Boyd, R., Fehr, E., 2003. Explaining altruistic behavior in humans. Evol. Hum. Behav. 24, 153–172.
Gneezy, U., Rustichini, A., 2000. Pay enough or don't pay at all. Q. J. Econ. 115 (3), 791–810.
Gneezy, U., Meier, S., Rey-Biel, P., 2011. When and why incentives (don't) work to modify behavior. J. Econ. Perspect. 25, 191–210.
Haber, S.N., Knutson, B., 2010. The reward circuit: linking primate anatomy and human imaging. Neuropsychopharmacology 35, 4–26.
Harackiewicz, J.M., 1979. The effects of reward contingency and performance feedback on intrinsic motivation. J. Pers. Soc. Psychol. 37, 1352–1363.
Harlow, B.Y.H.F., Harlow, M.K., Meyer, D.R., 1950. Learning motivated by a manipulation drive. J. Exp. Psychol. 40, 228–234.
Heinrich, C.J., Marschke, G., 2010. Incentives and their dynamics in public sector performance management systems. J. Pol. Anal. Manage. 29, 183–208.
Helson, H., 1948. Adaptation level as a basis for quantitative theory of frames of references. Psychol. Rev. 55, 297–313.
Jenkins, G.D., Mitra, A., Gupta, N., Shaw, J.D., 1998. Are financial incentives related to performance? A meta-analytic review of empirical research. J. Appl. Psychol. 83, 777–787.
Latham, G.P., Dossett, D.L., 1978. Designing incentive plans for unionized employees: a comparison of continuous and variable ratio reinforcement. Pers. Psychol. 31, 47–61.
London, M., Oldham, G.R., 1977. Comparison of group and individual incentive plans. Acad. Manage. J. 20, 34–41.
Luthans, F., Paul, R., Baker, D., 1981. An experimental analysis of the impact of contingent reinforcement on salespersons' performance behavior. J. Appl. Psychol. 66, 314–323.
Meier, S., 2007. Do subsidies increase charitable giving in the long run? Matching donations in a field experiment. J. Eur. Econ. Assoc. 5, 1203–1222.
Mellström, C., Johannesson, M., 2008. Crowding out in blood donation: was Titmuss right? J. Eur. Econ. Assoc. 6, 845–863.
Mobbs, D., Hassabis, D., Seymour, B., Marchant, J.L., Weiskopf, N., Dolan, R.J., Frith, C.D., 2009. Choking on the money. Psychol. Sci. 20, 955–962.
Murayama, K., Matsumoto, M., Izuma, K., Matsumoto, K., 2010. Neural basis of the undermining effect of monetary reward on intrinsic motivation. Proc. Natl. Acad. Sci. 107, 20911–20916.
Park, S.Q., Kahnt, T., Talmi, D., Rieskamp, J., Dolan, R.J., Heekeren, H.R., 2012. Adaptive coding of reward prediction errors is gated by striatal coupling. Proc. Natl. Acad. Sci. 109, 4285–4289.
Pritchard, R.D., DeLeo, P.J., Von Bergen, C.W., 1976. A field experimental test of expectancy-valence incentive motivation techniques. Organ. Behav. Hum. Perform. 15, 355–406.
Rosenfield, D., Folger, R., Adelman, H.F., 1980. When rewards reflect competence: a qualification of the overjustification effect. J. Pers. Soc. Psychol. 39, 368–376.
Rothstein, R., 2008. Holding Accountability to Account: How Scholarship and Experience in Other Fields Inform Exploration of Performance Incentives in Education. Working Paper.
Rousseau, D.M., 1998. The "problem" of the psychological contract considered. J. Organ. Behav. 19, 665–671.

Rousseau, D.M., 2001. Schema, promise and mutuality: the building blocks of the psychological contract. J. Occup. Organ. Psychol. 74, 511–541.

Skinner, B.F., 1963. Operant behavior. Am. Psychol. 18, 503–515.

Strang, S., Park, S.Q., 2016. Human cooperation and its underlying mechanisms. Current Topics in Behavioral Neurosciences, Springer Berlin Heidelberg, Berlin, Heidelberg.

Strombach, T., Hubert, M., Kenning, P., 2015. The neural underpinnings of performance-based incentives. J. Econ. Psychol. 50, 1–12.

Terborg, J.R., Miller, H.E., 1978. Motivation, behavior, and performance: a closer examination of goal setting and monetary incentives. J. Appl. Psychol. 63, 29–39.

Titmus, R.M., 1971. The Gift Relationship: From Human Blood to Social Policy. Pantheon Books, New York.

Toppen, J.T., 1965. Effect of size and frequency of money reinforcement on human operant (work) behavior. Percept. Mot. Skills 20, 259–269.

Uhl, C.N., Young, A.G., 1967. Resistance to extinction as a function of incentive, percentage of reinforcement and number of reinforcement trials. J. Exp. Psychol. 73, 556–564.

Wang, Y.C., McPherson, K., Marsh, T., Gortmaker, S.L., Brown, M., 2011. Health and economic burden of the projected obesity trends in the USA and the UK. Lancet 378, 815–825.

Wimperis, B., Farr, J., 1979. The effects of task content and reward contingency upon task performance and satisfaction. J. Appl. Soc. Psychol. 9, 229–249.

Yukl, G.a., Latham, G.P., Pursell, E.D., 1976. The effectiveness of performance incentives under continuous and variable ratio schedules of reinforcement. Pers. Psychol. 29, 221–231.

Zinser, O., Young, J.G., King, P.E., 1982. The influence of verbal reward on intrinsic motivation in children. J. Gen. Psychol. 106, 85–91.

CHAPTER

Rewarding feedback promotes motor skill consolidation via striatal activity

13

M. Widmer*,†,‡,1, N. Ziegler§,✝, J. Held*,†, A. Luft*,†, K. Lutz*,†,¶

**University Hospital of Zurich, Zurich, Switzerland*

†Cereneo, Center for Neurology and Rehabilitation, Vitznau, Switzerland

‡Neural Control of Movement Lab, ETH Zurich, Zurich, Switzerland

§Institute of Human Movement Sciences and Sport, ETH Zurich, Zurich, Switzerland

¶Institute of Psychology, University of Zurich, Zurich, Switzerland

[1]Corresponding author: Tel.: +41 44 255 88 06; Fax: +41 44 255 12 80, e-mail address: widmemar@ethz.ch

Abstract

Knowledge of performance can activate the striatum, a key region of the reward system and highly relevant for motivated behavior. Using functional magnetic resonance imaging, striatal activity linked to knowledge of performance was measured during the training of a repetitive arc-tracking task. Knowledge of performance was given after a random selection of trials or after good performance. The third group received knowledge of performance after good performance plus a monetary reward. Skill learning was measured from pre- to post- (acquisition) and from post- to 24 h posttraining (consolidation). Our results demonstrate an influence of feedback on motor skill learning. Adding a monetary reward after good performance leads to better consolidation and higher ventral striatal activation than knowledge of performance alone. In turn, rewarding strategies that increase ventral striatal response during training of a motor skill may be utilized to improve skill consolidation.

Keywords

Motor skill learning, Monetary reward, Performance feedback, Knowledge of performance, fMRI, Striatum, Pointing task, Consolidation

✝Deceased July 26, 2014.

Progress in Brain Research, Volume 229, ISSN 0079-6123, http://dx.doi.org/10.1016/bs.pbr.2016.05.006

Abbreviations

fMRI functional magnetic resonance imaging
GLMM generalized linear mixed model

1 INTRODUCTION

Extrinsically motivated actions, are performed because they lead to an outcome, eg, to a reward (Ryan and Deci, 2000). By increasing the extrinsic subjective value, rewards augment the overall subjective benefit of a task, making people tolerate higher subjective costs, and are thus traditionally defined as stimuli an organism is willing to work for (Knutson and Cooper, 2005; Lutz and Widmer, 2014). Intrinsic motivation, on the other hand, refers to doing something because it is inherently interesting or enjoyable, which is influenced by factors such as the subject's perceived autonomy, competence for or relatedness to a task (Ryan and Deci, 2007). Similar to motivation, reward can be classified as extrinsic or intrinsic (Deci et al., 1999, 2001; Reitman, 1998). While extrinsic reward refers to the receipt of material (eg, food or money) for a specific activity, the term "intrinsic reward" refers to reward derived from task inherent stimulation (eg, information about an achieved performance, watching a self-painted picture, or feeling self-produced movements). Evidence from behavioral studies implies that extrinsic reward might undermine intrinsic motivation and thus may lead to a decrease in performance (Callan and Schweighofer, 2008; Deci et al., 1999; Kohn, 1999; Murayama et al., 2010; Spence, 1970). For instance, the time children spend drawing decreases below baseline after this behavior had been (externally) rewarded and the reward has then been withdrawn (Greene and Lepper, 1974).

In experiments using functional magnetic resonance imaging (fMRI), both intrinsic and extrinsic (performance-dependent) reward have been shown to increase the neural activity in the striatum (Lutz et al., 2012), a key locus of reward processing (Knutson et al., 2008). In these experiments, only the ventral striatum was active during performance feedback, while feedback plus monetary reward activated both ventral and dorsal parts of the striatum. However, other studies found activation elicited by feedback alone also in the dorsal striatum (Poldrack et al., 2001; Tricomi and Fiez, 2008; Tricomi et al., 2004, 2006). Furthermore, dorsal striatal activity was shown to be modulated by the subject's sense of agency for having achieved a goal (Han et al., 2010; Tricomi and Fiez, 2008).

Previous research has investigated the influence of feedback and reward on the acquisition of cognitive tasks, eg, decision-making paradigms (den Ouden et al., 2013; Frank et al., 2004; Robinson et al., 2010). Our animal studies suggest that dopaminergic signals originating in reward-coding brain regions (ventral tegmental area) are required for motor skill acquisition. In rodents, dopaminergic projections from the ventral tegmental area to the primary motor cortex enable motor learning and long-term potentiation in cortico-cortical projections (Hosp et al., 2011;

Molina-Luna et al., 2009). These projections are not necessary for task execution (Molina-Luna et al., 2009). We hypothesize that this system can be used to facilitate motor skill learning by amplification of rewarding stimuli.

Indeed, recent work suggests positive effects of monetary reward on procedural (Wachter et al., 2009) and skill motor learning (Abe et al., 2011) as well as on motor adaption (Galea et al., 2015). Notably, all of these studies reported dissociable effects of positive and negative reward, and the latter two found positive reward to impact task consolidation/retention. Moreover, the reward-related learning effect reported by Wachter et al. (2009) was found to be mediated by the dorsal striatum. However, these studies exclusively used money as an extrinsic reward, albeit, as illustrated earlier, also intrinsic rewards (eg, knowledge of performance) were shown to activate the human reward circuits and thereby possibly influence motor learning.

Dopaminergic neurons in the midbrain signal outcomes that are better than expected (positive prediction error (Schultz, 2000)). Being informed about unexpectedly good performance may thus cause a positive prediction error. Indeed, only being informed about positive task outcome resulted in better performance than being informed about the outcome of poorly solved trials (Chiviacowsky and Wulf, 2007). Whether these findings come along with higher reward activity after good performance feedback remains to be elucidated.

In the present study, a modified version of the arc-pointing task that involves a visually guided precision movement of the wrist (Shmuelof et al., 2012) was used to test the hypothesis that striatum activation is increased if knowledge of performance is given after good performance instead of a random selection of trials. Adding a performance-dependent monetary reward was expected to further increase this activation. In addition, we hypothesized that motor skill learning is improved in conditions with enhanced striatum activity.

2 METHODS

2.1 PARTICIPANTS

Forty-five healthy right-handed volunteers (22 females, 20–34 years of age, 24.5 years on average; Table 1) participated in this study that was approved by the cantonal ethics committee (KEK-LU 13054). Hand preference and dominance were

Table 1 Subject Characteristics

	Overall	KP_{random}	KP_{good}	KP_{good}+MR
N (Dropouts)	44 (1)	15 (–)	14 (1)	15 (–)
Age (SD)	24.5 (3.2)	25.9 (2.8)	25.1 (3.7)	22.9 (2.5)
Sex, male/female	23/21	7/8	7/7	9/6

N *reports the number of subjects per group with dropouts listed in brackets. SD is standard deviation. Note that groups were allocated randomly, not by matching any of the reported characteristics.*

assessed using the Edinburgh Handedness Inventory (Oldfield, 1971) and the Hand Dominance Test (Steingruber and Lienert, 1971), respectively, confirming that all participants were classified as right handed. Subjects were recruited from the University community or shared a similar educational status. They were not specifically skilled or trained in comparable motor tasks. All participants gave written informed consent before being randomly assigned to one of three groups. Allocation was according to a computer-generated random number sequence. Subjects were unaware of the other groups and the scientific rationale of the study. All subjects received financial compensation in comparable amounts, but only for one group payments depended on individual performance during the training of the motor task.

2.2 STUDY DESIGN

Subjects participated in the study for 2 consecutive days. Neutral (group-independent) test sessions were performed to assess momentary performance on day 1, before and after the group-specific training. To assess overnight task consolidation, subjects returned 20–28 h after finishing day 1 training.

2.3 MOTOR TASK

Originally, the arc-pointing task (Shmuelof et al., 2012) was developed to investigate the speed-accuracy trade-off function during motor skill learning. To examine the influence of knowledge of performance with or without monetary reward on brain activity and motor skill learning, we modified the task. Here, the task required subjects to perform wrist movements to steer a cursor on a computer screen through a semicircular channel (Fig. 1). To maximize the dynamic range of learning the non-dominant left rather than the right wrist was chosen assuming that initial performance would be worse with the left. Ideally, the cursor had to be guided along the middle of an arc-channel with the nominal movement speed dictated by a clock hand pointing at the current nominal position (Fig. 2A). For each frame (at 60 frames per second), the absolute distance from the actual to the nominal position was calculated and the average over the whole movement was used as performance measure to determine a score (with or without monetary consequences, Fig. 2B).

Prior to each new block of movements, subjects viewed a computer-generated demonstration of the clock hand moving along the channel in the required movement time. At the beginning of each trial, subjects placed the cursor in the red starting box. After a variable delay (800–1600 ms), the box turned green as an "ok-to-go" signal (reaction time was not a measure of performance and subjects were told to start any time after the box turned green). As soon as the cursor had left the box in positive y-direction (=upward), the clock hand started to move with uniform angular velocity continuously pointing at the nominal cursor position that subjects tried to adhere to. The cursor was visible throughout the movement (online feedback; Fig. 2A) and the trial automatically ended when the clock hand arrived at the end of the channel. Then the screen froze for a variable period of time (500–4500 ms). During test sessions,

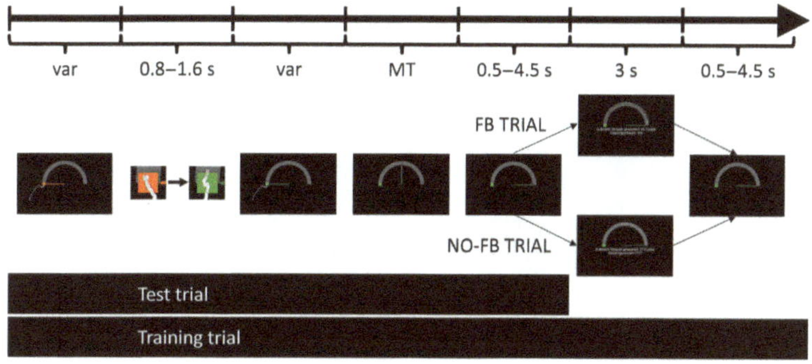

FIG. 1

Trial sequence. After placing the cursor in the start box, the box eventually turned *green* ("ok-to-go" signal) and subjects were free to start the movement whenever ready. The placing of the cursor in the start box, as well as the period from "ok-to-go" to the actual start of the movement were self-paced and hence of variable length (var). A specific movement time (MT) according to the speed requirements of the current block of trials was allowed to steer the cursor through the semicircular channel. As soon as movement time elapsed, the screen froze. During test sessions, the next trial directly followed. In case of a training trial, a group-specific knowledge of performance feedback was presented after feedback trials (FB TRIAL), or subjects were shown a neutral visual control stimulus after no-feedback trials (NO-FB TRIAL). Either way, the next training trial began after another delay period.

the subsequent trial directly followed. For training trials knowledge of performance or knowledge of performance plus monetary reward was presented for 3000 ms at this point, followed by another variable delay period (500–4500 ms) before the subsequent trial began. Fig. 1 shows a schematic summary of the paradigm.

To assess skill level in the absence of knowledge of performance and monetary reward, participants had to perform the arc-pointing task at five different movement speeds defined by the movement time that was allowed to move the cursor through the arc-channel (the clock hand uniformly travelled along the arc in exactly that time). Per test session, seven consecutive trials were performed as blocks with one of five movement times (movement time in ms: 800, 1000, 1200, 1400, and 1800) and these blocks were randomly ordered with 15 s breaks in between. Ten familiarization trials were allowed prior to the very first test session (ie, pretraining test) and, as already mentioned, a demonstration of the movement time was shown at the beginning of each movement time block. All in all, participants performed 35 trials per test session.

The training, on the other hand, was composed of five blocks of 50 trials each with 15 s breaks after 25 movements (within blocks) and 2 min breaks between the blocks. All 250 training trials were performed at one single movement time (ie, 1200 ms). After a movement, subjects received a terminal feedback by ~50% chance. Here, the three groups differed in terms of the selection of feedback trials

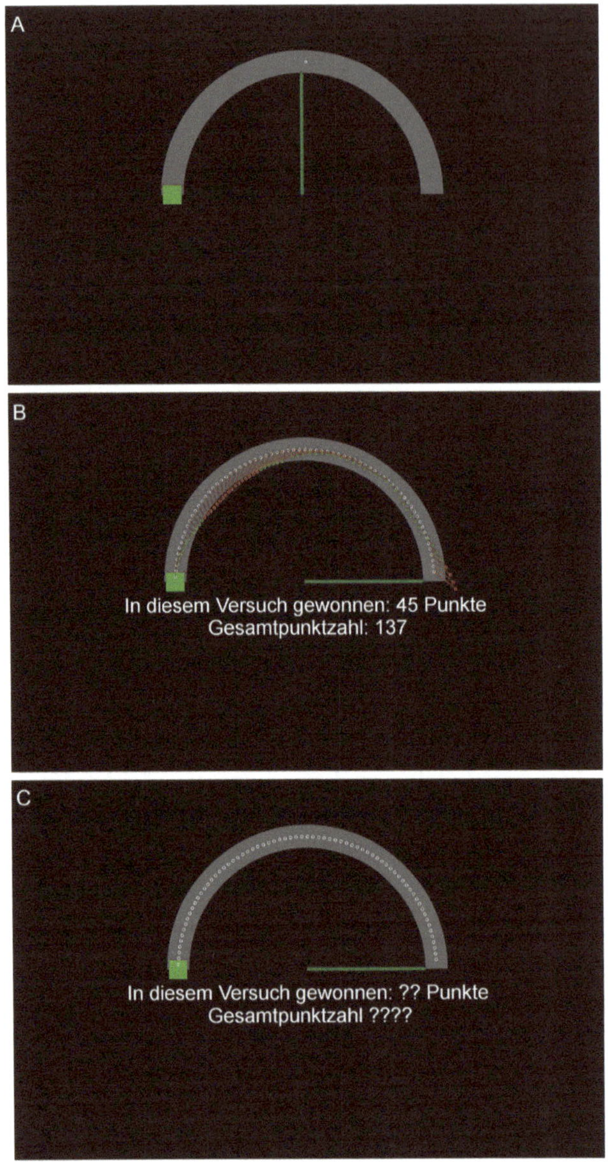

FIG. 2

(A) During the movement, the position of the cursor was indicated with a *white circle* (online feedback) and a clock hand continuously pointed at the current nominal position, which was defined to be in the middle of the semicircular channel. (B) A knowledge of performance feedback was presented after feedback trials, including the trajectory traveled by the cursor (series of *green* (inside of channel) or *red* (outside of channel) *colored circles*), as well as the nominal trajectory (series of uniformly distributed *white circles*). A *red line* and in terms of the type of feedback they were given. While the first group received knowledge of performance after randomly selected trials (KP_{random}), the other groups got either knowledge of performance only (KP_{good}) or knowledge of performance signifying a monetary reward ($KP_{good}+MR$) after relatively good performance, ie, when they performed better than the moving median over their performance in the last 10 trials. As described earlier, the tip of the clock hand pointed at the nominal position for each frame during a trial and the cursor's mean distance ($\bar{d}$) to the corresponding nominal position over all 72 frames per training trial (1200 ms at 60 frames per second) was used as measure to quantify performance.

$$\bar{d}_t = \frac{\sum_{f=1}^{72} d_f}{72},$$

where t is the number of the current trial and f stands for frame number. For members of KP_{good} and $KP_{good}+MR$, hence, a feedback was delivered from the 11th trial on, if $\bar{d}_t < \tilde{d}\left\{\bar{d}_{t-1}, \bar{d}_{t-2}, \ldots \bar{d}_{t-10}\right\}$, where $\tilde{d}$ is the median value. If selected as feedback trial, the feedback included, as a still image, the presentation of the trajectory traveled by the cursor as a series of circles that were colored according to their positions with respect to the channel (green if inside and red if outside of the channel). Moreover, the nominal trajectory was drawn as a series of equally spaced white circles along the middle of the channel and circles of the trajectory traveled by the cursor were linked to the corresponding nominal position by red lines (line width = 2 pixels $\approx$ 0.02 degree visual angle) to visualize d_f (Fig. 2B). Additionally, a score-feedback, for KP_{random} and KP_{good}, and a monetary reward, for $KP_{good}+MR$, was calculated based on $\bar{d}_t$. The relation between $\bar{d}_t$ and the monetary reward was chosen, based on pilot measurements, to allow members of $KP_{good}+MR$ to earn approximately 50 Swiss Francs (CHF; approx. 50 US Dollars) over the course of the experiment, since their minimal financial compensation was fixed to be 50 CHF less than that of KP_{random} and KP_{good}, if performance-related monetary rewards are not considered. Therefore, the monetary reward in Rappen (1 Rappen = 0.01 CHF; approx. 0.01 US Dollars) was set to be equal to $100 - \bar{d}_t/2$, if $\bar{d}_t < 200$ pixels, and 0 if $\bar{d}_t \geq 200$ pixels. Accordingly, a maximum of 1 CHF per trial could be won in the unrealistic case of perfect performance (ie, $\bar{d}_t = 0$). Note that no money was deducted after poor

linked each point of the cursor's trajectory to its corresponding nominal position and the average length of these lines was used to determine a score (for KP_{random}- and KP_{good}-groups) or a monetary reward (for $KP_{good}+MR$). "In diesem Versuch gewonnen: 45 Punkte" is the German expression explaining that the subject has won 45 points in the preceding trial, which, in this example, sums up to a total score of 137 over the whole experiment (Gesamtpunktzahl: 137). The neutral visual control stimulus presented after no-feedback trials is shown in (C). Note that the traveled trajectory was omitted and numbers specifying the score or monetary reward were replaced by question marks.

performance. Knowledge of performance for KP_{random} and KP_{good} was equally calculated, but its unit was points instead of Rappen, and for all groups the result of the current trial as well as the sum over the whole course of the experiment (money in CHF) was presented after feedback trials (all in letters and digits of ≈0.38° visual angle; Fig. 2B). In case of no-feedback trials, subjects were shown a similar screen in which scores or monetary rewards were replaced by question marks and only the nominal trajectory was presented. This ensured a comparable visual stimulus to the feedback conditions (no-feedback screen; Fig. 2C).

2.4 fMRI MEASUREMENTS

During the experiment, subjects lay supine in the MR scanner having their left forearm fixated with a customized armrest that was screwed on to the scanner table. A spherical reflective marker was attached to the proximal interphalangeal joint (knuckle) of their left index finger using surgical double-sided adhesive tape. An MRI-compatible motion capture camera set (Oqus MRI, Qualisys AB, Gothenburg, Sweden) consisting of eight cameras was used to continuously track the marker position at a frequency of 400 Hz. This information was imported online into Matlab R2012b (Mathworks Inc., Natick, MA, USA) using the Qualisys Matlab plug-in. A computer program written in "Presentation 16.3" software (Neurobehavioral Systems, Inc., Albany, USA) sampled the Qualisys-marker position via Matlab interface and transformed it into screen coordinates. To do so, in a calibration step, subjects were asked to move their wrist maximally in all directions having arm movements prevented by the aforementioned armrest. During this step, extreme x- (left-right) and y-positions (up–down) were logged and the screen was adjusted to display, in x- and y-direction, the middle 60% of each subject's individual range of motion. This procedure ensured that all participants were able to perform the required movements within a comfortable movement range.

The computer program also controlled stimulus presentation. While moving within the calibrated area, the marker position was displayed as a circle (≈0.13° visual angle) on a screen (0.64 × 0.4 m; 1920 × 1200 pixels) visible via mirrors to the subject inside the scanner (distance mirror − screen ≈ 1.90 m). The arc was centered around the middle of the screen (origin of ordinates) with an inner and outer radius of 384 (≈3.86° visual angle) and 456 pixels (≈4.58° visual angle), respectively. However, only the upper arc was used for task execution and all task movements were performed in clockwise direction. To indicate the start position, a red square with the side length equaling the width of the channel (72 pixels ≈ 0.72° visual angle) was placed at the beginning of the arc (box-center coordinates: $x = -420$ pixels, $y = 0$ pixels). Finally, a clock hand used to point at the nominal position for each frame during a trial, starting at the origin of ordinates with a length of 384 pixels (≈3.86° visual angle) and a width of 10 pixels (≈0.10 degree visual angle), completed the visual stimulus presented during each trial.

A Philips Ingenia 3.0T MRI scanner (Philips Healthcare, Best, The Netherlands) equipped with a Philips 32-channel dS head coil was used. During scanning sessions, head movement was minimized using a cushion and foam material parts. Three-dimensional anatomical images of the entire brain were obtained by using a T1-weighted three-dimensional spoiled gradient echo pulse sequence (180 slices, TR = 20 ms, TE = 2.3 ms, flip angle = 20°, FOV = 220 mm × 220 mm × 135 mm, matrix size = 224 × 187, voxel size = 0.98 mm × 1.18 mm × 0.75 mm). Functional data were obtained in 150 scans per testing session and 317 scans per training block, all consisting of 40 slices (slice thickness 3.5 mm, ascending acquisition order, no interslice-gap) covering the whole brain in oblique acquisition orientation. We used a sensitivity encoded (SENSE, factor 1.8) single-shot echo planar imaging technique (FEEPI; TR = 2.35 s; TE = 32 ms; FOV = 240 mm × 240 mm × 140 mm; flip angle = 82°; matrix size = 80 × 80; voxel size = 3 mm × 3 mm × 3.5 mm) with three dummy scans acquired at the beginning of each run and discarded in order to establish a steady state in T1 relaxation for all functional scans to be analyzed. Moreover, cardiac and respiratory cycles were continuously recorded (Invivo Essential MRI Patient Monitor, Invivo Corporation, Orlando, FL, USA) to allow correction of fMRI data for physiological noise (see Section 2.5).

2.5 ANALYSIS OF IMAGING DATA

Artifact minimization and MRI data analysis were performed using Matlab R2013b and the SPM8 software package (Institute of Neurology, London, UK; http://fil.ion.ucl.ac.uk/spm). All images were realigned to the first volume, normalized into standard stereotactic space (using the EPI-template provided by the Montreal Neurological Institute, MNI brain), resliced to 3 mm × 3 mm × 3 mm voxel size and smoothed using a 6 mm full-width-at-half-maximum Gaussian kernel. Since the interest of this study lay in the activation of rather small brain areas, a 6-mm rather than a larger Gaussian kernel was chosen, providing higher spatial resolution of resulting images and thus smaller partial volume effects in region of interest (ROI) analyses. Correction for physiological noise was performed via RETROICOR (Glover et al., 2000; Hutton et al., 2011) using Fourier expansions of different order for the estimated phases of cardiac pulsation (third order), respiration (fourth order), and cardio–respiratory interactions (first order) (Harvey et al., 2008). The corresponding confound regressors were created using the Matlab physIO Toolbox (Kasper et al., 2009, open source code available as part of the TAPAS software collection: http://www.translationalneuromodeling.org/tapas/). For first level data analysis of the arc-pointing task training, after highpass filtering (cut-off 128 s), an individual statistical general linear model was set up for each subject (Friston et al., 1995) by defining six regressors, corresponding to six recurring conditions per training block. Onsets and durations (in seconds) for each condition were extracted from Presentation-log-files using custom Matlab routines. The first regressor defined, for each trial, the time interval needed to place the cursor into the start box. The second condition started immediately after reaching the box and thus the corresponding regressor included both the planning and the execution of the complete movement (movement phase). This was followed by a period of variable length where subjects were looking at a still image of the arc waiting to either be shown the feedback screen after feedback trials or the no-feedback screen after no-feedback trials. Feedback screens then have been presented for 3 s and were modeled as separate regressors (feedback presentation and no-feedback presentation). The sixth regressor was a parametric modulation of the feedback regressor by the number of points (when KP_{random} or KP_{good} was presented) or the magnitude of the monetary reward (when KP_{good} + MR was presented) presented on the feedback screen in case of a feedback trial. Delays were not modeled and thus were used as baseline.

Based on our hypothesis of improving motor skill learning by reward-induced striatal upregulation, we focused the fMRI analysis on the striatum. To separate the signal change due to knowledge of performance and monetary reward from irrelevant visual input, the linear contrast "feedback vs no-feedback presentation" was specified. Thus, the relative signal increase during reward presentation after feedback trials relative to the signal elicited when looking at a visual control stimulus after no-feedback trials (both with respect to baseline signal during break periods) was calculated and represented as beta weights. These contrast values were then averaged over two ROIs (ventral and dorsal striatum) using an in-house Matlab ROI analysis routine. The striatum was partitioned into ventral and dorsal parts according to Lutz et al. (2012). To test for significant activation of the ROI, average effect sizes per participant were tested against null by one-tailed one-sample *t*-tests. All statistical analyses (imaging and behavioral data) were performed using SAS Enterprise Guide (5.1, SAS Institute, Cary, NC, USA). Moreover, beta values from the contrast "feedback vs no-feedback presentation" were subjected to a one-way ANOVA with the between-subject factor "group" (KP_{random}, KP_{good}, and KP_{good} + MR), and results were Bonferroni-corrected for performing multiple ANOVAs (two ROIs). Dunnett's two-tailed *t*-tests were then used to locate eventual influences of reward type (KP_{good} + MR vs KP_{good}) and/or feedback schedule (KP_{random} vs KP_{good}), where applicable (ie, in case of a significant main effect "group").

2.6 ANALYSIS OF BEHAVIOR

Boolean cursor position with respect to the arc-channel and d_f were calculated online for each frame and logged together with all other relevant experimental information. Data were extracted and $\bar{d}_t$ (according to the formula presented earlier) and ratios of data points lying within the arc-channel were determined using custom Matlab routines. Data were then corrected for outlier trials ($\bar{d}_t >$ average $\bar{d}_t$ of the corresponding block of trials + 2 standard deviations (SDs) or $\bar{d}_t >$ average $\bar{d}_t$ of the corresponding block of trials − 2 SD) using SAS Enterprise Guide 5.1. For statistical analysis of absolute movement errors during arc-pointing task training, the absolute error was logarithmically transformed in order to fulfill requirements for statistical tests. Performance changes between sessions were calculated, for each subject and movement time, as percentage changes relative to the corresponding baseline. That is, relative to the individual pretraining $\bar{d}_t$ for task acquisition and relative to post-training $\bar{d}_t$ for quantification of task consolidation. Generalized linear mixed models

(GLMM) for repeated measures were applied using SAS proc mixed. GLMM1: Analysis of absolute errors during training included the main factors "group" (levels: KP_{random}, KP_{good}, and KP_{good}+MR) and "training block" (levels 1–5). GLMM2: Analysis of percentage change in performance comprised the main factors "group" (levels: KP_{random}, KP_{good}, and KP_{good}+MR), "learning phase" (levels: acquisition and consolidation) and "movement time" (levels: 0.8, 1.0, 1.2, 1.4, and 1.6 ms). For posthoc analysis, Dunnett's *t*-tests, with KP_{good} acting as control condition, were used to locate whether differential skill development can be attributed to either the usage of different feedback schedules (KP_{random} vs KP_{good}) or different types of reward (KP_{good}+MR vs KP_{good}). One-tailed (hypothesis driven) Dunnett's *t*-tests were performed, where differences in striatal activations between two conditions reached significance. Moreover, one-sample *t*-tests were used to examine whether the groups' skill level changed during either of the learning phases, ie, whether percentage changes were significantly different from zero.

3 RESULTS

Data from one subject had to be excluded due to a software crash during the training of the task, which required recalibration and a restart of the experiment thus hampering comparability to the data of other participants.

3.1 fMRI

Using the contrast "feedback vs no-feedback presentation," one-tailed one-sample *t*-tests revealed significant activations of the ventral striatum for KP_{random} and KP_{good}+MR ($t=2.40$, $p=0.0153$ and $t=4.57$, $p=0.0002$, respectively) and of the dorsal striatum for KP_{good}+MR exclusively ($t=3.11$, $p=0.0077$; Fig. 3). The reward condition (main effect "group") significantly influenced the relative signal increase in the ventral striatum ($F=5.04$, $p=0.0220$), but less clearly in the dorsal striatum ($F=2.56$, $p=0.179$). In the ventral striatum, KP_{good}+MR showed significantly higher activation than KP_{good} ($t=2.98$, $p_{\mathrm{Dunnett}}=0.0093$).

3.2 BEHAVIORAL RESULTS

Task performance is expressed as $\bar{d}_t$, which was the measure determining knowledge of performance and monetary rewards. As a result of our experimental manipulation (ie, selecting well-solved trials for feedback), average performance during feedback trials was better ($\bar{d}_t$ was smaller) in KP_{good} (54.16 ± 18.12 pixels, $t=3.77$, $p=0.0005$) and KP_{good}+MR (50.07 ± 8.851 pixels, $t=4.47$, $p<0.0001$) compared with KP_{random} (75.11 ± 17.22 pixels; GLMM1: main effect "group": $F=11.58$, $p<0.0001$). As a consequence, these subjects were shown higher average scores per feedback trial (41.03 ± 7.158 points and 42.59 ± 13.97 Rappen vs 32.98 ± 5.421 points) and reached higher total scores over the course of the experiment

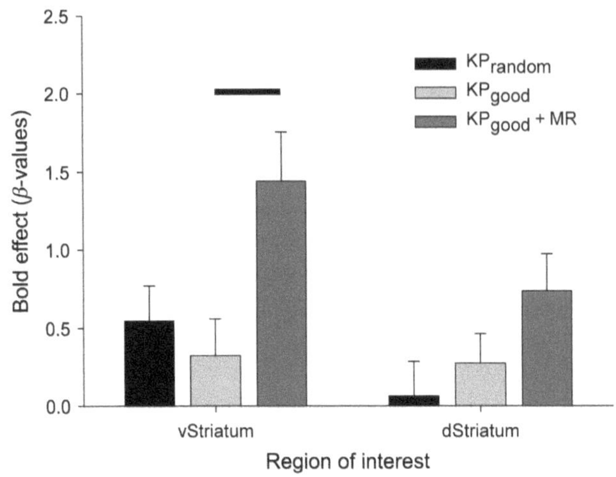

FIG. 3

Striatal activations (β-values) for the "feedback vs no-feedback presentation" contrast. Group effects were found to be significant in the ventral Striatum (vStriatum), but not significant in the dorsal Striatum (dStriatum). Means $\pm$ standard error of the mean (SEM). —, Significant pairwise comparison ($p < 0.05$). $N = 44$.

(5203 ± 908.7 points and 5358 ± 572.5 Rappen vs 4122 ± 677.6 points), all KP_{good} and $KP_{good}+MR$ vs KP_{random}.

Considering all trials, including no-feedback trials, overall performance increased (ie, $\bar{d}_t$ decreased) over the course of training (Fig. 4; GLMM1: main effect "training block": $F = 28.02$, $p < 0.0001$). No difference in overall $\bar{d}_t$ was found between groups (GLMM1: main effect "group": $F = 0.58$, $p = 0.5599$), but performance development over the course of the training was influenced by the group-specific reward condition (GLMM1: interaction "group*training block": $F = 2.20$, $p = 0.0247$).

Performance in our version of the arc-pointing task has been assessed before, right after and 24 h after the training of the arc-pointing task without providing additional terminal feedback in these testing sessions. The evolution of absolute errors, ie, of $\bar{d}_t$, across the different test sessions is presented in Fig. 5 (top). Of greater relevance than absolute error values, however, are performance changes between pre- and post- (due to task acquisition), as well as between post- and 24 h posttraining tests (due to task consolidation processes). Fig. 5 (bottom) displays percentage changes relative to the corresponding baseline value (ie, relative to pretraining $\bar{d}_t$ for acquisition and relative to posttraining $\bar{d}_t$ for consolidation). Online learning and consolidation differentially influenced performance (GLMM2: main effect "learning phase": $F = 81.80$, $p < 0.0001$), with greater changes caused by online learning. This change was influenced by task difficulty (GLMM2: interaction "learning phase*movement time": $F = 11.15$, $p < 0.0001$). Performance improved

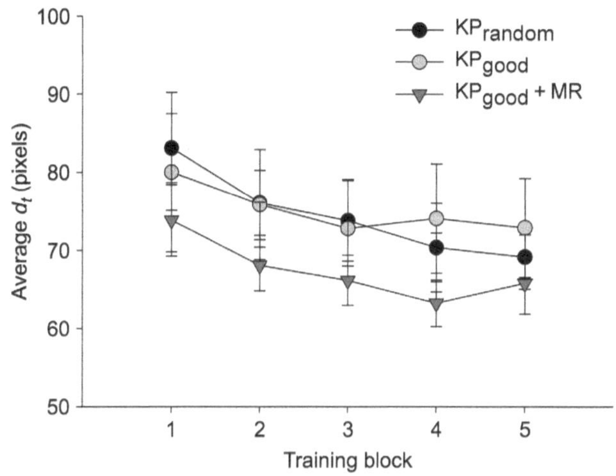

FIG. 4

Development of absolute errors ($\bar{d}_t$) in pixels for all trials (feedback and no-feedback trials) averaged over each training block (1–5) for all three study groups. Means ± SEM. $N=44$.

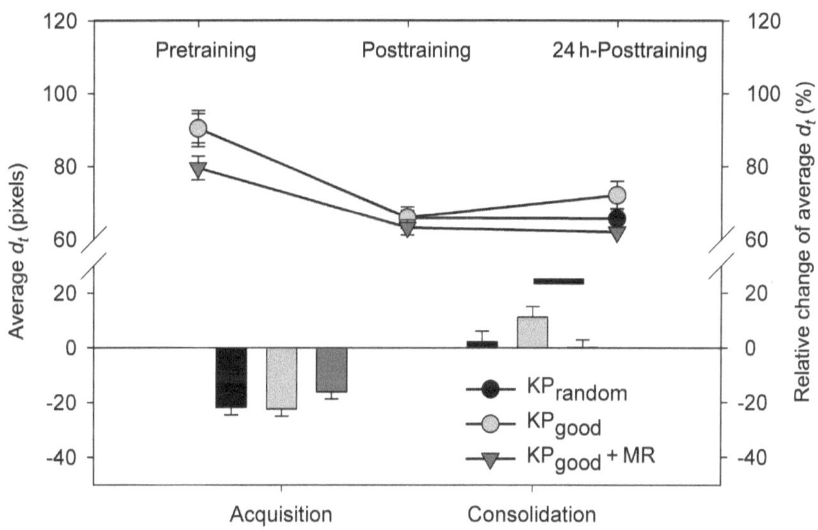

FIG. 5

Absolute performance ($\bar{d}_t$) during test sessions (*top*, upper *x*-axis, left *y*-axis) and relative performance change (in %) compared to the preceding test session (*bottom*, lower *x*-axis, right *y*-axis), ie, to pretraining $\bar{d}_t$ for task acquisition and to posttraining $\bar{d}_t$ for consolidation. All data are presented as Means ± SEM. –, Significant posthoc comparison ($p<0.05$). $N=44$.

due to online learning at all movement times, while, on the other hand, performance at 24 h could be maintained for movement times ≥ 1.2 but significantly suffered from "forgetting" at shorter movement times (ie, at higher task difficulty). Furthermore, "learning phase" significantly interacted with the "group" factor (GLMM2: $F=3.69$, $p=0.0259$). While all groups profited similarly from arc-pointing task training, only KP_{random} and KP_{good}+MR consolidated their performance overnight. KP_{good}'s performance decreased significantly ($t=3.39$, $p=0.0008$) and this worsening was significantly different compared with KP_{good}+MR ($t=2.42$, $p_{Dunnett}=0.0324$), and by tendency different compared with KP_{random} ($t=2.09$, $p_{Dunnett}=0.1399$).

4 DISCUSSION

Our results demonstrate that both striatal response and motor skill learning, measured as relative change of error from pre- to posttraining (=acquisition) and from posttraining to 24 h thereafter (=consolidation), are influenced by manipulations of the schedule for performance feedback and/or the type of reward. Specifically, adding an extrinsic (monetary) reward increases ventral striatal activation to performance feedback, which is associated with better motor skill consolidation overnight.

4.1 TRAINING AND MOTOR SKILL ACQUISITION

All groups practiced in identical intensity. Interventions only differed in terms of which trials were selected for KP and whether performance had monetary consequences or not. Higher subjective benefit through additional extrinsic (monetary) reward at stable cost should raise the motivation for a specific exercise. Motivation may rely on dopaminergic activity in the nucleus accumbens, as animal studies have shown that dopamine depletion in nucleus accumbens or low doses of dopamine antagonists reduce the willingness to work for extrinsic rewards (reviewed by Salamone and Correa, 2002). Enclosing nucleus accumbens, ventral striatum activations observed during our experiment (Fig. 3) could thus be an indication that groups invested varying amounts of effort into training. But, MR neither improved performance during training nor skill acquisition. This supports the results from Abe et al. (2011), who also found no difference in acquisition between reward, punishment, and control groups. However, other studies showed that punishment, but not reward improved the acquisition of a motor adaption paradigm (Galea et al., 2015) and induced a performance effect in a procedural motor task (Wachter et al., 2009). But, Wachter et al. (2009) also found that the acquisition of an implicit motor learning task profited from reward but not from punishment. This apparent inconsistency should be taken as indication that conclusions across different (motor) learning modalities like procedural, skill, or adaption learning must be drawn with caution (Shmuelof et al., 2012).

4.2 CONSOLIDATION

Our study design allows investigating the influence of using different schedules for intrinsic reward on neural activity and motor skill learning by comparing KP_{good} and KP_{random} conditions. While feedback trials were randomly selected in case of KP_{random}, subjects in KP_{good} were only informed about trials with good performance. Interestingly and against our hypothesis, striatal activation was only observed in KP_{random} but not KP_{good}. Behaviorally, this resulted in successful task consolidation for KP_{random} and significant overnight forgetting in KP_{good} with a between-group difference close to significance. Thus, ventral striatal activation during training supports successful consolidation of a newly learned motor skill.

Poor performance and striatal underactivation in KP_{good} were unexpected. This result is in contrast to findings from Chiviacowsky and Wulf (2007), who studied two experimental groups, one receiving knowledge of result after good (KR_{good}) and the other after bad performance (KR_{poor}), in a ballistic task that required subjects to throw beanbags at a target with their eyes covered. In their experiment, the KR_{good}-group significantly outperformed the KR_{poor}-group when subjects repeated the task 1 day after the training without knowledge of result. Therefore, the authors proposed motivational properties of positive feedback to have a direct effect on learning. On the contrary, the guidance hypothesis of feedback suggests that feedback is more beneficial if presented after larger rather than smaller errors because it then better guides the learner to the correct response (Salmoni et al., 1984; Schmidt, 1991). Relating this controversy to our finding of a tendency towards better consolidation in KP_{random} compared with KP_{good}, it appears that KP_{random} combines the best of both theories. That is, adequate error information guiding subject's response towards better performance, but still keeping subjects motivated by frequently including knowledge of performance after good performance. A positive motivational status might be indicated by the observed activation of the ventral striatum in KP_{random}, as motivation may rely on dopaminergic activity in the nucleus accumbens (Salamone and Correa, 2002). However, the question remains why knowledge of performance after average performance (KP_{random}) lead to striatal activation, while knowledge of performance after good performance did not. Attentively steering the cursor along the arc-channel under visual control may have enabled subjects to evaluate their performance online and thus to make predictions about the feedback. This, in turn, may have allowed KP_{good}-group to predict the reception of knowledge of performance, as for them the selection of feedback trials depended on performance. We know from experiments in primates that dopamine neurons appear to emit an alerting message about the surprising presence or absence of rewards and that response to rewards and reward-predicting stimuli depend on event predictability (Schultz, 1998). It therefore seems to be the unpredictable selection of feedback trials in KP_{random}, rather than the magnitude of the score that made up the activation in the ventral striatum. This finding is supported by the absence of significant activations to a parametric modulation of the "feedback presentation" contrast by the amount of points won during a trial.

Interestingly, although KP_{good} failed to induce any striatal activation and was accompanied by overnight forgetting, knowledge of performance after good performance lead to highest ventral striatum response and also activated the dorsal striatum when knowledge of performance signified a monetary outcome. Both ventral striatum activation and overnight task consolidation were significantly higher/better in KP_{good}+MR compared with KP_{good}. A beneficial influence of increased motivation due to higher subjective benefit (induced by extrinsic reward) on the consolidation component of motor skill learning thus emerges from our results. This corroborates previous findings on motor skill learning (Abe et al., 2011) and motor adaption (Galea et al., 2015). The former experiment used an isometric pinch force tracking task to investigate motor skill learning under either monetarily rewarded, punished, or neutral control training conditions. While at 24 h posttraining, punishment, and control groups performed at a similar level as immediately after the training, the rewarded group experienced significant offline gains, which remained present at 30 days posttraining. In contrast, the neutral and punished groups showed substantial performance loss at 30 days. When comparing to the experiment of Abe et al. (2011), the beneficial effect of reward could be similarly demonstrated in the present study. Although, for practical reasons, we did not test further than 24 h posttraining. Some remaining discrepancies of performance changes at 24 h posttraining may be attributed to differential influences of task complexity or difficulty between the pinch force task and the arc-pointing task, as indicated by our finding of a significant "learning phase*movement time" interaction. That is, changes due to task consolidation highly depended on task difficulty (ie, movement time).

However, regarding the comparison between KP_{good}+MR and KP_{good}, observed striatal activations are in line with previous work, revealing that feedback related activity in the ventral striatum is increased if knowledge of performance has monetary consequences and that a monetary incentive is needed to elicit a neural response in the dorsal striatum (Lutz et al., 2012). The absence of a response of the dorsal striatum to performance feedback is, on the other hand, in contrast to findings from other studies (Poldrack et al., 2001; Tricomi and Fiez, 2008; Tricomi et al., 2004, 2006). Unfortunately, different approaches for defining striatal subdivisions hamper comparability between these results.

To summarize, training under a feedback condition, which elicited higher activation of the ventral striatum, positively influenced skill development via better task consolidation. Overall, it seems that training under a feedback condition that induces activation in the ventral striatum helps for successful task consolidation. It is known that, in a rewarded task, hemodynamic ventral striatal response correlates with dopamine release in the ventral striatum, which as well correlates with the reward-related neural activity in the substantia nigra/ventral tegmental area, the origin of the dopaminergic projection (Schott et al., 2008). Reward-related ventral striatal activity may thus be an indication for increased dopaminergic function in the midbrain. In rodents, the existence of direct pathways linking midbrain reward centers to the motor cortex has been demonstrated (Hosp et al., 2011). In the motor cortex, dopamine facilitates long-term potentiation (Molina-Luna et al., 2009), a form of synaptic plasticity discussed to be critically involved in skill learning (Rioult-Pedotti et al., 2000; Ziemann et al., 2004). In their experiment, Hosp et al. (2011) could demonstrate that destroying dopaminergic neurons in the ventral tegmental area prevented improvements in forelimb reaching, a state that was abolished on administration of levodopa into the primary motor cortex. Dopamine-dependent long-term potentiation develops gradually over hours (Huang and Kandel, 1995) and persists for days to weeks (Abraham, 2003). We thus propose increased dopamine release into the primary motor cortex in feedback conditions with significant activation of the ventral striatum to be the key factor facilitating motor skill learning via better task consolidation.

4.3 LIMITATIONS

The striatum is involved in fine motor control. Therefore, it is not surprising that both ventral and dorsal striatum activation was observed during movement execution in this experiment. These activations, however, did not differ between groups (data not shown) and the movement phase was well separated from feedback/no-feedback presentation through a variable delay period (Fig. 1). Hence, we do not expect striatal involvement in movement control to have an influence on our imaging results observed during reward processing.

Furthermore, the present study does not yield a double dissociation between the influence of feedback schedule (random selection/good performance) and type of reward (knowledge of performance only/knowledge of performance plus monetary reward), because we have not fully balanced the possible conditions (KP_{random}, KP_{good}, $KP_{random}+MR$, and $KP_{good}+MR$). Nevertheless, we can corroborate influences of monetary reward on striatal activity and can link these to consolidation of a motor skill. It also allows to discuss effects of performance feedback schedules on striatal activity and motor skill learning, but it does not allow to investigate interactions between these two factors.

Moreover, generalization of these findings to other types of motor or nonmotor learning is limited. In motor skill learning, motor learning is investigated in the absence of a perturbation and the main goal is to reduce a variable error (Deutsch and Newell, 2004; Guo and Raymond, 2010; Hung et al., 2008; Liu et al., 2006; Muller and Sternad, 2004; Ranganathan and Newell, 2010). Task difficulty limits performance, usually in the form of a trade-off between speed and accuracy. Learning consists of breaking through this limit (ie, improving the speed-accuracy trade-off) (Reis et al., 2009). In the original work introducing the arc-pointing task, the authors well defined and checked for fulfillment of speed requirements (ie, the movement time) and then investigated an isolated measure of accuracy (Shmuelof et al., 2012). In contrast, our main outcome measure, $\bar{d}_t$, is influenced by both speed and accuracy. A reduction in $\bar{d}_t$ can thus occur by improved accuracy, more accurate timing, or a combination of both. Although we refrained from defining a target zone and thus from strictly checking for observance of the movement time, we excluded outlier trials, where, for example, the trial was accidentally started. In conclusion, although we can demonstrate a shift in the speed-accuracy trade-off function for the entire subject population, comparing groups by means of a separable measure of either speed or accuracy is in our case not valid, as it was the combined measure $\bar{d}_t$ that determined group-specific feedback conditions. This might be viewed as a shortcoming, hampering clear definition of the behavior observed during our study as motor skill learning, but on the other hand it allowed effective investigation of learning of goal-oriented movements with clearly set goals and well-defined feedback on goal achievement.

5 CONCLUSION

Our results demonstrate that motor skill learning is influenced by different reward conditions applied during the training of a motor task. Particularly, linking performance feedback to a monetary outcome efficiently raises ventral striatum activation, which comes along with better overnight task consolidation of the corresponding study group. Notably, all groups showing a significant response of the ventral striatum to feedback during training could retain their performance from the first day at the 24 h posttraining test, whereas a lack of ventral striatal response in the other group was accompanied by significant overnight forgetting. This leads us to conclude that increasing ventral striatal activity during acquisition of a motor skill by using appropriate reward improves consolidation of the acquired skill.

ACKNOWLEDGMENTS

The authors are indebted to the volunteers for their dedicated participation in this study. Special thanks go to Benjamin Hertler for his support in the implementation of the study and Peter Rasmussen for his help with the statistical analysis of the data. This study was supported by the Clinical Research Priority Program Neuro-Rehab (CRPP) of the University of Zurich. We would like to dedicate this work to Nadja Ziegler who sadly passed away over the course of this project.

Conflict of Interest: The authors have no conflicts of interest to declare.

ClinicalTrials.gov Identifier: NCT02189564.

REFERENCES

Abe, M., Schambra, H., Wassermann, E.M., Luckenbaugh, D., Schweighofer, N., Cohen, L.G., 2011. Reward improves long-term retention of a motor memory through induction of offline memory gains. Curr. Biol. 21, 557–562.

Abraham, W.C., 2003. How long will long-term potentiation last? Philos. Trans. R. Soc. Lond. B Biol. Sci. 358, 735–744.

Callan, D.E., Schweighofer, N., 2008. Positive and negative modulation of word learning by reward anticipation. Hum. Brain Mapp. 29, 237–249.

Chiviacowsky, S., Wulf, G., 2007. Feedback after good trials enhances learning. Res. Q. Exerc. Sport 78, 40–47.

Deci, E.L., Koestner, R., Ryan, R.M., 1999. A meta-analytic review of experiments examining the effects of extrinsic rewards on intrinsic motivation. Psychol. Bull. 125, 627–668. discussion 692–700.

Deci, E.L., Koestner, R., Ryan, R.M., 2001. Extrinsic rewards and intrinsic motivation in education: reconsidered once again. Rev. Educ. Res. 71, 1–27.

den Ouden, H.E.M., Daw, N.D., Fernandez, G., Elshout, J.A., Rijpkema, M., Hoogman, M., Franke, B., Cools, R., 2013. Dissociable effects of dopamine and serotonin on reversal learning. Neuron 80, 1090–1100.

Deutsch, K.M., Newell, K.M., 2004. Changes in the structure of children's isometric force variability with practice. J. Exp. Child Psychol. 88, 319–333.

Frank, M.J., Seeberger, L.C., O'Reilly, R.C., 2004. By carrot or by stick: cognitive reinforcement learning in Parkinsonism. Science 306, 1940–1943.

Friston, K.J., Holmes, A.P., Poline, J.B., Grasby, P.J., Williams, S.C., Frackowiak, R.S., Turner, R., 1995. Analysis of fMRI time-series revisited. Neuroimage 2, 45–53.

Galea, J.M., Mallia, E., Rothwell, J., Diedrichsen, J., 2015. The dissociable effects of punishment and reward on motor learning. Nat. Neurosci. 18, 597–602.

Glover, G.H., Li, T.Q., Ress, D., 2000. Image-based method for retrospective correction of physiological motion effects in fMRI: RETROICOR. Magnet. Reson. Med. 44, 162–167.

Greene, D., Lepper, M.R., 1974. Effects of extrinsic rewards on children's subsequent intrinsic interest. Child Dev. 45, 1141–1145.

Guo, C.C., Raymond, J.L., 2010. Motor learning reduces eye movement variability through reweighting of sensory inputs. J. Neurosci. 30, 16241–16248.

Han, S., Huettel, S.A., Raposo, A., Adcock, R.A., Dobbins, I.G., 2010. Functional significance of striatal responses during episodic decisions: recovery or goal attainment? J. Neurosci. 30, 4767–4775.

Harvey, A.K., Pattinson, K.T.S., Brooks, J.C.W., Mayhew, S.D., Jenkinson, M., Wise, R.G., 2008. Brainstem functional magnetic resonance imaging: disentangling signal from physiological noise. J. Magn. Reson. Imaging 28, 1337–1344.

Hosp, J.A., Pekanovic, A., Rioult-Pedotti, M.S., Luft, A.R., 2011. Dopaminergic projections from midbrain to primary motor cortex mediate motor skill learning. J. Neurosci. 31, 2481–2487.

Huang, Y.Y., Kandel, E.R., 1995. D1/D5 receptor agonists induce a protein synthesis-dependent late potentiation in the CA1 region of the hippocampus. Proc. Natl. Acad. Sci. U.S.A. 92, 2446–2450.

Hung, Y.C., Kaminski, T.R., Fineman, J., Monroe, J., Gentile, A.M., 2008. Learning a multi-joint throwing task: a morphometric analysis of skill development. Exp. Brain Res. 191, 197–208.

Hutton, C., Josephs, O., Stadler, J., Featherstone, E., Reid, A., Speck, O., Bernarding, J., Weiskopf, N., 2011. The impact of physiological noise correction on fMRI at 7T. Neuroimage 57, 101–112.

Kasper, L., Marti, S., Vannesjö, S., Hutton, C., Dolan, R., Weiskopf, N., Stephan, K., Prüssmann, K., 2009. Cardiac artefact correction for human brainstem fMRI at 7 Tesla. In: Proceedings of the Organization for Human Brain Mapping, Vol. 15, San Francisco.

Knutson, B., Cooper, J.C., 2005. Functional magnetic resonance imaging of reward prediction. Curr. Opin. Neurol. 18, 411–417.

Knutson, B., Delgado, M.R., Phillips, P.E., 2008. Representation of subjective value in the striatum. In: Glimcher, P.W., Camerer, C.F., Fehr, E., Poldrack, R.A. (Eds.), Neuroeconomics: Decision Making and the Brain. Academic Press, London, pp. 398–406.
Kohn, A., 1999. Punished by Rewards: The Trouble with Gold Stars, Incentive Plans, A's, Praise, and Other Bribes. Houghton Mifflin Harcourt, Boston.
Liu, Y.T., Mayer-Kress, G., Newell, K.M., 2006. Qualitative and quantitative change in the dynamics of motor learning. J. Exp. Psychol. Hum. Percept. Perform. 32, 380–393.
Lutz, K., Widmer, M., 2014. What can the monetary incentive delay task tell us about the neural processing of reward and punishment? Neurosci. Neuroecon. 4, 33–45.
Lutz, K., Pedroni, A., Nadig, K., Luechinger, R., Jancke, L., 2012. The rewarding value of good motor performance in the context of monetary incentives. Neuropsychologia 50, 1739–1747.
Molina-Luna, K., Pekanovic, A., Rohrich, S., Hertler, B., Schubring-Giese, M., Rioult-Pedotti, M.S., Luft, A.R., 2009. Dopamine in motor cortex is necessary for skill learning and synaptic plasticity. PLoS One 4, e7082.
Muller, H., Sternad, D., 2004. Decomposition of variability in the execution of goal-oriented tasks: three components of skill improvement. J. Exp. Psychol. Hum. Percept. Perform. 30, 212–233.
Murayama, K., Matsumoto, M., Izuma, K., Matsumoto, K., 2010. Neural basis of the undermining effect of monetary reward on intrinsic motivation. Proc. Natl. Acad. Sci. U.S.A. 107, 20911–20916.
Oldfield, R.C., 1971. The assessment and analysis of handedness: the Edinburgh inventory. Neuropsychologia 9, 97–113.
Poldrack, R.A., Clark, J., Pare-Blagoev, E.J., Shohamy, D., Creso Moyano, J., Myers, C., Gluck, M.A., 2001. Interactive memory systems in the human brain. Nature 414, 546–550.
Ranganathan, R., Newell, K.M., 2010. Influence of motor learning on utilizing path redundancy. Neurosci. Lett. 469, 416–420.
Reis, J., Schambra, H.M., Cohen, L.G., Buch, E.R., Fritsch, B., Zarahn, E., Celnik, P.A., Krakauer, J.W., 2009. Noninvasive cortical stimulation enhances motor skill acquisition over multiple days through an effect on consolidation. Proc. Natl. Acad. Sci. U.S.A. 106, 1590–1595.
Reitman, D., 1998. The real and imagined harmful effects of rewards: implications for clinical practice. J. Behav. Ther. Exp. Psychiatry 29, 101–113.
Rioult-Pedotti, M.S., Friedman, D., Donoghue, J.P., 2000. Learning-induced LTP in neocortex. Science 290, 533–536.
Robinson, O.J., Frank, M.J., Sahakian, B.J., Cools, R., 2010. Dissociable responses to punishment in distinct striatal regions during reversal learning. Neuroimage 51, 1459–1467.
Ryan, R.M., Deci, E.L., 2000. Intrinsic and extrinsic motivations: classic definitions and new directions. Contemp. Educ. Psychol. 25, 54–67.
Ryan, R.M., Deci, E.L., 2007. Active human nature: self-determination theory and the promotion and maintenance of sport, exercise, and health. In: Hagger, M.S., Chatzisarantis, N.L.D. (Eds.), Intrinsic Motivation and Self-Determination in Exercise and Sport. Human Kinetics, Champaign, IL, pp. 1–19.
Salamone, J.D., Correa, M., 2002. Motivational views of reinforcement: implications for understanding the behavioral functions of nucleus accumbens dopamine. Behav. Brain Res. 137, 3–25.

Salmoni, A.W., Schmidt, R.A., Walter, C.B., 1984. Knowledge of results and motor learning: a review and critical reappraisal. Psychol. Bull. 95, 355–386.

Schmidt, R.A., 1991. Frequent augmented feedback can degrade learning: evidence and interpretations. In: Requin, J., Stelmach, G.E. (Eds.), Tutorials in Motor Neuroscience. Kluwer Academic Publishers, Dordrecht, pp. 59–75.

Schott, B.H., Minuzzi, L., Krebs, R.M., Elmenhorst, D., Lang, M., Winz, O.H., Seidenbecher, C.I., Coenen, H.H., Heinze, H.J., Zilles, K., Duzel, E., Bauer, A., 2008. Mesolimbic functional magnetic resonance imaging activations during reward anticipation correlate with reward-related ventral striatal dopamine release. J. Neurosci. 28, 14311–14319.

Schultz, W., 1998. Predictive reward signal of dopamine neurons. J. Neurophysiol. 80, 1–27.

Schultz, W., 2000. Multiple reward signals in the brain. Nat. Rev. Neurosci. 1, 199–207.

Shmuelof, L., Krakauer, J.W., Mazzoni, P., 2012. How is a motor skill learned? Change and invariance at the levels of task success and trajectory control. J. Neurophysiol. 108, 578–594.

Spence, J.T., 1970. The distracting effects of material reinforcers in the discrimination learning of lower- and middle-class children. Child Dev. 41, 103–111.

Steingruber, H., Lienert, G., 1971. Hand-Dominanz-Test (HDT). Hogrefe, Göttingen, Germany.

Tricomi, E., Fiez, J.A., 2008. Feedback signals in the caudate reflect goal achievement on a declarative memory task. Neuroimage 41, 1154–1167.

Tricomi, E.M., Delgado, M.R., Fiez, J.A., 2004. Modulation of caudate activity by action contingency. Neuron 41, 281–292.

Tricomi, E., Delgado, M.R., McCandliss, B.D., McClelland, J.L., Fiez, J.A., 2006. Performance feedback drives caudate activation in a phonological learning task. J. Cogn. Neurosci. 18, 1029–1043.

Wachter, T., Lungu, O.V., Liu, T., Willingham, D.T., Ashe, J., 2009. Differential effect of reward and punishment on procedural learning. J. Neurosci. 29, 436–443.

Ziemann, U., Ilic, T.V., Pauli, C., Meintzschel, F., Ruge, D., 2004. Learning modifies subsequent induction of long-term potentiation-like and long-term depression-like plasticity in human motor cortex. J. Neurosci. 24, 1666–1672.

CHAPTER

How motivation and reward learning modulate selective attention 14

A. Bourgeois*,[1], L. Chelazzi[†,‡], P. Vuilleumier*

*Laboratory for Behavioral Neurology and Imaging of Cognition, University of Geneva, Geneva, Switzerland

[†]University of Verona, Verona, Italy

[‡]National Institute of Neuroscience, Verona, Italy

[1]Corresponding author: Tel.: +41-22-379-09-90; Fax: +41-22-379-50-02, e-mail address: alx.bourgeois@gmail.com

Abstract

Motivational stimuli such as rewards elicit adaptive responses and influence various cognitive functions. Notably, increasing evidence suggests that stimuli with particular motivational values can strongly shape perception and attention. These effects resemble both selective top-down and stimulus-driven attentional orienting, as they depend on internal states but arise without conscious will, yet they seem to reflect attentional systems that are functionally and anatomically distinct from those classically associated with frontoparietal cortical networks in the brain. Recent research in human and nonhuman primates has begun to reveal how reward can bias attentional selection, and where within the cognitive system the signals providing attentional priority are generated. This review aims at describing the different mechanisms sustaining motivational attention, their impact on different behavioral tasks, and current knowledge concerning the neural networks governing the integration of motivational influences on attentional behavior.

Keywords

Motivation, Reward, Attentional selection, Dopamine systems

1 INTRODUCTION

Our actions can be triggered by intentions, habits, or purely external incentives. More than two decades of neuroscience research in humans and animals have been centered on *incentive* motivation, ie, what causes individuals to engage behaviors according to the magnitude of reward they expect. Much of this research has focused on processes related to decision-making, typically associated with conscious and

Progress in Brain Research, Volume 229, ISSN 0079-6123, http://dx.doi.org/10.1016/bs.pbr.2016.06.004

effortful executive functions. In parallel to this important framework, a novel line of research has emerged in recent years in order to more specifically gain insight on how motivational stimuli operate to influence attentional selection (Anderson, 2015a; Chelazzi et al., 2013). Our brain has indeed evolved efficient selection mechanisms that can bias perception in favor of salient or behaviorally relevant stimuli, while ignoring irrelevant or distracting information. Such biases in attention are particularly important to behave adaptively given the limited capacity of both sensory and response systems. Attentional selection framework has established that attention can be controlled either voluntarily (ie, endogenously) according to strategic goals, or involuntarily (ie, exogenously) through more reflexive capture by salient or novel events. Influential neuroimaging studies have linked these two distinct aspects of attention to distinct networks in dorsal and ventral frontoparietal cortical areas (Corbetta and Shulman, 2002). The dorsal attentional network is composed by the intraparietal sulcus/superior parietal lobule and the frontal eye field/dorsolateral prefrontal cortex, whereas the ventral attentional network includes the temporoparietal junction and the ventral frontal cortex (inferior and middle frontal gyri). Remarkably, however, recent studies have pointed out that attentional selection might be modulated by stimuli with particular emotional or motivational values (see Vuilleumier, 2015, for a review). A variety of emotional or motivational cues can thus elicit adaptive responses that contribute to guide attention and modify perception. These value-based mechanisms appear functionally and anatomically distinct from the attentional systems classically associated with frontoparietal cortical networks in the brain. Moreover, these observations highlight the intimate but often neglected links between cognitive functions controlling selective attention and less well-known processes responsible for the appraisal of affective and motivational information. The aim of this review is to describe the mechanisms sustaining motivational attention and discuss the neural systems that may underlie this influence of motivational signals on attentional selection.

2 MOTIVATIONAL SIGNALS MODULATE SELECTIVE VISUAL ATTENTION

Numerous studies have investigated how monetary incentives can affect attentional performance. These studies have consistently reported a beneficial effect of monetary reward on attentional performance, attested by faster response times or increased detection performances, which could be linked to an increased perceptual sensitivity of the brain for rewarded stimuli (ie, Engelmann et al., 2009; Mohanty et al., 2008; Small et al., 2005). In all these studies, however, performance is directly linked to the reward outcome, that is, individuals are motivated to improve attention capacity in order to gain more reward.

Early evidence suggesting that reward outcomes may have a direct influence on attentional selection when performance is not immediately rewarded has been provided by Della Libera and Chelazzi (2006). These authors investigated the influence of monetary rewards as arbitrary feedback on performance in a negative-priming paradigm, where one stimulus feature has to be selected and another feature has to be ignored. Negative priming was consistent and prolonged following highly rewarded selections, but this effect was eliminated after poorly rewarded selections. This finding suggests that attentional selection and its lingering effects were dynamically modulated by reward contingency, such that the suppression of irrelevant stimulus information was more efficient and more persistent after high rewards.

Subsequent studies also demonstrated robust effects of reward on visual attentional selection. For instance, Anderson and collaborators (see Anderson, 2015a for a recent review, Anderson et al., 2011a,b, 2013b) performed a comprehensive series of experiments based on a reward association paradigm (see Fig. 1, for an example) in order to characterize value-driven attentional capture. In this paradigm, a high or a low reward is first associated with a basic feature of a stimulus, such as its color, during a learning association phase. The previously rewarded stimuli then appear as distractors in a subsequent visual search task, in order to investigate how value-associated stimuli compete for attentional selection. Their results demonstrated that value-associated stimuli strongly interfered with performance, shedding new light on how reward learning shapes attentional selection. Results typically show that the presence of previously rewarded cues (eg, color) produces a substantial

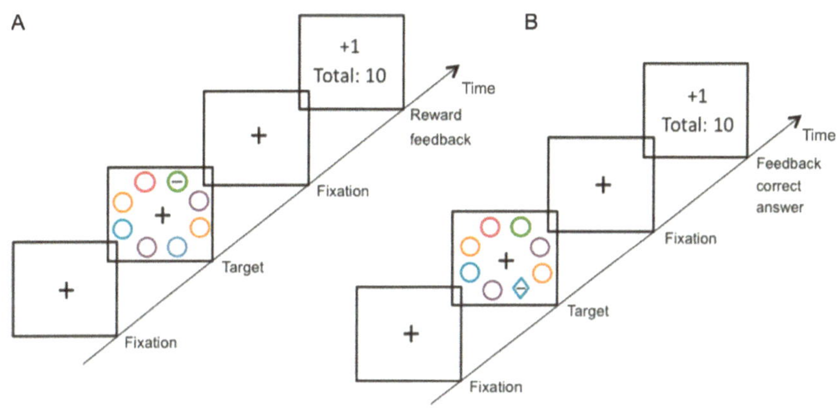

FIG. 1

Example of the reward association-based paradigm. (A) During the association phase, participants were asked to discriminate as fast and as accurately as possible a line, either horizontal or vertical, presented within a *red* or a *green* circle. In 80% of trials, one of the two targets (counterbalanced across participants) was followed by a high reward (+10), but by a low reward (+1) on the remaining 20%. (B) During the testing phase, participants were still required to discriminate a line, presented either horizontal or vertical. In order to investigate the attentional capture of previously high- or low-rewarded stimuli, one of the distractors was shown in *red* on 25% of trials, or *green* on another 25% trials.

Adapted from Bourgeois, A., Neveu, R., Bayle, D.J., et al., 2015. How does reward compete with goal-directed and stimulus-driven shifts of attention? Cogn. Emot. 24, 1–10.

slowing and diversion of attention away from the currently task-relevant targets. Thus, the motivational salience of visual stimuli may induce a strong bottom-up signal to guide attention and modulate sensory processing, which ultimately leads to competing choices that must be resolved.

Likewise, Hickey and collaborators (2014) designed a visual search task in which participants were required to select a target, while they ignored a salient distractor, and received a random-magnitude reward for correct performance. Response times were analyzed as a function of the magnitude of reward received in the preceding trial. Their results suggested that reward could guide attentional orienting by dynamically priming contextual locations of visual stimuli. Several studies also demonstrated that such reward effects operate even without any conscious awareness of the association contingency between rewards and particular stimulus features (eg, Anderson et al., 2011a,b, 2013b; Hickey et al., 2014). Finally, Della Libera and Chelazzi (2009) demonstrated not only that attentional processes are influenced by rewards but also that this effect is long-lasting, occurring several days after the end of the learning phase and when rewards are no longer at stake. These data open interesting perspectives for rehabilitation of patients with attention disorders (see Lucas et al., 2013; Olgiati et al., 2016) or abnormal reward-seeking behaviors (see Anderson et al., 2013a).

The latter study by Della Libera and Chelazzi (2009) further demonstrated that rewards can not only increase the salience of the associated stimuli, but can also enhance the suppression of distractors. Specifically, they showed that when during the learning phase a given stimulus was presented as a to-be-ignored distractor and was more often followed by high reward, then the system appeared to become relatively more efficient in ignoring the given item. This led the authors to suggest that, at least under the appropriate conditions, rewards act as teaching signals for learning and optimizing specific attentional operations, namely, selecting or ignoring, in relation to specific stimuli (Chelazzi et al., 2013). This idea is closely linked to notions of reinforcement learning applied to the attentional domain. Interestingly, in a subsequent study employing the exact same methodology as in the earlier study, except that participants were told that rewards were given on a random basis, ie, independently of their performance level, the effects of the reward treatment were different (Della Libera et al., 2011). In this case, stimuli that during learning were more often associated with high reward, and regardless of the role they played (target or distractor) when rewards were given, seemed to acquire increased salience, rendering them more easily selected when shown as targets and less easily ignored when shown as distractors. Therefore, it is of note that in this study the system did not appear to enhance distractor suppression for items more often associated with high reward, unlike what was found in the original study.

Going one step further, we recently examined where within the cognitive system the signal providing attentional priority may be generated when reward modulations compete with other mechanisms of spatial attention (Bourgeois et al., 2015). We designed a visual search task (see Fig. 1) based on the paradigm introduced by Anderson et al. (2011a). Spatial orienting of attention was manipulated across different exogenous and endogenous conditions, allowing us to pit reward effects and spatial-orienting effects against each other. Our results confirmed a robust effect of reward association on attentional capture. This effect occurred despite the concurrent attentional cues, either endogenous or exogenous, suggesting that reward is a powerful determinant of attentional selection that can mitigate the attentional orienting induced by other endogenous or exogenous signals. All together, these results suggest multiple, partly independent sources of modulation on visual orienting, which appear functionally and anatomically distinct from attentional systems classically associated with frontoparietal cortical networks in the brain (see also Pourtois et al., 2012, for similar modulations of attention by threat-related information).

Several elegant studies demonstrated that reward could create *oculomotor* salience by biasing not only perceptual mechanisms but also saccadic eye-movement systems. It is well known that there is a tight coupling between saccadic eye movements and shifts of spatial attention (Rizzolatti et al., 1987). In this context, it has been claimed that the oculomotor capture by rewarded stimuli might reflect exogenous process that are nonetheless influenced by top-down attentional set. Theeuwes and Belopolsky (2012) studied the oculomotor capture of previously rewarded stimuli in a subsequent visual search task. They found that eye movements tend to deviate toward a task-irrelevant but previously reward-associated stimulus in the search array, suggesting that the oculomotor capture was not driven by strategy. Using a reward association paradigm, Rothkirch et al. (2013) also demonstrated shorter latencies of voluntary saccades when they were directed toward faces previously associated with a high reward. Furthermore, Hickey and Van Zoest (2013) designed an oculomotor paradigm in which strategic attentional set was decoupled from the effect of reward and demonstrated that reward could guide visual selection independent of voluntary, strategic top-down control. Bucker and collaborators (2015) also examined the oculomotor capture of high, low, and not rewarded stimuli. However, unlike previous studies, the differently valued objects were presented simultaneously in close spatial proximity. Their results indicated that eyes were still biased toward the high value-associated stimulus. Moreover, this effect seemed to be robustly present even when rewards were no longer delivered.

2.1 REWARD-BASED LEARNING ALTERS PRIORITY MAPS OF SPACE

Chelazzi and colleagues (2014) addressed the possibility that reward-based learning alters the priority of stimuli appearing at various spatial locations, and cross-stimulus competition. Their paradigm comprised a learning phase during which participants performed a visual search task on stimulus arrays containing a target among a set of distractors. Correct responses were followed by reward, which could be high or low. The probability of earning high vs low reward was manipulated in a location-specific manner such that certain locations were associated more often with high reward, others with low reward and, finally, the remaining locations were associated with high and low reward in equal proportion. The influence of this reward treatment on attention was measured by comparing performance between a baseline session, occurring prior to learning, and a test session, occurring after learning. Importantly, rewards were not involved in these two sessions. During baseline and test, participants were asked to identify as many critical targets (letters and digits) as they could within briefly presented displays. Each display contained 1 or 2 targets accompanied, respectively, by 7 or 6 distractors. Of particular relevance for the given purposes were conditions in which two targets were presented but only one of them could be reported in the given trial, indicating that the target at one location had taken precedence over the competing one in entering short-term memory. Relative to the baseline session, it was found that at test, a target presented at a high-reward location (as established during learning) increased its priority when paired with a target at a low-reward location, and vice-versa. No reliable change instead occurred for locations associated with an equal probability of high vs low reward during learning. Based on this evidence the authors concluded that reward-based learning can alter the priority of spatial locations, presumably acting on those brain areas that are supposed to house priority maps of space for the sake of attentional guidance. Importantly, it was found that in the given context, effects of the reward-based treatment could be observed several days after the end of the learning phase and could generalize to new task and stimuli relative to those used during learning, further supporting the notion that the effect likely reflect plastic changes occurring at the level of priority maps of space.

2.2 MODULATION OF CONTEXTUAL CUEING

Among the different forms of attentional guidance, there is one, known as contextual cueing, that deserves to be kept separate from other forms of attentional guidance, as it is assumed to reflect unique cognitive operations and underlying brain mechanisms, in particular it is assumed to reflect an interaction between mechanisms controlling attentional deployment in space and those responsible for episodic and semantic memory (Chun, 2000; Goldfarb et al., 2016). The term contextual cueing refers to improved performance in searching for a given target stimulus within an array of distractors whenever the array contains some sort of consistent structure over repeated presentations (Chun et al., 2011). Specifically, spatial contextual cueing reflects an incidental form of learning that occurs when spatial distractor configurations are repeated in visual search displays (Chun and Jiang, 1998). Recently, it was reported that the efficiency of contextual cueing can be modulated by reward (Tseng and Lleras, 2013), a finding that was replicated in a subsequent fMRI investigation of reward-dependent enhancement of contextual cueing (Pollmann et al., 2016). Moreover, the latter study sought to characterize the neural underpinnings of reward-enhanced contextual cueing. In the reported experiment, reward value (high vs low) was associated with repeated displays in a learning session. The effect of reward value on context-guided visual search was assessed in a subsequent fMRI session without reward. Structures known to support explicit reward valuation, such as ventral frontomedial cortex and posterior cingulate cortex, were modulated by incidental reward learning. Furthermore, contextual cueing, leading to more efficient search, went along with decreased activation in the visual search (dorsal attention) network. Finally, the analysis revealed a special role of retrosplenial cortex, a brain area known to be involved in spatial navigation and spatial context memory (Miller et al., 2014), in that this cortical region showed both a main effect of reward and a reward × configuration interaction, raising the possibility that this region may be a central hub for the reward modulation of context-guided visual search.

2.3 CROSS-MODAL INTEGRATION OF VALUE-DRIVEN ATTENTIONAL CAPTURE

Studies of value-based attentional capture have been to date mainly focused on the visual sensory modality. Very recently, Anderson (2015b) examined the attentional capture of a previously rewarded auditory sound on the subsequent detection of a visual target. He demonstrated that value-associated auditory stimuli could interfere with performance in relation to the visual target. This result suggests that value-driven attentional capture may operate at a cross-modal level. Pooresmaeili and collaborators (2014) tested the effect of a previously rewarded sound on performance in a perceptual acuity task. These authors found that auditory stimuli that were previously paired with a high reward could subsequently enhance the sensitivity of visual perception. This effect occurred even when sounds and reward associations were task-irrelevant. Future studies should investigate how reward information is communicated and integrated across the different sensory modalities.

2.4 INDIVIDUAL DIFFERENCES IN REWARD SENSITIVITY

Behavioral studies in humans have demonstrated substantial variability between individuals in reward priming. Hickey and collaborators (2010b) reported, for instance, that healthy subjects with reward-seeking personalities, assessed by the BIS/BAS inventory, showed a larger modulation by reward of an intertrial priming effect. Using a reward association paradigm, Anderson et al. (2013a) demonstrated that the attentional capture by previously high-rewarded stimuli in a subsequent visual search task was greater in patients receiving methadone maintenance treatment for opioid dependence. This effect was also shown in patients with HIV, and correlated with prior HIV-related risk-taking behavior. Although previous studies have consistently demonstrated attentional biases in addictive behaviors (see, eg, Field and Cox, 2008, for a review), more studies are however needed to better understand the neural mechanisms and the neurobiology of value-based attentional processing in addiction.

3 DOES MOTIVATIONAL ATTENTION REQUIRE CONSCIOUSNESS?

The motivational value of stimuli, such as past reward history, may thus govern specific and powerful mechanisms that select information from the environment and promote swift responses to them. Interestingly, most studies reviewed earlier have reported attentional capture by previously rewarded stimuli occurring without explicit knowledge of the acquired stimulus value. However, recent results (Bourgeois et al., 2015) suggested that value-based attentional orienting might be modulated to some degree by conscious awareness of the reward association, with greatest distraction by rewarded stimuli when participants were aware of the previous reward contingencies. This enhanced distraction was more evident on trials where spatial attention was directed by competing valid cues. On the other hand, many studies already reported a strong influence of stimuli with emotional significance that are not perceived consciously to induce behavioral and neurophysiological responses (Seitz et al., 2009; Vuilleumier, 2005). Therefore, is explicit knowledge of contingency required for reward learning to occur?

Seitz et al. (2009) studied whether awareness is necessary for the formation of perceptual learning during conditioning. Participants were deprived of food and water, and passively viewed gratings while receiving occasional drops of water as rewards. One orientation of the grating was paired with reward and presented under conditions of suppressed awareness. Their results indicated improved perceptual discrimination for the orientation paired with reward, as compared with the unrewarded orientation. Bijleveld et al. (2010) designed a speed-accuracy paradigm in order to study how subliminal and supraliminal reward may impact this trade-off differently. Participants were required to solve a mathematical task in which they could earn money when prioritizing either accuracy or speed to solve equations. Supraliminal, but not subliminal rewards influenced task strategy, inducing a change in speed-accuracy trade-off. Thus, in this context, awareness of the reward outcome was necessary to influence performance. Zedelius et al. (2011) (see also Zedelius et al., 2014 for a review) also studied the role of consciously and unconsciously perceived rewards on the active maintenance of goal-relevant information. They used a word span task and presented either supraliminal or subliminal reward value before or after participants processed the target words. When supraliminal or subliminal rewards were presented before the target words, high reward led to enhanced performance. However, when rewards were presented after the word, that is, during the active maintenance processing, performance increased when high rewards were presented subliminally, but decreased when they were presented supraliminally. Finally, Bijleveld et al. (2014) studied the neural bases of subliminal and supraliminal monetary reward during a mental-rotation task fMRI study. They found that delivery of conscious and unconscious rewards produced the same behavioral outcomes. However, supraliminal, but not subliminal rewards engaged brain areas usually involved in reward processing such as the ventral striatum.

To sum up, there is still controversial evidence demonstrating the influence of subliminal vs supraliminal processing of reward information on behavior. Moreover, much of the extant literature is based on paradigms in which stimuli that predict reward, either supraliminal or subliminal, have an inherent value of motivational significance in the task itself. This does not allow a clear differentiation of the specific role of reward value on attentional selection outside of awareness from the well-known role of reward in the strategic establishment of attentional sets.

3.1 REWARD EFFECTS IN NEGLECT PATIENTS

A few neuropsychological studies have begun to investigate the impact of reward on attentional orienting in brain-damaged patients with spatial neglect after stroke. The neglect syndrome is a neurological disorder typically following right hemispheric lesion, and characterized by a loss of awareness for stimuli in the opposite side of space (typically left), despite intact visual pathways in early visual areas in the occipital cortex. The attentional imbalance in neglect seems to be a consequence of deficits in exogenous attentional orienting, whereas endogenous orienting appears relatively spared, even if slowed (Bartolomeo et al., 2001). Recent studies revealed that rewards associated with left-sided targets could progressively bias visual exploration toward the left hemispace, in patients suffering from left neglect after damage to right frontoparietal attention networks (Lucas et al., 2013; Malhotra et al., 2013). Interestingly, this spatially specific bias seems to occur without conscious awareness of asymmetric rewards contingencies. Likewise, Lecce et al. (2015) demonstrated in a spatial reward-learning task that despite defective allocation of attention toward the contralesional left hemispace, right brain-damaged patients with neglect showed preserved contralesional reward learning compared to right brain-damaged patients without neglect. A more detailed review of this topic is provided by Olgiati et al. (2016).

4 NEURAL BASES OF VALUE-DRIVEN ATTENTIONAL SELECTION

The precise mechanisms and neural bases underlying value-driven attentional capture remain partly unresolved. Specifically, the source of this enhancement is still unclear, as many studies did not clearly differentiate attention and reward expectations, giving rise to uncertainty about how reward modulates attentional selection in the brain (Maunsell, 2004). However, an increasing number of recent studies have been conducted in both human and nonhuman primates to investigate how reward information can influence early visual processing.

4.1 VALUE-BASED MODULATION IN VISUAL CORTEX OF HUMAN AND NONHUMAN PRIMATES

Early functional MRI investigations reported value-related modulations throughout specific areas of the human visual system. Serences (2008) conducted a pioneer fMRI study using a paradigm in which participants were asked to maximize their gain by choosing one of two targets varying in value across the course of the experimental session. The results indicated that current stimulus value influences the magnitude of cortical responses within early sensory regions of the visual cortex including V1 (see also Serences, 2008; Weil et al., 2010).

Value-based attentional selection may amplify sensory processing through mechanisms which seemed to be at least partly independent of attentional frontoparietal networks. One study in monkeys found that V1 neurons with a strong value effect also exhibited a strong attention effect, suggesting overlap between relative value and top-down attention (Stanisor et al., 2013). Interestingly, Baruni and collaborators (2015) conducted a study in monkeys in order to differentiate reward expectation from attentional source of modulation in visual area V4. Neurophysiological studies have indeed repeatedly found a modulation of V4 by selective attention (Fries et al., 2001; Moran and Desimone, 1985). The authors found that neurons in V4 were specifically modulated by the absolute reward value associated with visual stimuli. Enhanced neuronal responses allowing more efficient selection might be a critical mechanism linking visual modulations in V4 by reward value to attentional competition and selection.

4.2 ROLE OF DOPAMINERGIC SIGNALS

A comprehensive theoretical framework has been well established to support a link between dopamine systems and rewards across a variety of domains, including decision-making and memory (Schultz, 1997). Interestingly, Arsenault and collaborators (2013) measured fMRI activity of stimulus-reward coupling in monkey visual cortex. Reward-associated stimuli were found to induce specific perceptual learning effects, suggesting that rewards could induce selective plasticity within the visual cortex. This effect may depend on dopamine signaling, as attested by a deactivation of uncued reward activity within the visual cortex during the postinjection of dopamine antogonist. The recruitment of the dopaminergic reward system might be responsible for a boosting of the perceptual representation of stimuli paired with reward value, such that these stimuli become salient and attention drawing (Berridge and Robinson, 1998).

Beyond sensory cortex, other brain areas have been reported to exhibit reward-based modulation. Some of these regions may influence directly or indirectly visual processing, underlying reward-related modulations. Using a reward association-based paradigm, Anderson et al. (2014) demonstrated a selective activation of the extrastriate cortex but also of the caudate nucleus when previously high-rewarded visual distractor were presented in a visual search task. The caudate is a part of the basal ganglia receiving strong projections from dopaminergic pathways and connected with several associative cortical areas.

In a similar vein, Yamamoto and collaborators (2013) recorded neurons in the tail of the caudate nucleus (CDt) of three rhesus monkeys who were presented with various objects, of which half were associated with a large reward. Over time monkeys developed a consistent gaze bias, together with a stronger response of CDt neurons, for the high-valued objects compared to the low-valued objects. Going one step further, Hikosaka and collaborators (2014) demonstrated that the head and the CDt code differentially reward value of objects in short- and long-term value memories, respectively. Indeed, the CDt may encode values of visual objects stably, while the head of the caudate nucleus could encode values more flexibly. Furthermore, both stable and flexible value signals appear to be sent to the superior colliculi through different parts of the substantia nigra, thereby biasing gaze to high-valued objects (Yasuda and Hikosaka, 2015). Indeed, studies in nonhuman primates demonstrated that, in overt approach behaviors, reward expectations do not only recruit the dopaminergic system but also produce a concomitant increase of neuronal activity in several brain regions controlling attention and/or eye movements (Ding and Hikosaka, 2006; Maunsell, 2004; Platt and Glimcher, 1999; Weldon et al., 2008). Midbrain regions may assign priority to sensory sources of information, and then transmit this reward-associated signal to oculomotor regions such as the superior colliculi (Ikeda and Hikosaka, 2007), or to the frontal eye field (Ding and Hikosaka, 2006). This may allow to move the eyes automatically to value-associated stimuli, and thus promote faster/stronger accumulation of evidence for upcoming actions (see Fig. 2), but also result in selective top-down effects modulating activity in sensory areas (Dominguez-Borras and Vuilleumier, 2013; Moore and Fallah, 2004).

To sum up, dopamine modulation of midbrain neurons (striatum/caudate nucleus) may signal the difference between expected and actual reward, and then influence various brain systems involved in attention and motivation as well as decision-making (Hikosaka, 2007; Nakamura and Hikosaka, 2006; Pessiglione et al., 2006; Yamamoto et al., 2013). Stable and flexible representations of values encoded in the caudate nucleus might be transmitted to the superior colliculi via different parts of the substantia nigra, in order to bias sensorimotor behaviors toward rewarded information. The reward signal may be then further sent to cortical brain regions, such as orbital and medial prefrontal cortices (O'Doherty, 2004) or the anterior cingulate gyrus (Bush et al., 2002; Chudasama et al., 2013) which act to integrate and utilize the reward signal to dynamically modify behavior and response selection.

Contrastingly, most studies in humans used *covert* behavioral paradigms, but to date only few have evidenced direct links between the dopaminergic system and other brain regions mediating changes in perception and attention. Interestingly, Hickey and Peelen (2015) conducted an fMRI study in humans to identify the neural bases for the encoding of task-irrelevant reward-associated stimuli in naturalistic environments (see Fig. 3). Their results demonstrated first that reward could impact representations at the level of semantic category, composed of visually heterogeneous objects. More importantly, their results indicated that the strength of modulations by reward-associated distractors in object-selective visual cortex was predicted by a distributed network of brain areas, including frontal regions (orbitofrontal cortex and dorsolateral prefrontal cortex), the anterior cingulate, the parietal lobe, and notably dopaminergic midbrain areas.

Other studies also implicated the cingulate cortex in reward processing. For instance, Lecce et al. (2015) tested right brain-damaged patients with and without neglect in a spatial reward-learning task. Monetary rewards were displayed more frequently either in a box situated on the left side or in a box situated on the right side in two different sessions. Despite defective allocation of attention toward the contralesional left hemispace, neglect patients showed preserved contralesional reward learning compared to right brain-damaged patients without neglect. Notably, however, this reward-learning effect was not present in one neglect patient

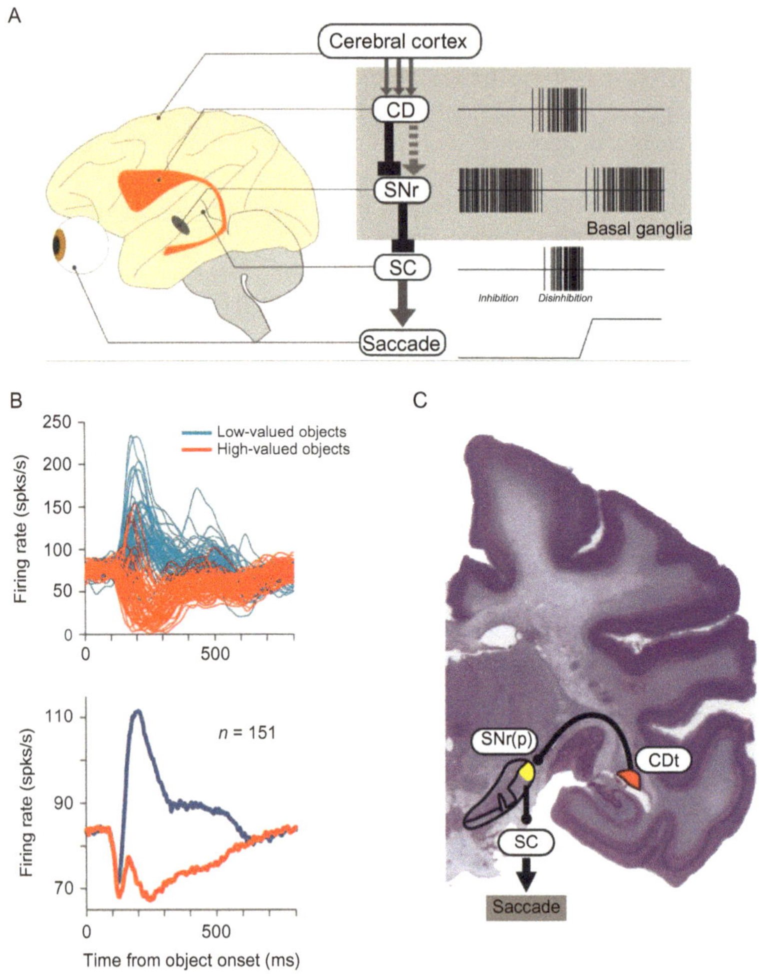

FIG. 2

(A) Basal ganglia circuit controlling the initiation of saccadic eye movements. Some neurons in the monkey caudate nucleus (CD) are activated by visual inputs which originate from the cerebral cortices and other areas. The CD neurons can inhibit the tonic activity of substantia nigra pars reticulata (SNr) neurons through direct connections or enhance the tonic activity of SNr neurons through indirect connections. (B) The responses of an SC-projecting SNr neuron to 120 well-learned objects (B-*top*); average responses of 151 SNr neurons to high-valued objects (*red*) and low-valued objects (*blue*) which were chosen randomly from about 300 well-learned objects (B-*bottom*). (C) The locations of the tail of the CD and SNr shown on a coronal section. The tail of the CD (*red*) has a direct inhibitory connection to the dorsolateral SNr (*yellow*) which then inhibits presaccadic neurons in the SC.

Adapted from Hikosaka, O., Kim, H.F., Yasuda, M., et al., 2014. Basal ganglia circuits for reward value-guided behavior. Annu. Rev. Neurosci. 37, 289–306.

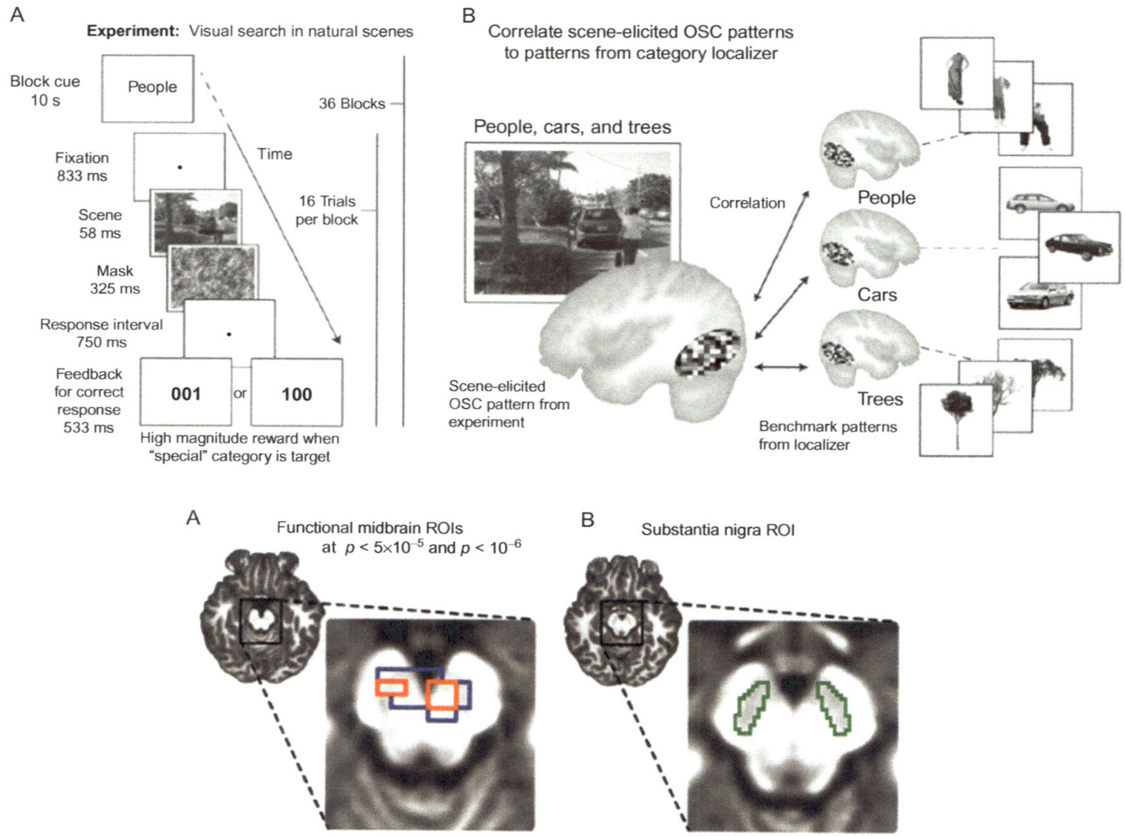

FIG. 3

Top of the panel (A) Experimental paradigm. Three different scene categories were used (people, car, trees). One target category was "special": when cued, correct detection of these objects garnered 100 points. (B) Analytic approach. Scene-evoked activity patterns in object-selective visual cortex (OSC) were cross-correlated with benchmark patterns identified in a separate localizer experiment. Strong correlations indicate increased category information in visual cortex during scene perception. *Bottom of the panel* (A). Functionally defined reward-sensitive region of interest (ROI). (B) Anatomical ROI in substantia nigra.

Adapted from Hickey, C., Peelen, M.V., 2015. Neural mechanisms of incentive salience in naturalistic human vision. Neuron 85, 512–518.

presenting a lesion affecting the anterior components of the parietofrontal attentional network, as well as the medial anterior cingulate cortex. Interestingly, Tosoni et al. (2013) demonstrated using fMRI that the posterior cingulate was modulated by cues signaling an increase in expected reward magnitude during an attentional orienting task, but not by cues instructing to shift and maintain attention, suggesting different networks underlying the distribution of spatial attention and its control by reward-related information. Finally, Hickey and collaborators (2010a) conducted an event-related potential (ERP) study on the impact of reward-associated stimuli on visual attention during a search task. They demonstrated that rewards could change visual salience of targets and distractors, independently of strategic set, an effect that was accompanied by distinctive neural responses located in the anterior cingulate.

Eventually, Vaidya and Fellows (2015) recently demonstrated a specific role of the ventromedial frontal (VMF) cortex for guiding attention in reward-predictive visual features. Participants were primed to attend to task-irrelevant colors associated with a probabilistic high or low reward. Healthy participants and patients with prefrontal damage sparing the VMF cortex demonstrated a larger color priming effect for high-reward distractors than low-reward distractors. This reward-related attentional modulation was, however, absent in patients with VMF damage. These results suggest a key role of the VMF for directing attention to reward-associated features, perhaps through its connections with higher-order sensory regions, which may in turn facilitate stimulus-value associations learning.

4.3 A ROLE FOR THE AMYGDALA IN MOTIVATIONAL ATTENTION?

Numerous studies have demonstrated a key role for the amygdala in reinforcement learning during the processing of both positive and negative emotions (Sergerie et al., 2008; Sander et al., 2003, for a review), and in particular for learning the positive or negative value of external stimuli through their association with rewards and punishments (Paton et al., 2006). Peck and collaborators (2013) tested three rhesus monkey to determine how the amygdala could integrate spatial-visual and motivational information, and thus possibly influence the allocation of spatial attention. These authors used reward-predictive cues in different spatial configurations and assessed how these cues could modulate amygdala neuron responses according to their location in the visual field. Their results indicated that both the spatial location and predicted reward magnitude could modulate the activity of the amygdala, suggesting that the latter might encode the value associated with visual information presented in particular visual locations in space. More studies in humans and nonhumans primates will be required in order to fully understand the role of this structure in motivational attention.

5 CONCLUSION

Converging evidence has accumulated in recent years to reveal a strong impact of motivational-related information, such as reward, on attentional selection. These effects seem to be functionally and anatomically independent from, but closely interacting with and complementary to other attentional systems mediating goal-directed and stimulus-driven orienting, classically associated with frontoparietal cortical networks in the brain. However, if the neuroanatomical basis of "motivated attention" has begun to be uncovered in monkeys, the exact mechanisms underlying this processing in the human brain and the specific role of different brain pathways still remain partly unresolved. More studies are needed to understand where within the cognitive system the signal providing attentional priority is generated and integrated, during both overt and covert behaviors. A better understanding of these neural mechanisms may be usefully exploited; for instance, in neurological rehabilitations strategies in patients suffering from attentional disorders, such as neglect patients, or to treat addicted populations in whom attention is captured by irrelevant but rewarding information.

REFERENCES

Anderson, B.A., 2015a. The attention habit: how reward learning shapes attentional selection. Ann. N. Y. Acad. Sci. 1369, 24–39.

Anderson, B.A., 2015b. Value-driven attentional capture in the auditory domain. Atten. Percept. Psychophys. 78, 242–250.

Anderson, B.A., Laurent, P.A., Yantis, S., 2011a. Learned value magnifies salience-based attentional capture. PLoS One 6, e27926.

Anderson, B.A., Laurent, P.A., Yantis, S., 2011b. Value-driven attentional capture. Proc. Natl. Acad. Sci. U.S.A. 108, 10367–10371.

Anderson, B.A., Faulkner, M.L., Rilee, J.J., et al., 2013a. Attentional bias for nondrug reward is magnified in addiction. Exp. Clin. Psychopharmacol. 21, 499–506.

Anderson, B.A., Laurent, P.A., Yantis, S., 2013b. Reward predictions bias attentional selection. Front. Hum. Neurosci. 7, 262.

Anderson, B.A., Laurent, P.A., Yantis, S., 2014. Value-driven attentional priority signals in human basal ganglia and visual cortex. Brain Res. 1587, 88–96.

Arsenault, J.T., Nelissen, K., Jarraya, B., et al., 2013. Dopaminergic reward signals selectively decrease fMRI activity in primate visual cortex. Neuron 77, 1174–1186.

Bartolomeo, P., Sieroff, E., Decaix, C., et al., 2001. Modulating the attentional bias in unilateral neglect: the effects of the strategic set. Exp. Brain Res. 137, 432–444.

Baruni, J.K., Lau, B., Salzman, C.D., 2015. Reward expectation differentially modulates attentional behavior and activity in visual area V4. Nat. Neurosci. 18, 1656–1663.

Berridge, K.C., Robinson, T.E., 1998. What is the role of dopamine in reward: hedonic impact, reward learning, or incentive salience? Brain Res. Brain Res. Rev. 28, 309–369.

Bijleveld, E., Custers, R., Aarts, H., 2010. Unconscious reward cues increase invested effort, but do not change speed-accuracy tradeoffs. Cognition 115, 330–335.

Bijleveld, E., Custers, R., Van Der Stigchel, S., et al., 2014. Distinct neural responses to conscious versus unconscious monetary reward cues. Hum. Brain Mapp. 35, 5578–5586.

Bourgeois, A., Neveu, R., Bayle, D.J., et al., 2015. How does reward compete with goal-directed and stimulus-driven shifts of attention? Cogn. Emot., 24, 1–10.

Bucker, B., Silvis, J.D., Donk, M., et al., 2015. Reward modulates oculomotor competition between differently valued stimuli. Vis. Res. 108, 103–112.

Bush, G., Vogt, B.A., Holmes, J., et al., 2002. Dorsal anterior cingulate cortex: a role in reward-based decision making. Proc. Natl. Acad. Sci. U.S.A. 99, 523–528.

Chelazzi, L., Perlato, A., Santandrea, E., et al., 2013. Rewards teach visual selective attention. Vis. Res. 85, 58–72.
Chelazzi, L., Estocinova, J., Calletti, R., et al., 2014. Altering spatial priority maps via reward-based learning. J. Neurosci. 34, 8594–8604.
Chudasama, Y., Daniels, T.E., Gorrin, D.P., et al., 2013. The role of the anterior cingulate cortex in choices based on reward value and reward contingency. Cereb. Cortex 23, 2884–2898.
Chun, M.M., 2000. Contextual cueing of visual attention. Trends Cogn. Sci. 4, 170–178.
Chun, M.M., Jiang, Y., 1998. Contextual cueing: implicit learning and memory of visual context guides spatial attention. Cogn. Psychol. 36, 28–71.
Chun, M.M., Golomb, J.D., Turk-Browne, N.B., 2011. A taxonomy of external and internal attention. Annu. Rev. Psychol. 62, 73–101.
Corbetta, M., Shulman, G.L., 2002. Control of goal-directed and stimulus-driven attention in the brain. Nat. Rev. Neurosci. 3, 201–215.
Della Libera, C., Chelazzi, L., 2006. Visual selective attention and the effects of monetary rewards. Psychol. Sci. 17, 222–227.
Della Libera, C., Chelazzi, L., 2009. Learning to attend and to ignore is a matter of gains and losses. Psychol. Sci. 20, 778–784.
Della Libera, C., Perlato, A., Chelazzi, L., 2011. Dissociable effects of reward on attentional learning: from passive associations to active monitoring. PLoS One 6, e19460.
Ding, L., Hikosaka, O., 2006. Comparison of reward modulation in the frontal eye field and caudate of the macaque. J. Neurosci. 26, 6695–6703.
Dominguez-Borras, J., Vuilleumier, P., 2013. Affective biases in attention and perception. In: Armony, J.L., Vuilleumier, P. (Eds.), Handbook of Human Affective Neuroscience. Cambridge University Press, New-York.
Engelmann, J.B., Damaraju, E., Padmala, S., et al., 2009. Combined effects of attention and motivation on visual task performance: transient and sustained motivational effects. Front. Hum. Neurosci. 3, 4.
Field, M., Cox, W.M., 2008. Attentional bias in addictive behaviors: a review of its development, causes, and consequences. Drug Alcohol Depend. 97, 1–20.
Fries, P., Neuenschwander, S., Engel, A.K., et al., 2001. Rapid feature selective neuronal synchronization through correlated latency shifting. Nat. Neurosci. 4, 194–200.
Goldfarb, E.V., Chun, M.M., Phelps, E.A., 2016. Memory-guided attention: independent contributions of the hippocampus and striatum. Neuron 89, 317–324.
Hickey, C., Peelen, M.V., 2015. Neural mechanisms of incentive salience in naturalistic human vision. Neuron 85, 512–518.
Hickey, C., Van Zoest, W., 2013. Reward-associated stimuli capture the eyes in spite of strategic attentional set. Vis. Res. 92, 67–74.
Hickey, C., Chelazzi, L., Theeuwes, J., 2010a. Reward changes salience in human vision via the anterior cingulate. J. Neurosci. 30, 11096–11103.
Hickey, C., Chelazzi, L., Theeuwes, J., 2010b. Reward guides vision when it's your thing: trait reward-seeking in reward-mediated visual priming. PLoS One 5, e14087.
Hickey, C., Chelazzi, L., Theeuwes, J., 2014. Reward-priming of location in visual search. PLoS One 9, e103372.
Hikosaka, O., 2007. Basal ganglia mechanisms of reward-oriented eye movement. Ann. N. Y. Acad. Sci. 1104, 229–249.
Hikosaka, O., Kim, H.F., Yasuda, M., et al., 2014. Basal ganglia circuits for reward value-guided behavior. Annu. Rev. Neurosci. 37, 289–306.

Ikeda, T., Hikosaka, O., 2007. Positive and negative modulation of motor response in primate superior colliculus by reward expectation. J. Neurophysiol. 98, 3163–3170.

Lecce, F., Rotondaro, F., Bonni, S., et al., 2015. Cingulate neglect in humans: disruption of contralesional reward learning in right brain damage. Cortex 62, 73–88.

Lucas, N., Schwartz, S., Leroy, R., et al., 2013. Gambling against neglect: unconscious spatial biases induced by reward reinforcement in healthy people and brain-damaged patients. Cortex 49, 2616–2627.

Malhotra, P.A., Soto, D., Li, K., et al., 2013. Reward modulates spatial neglect. J. Neurol. Neurosurg. Psychiatry 84, 366–369.

Maunsell, J.H., 2004. Neuronal representations of cognitive state: reward or attention? Trends Cogn. Sci. 8, 261–265.

Miller, A.M., Vedder, L.C., Law, L.M., et al., 2014. Cues, context, and long-term memory: the role of the retrosplenial cortex in spatial cognition. Front. Hum. Neurosci. 8, 586.

Mohanty, A., Gitelman, D.R., Small, D.M., et al., 2008. The spatial attention network interacts with limbic and monoaminergic systems to modulate motivation-induced attention shifts. Cereb. Cortex 18, 2604–2613.

Moore, T., Fallah, M., 2004. Microstimulation of the frontal eye field and its effects on covert spatial attention. J. Neurophysiol. 91, 152–162.

Moran, J., Desimone, R., 1985. Selective attention gates visual processing in the extrastriate cortex. Science 229, 782–784.

Nakamura, K., Hikosaka, O., 2006. Role of dopamine in the primate caudate nucleus in reward modulation of saccades. J. Neurosci. 26, 5360–5369.

O'Doherty, J.P., 2004. Reward representations and reward-related learning in the human brain: insights from neuroimaging. Curr. Opin. Neurobiol. 14, 769–776.

Olgiati, E., Russel, C., Soto, D., et al., 2016. Chapter 15—Motivation and attention following Hemispheric stroke. In: Studer, B., Knecht, S (Eds.), Progress in Brain Research, vol. 229. Elsevier, Amsterdam, pp. 343–366.

Paton, J.J., Belova, M.A., Morrison, S.E., et al., 2006. The primate amygdala represents the positive and negative value of visual stimuli during learning. Nature 439, 865–870.

Peck, C.J., Lau, B., Salzman, C.D., 2013. The primate amygdala combines information about space and value. Nat. Neurosci. 16, 340–348.

Pessiglione, M., Seymour, B., Flandin, G., et al., 2006. Dopamine-dependent prediction errors underpin reward-seeking behaviour in humans. Nature 442, 1042–1045.

Platt, M.L., Glimcher, P.W., 1999. Neural correlates of decision variables in parietal cortex. Nature 400, 233–238.

Pollmann, S., Estocinova, J., Sommer, S., et al., 2016. Neural structures involved in visual search guidance by reward-enhanced contextual cueing of the target location. Neuroimage 124, 887–897.

Pooresmaeili, A., Fitzgerald, T.H., Bach, D.R., et al., 2014. Cross-modal effects of value on perceptual acuity and stimulus encoding. Proc. Natl. Acad. Sci. U.S.A. 111, 15244–15249.

Pourtois, G., Schettino, A., Vuilleumier, P., 2012. Brain mechanisms for emotional influences on perception and attention: what is magic and what is not. Biol. Psychol. 92, 492–512.

Rizzolatti, G., Riggio, L., Dascola, I., et al., 1987. Reorienting attention across the horizontal and vertical meridians: evidence in favor of a premotor theory of attention. Neuropsychologia 25, 31–40.

Rothkirch, M., Ostendorf, F., Sax, A.L., et al., 2013. The influence of motivational salience on saccade latencies. Exp. Brain Res. 224, 35–47.

Sander, D., Grafman, J., Zalla, T., 2003. The human amygdala: an evolved system for relevance detection. Rev. Neurosci. 14, 303–316.

Schultz, W., 1997. Dopamine neurons and their role in reward mechanisms. Curr. Opin. Neurobiol. 7, 191–197.

Seitz, A.R., Kim, D., Watanabe, T., 2009. Rewards evoke learning of unconsciously processed visual stimuli in adult humans. Neuron 61, 700–707.

Serences, J.T., 2008. Value-based modulations in human visual cortex. Neuron 60, 1169–1181.

Sergerie, K., Chochol, C., Armony, J.L., 2008. The role of the amygdala in emotional processing: a quantitative meta-analysis of functional neuroimaging studies. Neurosci. Biobehav. Rev. 32, 811–830.

Small, D.M., Gitelman, D., Simmons, K., et al., 2005. Monetary incentives enhance processing in brain regions mediating top-down control of attention. Cereb. Cortex 15, 1855–1865.

Stanisor, L., Van Der Togt, C., Pennartz, C.M., et al., 2013. A unified selection signal for attention and reward in primary visual cortex. Proc. Natl. Acad. Sci. U.S.A. 110, 9136–9141.

Theeuwes, J., Belopolsky, A.V., 2012. Reward grabs the eye: oculomotor capture by rewarding stimuli. Vis. Res. 74, 80–85.

Tosoni, A., Shulman, G.L., Pope, A.L., et al., 2013. Distinct representations for shifts of spatial attention and changes of reward contingencies in the human brain. Cortex 49, 1733–1749.

Tseng, Y.C., Lleras, A., 2013. Rewarding context accelerates implicit guidance in visual search. Atten. Percept. Psychophys. 75, 287–298.

Vaidya, A.R., Fellows, L.K., 2015. Ventromedial frontal cortex is critical for guiding attention to reward-predictive visual features in humans. J. Neurosci. 35, 12813–12823.

Vuilleumier, P., 2005. How brains beware: neural mechanisms of emotional attention. Trends Cogn. Sci. 9, 585–594.

Vuilleumier, P., 2015. Affective and motivational control of vision. Curr. Opin. Neurol. 28, 29–35.

Weil, R.S., Furl, N., Ruff, C.C., et al., 2010. Rewarding feedback after correct visual discriminations has both general and specific influences on visual cortex. J. Neurophysiol. 104, 1746–1757.

Weldon, D.A., Patterson, C.A., Colligan, E.A., et al., 2008. Single unit activity in the rat superior colliculus during reward magnitude task performance. Behav. Neurosci. 122, 183–190.

Yamamoto, S., Kim, H.F., Hikosaka, O., 2013. Reward value-contingent changes of visual responses in the primate caudate tail associated with a visuomotor skill. J. Neurosci. 33, 11227–11238.

Yasuda, M., Hikosaka, O., 2015. Functional territories in primate substantia nigra pars reticulata separately signaling stable and flexible values. J. Neurophysiol. 113, 1681–1696.

Zedelius, C.M., Veling, H., Aarts, H., 2011. Boosting or choking—how conscious and unconscious reward processing modulate the active maintenance of goal-relevant information. Conscious. Cogn. 20, 355–362.

Zedelius, C.M., Veling, H., Custers, R., et al., 2014. A new perspective on human reward research: how consciously and unconsciously perceived reward information influences performance. Cogn. Affect. Behav. Neurosci. 14, 493–508.

CHAPTER

Motivation and attention following hemispheric stroke 15

E. Olgiati*, C. Russell[†], D. Soto[‡,§], P. Malhotra*[,1]

*Imperial College London, Charing Cross Hospital, London, United Kingdom

[†]Institute of Psychiatry, Psychology and Neuroscience, King's College London, London, United Kingdom

[‡]Basque Center on Cognition, Brain and Language, San Sebastian, Spain

[§]Ikerbasque, Basque Foundation for Science, Bilbao, Spain

[1]Corresponding author: Tel.: +44-20-33117286; Fax: +44-20-33117577, e-mail address: p.malhotra@imperial.ac.uk

Abstract

Spatial neglect (SN) is an extremely common disorder of attention; it is most frequently a consequence of stroke, especially to the right cerebral hemisphere. The current view of SN is that it is not a unitary deficit but a multicomponent syndrome. Crucially, it has been repeatedly shown that it has a considerable negative impact on rehabilitation outcome. Although a number of behavioral and pharmacological therapies have been developed, none of these appears to be applicable to all patients with SN or has proved unequivocally successful in clinical trials.

One potential avenue for therapeutic intervention in neglect relates to the interaction between motivation and attention. A number of investigators, including ourselves, have observed a possible motivational component to the syndrome and showed that motivational stimulation can temporarily improve attention in patients with SN.

In this chapter we review previous work looking at how motivation can modulate attention in healthy individuals and how it may be affected by neurological disease before discussing how motivational impairments may contribute to neglect, and how motivation has been used to modulate neglect. In the final section, we present recent experimental work examining how reward interacts with attentional biases in patients with SN. In this study, we adapted the classic Landmark task to explore the mechanisms behind the effect of reward in SN, and found that centrally located stimuli that were explicitly associated with reward appeared to improve neglect and reduce rightward bias. Our results suggest that positive motivation, in the form of anticipated monetary reward, may influence attentional bias via more general mechanisms, such as alerting and task engagement, rather than directly increasing salience of items in contralesional space. We conclude by discussing how motivation might be practically integrated into the rehabilitation of patients with this debilitating disorder.

Progress in Brain Research, Volume 229, ISSN 0079-6123, http://dx.doi.org/10.1016/bs.pbr.2016.06.011

Keywords

Motivation, Stroke, Reward, Attention, Neglect

1 REWARD CAN MODULATE ATTENTION

Rewards (and punishments) are known to shape behavior by changing the likelihood of succeeding actions, as described in classical conditioning procedures (Thorndike, 1911). By acting as a key reference for behavioral decision-making, reward plays a central role in survival and adaptation to our environment. In the brain, specific neural mechanisms have evolved to signal and predict the occurrence of a rewarding event, and it has been shown that dopaminergic neurons in the striatum, prefrontal cortex, and amygdala are involved in reward-based processes (Schultz, 2002). As different brain regions are interconnected to each other in networks, dopaminergic neurons that respond to the delivery of a reward can also modulate the activity of other neural populations and thus influence cognition. A number of investigators have shown that reward can modulate selective attention; that is, the ability to select important information while ignoring distractors (Della Libera and Chelazzi, 2006; Small et al., 2005). Over the last decade, the effect of reward on attention has been extensively investigated in healthy controls using a variety of attentional paradigms (eg, Kiss et al., 2009; Kristjansson et al., 2010; Raymond and O'Brien, 2009 and also see Bourgeois et al., 2016). Through integration between motivational systems and the neural substrates of space representation and action, reward has been shown to have a profound and specific impact on our ability to orient our attention (eg, see recent PET study by Anderson et al., 2016).

2 MOTIVATION IN BRAIN DISEASE

Motivation requires an individual to be sensitive to the intrinsic value of environmental cues and their expected outcomes, as well as accurate weighting of any associated costs. For example, one clear cost relates to the effort required to perform the behavior. Stimuli signaling reward carry positive motivational value because they increase the perceived subjective benefit–cost ratio (Studer and Knecht, 2016). However, brain damage or disease can affect response to motivational value and therefore the ability to react to reward. Specifically, investigators have recently taken a particular interest in how the clinical syndrome of apathy in brain disease might relate to lack of response to reward. Apathy is viewed as a primary deficit of motivation, with patients lacking in goal-directed behavior (Chong and Husain, 2016). The topic has been mainly investigated in the context of Parkinson's disease (PD), which is characterized by changes to dopaminergic pathways (eg, a loss of dopamine neurons in the substantia nigra pars compacta) thought to be critical to reward processing (Schultz, 2002; Wise, 1982). Although the clinical picture of

PD patients is dominated by the presence of movement disorders, it is often accompanied by cognitive problems, and a number of dysexecutive disorders and emotional changes, including depression, anxiety, and apathy (Owen et al., 1992). PD has therefore been extensively used as a model to explore the basis of motivated behavior (Lawrence et al., 2011; Martinez-Horta et al., 2014; Renfroe et al., 2016), using reward response as an index of motivation. For instance, a recent electrophysiological study suggested that behavioral deficits found in PD may be related to a general deficit in preparation for any action; however, patients demonstrated a benefit from incentives as well as healthy controls (Renfroe et al., 2016). Similar studies of reward response in PD have produced contradictory findings, eg, some studies found motivational influence on movement in patients "off medications" (Kojovic et al., 2014) and others did not (Shiner et al., 2012). It is likely that this contradiction is related to the heterogeneity and severity of the disease, effects of treatment, genotype, and different experimental paradigms that have been employed to investigate reward response in PD (Charbonneau et al., 1996; de la Fuente-Fernandez et al., 2001; Goerendt, 2004; Horvitz and Eyny, 2000; Kojovic et al., 2014; Rowe et al., 2008).

More recent work has examined reward processes in other clinical populations including patients with stroke. Blunted reward sensitivity has been reported in patients with striatal damage (Lucas et al., 2013; Malhotra et al., 2013; Schmidt et al., 2008), focal medial frontal lesions (Manohar and Husain, 2016), and right brain-damaged patients with cingulate cortex involvement (Lecce et al., 2015). Clinicians should be aware that a lack of response to motivational stimulation, which has been linked to clinical apathy in stroke populations (Adam et al., 2013; Rochat et al., 2013; Schmidt et al., 2008), may directly reduce their patients' potential for recovery following a stroke. However, it should also be noted that patients may experience a general loss of reward and motivation in their life following the onset of any acute brain disease (eg, jobs may end prematurely, social networks can be reduced, etc.), and therapists face the enormous challenge of engaging them into day-to-day rehabilitation activities (Robertson, 2013). Crucially, a decline in reward response can also contribute and exacerbate other underlying cognitive deficits (Robertson, 2013). A linked, but separate, concept relates to the importance of motivation during rehabilitation, and there is an increasing interest in how patients might directly benefit from the incorporation of motivation into rehabilitation following stroke (Cheng et al., 2015). Later, we discuss these issues in the context of an extremely common and heterogeneous disorder, which has a profound effect on rehabilitation outcome: the spatial neglect syndrome.

3 MOTIVATIONAL IMPAIRMENTS IN SPATIAL NEGLECT

The most common and disabling attention deficit caused by lesions to the right hemisphere of the brain is spatial neglect (SN), whereby patients fail to report sensory events occurring on the contralesional side of the space and to perform actions in the neglected portion of space (Vallar and Bolognini, 2014). SN is usually caused by large strokes in the middle cerebral artery (MCA) territory and its clinical manifestations can be heterogeneous, so that most patients do not manifest every single feature of the syndrome (Li and Malhotra, 2015). The typical patient often struggles in daily activities: patients may not attend to people approaching them from the left-hand side (even if they are speaking), eat food from only the right half of the plate, apply makeup to or shave the right-hand side of their face, and, when walking or using a wheelchair, bump into left-sided objects. Aberrant behaviors like these are often clearly visible to relatives and hospital staff, but a standard psychometric assessment is essential to make a diagnosis and monitor severity and recovery, as in the chronic poststroke phase these symptoms may not be as easy to observe. Traditionally, clinicians employ pen-and-paper tasks such as cancelation tests (eg, star cancelation (Wilson et al., 1987)), line bisection as well as copying and drawing objects (eg, five-element complex drawing (Gainotti et al., 1972; Wilson et al., 1987)). It is very important to detect the presence of SN, as it profoundly affects functional outcome (Suhr and Grace, 1999) and poses a considerable obstacle to successful rehabilitation (Di Monaco et al., 2011; Farne et al., 2004; Katz et al., 1999; Paolucci et al., 2001). Nonetheless, widely approved treatments for this condition are lacking. SN is such a challenging condition to treat partly because it is a multicomponent syndrome (Barbieri and De Renzi, 1989; Vallar, 1998). Whereas the core deficit in SN is typically an attention bias toward the right side of the visual field, a combination of nonlateralized associated cognitive deficits may vary across patients and can shape and exacerbate neglect (Corbetta and Shulman, 2002; Husain and Rorden, 2003; Vallar and Bolognini, 2014). These might also persist chronically even after apparent recovery on standard clinical assessments (see Parton et al., 2004 for a review). Important nonlateralized impairments include deficits in vigilance (Heilman et al., 1978; Husain and Rorden, 2003) and in spatial working memory (Malhotra et al., 2005). Additionally, evidence suggests long-lasting impairments in attention capacity when more challenging tasks are used (Bonato and Deouell, 2013; Russell et al., 2013b).

Early accounts of SN focussed on the importance of motivation mainly in the genesis of another associated feature of the syndrome, a disturbance of spontaneous movement known as motor neglect (MN). Following a lesion within the right hemisphere of the brain, patients with MN underutilize their contralesional limbs, even though they have normal strength and dexterity (Laplane and Degos, 1983). As movements in response to strong prompts are typically preserved (Laplane and Degos, 1983), MN may be considered a (unilateral) deficit of motor motivation. In support of this account, a recent lesion study performed by Migliaccio et al. (2014) showed that the only consistently damaged structure across MN patients was the cingulum, a major pathway of the medial motor system involved in motor initiative and motivational aspects of actions through connection with limbic structures. It has also been suggested that there is a motivational component to SN (Mesulam, 1985; Russell et al., 2013a; Vuilleumier, 2015) and, interestingly, evidence from animal studies suggests that persistent and more severe neglect could be linked to lesions to dopaminergic pathways (Christakou et al., 2005; Van Vleet et al., 2003). Conversely, motivational modulation may have more to contribute in those individuals without lesions affecting these pathways.

4 MOTIVATIONAL MODULATION OF ATTENTION DEFICITS

Directly related to the discussion earlier, understanding motivation in neurological disease is critical for successful rehabilitation (Robertson, 2013). Patients' volitional efforts in working toward recovery are a vital part of the rehabilitation process and could potentially be enhanced by ensuring a stimulating rehabilitation setting. Here, our focus is on motivation's potential applications as a therapeutic tool in attention impairments following stroke. In particular, we will describe how SN symptoms can be alleviated by boosting motivation. This opens up the intriguing possibility that incorporating motivational processes into the clinical management of SN and other poststroke disorders may be of significant therapeutic relevance. Investigators have used several methods to modulate motivational levels in patients with neglect. We summarize these later, before focussing on one particular mode of motivational stimulation–anticipation of reward.

4.1 FRAMING EXERCISES

Ishiai et al. (1990) clinically observed that numbering targets while searching for them improved SN patients' performance in a cancelation task, proposing that the act of numbering, and the ongoing expectation of the subsequent number may boost motivation to find more. Patients have also shown improved performance on a task that they have previously failed, when task instructions are manipulated such that they are implicitly provided with a strategy (ie, to arrange circles around rather than drawing a flower) (Ishiai et al., 1997). This improvement was explained by the authors in terms of a boost of motivation that led to a better arrangement of the elements in leftward space. Following these initial empirical observations, the possibility that a motivational component may emerge when one is actively completing a continuous, regular sequence has been further explored in the domain of music.

Merely listening to sequences of sounds structured in a musical form is known to automatically trigger motor responses in healthy individuals (see Maes et al., 2014 for an overview). This works as a function of previously established associations, developed through systematic, and repeated passive exposure, enabling one to predict the auditory consequences of one's own actions. The framing provided by the production of a predictable sequence again may give the exercise an organized structure that boosts participants' engagement. Following on from this theoretical base, Bodak et al. (2014) and Bernardi et al. (2015) have shown that the improvement obtained by playing a music scale on a horizontally aligned instrument can translate to clinical tests and may outlast the duration of the musical sessions.

4.2 SOUND ENVIRONMENT

The effect of music on cognition and recovery can also arise from general nonspatial effects, ie, the increased arousal and positive emotions triggered by merely listening to music. Sarkamo et al. (2008) examined the effect of an enriched sound environment on general cognitive recovery and mood in a sample of acute MCA stroke patients. Patients were randomly assigned to one out of three groups and were invited, for the following 2 months, to listen daily (minimum 1 h/day) to their preferred music, an audiobook, or no listening material. A neuropsychological assessment performed pre- and postintervention showed a significant improvement in general cognitive recovery (ie, verbal memory capacity and focused attention) and mood in the music group only. This is in keeping with previous findings (Husain et al., 2002), and with the arousal and mood hypothesis (Thompson et al., 2001). This states that any enjoyable stimuli, for instance in the form of high tempo or preferred music, can induce a general positive affective state or enhanced arousal, which in turn can lead to improved performance on cognitive tasks. It should be noted that in the study by Sarkamo et al. (2008), the sample was heterogeneous and although SN patients were included they were not examined separately. Therefore, a clear conclusion on whether music was able to improve SN cannot be drawn from this study.

However, other investigators have directly examined this question and explored whether positive mood, induced by listening to music, can modulate spatial attention deficits and improve SN in stroke populations. Initial work has been encouraging, with mere listening to music of preference reducing SN manifestations as assessed by standard neuropsychological assessment (Chen et al., 2013; Guilbert et al., 2014; Soto et al., 2009; Tsai et al., 2013). For instance, in the study by Soto et al. (2009), the authors assessed three patients with SN as they listened to their favorite music, and found significant improvement in their ability to complete perceptual and visual detection tasks as well as preliminary evidence of improvements in standard clinical tests such as line bisection, star cancelation, and reading tests. Hence, motivational stimulation, in the form of preferred music, seems to be a promising tool to induce positive emotions and consequently to reduce SN manifestations, possibly through activation of the mesolimbic dopaminergic reward system (Chen et al., 2013; Salimpoor et al., 2011; Soto et al., 2009).

4.3 POSITIVE AND AVERSIVE MOTIVATION

By and large, similar beneficial effects for attention are obtained by other emotion-evoking stimuli. Researchers have been studying how attention is enhanced by emotional events by looking at differences in performance following presentation of emotive faces, words, voices, and pictures of scenes. Emotionally valent stimuli have been used to successfully modulate attention in healthy controls (for a review, see Vuilleumier, 2015) and patient groups (for a review, see Dominguez-Borras et al., 2012). In patients, preliminary lesion studies have focused on the modulatory effects of emotional stimuli on the spatial distribution of extinction phenomena (ie, difficulty in reporting a contralesional target when it occurs simultaneously with an ipsilesional one) in SN patients (eg, Fox, 2002; Vuilleumier and Schwartz, 2001). Two further lesion studies found an improvement in a standard clinical task, ie, a reduction in rightward bias in a line bisection task (Tamietto et al., 2005) and a facilitation in a visual search paradigm (Lucas and Vuilleumier, 2008). The effect of emotional motivation is likely to be due to the modulatory influence of limbic regions, especially the amygdala (Morris et al., 2001; Vuilleumier et al., 2002). More recent work has looked specifically at negative motivation elicited, for example, by aversive Pavlovian conditioning, and shown that this can also enhance attention in patients (Dominguez-Borras et al., 2013). Defensive behavior in response to negative emotional signals could theoretically be exploited in the rehabilitation setting, although any potential therapeutic use of aversive stimulation would clearly have to be ethically sound and integrated into a wider treatment program.

4.4 MONETARY INCENTIVE

Over 30 years ago, Mesulam (1985) presented anecdotal observation from a single case of a patient with SN, showing that motivational factors (the provision of 1 penny for each detected target) may improve performance on a cancelation task. Following Mesulam's initial observation and the work examining motivational modulation of attention in healthy volunteers as mentioned earlier and described in detail elsewhere in this volume (Bourgeois et al., 2016), monetary incentive has been used to experimentally manipulate attention in pathological populations, with the ultimate aim of improving their performance and reducing their disability. It has now been shown in a number of empirical studies that it is possible to reduce neglect experimentally through administration of a monetary reward.

Anticipated reward, in the form of monetary incentive, can modulate attention at the clinical level, as indexed by standard neuropsychological assessments. The first experimental demonstration of this was by Malhotra et al. (2013), who showed that anticipation of monetary reward reduced clinical severity of the SN syndrome: there was a significant improvement on a modified standard (cancelation) clinical task in a sample of 10 patients affected by SN. Patients carried out one version of the task where targets were explicitly associated with monetary reward (1 pound coins), and a control condition where targets were not associated with reward (Brass buttons). No difference was observed between the two conditions at baseline, after which patients were given monetary reward, and informed that the amount received was solely related to their performance in the reward condition. In the second session, patients showed significantly improved performance in the reward condition without any such improvement in the nonrewarded condition. That this enhancement in the reward condition was present in the second session and not at baseline suggests that it was due to explicit association between the stimuli in that condition and with receipt of reward. Analysis of patients' lesions revealed that positive reward response required intact structures within the striatum and that patients with lesions affecting the striatum were unable to benefit from reward related performance enhancement. These results of reward exposure have potentially powerful clinical implications for rehabilitation of cognitive functions following a brain insult (Robertson, 2013).

Reward may also be effective in reducing nonspatial attentional deficits in right brain-damaged patients, such as temporal-based selection in an attentional blink (AB) paradigm (Li et al., 2016). The AB phenomenon is observed in healthy individuals when two visual stimuli are presented in close temporal proximity, but patients with neglect typically show a pathological prolongation of the AB, such that they are unable to detect a second target over a much greater time period (Husain et al., 1997). However, Li and colleagues (2016) showed that when reward is incorporated into an AB paradigm, it can facilitate identification of the second target. Interestingly, this effect was most prominent in those who had recovered from SN on standard clinical tests, suggesting a possible role for motivational responsiveness in recovery from attention deficits following stroke. Indeed, the study described earlier (Malhotra et al, 2013) suggests that this responsiveness might require intact striatal structures.

In these studies, reward was linked to an abstract rule, such that patients were informed that performance would be rewarded at the end of the session, or it was explicitly associated with targets that were equally distributed on both sides of the midline. However, other studies have used lateralized monetary incentives to examine reward learning in SN patients. Lecce et al. (2015) showed that SN patients can explicitly learn and take advantage of reward when it is presented in the contralesional hemispace. Likewise, Lucas et al. (2013) presented SN patients with a gambling task, whereby they had to search for the most rewarding target in a visual array. They found that space exploration could be biased by asymmetrical presentation of reward incentives, with rewarded left-sided (but not right-sided) targets leading to an improvement in SN manifestations on standard cancelation tasks (without reward), which were carried out separately after the reward session.

5 DISSECTING THE MECHANISMS UNDERLYING REWARD'S EFFECTS ON NEGLECT

The precise mechanism underlying the effect of reward on attentional performance has not yet been determined. Experiments with healthy volunteers suggest that reward may boost attention via an increase in target salience and/or a generalized (nonlateralized) increase in arousal levels (for a review, see Chelazzi et al., 2013). The effect could arise from reward enhancing behavioral salience and meaningfulness of a relevant stimulus, thus increasing its weight in the competition for attention. Previous work has suggested that salient targets presented in the neglected hemispace can improve visual search in SN patients, as if they are pulling attention toward the impaired side (Bays et al., 2010). It is therefore possible that when targets are explicitly associated with a reward, they become more salient and therefore potentially able to reduce rightward bias. Evidence to support this salience account of the effect comes from studies in healthy individuals, which have shown that performance can be *worsened* if distractors rather than targets are associated with reward (Anderson et al., 2011; Della Libera and Chelazzi, 2006). Rewarded stimuli can capture attention and affect performance even if they are irrelevant to the task, suggesting that the increased salience following reward learning is involuntary and automatic. In contrast, the very same effect has also been observed when no monetary reward is used during a training phase, suggesting that this might reflect a general attentional capture induced by previous targets and may not be specifically linked to reward-based processes (Sha and Jiang, 2016).

Alternatively, reward might affect fronto-parietal attention networks via modulation of ascending reticular input associated with the regulation of levels of arousal/alertness. SN has previously been shown to be efficiently modulated by the presence of alerting auditory cues, presented before or during appearance of a target (Finke et al., 2012; Robertson et al., 1998). Reward may share similar arousing effects, enhancing the strategic control of attention and general effort in a top-down manner (Chelazzi et al., 2013; Hubner and Schlosser, 2010). In support of this account, monetary reward has been associated with increased galvanic skin response (Pessiglione et al., 2007), and it has also been shown that expected value and attentional demands are, to some extent, integrated in cortico-striatal-thalamic circuits (Krebs et al., 2012). To explore these issues further and to directly examine these two putative mechanisms for reward's effects on neglect, we compared patients' performances in two adapted versions of a well-known standard clinical task (ie, the Landmark task, see below). These adapted versions were intended to induce either a generalized boost in arousal or an increase in targets' relative salience.

5.1 AN INVESTIGATION USING THE LANDMARK TASK

5.1.1 Aim

We conducted a study to explore how reward modulates attention in a sample of stroke patients affected by attention deficits. Our key aim was to differentiate between the salience and arousal accounts (see above) of anticipated monetary reward's effects on neglect, by means of an adapted version of the classic Landmark task (Milner et al., 1992). In the standard version of this task, participants are presented with a set of horizontal lines that are prebisected at the midpoint or to the left/right of true center. Subject are usually falsely informed that none of the bisections are placed in the exact center and asked to report which is the shortest half of the line. As patients with SN are clinically "over-oriented" toward the right side of the space, evenly prebisected lines are typically perceived as bisected closer to the left end of the line—a perceptual bias which indicates a leftward underestimation of the line's length (Milner et al., 1993). For the same reason, errors with noncentral landmarks are usually more frequent when lines are prebisected toward the right side.

We adapted the task by systematically positioning rewarding and nonrewarding cues centrally or above the extreme ends of each line. We hypothesized that if the effects of reward were explained by the increased salience of items explicitly associated with monetary value, then these would increase rightward bias if positioned above the rightward end of the line, and reduce rightward bias if placed at the leftward, neglected, end of the line. On the other hand, if more general attention mechanisms, including arousal/alertness, were responsible for the effects of reward, then even nonlateralized centrally presented reward cues might modulate bias when displayed immediately before the horizontal line display.

5.1.2 Methods

Eight brain-damaged patients (all right handed) with left SN (see Table 1 for details) following a right hemispheric ischemic stroke were recruited from Imperial College Healthcare NHS Trust (London).

5.1.2.1 Experimental task

Baseline: Patients sat in a dimly lit, quiet room, at a distance of 57 cm in front of a laptop computer (HP EliteBook 846p, Windows 7, Intel i5 core processor). We presented them with a computerized version of the classic Landmark paradigm (Milner et al., 1992), whereby they were asked to state which section of prebisected lines appeared shorter (baseline). In order to ensure that patients understood the instructions correctly, in a training phase they were presented with lines that had been bisected grossly to the right or to the left. After four consecutive correct responses, the experiment was started. Each trial began with the presentation of a central fixation circle for 2000 ms. Following the presentation of a visual mask (50 ms), a prebisected line appeared in the middle of the screen, and patients were asked to judge which was the shorter half of the line by pressing a response key using their right hand with their index finger (if left side appeared shorter) and middle finger (if right side appeared shorter). Unlike other studies (Harvey et al., 1995), we did not ask patients to point toward the shorter half of the line but used a response key to minimize manual orienting behavior. Lines were always shown one at a time, for a total of 36 trials. There were 12 evenly and 24 unevenly prebisected lines, all of which were 17 cm long. The landmark was positioned 2, 4, or 6 mm to the left or the right of the true center for the unevenly bisected condition.

Trials with unevenly bisected lines allowed us to compute an accuracy score. Trials with evenly bisected lines allowed the assignment of rightward perceptual biases by calculating the proportion of "right" responses for each line. The task was always conducted in free viewing (ie, there was no time limit imposed), but participants were encouraged to give their first impression when experiencing uncertainty.

5.1.2.2 Tasks A (central cue—arousal) and B (lateralized cue—behavioral salience)

In the two subsequent sessions (the order of these was counterbalanced across subjects) task conditions were identical to those in baseline, except for the trials when evenly bisected lines were shown. As in our baseline task and previous studies (Milner et al., 1993; Olk and Harvey, 2002) asymmetrical stimuli were always presented without any cue; however, these lines were interspersed with evenly bisected *cued* lines where images of rewarding and neutral objects were also presented at

Table 1 Patient Information and Scores on Standard Neuropsychological Tests at Time of Participation

Patient	Sex/ Age	Time Since Stroke (Months)	BIT Copying	BIT Star Cancelation	Mesulam Shape Cancelation	Objects in Room	Clock Drawing	Line Bisection
1	F, 75	96	5/6	54/54	60/60	+20, −60	12/12	+8.6
2	M, 71	62	5.5/6	54/54	53/60	+20, −65	12/12	+4.2
3	M, 66	89	4/6	51/54	48/60	+60, −60	12/12	+33.4
4	M, 86	5	4.5/6	52/54	48/60	[a]	9/12	−5.4
5	M, 75	25	6/6	54/54	48/60	+90, +35	12/12	+1.8
6	F, 75	5	3.5/6	48/54	55/60	+80, −80	3/12	+2.2
7	M, 61	1	3.5/6	49/54	37/60	+90, −45	12/12	+10.8
8	F, 22	5	6/6	53/54	54/60	+90, −70	12/12	+4.2

BIT, *behavioral inattention test (Thames Valley Test Company, 1987).*

[a]*Indicates where test was not carried out. For the star/shape cancellation and drawing, the total raw accuracy score is reported; for the naming of 10 objects in the room, degrees of the most peripheral objects named by the patient are reported; for the line bisection, the patient's mean percentage of the rightward bias is reported. +, rightward; −, leftward.*

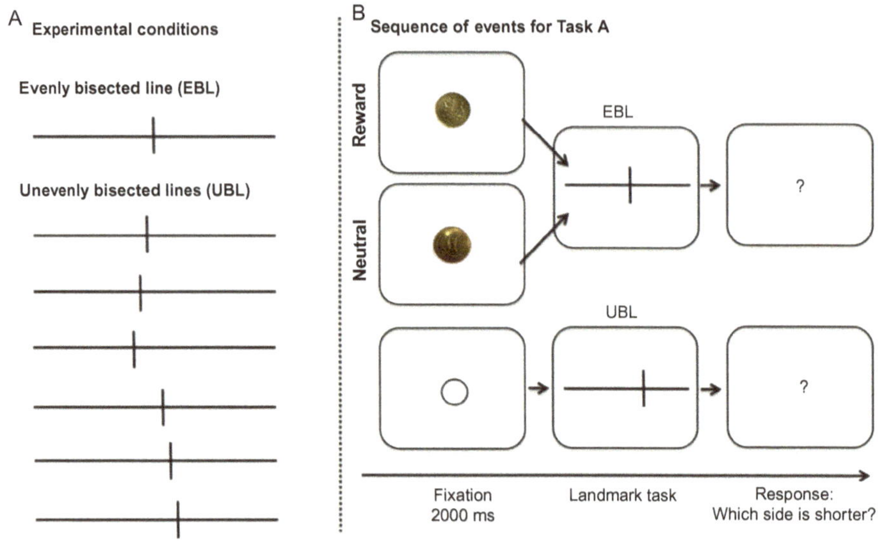

FIG. 1

(A) Experimental conditions and (B) schematic representation of the sequence of events in Task A (Arousal). Note that the baseline consisted of a mixture of evenly (EBL) and unevenly bisected lines (UBL) that were always preceded by a *circle*.

fixation immediately preceding the appearance of the line (Task A) or simultaneously with the line at one or both ends (Task B). The reward and neutral conditions employed images of the same size displaying pound coins (£1.00) and brass buttons, respectively (see Fig. 1). Prior to inclusion, we ensured that patients could distinguish between the two objects. In *Task A (arousal)*, reward/neutral objects were presented at fixation (2000 ms), followed by a mask and then an evenly bisected line (80 in total— 40 lines were preceded by a reward and 40 lines by a neutral object). Twenty-four trials with unevenly bisected lines were also included and, as in the baseline task, they were always preceded by a circle fixation point (2000 ms).

In *Task B (behavioral salience)*, following presentation of a fixation circle (2000 ms) and a subsequent mask, reward or neutral stimuli were placed above the left, the right, or both ends of each of the 96 evenly bisected lines (24 lines had a single reward; 12 lines had bilateral rewards; 24 lines had a single neutral object, 12 lines had bilateral neutral objects; 12 trials had mixed reward/neutral objects with reward on the right and vice versa for 12 trials) and remained on screen until a response was made (see Fig. 2). Twenty-four trials with unevenly bisected lines were also included and, as in baseline, they were not accompanied by any reward/neutral objects. As described earlier, it was thought that the central presentation of reward in Task A might induce a generalized boost in arousal, whereas the lateralized cues in Task B would increase the relative salience of either end of the horizontal line.

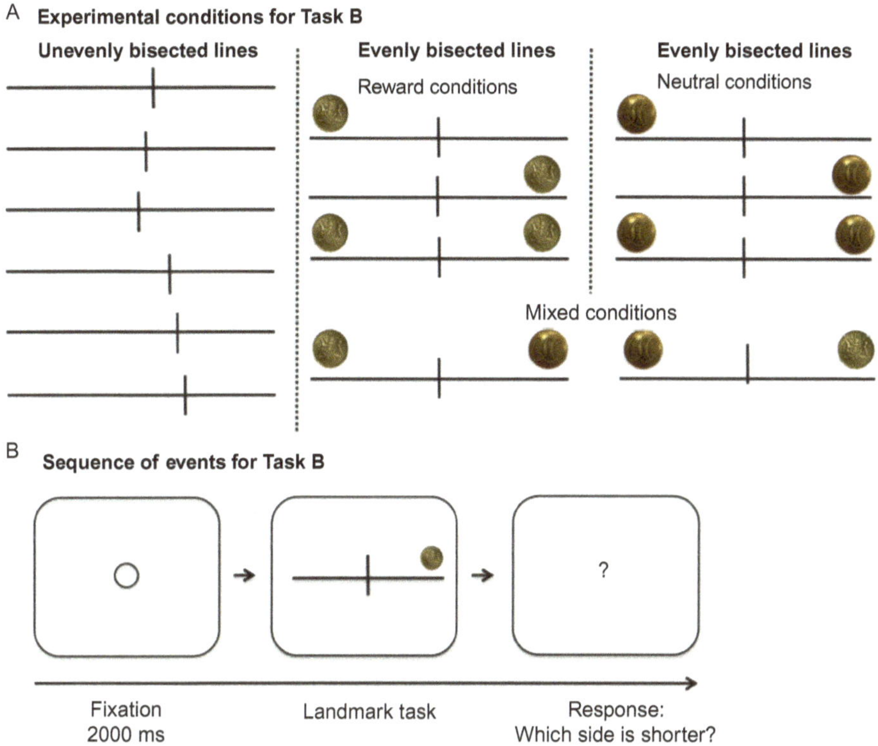

FIG. 2

(A) Experimental conditions and (B) schematic representation of the sequence of events in Task B (behavioral salience).

Patients were informed that the money they would receive would be calculated from performance in rewarded trials only. However, as requested by our local Ethics committee, all patients actually received an identical amount of money at the end of the experiment (£20 in vouchers) regardless of their performance.

Previous studies in healthy individuals and patient populations have demonstrated that reward only appears to affect attention when participants are given sufficient time and/or feedback to learn the association between target and reward (Kiss et al., 2009; Lucas et al., 2013; Malhotra et al., 2013). Accordingly, following administration of the Landmark Task in the baseline session, we asked patients to complete the pound cancelation task (as in Malhotra et al., 2013) and rewarded their performance (£5 in vouchers). The use of incentive at this stage was to induce positive motivation and trigger the effect of reward in the two subsequent sessions.

5.1.3 Results

For unevenly bisected lines only (ie, the Landmark was asymmetrically located toward the left/right end of the line), we were able to compute the number of correct responses. A within-participant ANOVA was used to investigate accuracy across tasks, with Task (baseline vs Task A vs Task B) and Landmark position (left vs right) as factors; series of paired t tests were then used to examine the effects. Note that no reward or neutral cues were presented with the unevenly bisected lines.

We also looked at the rightward perceptual bias for evenly bisected lines, computed as the proportion of "right" responses for each line. In order to compare the effect of the reward/neutral cues in Task A and Task B, a within-participant ANOVA with condition (reward vs neutral) and side (central vs left vs right vs bilateral vs mixed) was used. We then conducted a series of paired t tests to directly compare the rightward shift manifested in Task A and Task B to that of the baseline session. Finally, we used paired t tests to compare performances against chance level (probability of 50%). The partial Eta squared ($p\eta^2$) of significant effects, which measures the proportion of the total variance that is attributable to a main factor or to an interaction (Cohen, 1988), was also computed in order to detect effect sizes. For paired samples t tests, Cohen's d_z and Cohen's d_s of significant effects were also computed (Cohen, 1988).

Accuracy in nonsymmetrically bisected lines—No main effect of Task was found ($F(2,6)=0.108$, $p=0.9$). There was however a significant main effect of Landmark position ($F(1,7)=9.513$, $p=0.018$, $p\eta^2=0.576$); in keeping with previous studies, accuracy was lowest for trials when lines had been prebisected toward the right—the most challenging condition for patients affected by SN (see Fig. 3). A significant Task by Landmark position interaction also emerged ($F(2,6)=7.027$, $p=0.027$, $p\eta^2=0.701$) and was further analyzed via a series of paired t tests; these showed that when the line was bisected to the right, patients were significantly more accurate in Task A as compared to baseline ($t(7)=-4.029$, $p=0.005$, $d_z=1.46$) but were not more accurate in Task B as compared to baseline ($t(7)=-0.397$, $p=0.703$). As Task A and Task B were administered in counterbalanced order across participants, a practice effect can be ruled out. When the line was bisected to the left, patients were significantly less accurate in Task A as compared to baseline ($t(7)=3.228$, $p=0.014$, $d_z=1.13$), with a borderline significant difference in Task B as compared to baseline ($t(7)=2.304$, $p=0.055$).

It should be noted that the stimuli being responded to (ie, uncued lines prebisected toward the right) in this analysis were *exactly the same* across the three tasks. That is for these unevenly bisected stimuli *no* cues (reward or neutral) were used. Therefore, it is possible that the improved accuracy that we observed for lines bisected to the right during Task A may have been secondary to an increase in general arousal, possibly induced by cues associated with preceding trials. To address this directly, we examined successive trials in each condition by comparing accuracy rates in those trials *that followed* rewarded trials vs nonrewarded trials in Task A (Fig. 3, lower panel). Interestingly, a significant difference between the two types of cues emerged in Task A, with performance being significantly more accurate for uncued lines

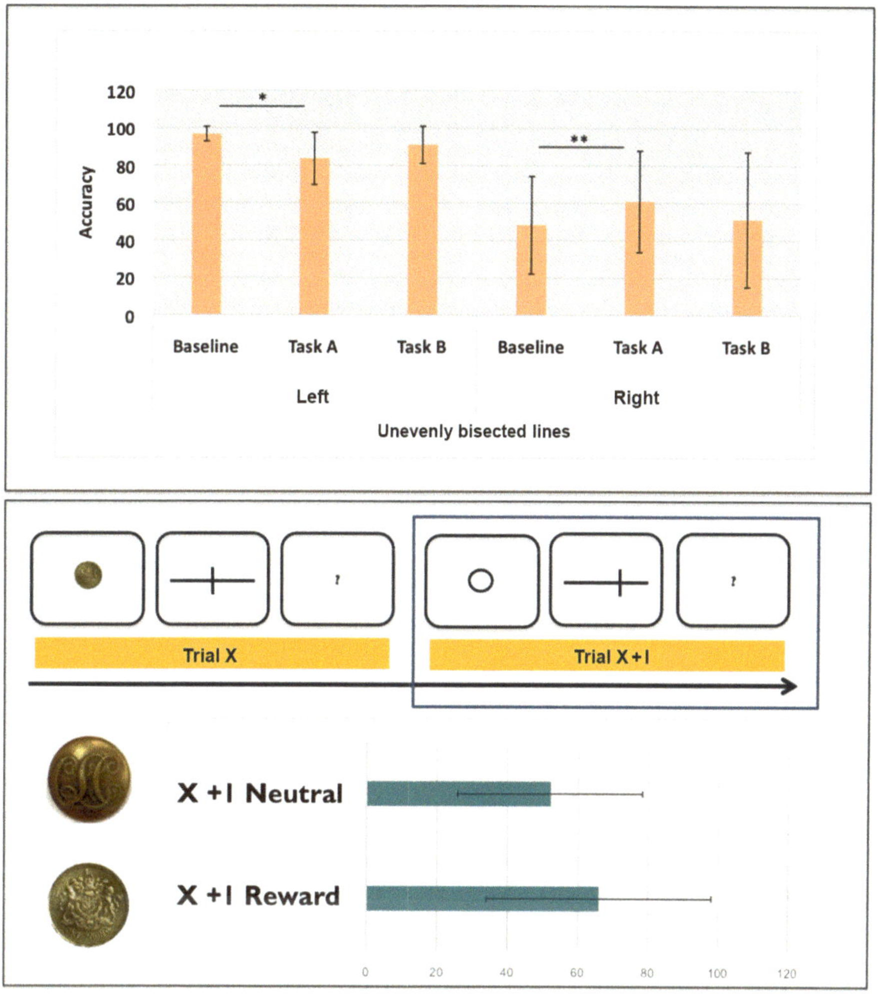

FIG. 3

Accuracy results. Top panel shows mean accuracy (percentage) in baseline, Task A (arousal), and B (behavioral salience) for lines asymmetrically bisected to the left or right. Accuracy for the most challenging condition for patients with neglect (ie, lines prebisected to the right) was significantly greater in Task A than in baseline. Error bars = standard deviation (SD); *p = 0.014; **p = 0.005 significant difference. Lower panel shows how mean accuracy (percentage) in Task A for lines bisected to the right (Trial X+1) was affected by the nature of the preceding trial (Trial X). Trials that followed presentation of a reward are compared to trials that followed presentation of a neutral object. Error bars = standard deviation (SD).

prebisected to the right (ie, the most difficult condition for SN patients) following previously rewarded trials vs trials that followed nonrewarded trials (66% vs 52%, respectively; $t(7) = 2.554$, $p = 0.038$). This is an indicator that anticipated reward affected arousal levels during Task A.

Assessing rightward bias in evenly bisected lines—When judging evenly bisected lines in the baseline session, patients showed an overall systematic bias favoring the right side of the line (ie, 89% of the times they perceived the right side of the lines as being longer).

In order to compare the effect of the reward/neutral cues in Task A and Task B, a within-participant ANOVA with condition (reward vs neutral) and side (central vs left vs right vs bilateral vs mixed) was used and showed no main effect of condition ($F(1,7)=0.267$, $p=0.621$), side ($F(4,4)=0.709$, $p=0.626$), or condition by side interaction ($F(4,4)=0.245$, $p=0.899$). Given the absence of significant effects of the factor side in Task B, spatial positions (left/right/bilateral) were collapsed for further analyses on the effect of reward/neutral objects.

Paired *t* tests were then used to directly compare rightward bias in Task A and Task B to the baseline session: a significantly smaller rightward deviation (see Fig. 4) emerged in Task A (reward condition: $t(7)=3.812$, $p=0.007$, $d_z=1.339$; neutral condition: $t(7)=3.321$, $p=0.013$, $d_z=1.141$) and Task B (reward condition: $t(7)=3.355$, $p=0.012$, $d_z=1.217$; neutral condition: $t(7)=3.465$, $p=0.010$, $d_z=1.226$), both compared to baseline, without any difference between the two types of cue (Task A: $t(7)=-0.638$, $p=0.544$; Task B: $t(7)=0.253$, $p=0.807$).

At baseline, a one sample *t*-test against chance value (a probability of 0.5) revealed that patients judged evenly bisected lines as "left shorter" above chance, ie, they were strongly biased to the right, as expected ($t(7)=8.46$, $p<0.01$, $d_s=5.982$). However, when the line was preceded (Task A) or associated (Task B) with a reward/neutral object, whether it was a pound coin (ie, reward condition, Task A: $t(7)=1.676$, $p=0.138$; Task B: $t(7)=1.744$, $p=0.125$) or a brass button (ie, neutral condition, Task A: $t(7)=1.920$, $p=0.96$; Task B: $t(7)=1.566$, $p=0.161$), performance did not differ from chance level and patients did not systematically judge the left side as being shorter (ie, rightward bias) anymore. Remarkably,

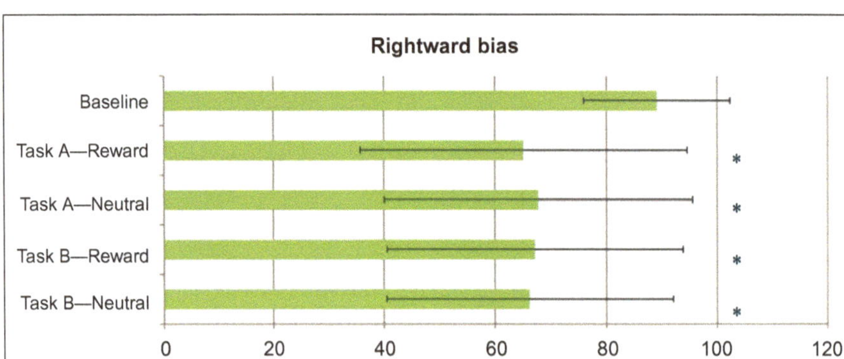

FIG. 4

Rightward bias results. Mean rightward bias (percentage) in baseline, Task A (arousal), and B (behavioral salience). Error bars = standard deviation (SD); * = significant difference between baseline and reward/neutral conditions in both Task A and Task B.

centrally presented (Task A) and lateralized cues (Task B) proved equally able to improve performance and reduce the rightward shift, with no increased effect of reward.

5.1.4 Discussion

Our results show that in the Landmark paradigm, stimuli explicitly associated with reward do not redirect spatial attention any more than neutral stimuli, with both cues being equally effective in reducing the rightward bias in SN patients. Our results do not rule out a salience effect occurring with stimuli that are explicitly associated with reward. However, here, when reward was centrally located and presented before the line, it boosted performance on subsequent trials: it triggered carry-over effects that were likely associated with reward expectations across trials, leading to a reduction in rightward bias (as suggested by an increase in accuracy for lines prebisected to the right, and a reduction in accuracy for lines prebisected toward the left (see below)) on trials that followed rewarded ones. This effect can be regarded as evidence supporting the theory that generalized arousal is the main contributor to reward's effects on clinical tasks in SN patients. Rewarded targets have been shown to induce changes in galvanic skin response (Pessiglione et al., 2007) and pupillary diameter (Manohar and Husain, 2015), both of which are indices of physiological arousal. In the auditory domain, increased arousal obtained through an alerting sound presented before or during the task has been shown to ameliorate neglect (Hommel et al., 1990; Robertson et al., 1998).

We would suggest that, in the current experiment, nonspatial alerting induced by reward presentation effectively produced a leftward shift in patients' performance. This effect was also evident when they were asked to evaluate which half of left-bisected lines appeared shorter—the easier condition for neglect patients, as they tend to perceive the left side of the line as shorter. That is, the finding that accuracy was diminished for those trials following presentation of a nonspatial rewarding cue in Task A could actually be the result of a leftward shift in spatial attention (inducing an expansion of the left side of the line). This is similar to the finding by Robertson and co-workers who observed that phasic alerting induced by warning tones paradoxically induced an advantage for left visual events in patients (Robertson et al., 1998). The authors explained the reversal in terms of a leftward shift in attention that exceeded the patient's chronic rightward bias, thus extending further than the center.

In our data there is evidence that attentional capture driven by rewarding stimuli outlasts (albeit briefly) the rewarded trial. The effect of immediate preceding rewards was recently explored in an fMRI study conducted by Serences (2008), who found that value also influences activation levels within early regions of visual cortex (V1) and that these modulations appear to be influenced primarily by the history of recent rewards as opposed to generalized biases in the subjective value of the selected stimulus that occurred on either a trial-by-trial or a scan-by-scan basis.

The cueing effect of both rewarded and neutral targets found in the current experiment was greater than that found in previous studies (Harvey et al., 1995; Olk and Harvey, 2002) which showed no clear effect or a tendency to be less biased when a cue was displayed on the right-hand side, but left/right judgement ratios differed significantly from chance in all cueing conditions. It should be noted that, unlike previous studies, our paradigm did not require patients to respond by pointing to the line; instead, we asked patients to press a key with their unaffected hand. Also, our sample may differ in the severity of their SN.

In our study, Task A and Task B differed in the spatial location of the reward/neutral objects (centrally presented vs lateralized incentives) and duration of the reward/cue on the screen (2000 ms in Task A vs until response in Task B). Despite the fact that in Task B cueing could potentially have had higher chances to be processed because it remained on the screen for longer, an improvement in accuracy for lines bisected to the right was evident in Task A only. It could be argued that the effect of reward was more prominent in Task A because it was presented at fixation and hence more clearly seen than peripheral cues in Task B. That is, left lateralized reward cues in Task B might not be attended to at all, and right-lateralized reward cues would lead to a worsening of the pathological bias toward the ipsilesional side of the line. However, our data showed that patients *implicitly* processed the left-sided reward just as well as the right-sided reward, and both lateralized reward and lateralized neutral cues improved performance equally. This strongly suggests that the association of monetary reward with a stimulus does not appear to have a direct effect on pathological attentional bias.

6 PRACTICAL IMPLICATIONS OF MOTIVATION-ATTENTION STUDIES IN NEGLECT

These results regarding the effect of motivation on attention may potentially have significant implications for how we approach rehabilitation of patients with SN. As a rule, it is important to detect all possible modifiable factors and conditions to be exploited during the rehabilitation process. In the last few years, it has become clear that motivation can be considered a useful tool to enhance engagement in the rehabilitation setting, so that patients can participate and benefit fully from rehabilitation therapy. Different motivators have been used and proved able to successfully modulate attention deficits following stroke, reducing SN manifestations. This opens up the possibility to incorporate reward into the therapeutic setting and create individually *tailored* care plans, choosing and varying incentives depending on subjects' interests. For instance, the use of preferred tunes to trigger positive emotional states can aid and stimulate recovery following a brain insult.

As described earlier, different modalities of reward delivery have been tested and proved able to modulate SN, possibly by acting as natural dopaminergic stimulants. Dopaminergic drugs have been used in previous trials in SN, with conflicting results (Barrett et al., 1999; Fleet et al., 1987; Gorgoraptis et al., 2012), and it could be that the individual differences in motivational response which have been described may explain some of this variability in the response and that this might be associated with lesion anatomy. Further work is necessary to fully understand the interaction between response to motivational stimulation and the effects of dopaminergic stimulation in patients with attentional deficits (Li et al., 2013).

7 OUTSTANDING ISSUES

There are several outstanding issues in the literature on reward, and therefore the complex interaction between motivation and SN deserves further examination. To begin with, the variability of response across subjects seems remarkable but is still poorly understood. In addition, the current study is relatively small, and thus it is possible that it did not have the statistical power to detect weaker reward effects on task performance. Another issue regards the duration of the effect. Li et al. (2016) and Lucas et al. (2013) showed that once the association has been made, there is evidence for the effect in the following experimental session, even if this does not involve reward at all. However, in our experience the effect of monetary incentives usually appears to decline over time, which makes it less practical to use in the clinical setting. However, monetary reward is a useful research tool that can be translated into more personally relevant motivation by clinicians. It will also be important to determine which improvements in standard clinical tests are transferrable to improvements in everyday life. Another issue concerns clinical validity: it will be important to develop meaningful predictors of clinical response to reward, in order to target rehabilitation and predict outcome. In addition it is worth investigating individual variability in reward response and whether some patients have greater potential to benefit from motivational influences. These differences might be due to many factors, for example, the length of their illness, the presence of concurrent apathy and/or other mood disorders, and site of lesion.

ACKNOWLEDGMENTS

This work was supported by the National Institute for Health Research (NIHR) Imperial Biomedical Research Centre. D.S. acknowledges financial support from the Spanish Ministry of Economy and Competitiveness, through the "Severo Ochoa" Programme for Centres/Units of Excellence in R&D" (SEV-2015-490).

REFERENCES

Adam, R., Leff, A., Sinha, N., Turner, C., Bays, P., Draganski, B., Husain, M., 2013. Dopamine reverses reward insensitivity in apathy following globus pallidus lesions. Cortex 49, 1292–1303.

Anderson, B.A., Laurent, P.A., Yantis, S., 2011. Value-driven attentional capture. Proc. Natl. Acad. Sci. U. S. A. 108, 10367–10371.

Anderson, B.A., Kuwabara, H., Wong, D.F., Gean, E.G., Rahmim, A., Brasic, J.R., George, N., Frolov, B., Courtney, S.M., Yantis, S., 2016. The role of dopamine in value-based attentional orienting. Curr. Biol. 26 (4), 550–555. doi:http://dx.doi.org/10.1016/j.cub.2015.12.062.

Barbieri, C., de Renzi, E., 1989. Patterns of neglect dissociation. Behav. Neurol. 2, 13–24.

Barrett, A.M., Crucian, G.P., Schwartz, R.L., Heilman, K.M., 1999. Adverse effect of dopamine agonist therapy in a patient with motor-intentional neglect. Arch. Phys. Med. Rehabil. 80, 600–603.

Bays, P.M., Singh-Curry, V., Gorgoraptis, N., Driver, J., Husain, M., 2010. Integration of goal- and stimulus-related visual signals revealed by damage to human parietal cortex. J. Neurosci. 30, 5968–5978.

Bernardi, N.F., Cioffi, M.C., Ronchi, R., Maravita, A., Bricolo, E., Zigiotto, L., Perucca, L., Vallar, G., 2015. Improving left spatial neglect through music scale playing. J. Neuropsychol. doi:http://dx.doi.org/10.1111/jnp.12078.

Bodak, R., Malhotra, P., Bernardi, N.F., Cocchini, G., Stewart, L., 2014. Reducing chronic visuo-spatial neglect following right hemisphere stroke through instrument playing. Front. Hum. Neurosci. 8, 413.

Bonato, M., Deouell, L.Y., 2013. Hemispatial neglect: computer-based testing allows more sensitive quantification of attentional disorders and recovery and might lead to better evaluation of rehabilitation. Front. Hum. Neurosci. 7, 162.

Bourgeois, A., Chelazzi, L., Vuillleumier, P., 2016. Chapter 14—How motivation and reward learning modulate selective attention. In: Studer, B., Knecht, S. (Eds.), Progress in Brain Research, vol. 229. Elsevier, Amsterdam, pp. 325–342.

Charbonneau, D., Riopelle, R.J., Beninger, R.J., 1996. Impaired incentive learning in treated Parkinson's disease. Can. J. Neurol. Sci. 23, 271–278.

Chelazzi, L., Perlato, A., Santandrea, E., Della Libera, C., 2013. Rewards teach visual selective attention. Vision Res. 85, 58–72.

Chen, M.C., Tsai, P.L., Huang, Y.T., Lin, K.C., 2013. Pleasant music improves visual attention in patients with unilateral neglect after stroke. Brain Inj. 27, 75–82.

Cheng, D., Qu, Z., Huang, J., Xiao, Y., Luo, H., Wang, J., 2015. Motivational interviewing for improving recovery after stroke. Cochrane Database Syst. Rev. 6, CD011398.

Chong, T.T.-J., Husain, M., 2016. Chapter 17—The role of Dopamine in the pathophysiology and treatment of apathy. In: Studer, B., Knecht, S. (Eds.), Progress in Brain Research, vol. 229. Elsevier, Amsterdam, pp. 389–426.

Christakou, A., Robbins, T.W., Everitt, B.J., 2005. Prolonged neglect following unilateral disruption of a prefrontal cortical-dorsal striatal system. Eur. J. Neurosci. 21, 782–792.

Cohen, J., 1988. Statistical Power Analysis for the Behavioral Sciences. Routledge Academic, New York, NY.

Corbetta, M., Shulman, G.L., 2002. Control of goal-directed and stimulus-driven attention in the brain. Nat. Rev. Neurosci. 3, 201–215.

de la Fuente-Fernandez, R., Ruth, T.J., Sossi, V., Schulzer, M., Calne, D.B., Stoessl, A.J., 2001. Expectation and dopamine release: mechanism of the placebo effect in Parkinson's disease. Science 293, 1164–1166.

Della Libera, C., Chelazzi, L., 2006. Visual selective attention and the effects of monetary rewards. Psychol. Sci. 17, 222–227.

di Monaco, M., Schintu, S., Dotta, M., Barba, S., Tappero, R., Gindri, P., 2011. Severity of unilateral spatial neglect is an independent predictor of functional outcome after acute inpatient rehabilitation in individuals with right hemispheric stroke. Arch. Phys. Med. Rehabil. 92, 1250–1256.

Dominguez-Borras, J., Saj, A., Armony, J.L., Vuilleumier, P., 2012. Emotional processing and its impact on unilateral neglect and extinction. Neuropsychologia 50, 1054–1071.

Dominguez-Borras, J., Armony, J.L., Maravita, A., Driver, J., Vuilleumier, P., 2013. Partial recovery of visual extinction by pavlovian conditioning in a patient with hemispatial neglect. Cortex 49, 891–898.

Farne, A., Buxbaum, L.J., Ferraro, M., Frassinetti, F., Whyte, J., Veramonti, T., Angeli, V., Coslett, H.B., Ladavas, E., 2004. Patterns of spontaneous recovery of neglect and associated disorders in acute right brain-damaged patients. J. Neurol. Neurosurg. Psychiatry 75, 1401–1410.

Finke, K., Matthias, E., Keller, I., Müller, H.J., Schneider, W.X., Bublak, P., 2012. How does phasic alerting improve performance in patients with unilateral neglect? A systematic analysis of attentional processing capacity and spatial weighting mechanisms. Neuropsychologia 50, 1178–1189.

Fleet, W.S., Valenstein, E., Watson, R.T., Heilman, K.M., 1987. Dopamine agonist therapy for neglect in humans. Neurology 37, 1765–1770.

Fox, E., 2002. Processing emotional facial expressions: the role of anxiety and awareness. Cogn. Affect. Behav. Neurosci. 2, 52–63.

Gainotti, G., Messerli, P., Tissot, R., 1972. Qualitative analysis of unilateral spatial neglect in relation to laterality of cerebral lesions. J. Neurol. Neurosurg. Psychiatry 35, 545–550.

Goerendt, I.K., 2004. Reward processing in health and Parkinson's disease: neural organization and reorganization. Cereb. Cortex 14, 73–80.

Gorgoraptis, N., Mah, Y.H., Machner, B., Singh-Curry, V., Malhotra, P., Hadji-Michael, M., Cohen, D., Simister, R., Nair, A., Kulinskaya, E., Ward, N., Greenwood, R., Husain, M., 2012. The effects of the dopamine agonist rotigotine on hemispatial neglect following stroke. Brain 135, 2478–2491.

Guilbert, A., Sylvain, C., Moroni, C., 2014. Hearing and music in unilateral spatial neglect neuro-rehabilitation. Front. Psychol. 5, 1503.

Harvey, M., Milner, A.D., Roberts, R.C., 1995. An investigation of hemispatial neglect using the Landmark Task. Brain Cogn. 27, 59–78.

Heilman, K.M., Schwartz, H.D., Watson, R.T., 1978. Hypoarousal in patients with the neglect syndrome and emotional indifference. Neurology 28, 229–232.

Hommel, M., Peres, B., Pollak, P., Memin, B., Besson, G., Gaio, J.M., Perret, J., 1990. Effects of passive tactile and auditory stimuli on left visual neglect. Arch. Neurol. 47, 573–576.

Horvitz, J.C., Eyny, Y.S., 2000. Dopamine D2 receptor blockade reduces response likelihood but does not affect latency to emit a learned sensory-motor response: implications for Parkinson's disease. Behav. Neurosci. 114, 934–939.

Hubner, R., Schlosser, J., 2010. Monetary reward increases attentional effort in the flanker task. Psychon. Bull. Rev. 17, 821–826.

Husain, M., Rorden, C., 2003. Non-spatially lateralized mechanisms in hemispatial neglect. Nat. Rev. Neurosci. 4, 26–36.

Husain, M., Shapiro, K., Martin, J., Kennard, C., 1997. Abnormal temporal dynamics of visual attention in spatial neglect patients. Nature 385, 154–156.

Husain, G., Thompson, W.F., Schellenberg, E.G., 2002. Effects of musical tempo and mode on arousal, mood, and spatial abilities. Music Percept. 20, 151–171.

Ishiai, S., Sugishita, M., Odajima, N., Yaginuma, M., Gono, S., Kamaya, T., 1990. Improvement of unilateral spatial neglect with numbering. Neurology 40, 1395–1398.

Ishiai, S., Seki, K., Koyama, Y., Izumi, Y., 1997. Disappearance of unilateral spatial neglect following a simple instruction. J. Neurol. Neurosurg. Psychiatry 63, 23–27.

Katz, N., Hartman-Maeir, A., Ring, H., Soroker, N., 1999. Functional disability and rehabilitation outcome in right hemisphere damaged patients with and without unilateral spatial neglect. Arch. Phys. Med. Rehabil. 80, 379–384.

Kiss, M., Driver, J., Eimer, M., 2009. Reward priority of visual target singletons modulates event-related potential signatures of attentional selection. Psychol. Sci. 20, 245–251.
Kojovic, M., Mir, P., Trender-Gerhard, I., Schneider, S.A., Parees, I., Edwards, M.J., Bhatia, K.P., Jahanshahi, M., 2014. Motivational modulation of bradykinesia in Parkinson's disease off and on dopaminergic medication. J. Neurol. 261, 1080–1089.
Krebs, R.M., Boehler, C.N., Roberts, K.C., Song, A.W., Woldorff, M.G., 2012. The involvement of the dopaminergic midbrain and cortico-striatal-thalamic circuits in the integration of reward prospect and attentional task demands. Cereb. Cortex 22, 607–615.
Kristjansson, A., Sigurjonsdottir, O., Driver, J., 2010. Fortune and reversals of fortune in visual search: reward contingencies for pop-out targets affect search efficiency and target repetition effects. Atten. Percept. Psychophys. 72, 1229–1236.
Laplane, D., Degos, J.D., 1983. Motor neglect. J. Neurol. Neurosurg. Psychiatry 46, 152–158.
Lawrence, A.D., Goerendt, I.K., Brooks, D.J., 2011. Apathy blunts neural response to money in Parkinson's disease. Soc. Neurosci. 6, 653–662.
Lecce, F., Rotondaro, F., Bonni, S., Carlesimo, A., Thiebaut de Schotten, M., Tomaiuolo, F., Doricchi, F., 2015. Cingulate neglect in humans: disruption of contralesional reward learning in right brain damage. Cortex 62, 73–88.
Li, K., Malhotra, P.A., 2015. Spatial neglect. Pract. Neurol. 15, 333–339.
Li, K., Soto, D., Russell, C., Balaji, C., Malhotra, P., 2013. The effects of l-dopa on the interaction between reward and attention in patients with right hemisphere stroke. Poster Presented at the Rovereto Attention Workshop, 24–26. October.
Li, K., Russell, C., Balaji, N., Saleh, Y., Soto, D., Malhotra, P., 2016. The effects of motivational reward on the pathological attentional blink following right hemisphere stroke. Neuropsychologia doi:http://dx.doi.org/10.1016/j.neuropsychologia.2016.03.037. pii:S0028-3932(16)30108-7.
Lucas, N., Vuilleumier, P., 2008. Effects of emotional and non-emotional cues on visual search in neglect patients: evidence for distinct sources of attentional guidance. Neuropsychologia 46, 1401–1414.
Lucas, N., Schwartz, S., Leroy, R., Pavin, S., Diserens, K., Vuilleumier, P., 2013. Gambling against neglect: unconscious spatial biases induced by reward reinforcement in healthy people and brain-damaged patients. Cortex 49, 2616–2627.
Maes, P.J., Leman, M., Palmer, C., Wanderley, M.M., 2014. Action-based effects on music perception. Front. Psychol. 4, 1–14.
Malhotra, P., Jager, H.R., Parton, A., Greenwood, R., Playford, E.D., Brown, M.M., Driver, J., Husain, M., 2005. Spatial working memory capacity in unilateral neglect. Brain 128, 424–435.
Malhotra, P.A., Soto, D., Li, K., Russell, C., 2013. Reward modulates spatial neglect. J. Neurol. Neurosurg. Psychiatry 84, 366–369.
Manohar, S.G., Husain, M., 2015. Reduced pupillary reward sensitivity in Parkinson's disease. NPJ Parkinson's Dis. 1, 15026.
Manohar, S.G., Husain, M., 2016. Human ventromedial prefrontal lesions alter incentivisation by reward. Cortex 76, 104–120.
Martinez-Horta, S., Riba, J., De Bobadilla, R.F., Pagonabarraga, J., Pascual-Sedano, B., Antonijoan, R.M., Romero, S., Mananas, M.A., Garcia-Sanchez, C., Kulisevsky, J., 2014. Apathy in Parkinson's disease: neurophysiological evidence of impaired incentive processing. J. Neurosci. 34, 5918–5926.
Mesulam, M., 1985. Principles of Behavioral Neurology. F.A. Davis, Philadelphia, PA.
Migliaccio, R., Bouhali, F., Rastelli, F., Ferrieux, S., Arbizu, C., Vincent, S., Pradat-Diehl, P., Bartolomeo, P., 2014. Damage to the medial motor system in stroke patients with motor neglect. Front. Hum. Neurosci. 8, 408.

Milner, A.D., Brechmann, M., Pagliarini, L., 1992. To halve and to halve not: an analysis of line bisection judgements in normal subjects. Neuropsychologia 30, 515–526.

Milner, A.D., Harvey, M., Roberts, R.C., Forster, S.V., 1993. Line bisection errors in visual neglect: misguided action or size distortion? Neuropsychologia 31, 39–49.

Morris, J.S., Degelder, B., Weiskrantz, L., Dolan, R.J., 2001. Differential extrageniculostriate and amygdala responses to presentation of emotional faces in a cortically blind field. Brain 124, 1241–1252.

Olk, B., Harvey, M., 2002. Effects of visible and invisible cueing on line bisection and Landmark performance in hemispatial neglect. Neuropsychologia 40, 282–290.

Owen, A.M., James, M., Leigh, P.N., Summers, B.A., Marsden, C.D., Quinn, N.P., Lange, K.W., Robbins, T.W., 1992. Fronto-striatal cognitive deficits at different stages of Parkinson's disease. Brain 115 (Pt. 6), 1727–1751.

Paolucci, S., Antonucci, G., Grasso, M.G., Pizzamiglio, L., 2001. The role of unilateral spatial neglect in rehabilitation of right brain-damaged ischemic stroke patients: a matched comparison. Arch. Phys. Med. Rehabil. 82, 743–749.

Parton, A., Malhotra, P., Husain, M., 2004. Hemispatial neglect. J. Neurol. Neurosurg. Psychiatry 75, 13–21.

Pessiglione, M., Schmidt, L., Draganski, B., Kalisch, R., Lau, H., Dolan, R.J., Frith, C.D., 2007. How the brain translates money into force: a neuroimaging study of subliminal motivation. Science 316, 904–906.

Raymond, J.E., O'Brien, J.L., 2009. Selective visual attention and motivation: the consequences of value learning in an attentional blink task. Psychol. Sci. 20, 981–988.

Renfroe, J.B., Bradley, M.M., Okun, M.S., Bowers, D., 2016. Motivational engage ment in Parkinson's disease: preparation for motivated action. Int. J. Psychophysiol. 99, 24–32.

Robertson, I.H., 2013. The neglected role of reward in rehabilitation. J. Neurol. Neurosurg. Psychiatry 84, 363.

Robertson, I.H., Mattingley, J.B., Rorden, C., Driver, J., 1998. Phasic alerting of neglect patients overcomes their spatial deficit in visual awareness. Nature 395, 169–172.

Rochat, L., Van Der Linden, M., Renaud, O., Epiney, J.B., Michel, P., Sztajzel, R., Spierer, L., Annoni, J.M., 2013. Poor reward sensitivity and apathy after stroke: implication of basal ganglia. Neurology 81, 1674–1680.

Rowe, J.B., Hughes, L., Ghosh, B.C., Eckstein, D., Williams-Gray, C.H., Fallon, S., Barker, R.A., Owen, A.M., 2008. Parkinson's disease and dopaminergic therapy—differential effects on movement, reward and cognition. Brain 131, 2094–2105.

Russell, C., Li, K., Malhotra, P.A., 2013a. Harnessing motivation to alleviate neglect. Front. Hum. Neurosci. 7, 230.

Russell, C., Malhotra, P., Deidda, C., Husain, M., 2013b. Dynamic attentional modulation of vision across space and time after right hemisphere stroke and in ageing. Cortex 49, 1874–1883.

Salimpoor, V.N., Benovoy, M., Larcher, K., Dagher, A., Zatorre, R.J., 2011. Anatomically distinct dopamine release during anticipation and experience of peak emotion to music. Nat. Neurosci. 14, 257–262.

Sarkamo, T., Tervaniemi, M., Laitinen, S., Forsblom, A., Soinila, S., Mikkonen, M., Autti, T., Silvennoinen, H.M., Erkkila, J., Laine, M., Peretz, I., Hietanen, M., 2008. Music listening enhances cognitive recovery and mood after middle cerebral artery stroke. Brain 131, 866–876.

Schmidt, L., D'Arc, B.F., Lafargue, G., Galanaud, D., Czernecki, V., Grabli, D., Schupbach, M., Hartmann, A., Levy, R., Dubois, B., Pessiglione, M., 2008. Disconnecting force from money: effects of basal ganglia damage on incentive motivation. Brain 131, 1303–1310.

Schultz, 2002. Getting formal with Dopamine and Reward. Neuron 36, 241–263.

Serences, J.T., 2008. Value-based modulations in human visual cortex. Neuron 60, 1169–1181.

Sha, L.Z., Jiang, Y.V., 2016. Components of reward-driven attentional capture. Atten. Percept. Psychophys. 78, 403–414.

Shiner, T., Seymour, B., Symmonds, M., Dayan, P., Bhatia, K.P., Dolan, R.J., 2012. The effect of motivation on movement: a study of bradykinesia in Parkinson's disease. PLoS One 7, e47138.

Small, D.M., Gitelman, D., Simmons, K., Bloise, S.M., Parrish, T., Mesulam, M.M., 2005. Monetary incentives enhance processing in brain regions mediating top-down control of attention. Cereb. Cortex 15, 1855–1865.

Soto, D., Funes, M.J., Guzman-Garcia, A., Warbrick, T., Rotshtein, P., Humphreys, G.W., 2009. Pleasant music overcomes the loss of awareness in patients with visual neglect. Proc. Natl. Acad. Sci. U. S. A. 106, 6011–6016.

Studer, B., Knecht, S., 2016. Chapter 2—A benefit–cost framework of motivation for a specific activity. In: Studer, B., Knecht, S. (Eds.), Progress in Brain Research, vol. 229. Elsevier, Amsterdam, pp. 25–47.

Suhr, J.A., Grace, J., 1999. Brief cognitive screening of right hemisphere stroke: relation to functional outcome. Arch. Phys. Med. Rehabil. 80, 773–776.

Tamietto, M., Latini Corazzini, L., Pia, L., Zettin, M., Gionco, M., Geminiani, G., 2005. Effects of emotional face cueing on line bisection in neglect: a single case study. Neurocase 11, 399–404.

Thompson, W.F., Schellenberg, E.G., Husain, G., 2001. Arousal, mood, and the Mozart effect. Psychol. Sci. 12, 248–251.

Thorndike, E.L., 1911. Animal Intelligence. Macmillan, New York, NY.

Tsai, P.L., Chen, M.C., Huang, Y.T., Lin, K.C., Chen, K.L., Hsu, Y.W., 2013. Listening to classical music ameliorates unilateral neglect after stroke. Am. J. Occup. Ther. 67, 328–335.

Vallar, G., 1998. Spatial hemineglect in humans. Trends Cogn. Sci. 2, 87–97.

Vallar, G., Bolognini, N., 2014. Unilateral spatial neglect. In: Nobre, A.C.K., Kastner, S. (Eds.), The Oxford Handbook of Attention. Oxford University Press. Oxford Handbooks Online, Oxford.

van Vleet, T.M., Heldt, S.A., Pyter, B., Corwin, J.V., Reep, R.L., 2003. Effects of light deprivation on recovery from neglect and extinction induced by unilateral lesions of the medial agranular cortex and dorsocentral striatum. Behav. Brain Res. 138, 165–178.

Vuilleumier, P., 2015. Affective and motivational control of vision. Curr. Opin. Neurol. 28, 29–35.

Vuilleumier, P., Schwartz, S., 2001. Modulation of visual perception by eye gaze direction in patients with spatial neglect and extinction. Neuroreport 12, 2101–2104.

Vuilleumier, P., Armony, J.L., Clarke, K., Husain, M., Driver, J., Dolan, R.J., 2002. Neural response to emotional faces with and without awareness: event-related fMRI in a parietal patient with visual extinction and spatial neglect. Neuropsychologia 40, 2156–2166.

Wilson, B., Cockburn, J., Halligan, P., 1987. Development of a behavioral test of visuospatial neglect. Arch. Phys. Med. Rehabil. 68, 98–102.

Wise, R.A., 1982. Neuroleptics and operant behavior: the anhedonia hypothesis. Behav. Brain Sci. 5, 39–53.

CHAPTER

Increasing self-directed training in neurorehabilitation patients through competition

16

B. Studer*,†,1, H. Van Dijk*, R. Handermann†, S. Knecht*,†

**Institute of Clinical Neuroscience and Medical Psychology, Medical Faculty, Heinrich-Heine-University Düsseldorf, Düsseldorf, Germany*

†Mauritius Hospital, Meerbusch, Germany

[1]Corresponding author: Tel.: +49-2159-679-5114; Fax: +49-2159-679-1535, e-mail address: bettina.studer@stmtk.de

Abstract

This proof-of-concept study aimed to test whether competition could be a useful tool to increase intensity and amount of self-directed training in neurorehabilitation. Stroke patients undergoing inpatient neurorehabilitation ($n=93$) conducted self-directed endurance training on a (wheelchair-compatible) bicycle trainer under three experimental conditions: a "Competition" condition and two noncompetition control conditions (repeated randomized within-subject design). Training performance and perceived exertion were recorded and statistically analyzed. Three motivational effects of competition were found. First, competition led to an increase in self-directed training. Patients exercised significantly more intensively under competition than in the two noncompetition control conditions. Second, (winning a) competition had a positive influence on performance in the subsequent training session. Third, training performance was particularly high during rematch competitions; that is to say, during second encounter competitions against an opponent that the patient had just beaten. No systematic effect of competition upon perceived exertion (controlled for training performance) was found. Together, our results demonstrate that competition is a potent motivational tool to increase self-directed training in neurorehabilitation.

Keywords

Motivation, Training, Competition, Enrichment, Effort, Perceived exertion, Recovery, Stroke

1 INTRODUCTION

Every year, approximately 16 million people globally suffer a first-time stroke (Strong et al., 2007), and brain damage through stroke, trauma, or other causes is the leading cause of acquired disability in the developed world (eg, Mukherjee

Progress in Brain Research, Volume 229, ISSN 0079-6123, http://dx.doi.org/10.1016/bs.pbr.2016.06.012

and Patil, 2011). Fortunately, impairments of physical and cognitive functions as a result of brain injury can be significantly reduced through neurorehabilitation with specialized multidisciplinary training. Studies in animal models as well as analyses of human clinical and experimental datasets show that the intensity of neurorehabilitative training is a critical factor of its success; the more training and repetitions patients undergo, the larger and faster is the functional recovery (Albert and Kesselring, 2012; Kleim and Jones, 2008; Knecht et al., 2016; Krakauer et al., 2012; Lohse et al., 2014; Zeiler and Krakauer, 2013). However, in clinical practice, the quest to maximize training intensity faces two major challenges. The first challenge is motivational: learning and relearning of physical and cognitive function is an active, demanding, and effortful process. Therefore, neurorehabilitative training requires high drive and persistence. Overcoming such motivational demands is not an easy task even for healthy adults, and patients with brain damage often present with reduced drive, effort, and perseverance, and can be easily frustrated and lose motivation if challenged (Nicholson et al., 2013). Indeed, apathy—a disorder characterized by diminished motivation, reduced goal-directed behavior and cognitive activity, and impoverished emotions (Robert et al., 2009)—is observed in approximately one-third of stroke patients (see Caeiro et al., 2013 for meta-analysis) and negatively impacts the degree of functional recovery following stroke (Caeiro et al., 2012; Harris et al., 2014; Mayo et al., 2009; Santa et al., 2008; Van Dalen et al., 2013). The second challenge in the quest to maximize the amount and intensity of neurorehabilitative training patients undergo is of a logistical nature: personnel resources are limited and the amount of therapist-guided training administered in neurorehabilitation cannot be increased without also enhancing financial costs. A solution to this challenge, which is increasingly applied in current neurorehabilitation practice, is to complement therapist-guided training with self-directed training. In self-directed training, also referred to as "patient-led training," patients carry out exercises outside of therapy sessions and without direct supervision. Unfortunately, however, self-directed training is likely to be particularly vulnerable to motivational impairments given that it lacks the direct and proximal social encouragement and reinforcement therapists provide. Indeed, a recent randomized controlled trial testing adherence to, and feasibility and acceptability of, self-directed upper and lower limb training in stroke patients undergoing inpatient neurorehabilitation found that self-directed training was conducted far less often and for shorter durations than recommended (Tyson et al., 2015). This occurred despite patients reporting a positive attitude towards self-directed training. Loss of motivation was identified as one reason for this lack of actually executed self-directed training (Horne et al., 2015). Finding novel ways to increase patients' motivation, effort, and persistence during self-directed and therapist-guided training is thus crucial to optimizing training intensity in neurorehabilitation and utilizing the full recovery potential of each patient.

We propose that an effective approach to increase motivation and training intensity in neurorehabilitation patients could be to add an element of competition to the exercise environment. Competition has been shown to enhance performance and goal-directed behavior in a range of contexts, from sports to economic behavior. A meta-analysis of 64 studies using a variety of sport and laboratory motor tasks found that competition led to a significant increase in performance when compared to individualist task execution (Stanne et al., 1999). More recent studies further confirmed these results: for instance, competition was associated with a significant improvement in performance on cycling time trials in both regular students (Corbett et al., 2012) and trained cyclists (Williams et al., 2015). And, two recent laboratory studies demonstrated that competition evoked higher (Le Bouc and Pessiglione, 2013) and more sustained (Cooke et al., 2011) physical effort on handgrip force tasks. In the field of economics, a well-described phenomenon is that people frequently overbid in (private-value) auctions (eg, Cooper and Fang, 2008; Cox et al., 1982, 1988; Goeree et al., 2002; Kagel and Levin, 1993; Kagel et al., 1987). That is to say, bids placed in an auction are often higher than the bidder's subjective value of the same item assessed outside of a competitive situation and higher than the objectively optimal bid (the (risk-neutral) Nash-equilibrium). In conclusion, these converging experimental findings imply that competition is a powerful motivator.

What might then be the mechanism by which competition increase motivation? One proposal is that competition raises the enjoyment of an activity (Tauer and Harackiewicz, 2004); in other words, it increases the intrinsic benefit of the activity. An alternative explanation is that the anticipation of a "joy of winning" (Cooper and Fang, 2008; Grimm and Engelmann, 2005; Roider and Schmitz, 2012) increases the subjective benefit of a high effort (ie, of giving it your all) under competition. In line with this proposition, two recent neuroimaging studies that found that participating in or winning a competition is associated with increased activation of the medial prefrontal cortex (Fareri and Delgado, 2014; Le Bouc and Pessiglione, 2013), a key area of the brain valuation/reward system (see, eg, Kennerley and Walton, 2011; Noonan et al., 2012; O'Doherty, 2004; Rushworth et al., 2011; Studer et al., 2015; Volz et al., 2006). Two other proposed explanations are that anticipation of losers' regret (Engelbrecht-Wiggans and Katok, 2006; Filiz-Ozbay and Ozbay, 2007) or a wish to avoid a social loss (Delgado et al., 2008) and associated reduction in social status (Mazur and Booth, 1998) increase motivation under competition. In addition, some have argued that competition might induce anxiety and pressure, which in turn can boost (or also weaken) behavior and performance (eg, Baumeister and Showers, 1986; Cooke et al., 2011; Dimenichi and Tricomi, 2015). This last notion would imply that competition might not be appropriate for a neurorehabilitation setting, as competition might frustrate, rather than entice, patients with functional impairments. However, previous theoretical and experimental work suggests that whether a competition is perceived as motivating or threatening does not depend on an individual's performance level per se, but rather on whether they believe to have a reasonable chance of winning and that the competition outcome is dependent on his or her effort (Johnson and Johnson, 1974; Salvador, 2005; Salvador and Costa, 2009; Stanne et al., 1999). Other characteristics of a positively perceived competition are that winning should not be too important, that the rules of the competition are fixed, and that monitoring of the competition outcome is possible (Johnson and Johnson, 1974; Stanne et al., 1999). Therefore, we argue that—when carefully controlled—competition could be a viable and potent motivator to increase the training efforts and performance of neurorehabilitation patients.

In this proof-of-concept study, we tested whether competition leads to an increase in the intensity of self-directed endurance training on (wheelchair-compatible) bicycle trainers in 93 stroke patients undergoing inpatient neurorehabilitation. Cardiorespiratory endurance training is indicated in stroke patients not only due to its beneficial impact on general physical health and fitness, but also because it can augment patients' participation capacities in function specific, therapist-guided training, improve walking speed and mobility, and enhance quality of life (Brazzelli et al., 2011; Pang et al., 2013; Tang et al., 2009). To assess the effectiveness of competition to increase self-directed training in our neurorehabilitation sample, we recorded patients' training amount and intensity under three experimental conditions using a repeated within-subject design: the main intervention condition "Competition" and two control conditions termed "Baseline" and "Feedback." In the "Competition" condition, participants competed against an anonymous same-sex opponent of a similar training level and were told that both their own and the competitor's training performance would be recorded, analyzed, and reported back to them (together with the competition outcome). In the "Baseline" control condition, training was performed without feedback or external monitoring (as in current clinical practice), and training performance was recorded covertly. Meanwhile, in the "Feedback" condition, patients were informed that their performance would be recorded, analyzed, and reported back to them. This high-level control condition was included to allow isolating the effect of competition per se from potential motivation enhancement caused by the knowledge of being monitored and receiving feedback (common to both the "Feedback" and "Competition" conditions). Training performance served as a primary outcome measure, and we hypothesized that performance would be highest in the "Competition" condition. Subjective training effort (assessed on the Borg Rating of Perceived Exertion scale (Borg, 1982)) served as a secondary outcome measure, with no a priori predictions about the direction of a potential effect of competition.

2 METHODS

2.1 STUDY DESIGN AND SETTING

This proof-of-concept study was conducted at the Mauritius Hospital Meerbusch, a specialized center for inpatient neurorehabilitation with 200 beds and a catchment area of approximately 2.8 million people. A cross-over controlled within-subject design was used: each participant underwent each of the three experimental conditions ("Baseline," "Feedback," and "Competition") repeatedly (minimum: twice, maximum: unrestricted), and the order of conditions was randomized across participants. Training performance was recorded for each training session and compared across the experimental conditions.

2.2 ETHICAL APPROVAL AND CONSENT

This research was approved by the independent Ethics Committee of the Medical Faculty of the Heinrich Heine University Düsseldorf, Germany (protocol no 4835) and conducted in accordance with the revised Declaration of Helsinki. All participants provided written informed consent.

2.3 PARTICIPANTS

Eligible adult stroke patients undergoing inpatient neurorehabilitation at the Mauritius Hospital Meerbusch were prospectively recruited for this study over a period of 14 months (12/01/2014 to 01/05/2016). Inclusion criteria were (i) suffered from an ischemic or hemorrhagic stroke at least two and no more than 20 weeks prior to study inclusion, (ii) German speaking, (iii) sufficient preserved leg function for training on wheel-chair compatible or conventional bicycle trainer, (iv) able to provide informed consent, and (v) stable medical condition. Exclusion criteria were (i) moderate or severe cognitive impairment, (ii) moderate or severe aphasia, (iii) predicted remaining inpatient stay shorter than 2 weeks, (iv) in single-room isolation due to multiresistant bacteria, and (v) unsupervised cardiorespiratory fitness training contraindicated due to comorbidity associated with increased risk of cardiovascular overstressing (for instance acute myo- or endocarditis, coronary heart disease or cardiac insufficiency NYHA Class IV, or acute infection with fever). Patients were only eligible for participation when all inclusion criteria and none of the exclusion criteria were met, and eligibility was confirmed for each patient by their treating physician at the Mauritius Hospital Meerbusch. Two different types of bicycle trainers were used in this study, a conventional bicycle trainer (suitable for patients with unaided walking ability and good balance) and a wheel-chair compatible bicycle trainer (see Fig. 1); and the decision which device type was more suitable for an individual patient was made by a qualified physiotherapist or sports therapist familiar with the patient. Details of the enrollment process are provided in Fig. 2.

A total of 93 patients took part in this study; 58 patients performed the self-directed training on a wheel-chair compatible bicycle trainer and 35 patients used a conventional bicycle trainer. Duration of study participation and number of (recorded) training days depended upon the length of each participants inpatient stay and thus differed between participants (range 0–37 recorded training sessions). Thirty-three of 93 included participants were excluded from statistical analysis because training performance was recorded on less than five training sessions (see Fig. 2 for reasons). Final statistical analysis was thus performed on a total sample of 60 patients and a total of 701 recorded training sessions. Thirty-six patients of this final sample used the wheel-chair compatible bicycle trainer (total number of measures = 526); the other 24 patients exercised on a conventional bicycle trainer (total number of measures = 175). Demographic characteristics of the final samples, information on stroke type, affected hemisphere and time since stroke onset, and average number of recorded training sessions are provided in Table 1.

FIG. 1

Wheel-chair compatible (A) and conventional (B) bicycle trainers used for self-directed training.

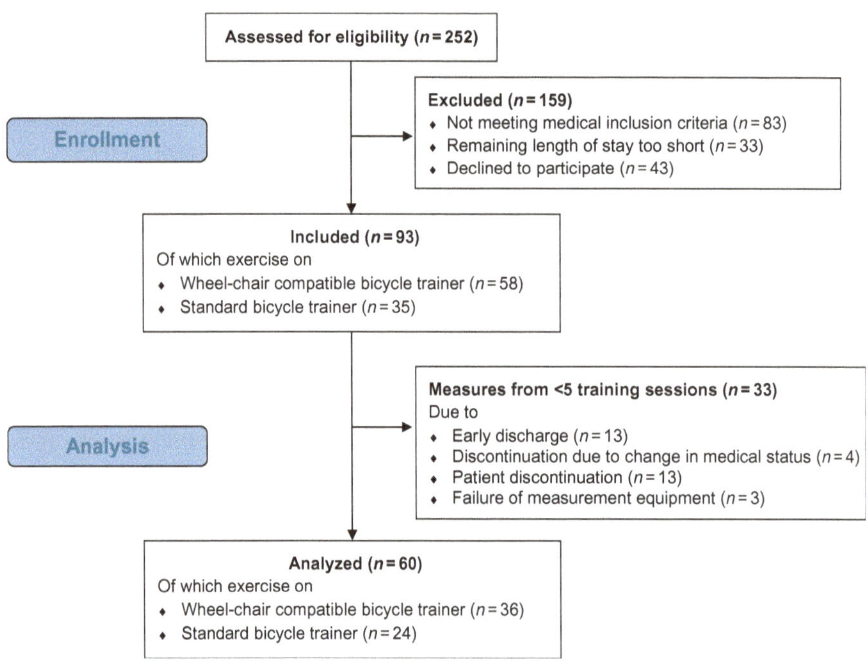

FIG. 2

Enrollment process, data collection phase, and data analysis.

Table 1 Characteristics of Patient Sample

	Subgroup Exercising on Wheel-Chair Compatible Bicycle Trainer	Subgroup Exercising on Conventional Bicycle Trainer
Sample size	36	24
Gender (M/F)	22/14	17/7
Stroke type (ischemic/hemorrhagic)	33/3	20/4
Affected hemisphere (left/right)	11/25	9/14[a]
Days since stroke—mean (±SD)	36.97 (±24.93)	34.17 (±16.15)
Age—mean (±SD)	71.92 (±7.91)	65.58 (±9.56)
Barthel index at time of first training—mean (±SD)	74.58 (±17.17)	96.45 (±6.67)
# recorded training sessions—mean (±SD)	15.06 (±8.02)	8.04 (±4.07)

[a]*In one case, the affected hemisphere was not known.*

2.4 PROCEDURE

Prior to the first training sessions, participants underwent a standardized step-test on the wheelchair-compatible or conventional bicycle trainer. In this test, the required intensity measured in Watts was raised every 4 min in steps of 25 W with continuous monitoring of heart rate. The test served as an estimate of patients' fitness levels and training capacities, and was also used to instruct patients on how to operate the device interface. Then, participants conducted a daily self-directed training on the standard or wheel-chair compatible bicycle trainer during a fixed prebooked time window (weekdays only). They were free to choose for how long and with which intensity (ie, speed and physical resistance) they wanted to exercise on each day, and the experimenter was not present during the training. Directly before and after each training, participants met with the experimenter. In the pretraining meeting, participants were told their recorded training performance on the preceding day (unless the previous training took place in the "Baseline" condition) in a commending manner. If the previous training took place under Competition, the training performance of the competitor, and the outcome of the competition were reported as well. Then, participants were informed about the condition (Baseline, Feedback, or Competition) for the upcoming training. In the posttraining meeting, the participant was asked to rate perceived exertion for the just completed training.

Following the last training day, patients underwent a posttrial interview and debriefing, which included two questions about the perceived effect of competition: (i) whether they felt particularly motivated in competition sessions and (ii) whether they believe to have performed better and/or increased training effort in the Competition sessions compared to other sessions.

2.5 CONDITIONS

Three experimental conditions were used in this experiment: a standard control condition termed "Baseline", the intervention condition termed "Competition" and a second, high-level control condition termed "Feedback."

i. Baseline = *Control Condition (covert recording, no feedback)*

   The "Baseline" condition was designed to reflect current clinical praxis, where self-directed training is usually not inspected, checked, or recorded. Therefore, training performance in this condition was recorded covertly: patients were told that data recording equipment had to be readjusted and that their training performance of this day would not be analyzed.

ii. Competition = *Intervention Condition (overt recording, feedback about own and opponent's performance)*

   In the "Competition" condition, patients were instructed that they would compete against an anonymous opponent and could win the competition by outperforming their opponent. Patients were advised that they would not exercise at the same time and thus not see the opponent, but that both their own and the competitors' training performance would be recorded and then compared offline. Patients were also informed prior to the training that they would receive feedback on their own performance, the opponent's performance and who won the competition the next day.

   Given that competition is expected to be most motivating when the opponent is well-matched (Stanne et al., 1999), we used a fictional opponent and cover story: patients were told that their opponent is of the same gender, similar age, similar functional and training level and has also suffered a stroke. This approach also allowed full experimental control over competition outcomes, which we deemed important given that winning, but not losing, a competition is expected to positively impact motivation, perceived competence and subsequent performance (eg, Reeve and Deci, 1996; Reeve et al., 1985; Vallerand et al., 1986). As a rule, the participant won the competition with a tight result. An exception from this rule was made only in the very rare case where a patients' training performance was markedly lower than his/her previous performances. In such cases, patients were told that the opponent won the competition to protect the plausibility of the cover story and competition feedback.

iii. Feedback = *High-Level Control Condition (overt recording, feedback about own performance)*

   The "Feedback" condition served as a high-level control condition and was included to allow isolating the effect of Competition per se from a potential effect caused by the knowledge of being recorded and receiving feedback (integral elements of the "Competition" condition). In this condition, participants were told that their training performance would be recorded and analyzed and that they would receive feedback on it the next day.

2.6 RANDOMIZATION OF ORDER OF EXPERIMENTAL CONDITIONS

The order of experimental conditions was predetermined for each participant by a pseudo-randomly created and randomly assigned sequence. In the pseudo-random creation of the sequences, two constraints were employed: a given condition was always repeated across two successive training days (eg, B, B, C, C, F, F) and each discrete subset of six training sessions (eg, sessions 1–6 and sessions 7–12) had to include all three conditions.

2.7 MEASUREMENT OF TRAINING PERFORMANCE

Training performance served as a primary outcome measure and was collected for each training session and participant. Because the recording abilities of the two used types of bicycle trainers differed, diverse measures of training performance had to be used in the two subgroups.

2.7.1 Subgroup using wheel-chair compatible bicycle trainer

Wheel-chair compatible bicycle trainers (THERA-Vital; medica Medizintechnik GmbH, Germany) allowed direct recording of the performed exercise onto a detachable memory stick. For each training session and patient, duration, and intensity (average Watts) of the performed training was recorded. These two key measures were then integrated to determine training performance, defined as performance [work; in joules] = average power [in Watts] × duration [in seconds]. Calculation of training performance was explained to all participants, and all participants were explicitly advised that they could increase training performance by exercising longer, cycling faster, or setting a higher level of resistance on the bicycle trainer.

2.7.2 Subgroup using conventional bicycle trainer

Conventional bicycle trainers (Ergo-Fit 407, ERGO-FIT GmbH & Co. KG) were not able to record performed exercise. Therefore, we used activity trackers (ViFit Activity Tracker, Medisana AG, Germany) as a measurement tool in this subgroup: activity trackers reliably recorded the total number of pedal revolutions performed in a given session, and this count served as a measure of training performance. All participants were instructed about how training performance was assessed. Participants were also advised that increases in training duration or cycling speed would lead to an increase in training performance, whereas augmenting the level of resistance on the bicycle trainer would not lead to an increase in the performance measure.

2.8 FURTHER COLLECTED MEASURES

Immediately following each training session, participants were asked to provide a rating of the subjective exercise intensity on the Borg Rating of Perceived Exertion (Borg RPE; Borg, 1982). On the Borg RPE respondents rate subjective effort and exertion associated with an exercise on a single item scale rating from 6 (no exertion) to 20 (maximal exertion). Test–retest reliability of the Borg RPE is high (Eston and Williams, 1988) and Borg RPE scores correlate strongly with heart rate (ie, a physiological measure of exertion) during cycling exercise in healthy adults (Borg, 1970, 1982).

2.9 DATA PROCESSING AND STATISTICAL ANALYSIS

Calculated session-by-session training performance measures were standardized (z-scored) for each patient prior to statistical analysis to enable comparison across the two training subgroups and control for interindividual differences in the overall performance level.

Generalized estimating equation (GEE) models were used to assess the effect of experimental condition (Baseline, Feedback, or Competition) upon patients' training performance (primary outcome) and perceived exertion (secondary outcome). GEE models account for nonindependence in the observation of a single subject and thereby allow calculating generalized linear regression models for datasets with repeated response measures (Liang and Zeger, 1986; Zeger and Liang, 1986; Zeger et al., 1988). GEE models test the effect of multiple hypothesized factors simultaneously, and—unlike analysis of variance—can handle missing observations and violations in sphericity.

A first GEE model assessed the effects of experimental condition (Baseline, Feedback, or Competition) on patients' training performance, while controlling for time-/practice-related effects. Z-scored training performance was entered as a linear response measure in this model, and two main predictors of interest were assessed: (i) *Experimental Condition* (categorical; reference category = Baseline) and (ii) *Time* (number of days since the first recorded training; continuous). In this and all other calculated GEE models, *Subject ID* and *Training Session Number* were entered as between- and within-subject identifier variables and an unstructured working correlation matrix was used. To increase power of detecting true effects, this GEE model was calculated for the pooled data from both subgroups, with *Training Type* entered as an additional categorical predictor. However, given that measurement of training performance for exercise on the conventional bicycle trainer differed from and was less precise than on the wheel-chair compatible bicycle trainer, we also report the results of an equivalent GEE model (without the factor *Training Type*) calculated for the data from the wheel-chair compatible bicycle trainer subgroup only.

A second GEE model tested whether perceived exertion as rated on the Borg Scale (linear response measure) differed between the experimental conditions. The model included three predictors: (i) *Experimental Condition*, (ii) *Time* (number of days since the first recorded training; continuous), and (iii) *z-scored Training Performance*. This model thus tested whether perceived exertion differed between the conditions after controlling for training-/practice-related effects as well as actual performance. Again, two models were calculated; one on the data from the pooled group (including the predictor *Training Type*), the other on the wheel-chair compatible bicycle trainer subgroup only.

Finally, we ran two sets of exploratory analyses assessing the effects of (winning a) competition upon subsequent training performance. The first analysis consisted of a GEE model which tested if (standardized) training performance on a given day was influenced by whether the preceding training occurred under competition or noncompetition (Baseline or Feedback), while controlling for the effects of *Experimental Condition*, *Time*, and *Training type*. This GEE model was identical to the first one described earlier except that it included an additional categorical predictor *Previous_Competition* (2 levels; yes, no). The second exploratory analysis tested whether participants were particularly motivated during a "rematch"/2nd encounter competitions, that is to say, in the case where they had just won against an opponent and now compete with the same opponent again. This was assessed by a GEE model calculated on (standardized performance in) competition trials only with two categorical predictors *Rematch_Competition* (2 levels; yes, no) and *Training Type*.

Statistical analyses were performed using SPSS Version 22 (IBM SPSS Inc., Chicago, IL). All analyses are reported two-tailed and alpha was set at 0.05.

3 RESULTS

3.1 TRAINING PERFORMANCE IN THE BASELINE, FEEDBACK, AND COMPETITION CONDITIONS

The GEE model for the pooled group ($n=60$) showed that patients' training performance was influenced by the *Experimental Condition* (main effect: Wald $\chi^2=11.85$, $p=0.003$) and *Time* (main effect: Wald $\chi^2=10.02$, $p=0.002$). The main effect for *Training Type* was also significant (Wald $\chi^2=11.90$, $p=0.001$). The parameter estimate for the predictor *Time* was $\beta=0.018$ (95% CI: 0.007 to 0.029, $p=0.002$), indicating that individuals' training performance significantly increased over the training period. The parameter estimates for the categorical predictor *Experimental Condition* showed that patients' training performance increased significantly during the "Competition" condition ($\beta=0.311$, 95% CI: 0.093 to 0.529, $p=0.005$), but not during the "Feedback" condition ($\beta=0.047$, 95% CI: −0.142 to 0.236, $p=0.63$), compared to the "Baseline" condition (see Fig. 3A). Direct comparison of the estimated average training performance in the Competition vs the "Feedback" condition (calculated for the mean time point of all recordings, which was 12.27 days after an individuals' first training session) also confirmed a significant increase in the "Competition" condition ($p=0.001$, see Fig. 3B).

Very similar results were found when the GEE model was calculated for the wheelchair compatible bicycle trainer subgroup ($n=36$) only. Training performance was systematically influenced by the *Experimental condition* (Wald $\chi^2=10.95$, $p=0.004$), with patients training more intensively in the "Competition" condition ($\beta=0.313$, 95% CI: 0.060 to 0.567, $p=0.015$), but not the "Feedback" condition ($\beta=0.022$, 95% CI: −0.209 to 0.252, $p=0.85$), than in the "Baseline" condition. Direct post hoc comparison of the estimated average training performance in the "Competition" vs "Feedback" conditions (estimated for 14.90 days after training start (=subsample mean)) confirmed that patients trained more intensively in the "Competition" condition ($p=0.002$). The main effect of *Time* was also significant

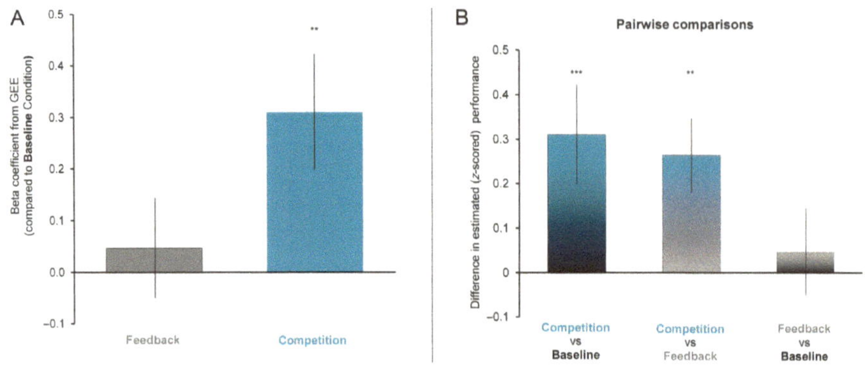

FIG. 3

Training performance in the Baseline, Feedback, and "Competition" condition. (A) Influence of the feedback (ns) and the "Competition" condition upon standardized training performance (beta coefficient from SEE model). Error bars represent standard error of the mean (SEM). (B) Pairwise comparison of model-estimated standardized training performance in the three experimental conditions (estimated for 12.27 days after the first training session). Error bars represent standard errors of the mean difference. Significant effects are marked by asterisks, $^{**}p<0.01$, $^{***}p<0.011$.

(Wald $\chi^2=8.446$, $p=0.004$), with training performance increasing slightly over the course of training ($\beta=0.017$, 95% CI: 0.005 to 0.028, $p=0.004$). Exemplary data from an individual patient are provided in Fig. 4.

In summary, competition led to a significant increase in self-directed training: patients exercised more intensively in the "Competition" condition than in the two control conditions.

3.2 PERCEIVED EXERTION IN THE BASELINE, FEEDBACK, AND COMPETITION CONDITIONS

The GEE model analyzing perceived exertion (scores on Borg scale) in the pooled group ($n=60$) revealed a significant main effect of (standardized) *Training Performance* (Wald $\chi^2=8.27$, $p=0.004$), with perceived exertion scaling positively with training performance ($\beta=0.198$, 95% CI: 0.068 to 0.333, $p=0.004$). No significant effects of *Experimental Condition* (Wald $\chi^2=0.772$, $p=0.68$), *Time* (Wald $\chi^2=0.00$, $p=0.99$) or *Training Type* (Wald $\chi^2=1.17$, $p=0.28$) were found.

The same analysis calculated with data from the wheel-chair compatible bicycle trainer group only ($n=36$) yielded qualitatively identical results. Perceived exertion again scaled positively with standardized *Training Performance* (Wald $\chi^2=9.485$, $p=0.002$), but was not systematically influenced by *Experimental Condition* (Wald $\chi^2=0.106$, $p=0.95$) or *Time* (Wald $\chi^2=0.074$, $p=0.79$).

In summary, perceived exertion covaried with objective training performance, but was not significantly altered by competition.

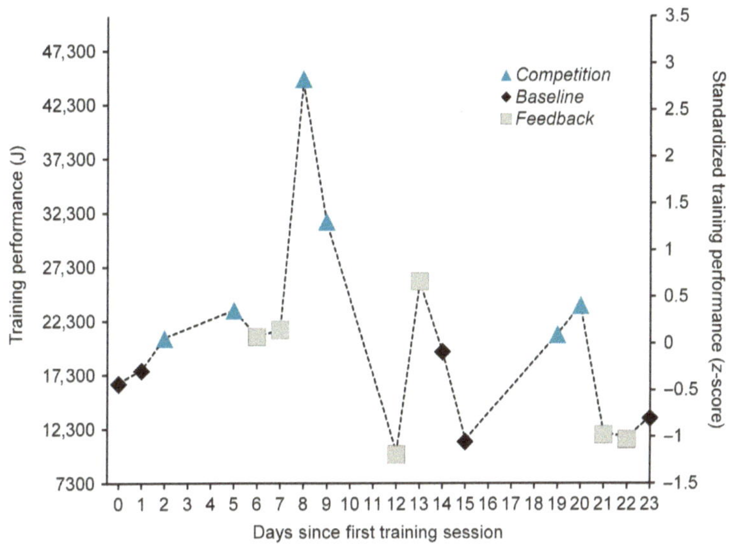

FIG. 4

Exemplary data from an individual patient showing a clear competition effect. Raw (*left y*-axis) and standardized (*right y*-axis) training performance of an individual patient is plotted. Note that training performance in the Competition sessions (denoted by *triangles*) was higher than training performance in Baseline (denoted by *diamonds*) and Feedback (denoted by *squares*) sessions.

3.3 EXPLORATORY ANALYSIS OF THE IMPACT OF COMPETITION UPON SUBSEQUENT TRAINING PERFORMANCE

First, we tested whether competition affected performance in the immediately following training session, regardless of the experimental condition of that subsequent training session. The GEE model indicated that patients' training performance tended to be higher following a competition than following a noncompetition session (parameter estimate for *Previous_Competition*: $\beta=0.111$, 95% CI: 0.014 to 0.236, $p=0.08$), although this effect was only marginally significant (Wald $\chi^2=3.023$, $p=0.08$). All other tested effects (ie, *Experimental Condition*, *Time*, and *Training Type*) were qualitatively unchanged from those reported earlier and statistically significant.

Next, we explored whether performance on rematch competitions, that is to say on occasions where a patient competed against the same opponent again after just having won, was systematically stronger than performance in first encounter competitions. The GEE model found that training performance was indeed significantly higher in rematch competition sessions compared to first encounter competitions (Wald $\chi^2=5.377$, $p=0.02$, parameter estimate for *Rematch_Competition*: $\beta=0.208$, 95% CI: 0.032 to 0.385, $p=0.02$, see Fig. 5). *Training Type* had no significant influence (Wald $\chi^2=0.227$, $p=0.63$).

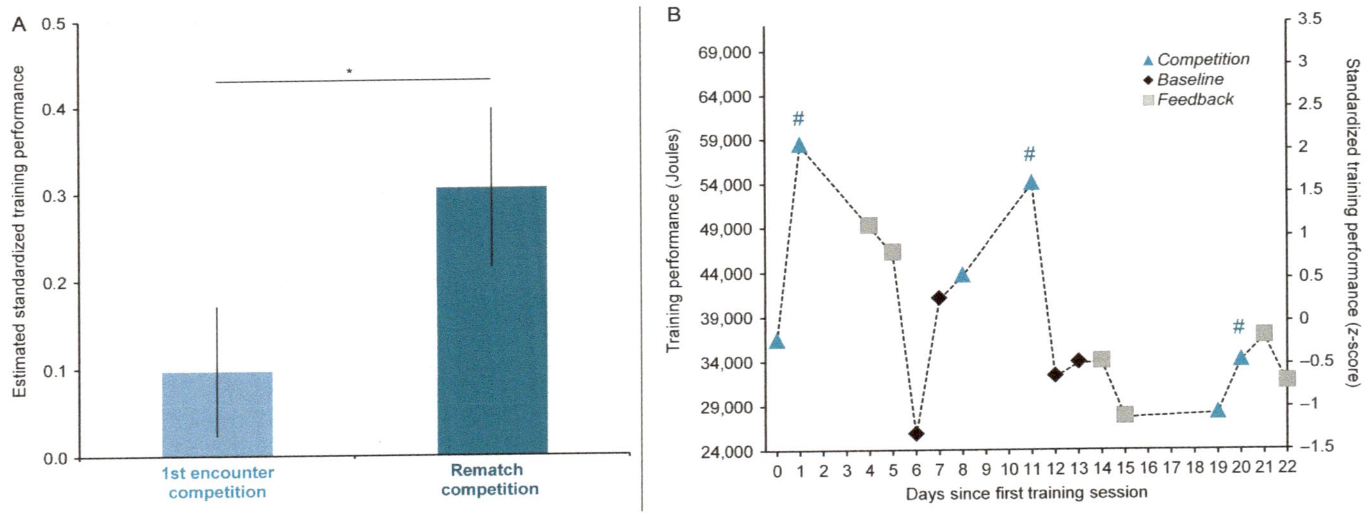

FIG. 5

Training performance in rematch vs 1st encounter competitions. (A) Model-estimated standardized training performance in 1st encounter vs rematch (2nd encounter) competitions. Rematch competitions were associated with a significantly higher training performance, $*p<0.05$. Error bars represent standard error of the mean. (B) Exemplary data from an individual patient. Raw (*left y*-axis) and standardized (*right y*-axis) training performance of an individual patient is plotted. Note the boost in the patient's training performance in rematch competitions (on days 1, 11, and 20, *triangles* marked with #) compared to both 1st encounter competitions (on days 0, 8, and 19, *triangles* without #) and noncompetition sessions (Baseline—*diamonds* and Feedback—*squares*).

In summary, competition also had an enhancing effect on subsequent training performance, and rematch competitions appeared particularly effective in increasing training intensity.

3.4 POSTTRIAL INTERVIEW

Posttrial interview answers were collected in 36 patients (data of $n=24$ were missing due to logistical reasons). Interestingly, patients' explicit report about the motivational impact of competition somewhat contrasted with the clear behavioral impact: while 53% of patients reported that they felt particularly motivated in competition sessions, only 29% believed to have performed better under competition (compared to noncompetition sessions).

4 DISCUSSION

This proof-of-concept study assessed the potential of competition to increase self-directed training in patients undergoing inpatient neurorehabilitation. Patients performed self-directed training on a wheelchair-compatible or conventional bicycle trainer under three experimental conditions (Baseline, Competition, and Feedback). Training performance and perceived exertion were recorded and analyzed. Our data revealed three motivational effects of competition: first, competition led to a significant increase in self-directed training. Patients exercised more intensively when competing against an anonymous same-sex opponent than in the two noncompetition control conditions. Second, patients' training performance was particularly high during rematch competitions, where a patient had just won a competition on the previous day and now competed against the same opponent again. Third, competition had a tentative effect upon subsequent training performance, with patients' training more intensively following a competition session.

Our findings demonstrate that competition can be a powerful tool to enhance self-directed training and complement previous findings of competition effects upon physical activity and performance in the healthy population. For instance, Cooke et al. (2011) asked healthy student volunteers to squeeze a handgrip dynamometer at 40% of their maximum contraction force for as long as possible, and found that participants maintained this isometric contraction for 22% longer in a competition than in an individualistic condition. Another recent laboratory study in students by Dimenichi and Tricomi (2015) revealed that competition led to a significant facilitation in reaction times on a simple paced keypress task. Competition has previously also been found to improve performance on sports trials such as basketball shooting (Giannini, 1988) and cycling time trials (Williams et al., 2015). And, three recent studies demonstrate that introducing competition-based games and assignments in university teaching courses improved students' learning and course performance (Burguillo, 2010; Cagiltay et al., 2015; Van Nuland et al., 2015). However, to the best of our knowledge, the potential of competition in neurological patients and for increasing recovery-relevant training has never been assessed to date.

In fact, the use of competition to increase motivation and performance in learning environments is even debated controversially in the extant (education) literature. While many have highlighted the benefits of competition, two main arguments against its use (in the context of learning interventions for and classroom education of children) have also been brought forth. The first is derived from Self-Determination Theory (Deci, 1980; Ryan and Deci, 2000) and states that adding "extrinsic" motivators (such as performance-based pay or competition) can under some circumstances undermine intrinsic motivation (Deci et al., 1981; Reeve and Deci, 1996). A potential antagonistic effect upon intrinsic motivation warrants consideration in situations where fostering intrinsic motivation and enjoyment of an activity is the intervention target. However, for applications such as ours, where exercise is presumably driven by its extrinsic benefits (ie, improving health, fitness, and strength) rather than enjoyment, and where intrinsic motivation appears insufficient to sustain behavior, this is arguably not very relevant. A second concern is that competition can increase anxiety and pressure (Baumeister and Showers, 1986; Beilock and Carr, 2001; Dimenichi and Tricomi, 2015)—negative emotional states associated with high attentional load demands that have sometimes been found to deteriorate performance (eg, Baumeister and Showers, 1986; Beilock and Carr, 2001; Bertrams et al., 2013; Jones and Hardy, 1988). In support of this argument, the aforementioned study testing the effect of competition on a grip force task by Cooke et al. (2011) found that self-reported anxiety did increase during competition, and that anxiety (negatively) modulated the performance-boost under competition. Note however, that competition was still successful in inducing a performance increase in that study. In the current study, two measures were taken to decrease the likelihood of competition-induced anxiety: first, competitors were always kept anonymous so that patients did not have to fear a decrease in social status upon losing (see Delgado et al., 2008; Mazur and Booth, 1998). Second, opponents were always described as a stroke patient matched in performance level, sex, and age so that participants would perceive their chance of winning as reasonable (Stanne et al., 1999; see Salvador and Costa, 2009). Since anxiety was not measured in this study, we cannot conclude whether these measures were successful in eradicating any competition-induced anxiety or pressure. However, the fact that competition did have a clear and significant boosting effect on training performance indicates that performance-*hampering* anxiety or pressure did not occur, or at least not in the majority of patients.

Instead, we propose that the best fitting explanation with regards to our results is that competition increased the subjective expected benefit of exercising (intensively) through anticipation of a "joy of winning" (Cooper and Fang, 2008; Grimm and Engelmann, 2005; Roider and Schmitz, 2012), that is to say by adding a new extrinsic benefit to the training (see also Studer and Knecht, 2016). This explanation is also congruent with our observation that (winning a) competition was associated with an increase in subsequent training performance. Previous work found that winning a competition enhances subsequent competition willingness and this "winner effect" is mediated by the release of the hormone testosterone in response to the competition win (for recent reviews, see Carre and Olmstead, 2015; Losecaat Vermeer et al., 2016; Zilioli and Watson, 2014). In addition, higher subsequent intrinsic motivation in competition winners compared to competition losers was reported (Vansteenkiste and Deci, 2003). Together, these findings demonstrate that winning a competition is a rewarding outcome (see also neuroimaging results by Fareri and Delgado, 2014; Le Bouc and Pessiglione, 2013) with motivational consequences, and it seems likely that anticipation of this positive experience and affective state would increase performance and motivation during competition. Finally, our data suggest that rematch competitions were particularly motivating, as training performance in these second encounter competitions (following a close competition win) was significantly higher than in 1st encounter competitions against the same opponent. This finding is consistent with a recent study by Mehta et al. (2015), which found that individuals who won a competition tightly (such as was always the case in this study) tended to rate the competition task as more fun than those who won a competition decisively.

We also assessed the influence of competition upon perceived exertion, as a measure of subjective training effort. Some previous research indicates that perception of effort might be affected by motivational context. In particular, a recent study by Fritz et al. (2013) in healthy volunteers found that perceived exertion was lower during work-out trials accompanied with movement-generated music than in a control condition where the music was not coupled to participants' movements. Meanwhile, objective performance did not differ significantly between the two work-out conditions. In addition, a recent model of subjective mental effort by Kurzban et al. (2013) postulates that perceived effort during performance of a cognitive task is high when the utility of the task is close to the utility of the most attractive alternative activity. Under the assumption that this model also applies to effort perception during physical activities, one could thus predict that increasing the utility (or expected benefit) of the training through competition would reduce perceived effort, because the difference between the expected benefit of the training and its opportunity costs is raised. In contrast to this prediction, we found no systematic effect of competition upon perceived exertion, when controlling for objective training performance (which was significantly correlated with ratings of perceived exertion). This could indicate that patients' perception of effort was unaffected by the competition manipulation, although caution is warranted in the interpretation of null findings given that alternative explanations (eg, a type II error or suboptimal choice of measure) are also plausible. Future research should further explore whether, and under which circumstances, motivation enhancement through competition or other mechanisms can attenuate perceived effort.

4.1 QUESTION FOR FUTURE (APPLICATION) RESEARCH

This study demonstrates that competition can be a powerful motivator and fruitful tool to increase training in neurorehabilitation. Future research can build upon this proof of concept and explore the potential of competition to boost motivation, training intensity and training amount in rehabilitation, and other therapeutic settings. One important question to answer is for whom competition is likely to be the most effective. Previous studies using self-report questionnaires show that competitiveness is subject to individual differences (Harris and Houston, 2010; Harris et al., 2015; Smither and Houston, 1992), and motivation theories predict that what (best) motivates a person will depend upon their personality (eg, Deci and Ryan, 2000; Elliot and Harackiewicz, 1994; Steele-Johnson et al., 2000). Knowledge about which individuals are likely to profit, together with versatile classification screening tools, would allow personalized treatment in (neuro-)rehabilitation. A second question which should be investigated in future research is which form of competition is the most motivating and performance-boosting. In addition to single head-to-head competitions as used in the current study, group-to-group competitions combining an element of cooperation (with team members) and competition (with the opponent team) might be interesting (see Tauer and Harackiewicz, 2004). A final open question is whether competition effects on motivation and (training) behavior could be further enhance by tapping into the neurobiological mechanisms underlying this effect, for instance through administration of testosterone (see also Losecaat Vermeer et al., 2016).

ACKNOWLEDGMENTS

The research presented in this chapter was funded by the Research Committee of the Medical Faculty at the Heinrich-Heine-University Düsseldorf (project grant number 23/2015 awarded to Bettina Studer). We are grateful to Heike Wittenberg and Helmut Krause for valuable discussion and assistance in study coordination, to Ulrich Rauf, Barbara Peilstöcker, Ute Wallner, Tanja Schill, Gabi Böhle, and the physiotherapists and physicians at the Mauritius Hospital Meerbusch for assistance in patient recruitment, and to Deborah Hunstiger, Franziska Hoffmann, and Karen Waterboer for help in data acquisition. We would also like to thank *Medisana AG* (Neuss, Germany) and *medica Medizintechnik GmbH* (Hochdorf, Germany) for technical support in data recording.

REFERENCES

Albert, S.J., Kesselring, J., 2012. Neurorehabilitation of stroke. J. Neurol. 259, 817–832.

Baumeister, R.F., Showers, C.J., 1986. A review of paradoxical performance effects: choking under pressure in sports and mental tests. Eur. J. Soc. Psychol. 16, 361–383.

Beilock, S.L., Carr, T.H., 2001. On the fragility of skilled performance: what governs choking under pressure? J. Exp. Psychol. Gen. 130, 701.

Bertrams, A., Englert, C., Dickhauser, O., Baumeister, R.F., 2013. Role of self-control strength in the relation between anxiety and cognitive performance. Emotion 13, 668–680.
Borg, G., 1970. Perceived exertion as an indicator of somatic stress. Scand. J. Rehabil. Med. 2, 92–98.
Borg, G.A., 1982. Psychophysical bases of perceived exertion. Med. Sci. Sports Exerc. 14, 377–381.
Brazzelli, M., Saunders, D.H., Greig, C.A., Mead, G.E., 2011. Physical fitness training for stroke patients. Cochrane Database Syst. Rev. 2011 (11), Cd003316.
Burguillo, J.C., 2010. Using game theory and competition-based learning to stimulate student motivation and performance. Comput. Educ. 55, 566–575.
Caeiro, L., Ferro, J.M., Figueira, M.L., 2012. Apathy in acute stroke patients. Eur. J. Neurol. 19, 291–297.
Caeiro, L., Ferro, J.M., Costa, J., 2013. Apathy secondary to stroke: a systematic review and meta-analysis. Cerebrovasc. Dis. 35, 23–39.
Cagiltay, N.E., Ozcelik, E., Ozcelik, N.S., 2015. The effect of competition on learning in games. Comput. Educ. 87, 35–41.
Carre, J.M., Olmstead, N.A., 2015. Social neuroendocrinology of human aggression: examining the role of competition-induced testosterone dynamics. Neuroscience 286, 171–186.
Cooke, A., Kavussanu, M., Mcintyre, D., Ring, C., 2011. Effects of competition on endurance performance and the underlying psychological and physiological mechanisms. Biol. Psychol. 86, 370–378.
Cooper, D.J., Fang, H., 2008. Understanding overbidding in second price auctions: an experimental study. Econ. J. 118, 1572–1595.
Corbett, J., Barwood, M.J., Ouzounoglou, A., Thelwell, R., Dicks, M., 2012. Influence of competition on performance and pacing during cycling exercise. Med. Sci. Sports Exerc. 44, 509–515.
Cox, J.C., Roberson, B., Smith, V., 1982. Theory and behavior of single object auctions. Res. Exp. Econ. 2, 1–43.
Cox, J.C., Smith, V.L., Walker, J.M., 1988. Theory and individual behavior of first-price auctions. J. Risk Uncertain. 1, 61–99.
Deci, E.L., 1980. The Psychology of Self-Determination. Heath, Lexington, MA.
Deci, E.L., Ryan, R.M., 2000. The "what" and "why" of goal pursuits: human needs and the self-determination of behavior. Psychol. Inquiry 11, 227–268.
Deci, E.L., Betley, G., Kahle, J., Abrams, L., Porac, J., 1981. When trying to win: competition and intrinsic motivation. Pers. Soc. Psychol. Bull. 7, 79–83.
Delgado, M.R., Schotter, A., Ozbay, E.Y., Phelps, E.A., 2008. Understanding overbidding: using the neural circuitry of reward to design economic auctions. Science 321, 1849–1852.
Dimenichi, B.C., Tricomi, E.M., 2015. The power of competition: effects of social motivation on attention, sustained physical effort, and learning. Front. Psychol. 6, 1–13.
Elliot, A.J., Harackiewicz, J.M., 1994. Goal setting, achievement orientation, and intrinsic motivation: a mediational analysis. J. Pers. Soc. Psychol. 66, 968.
Engelbrecht-Wiggans, R., Katok, E., 2006. Regret in auctions: theory and evidence. Econ. Theory 33, 81–101.
Eston, R.G., Williams, J.G., 1988. Reliability of ratings of perceived effort regulation of exercise intensity. Br. J. Sports Med. 22, 153–155.
Fareri, D.S., Delgado, M.R., 2014. Differential reward responses during competition against in- and out-of-network others. Soc. Cogn. Affect. Neurosci. 9, 412–420.

Filiz-Ozbay, E., Ozbay, E.Y., 2007. Auctions with anticipated regret: theory and experiment. Am. Econ. Rev. 97, 1407–1418.

Fritz, T.H., Hardikar, S., Demoucron, M., Niessen, M., Demey, M., Giot, O., Li, Y., Haynes, J.-D., Villringer, A., Leman, M., 2013. Musical agency reduces perceived exertion during strenuous physical performance. Proc. Natl. Acad. Sci. U.S.A. 110, 17784–17789.

Giannini, J.M., 1988. The effects of mastery, competitive, and cooperative goals on the performance of simple and complex basketball skills. J. Sport Exerc. Psychol. 10, 408417.

Goeree, J.K., Holt, C.A., Palfrey, T.R., 2002. Quantal response equilibrium and overbidding in private-value auctions. J. Econ. Theory 104, 247–272.

Grimm, V., Engelmann, D., 2005. Overbidding in first price private value auctions revisited: implications of a multi-unit auctions experiment. In: Schmidt, U., Traub, S. (Eds.), Advances in Public Economics: Utility, Choice and Welfare. Springer, Boston, MA.

Harris, P.B., Houston, J.M., 2010. A reliability analysis of the revised competitiveness index. Psychol. Rep. 106, 870–874.

Harris, A.L., Elder, J., Schiff, N.D., Victor, J.D., Goldfine, A.M., 2014. Post-stroke apathy and hypersomnia lead to worse outcomes from acute rehabilitation. Transl. Stroke Res. 5, 292–300.

Harris, N., Newby, J., Klein, R.G., 2015. Competitiveness facets and sensation seeking as predictors of problem gambling among a sample of university student gamblers. J. Gambl. Stud. 31, 385–396.

Horne, M., Thomas, N., Mccabe, C., Selles, R., Vail, A., Tyrrell, P., Tyson, S., 2015. Patient-directed therapy during in-patient stroke rehabilitation: stroke survivors' views of feasibility and acceptability. Disabil. Rehabil. 37, 2344–2349.

Johnson, D.W., Johnson, R.T., 1974. Instructional goal structure: cooperative, competitive, or individualistic. Rev. Educ. Res. 44, 213–240.

Jones, J.G., Hardy, L., 1988. The effects of anxiety upon psychomotor performance. J. Sports Sci. 6, 59–67.

Kagel, J.H., Levin, D., 1993. Independent private value auctions: bidder behaviour in first-, second- and third-price auctions with varying numbers of bidders. Econ. J. 103, 868–879.

Kagel, J.H., Harstad, R.M., Levin, D., 1987. Information impact and allocation rules in auctions with affiliated private values: a laboratory study. Econometrica 55, 1275–1304.

Kennerley, S.W., Walton, M.E., 2011. Decision making and reward in frontal cortex: complementary evidence from neurophysiological and neuropsychological studies. Behav. Neurosci. 125, 297–317.

Kleim, J.A., Jones, T.A., 2008. Principles of experience-dependent neural plasticity: implications for rehabilitation after brain damage. J. Speech Lang. Hear. Res. 51, S225–S239.

Knecht, S., Roßmüller, J., Unrath, M., Stephan, K.-M., Berger, K., Studer, B., 2016. Old benefit as much as young patients with stroke from high-intensity neurorehabilitation: cohort analysis. J. Neurol. Neurosurg. Psychiatry 87, 526–530.

Krakauer, J.W., Carmichael, S.T., Corbett, D., Wittenberg, G.F., 2012. Getting neurorehabilitation right: what can be learned from animal models? Neurorehabil. Neural Repair 26, 923–931.

Kurzban, R., Duckworth, A., Kable, J.W., Myers, J., 2013. An opportunity cost model of subjective effort and task performance. Behav. Brain Sci. 36, 661–679.

Le Bouc, R., Pessiglione, M., 2013. Imaging social motivation: distinct brain mechanisms drive effort production during collaboration versus competition. J. Neurosci. 33, 15894–15902.

Liang, K.-Y., Zeger, S.L., 1986. Longitudinal data analysis using generalized linear models. Biometrika 73, 13–22.

Lohse, K.R., Lang, C.E., Boyd, L.A., 2014. Is more better? Using metadata to explore dose-response relationships in stroke rehabilitation. Stroke 45, 2053–2058.

Losecaat Vermeer, A.B., Riečanský, I., Eisenegger, C., 2016. Chapter 9—Competition, testosterone, and adult neurobehavioral plasticity. In: Studer, B., Knecht, S. (Eds.), Progress in Brain Research, vol. 229. Elsevier, Netherlands, pp. 213–238.

Mayo, N.E., Fellows, L.K., Scott, S.C., Cameron, J., Wood-Dauphinee, S., 2009. A longitudinal view of apathy and its impact after stroke. Stroke 40, 3299–3307.

Mazur, A., Booth, A., 1998. Testosterone and dominance in men. Behav. Brain Sci. 21, 353–363.

Mehta, P.H., Snyder, N.A., Knight, E.L., Lassetter, B., 2015. Close versus decisive victory moderates the effect of testosterone change on competitive decisions and task enjoyment. Adapt. Hum. Behav. Physiol. 1, 291–311.

Mukherjee, D., Patil, C.G., 2011. Epidemiology and the global burden of stroke. World Neurosurg. 76, S85–S90.

Nicholson, S., Sniehotta, F.F., Van Wijck, F., Greig, C.A., Johnston, M., Mcmurdo, M.E.T., Dennis, M., Mead, G.E., 2013. A systematic review of perceived barriers and motivators to physical activity after stroke. Int. J. Stroke 8, 357–364.

Noonan, M.P., Kolling, N., Walton, M.E., Rushworth, M.F., 2012. Re-evaluating the role of the orbitofrontal cortex in reward and reinforcement. Eur. J. Neurosci. 35, 997–1010.

O'Doherty, J.P., 2004. Reward representations and reward-related learning in the human brain: insights from neuroimaging. Curr. Opin. Neurobiol. 14, 769–776.

Pang, M.Y., Charlesworth, S.A., Lau, R.W., Chung, R.C., 2013. Using aerobic exercise to improve health outcomes and quality of life in stroke: evidence-based exercise prescription recommendations. Cerebrovasc. Dis. 35, 7–22.

Reeve, J., Deci, E.L., 1996. Elements of the competitive situation that affect intrinsic motivation. Pers. Soc. Psychol. Bull. 22, 24–33.

Reeve, J., Olson, B.C., Cole, S.G., 1985. Motivation and performance: two consequences of winning and losing in competition. Motiv. Emotion 9, 291–298.

Robert, P., Onyike, C.U., Leentjens, A.F., Dujardin, K., Aalten, P., Starkstein, S., Verhey, F.R., Yessavage, J., Clement, J.P., Drapier, D., Bayle, F., Benoit, M., Boyer, P., Lorca, P.M., Thibaut, F., Gauthier, S., Grossberg, G., Vellas, B., Byrne, J., 2009. Proposed diagnostic criteria for apathy in Alzheimer's disease and other neuropsychiatric disorders. Eur. Psychiatry 24, 98–104.

Roider, A., Schmitz, P.W., 2012. Auctions with anticipated emotions: overbidding, underbidding, and optimal reserve prices. Scand. J. Econ. 114, 808–830.

Rushworth, M.F., Noonan, M.P., Boorman, E.D., Walton, M.E., Behrens, T.E., 2011. Frontal cortex and reward-guided learning and decision-making. Neuron 70, 1054–1069.

Ryan, R.M., Deci, E.L., 2000. Self-determination theory and the facilitation of intrinsic motivation, social development, and well-being. Am. Psychol. 55, 68.

Salvador, A., 2005. Coping with competitive situations in humans. Neurosci. Biobehav. Rev. 29, 195–205.

Salvador, A., Costa, R., 2009. Coping with competition: neuroendocrine responses and cognitive variables. Neurosci. Biobehav. Rev. 33, 160–170.

Santa, N., Sugimori, H., Kusuda, K., Yamashita, Y., Ibayashi, S., Iida, M., 2008. Apathy and functional recovery following first-ever stroke. Int. J. Rehabil. Res. 31, 321–326.

Smither, R.D., Houston, J.M., 1992. The nature of competitiveness: the development and validation of the competitiveness index. Educ. Psychol. Meas. 52, 407–418.

Stanne, M.B., Johnson, D.W., Johnson, R.T., 1999. Does competition enhance or inhibit motor performance: a meta-analysis. Psychol. Bull. 125, 133–154.

Steele-Johnson, D., Beauregard, R.S., Hoover, P.B., Schmidt, A.M., 2000. Goal orientation and task demand effects on motivation, affect, and performance. J. Appl. Psychol. 85, 724–738.

Strong, K., Mathers, C., Bonita, R., 2007. Preventing stroke: saving lives around the world. Lancet Neurol. 6, 182–187.

Studer, B., Knecht, S., 2016. Chapter 2—A benefit–cost framework of motivation for a specific activity. In: Studer, B., Knecht, S. (Eds.), Progress in Brain Research, vol. 229. Elsevier, Netherlands, pp. 25–47.

Studer, B., Manes, F., Humphreys, G., Robbins, T.W., Clark, L., 2015. Risk-sensitive decision-making in patients with posterior parietal and ventromedial prefrontal cortex injury. Cereb. Cortex 25, 1–9.

Tang, A., Sibley, K.M., Thomas, S.G., Bayley, M.T., Richardson, D., Mcilroy, W.E., Brooks, D., 2009. Effects of an aerobic exercise program on aerobic capacity, spatiotemporal gait parameters, and functional capacity in subacute stroke. Neurorehabil. Neural Repair 23, 398–406.

Tauer, J.M., Harackiewicz, J.M., 2004. The effects of cooperation and competition on intrinsic motivation and performance. J. Pers. Soc. Psychol. 86, 849–861.

Tyson, S., Wilkinson, J., Thomas, N., Selles, R., Mccabe, C., Tyrrell, P., Vail, A., 2015. Phase II pragmatic randomized controlled trial of patient-led therapies (mirror therapy and lower-limb exercises) during inpatient stroke rehabilitation. Neurorehabil. Neural Repair 29, 818–826.

Vallerand, R.J., Gauvin, L.I., Halliwell, W.R., 1986. Effects of zero-sum competition on children's intrinsic motivation and perceived competence. J. Soc. Psychol. 126, 465–472.

Van Dalen, J.W., Van Charante, E.P.M., Nederkoorn, P.J., Van Gool, W.A., Richard, E., 2013. Poststroke apathy. Stroke 44, 851–860.

Van Nuland, S.E., Roach, V.A., Wilson, T.D., Belliveau, D.J., 2015. Head to head: the role of academic competition in undergraduate anatomical education. Anat. Sci. Educ. 8, 404–412.

Vansteenkiste, M., Deci, E.L., 2003. Competitively contingent rewards and intrinsic motivation: can losers remain motivated? Motiv. Emotion 27, 273–299.

Volz, K.G., Schubotz, R.I., Von Cramon, D.Y., 2006. Decision-making and the frontal lobes. Curr. Opin. Neurol. 19, 401–406.

Williams, E.L., Jones, H.S., Andy Sparks, S., Marchant, D.C., Midgley, A.W., Mc Naughton, L.R., 2015. Competitor presence reduces internal attentional focus and improves 16.1 km cycling time trial performance. J. Sci. Med. Sport 18, 486–491.

Zeger, S.L., Liang, K.Y., 1986. Longitudinal data analysis for discrete and continuous outcomes. Biometrics 42, 121–130.

Zeger, S.L., Liang, K.-Y., Albert, P.S., 1988. Models for longitudinal data: a generalized estimating equation approach. Biometrics 44, 1049–1060.

Zeiler, S.R., Krakauer, J.W., 2013. The interaction between training and plasticity in the poststroke brain. Curr. Opin. Neurol. 26, 609–616.

Zilioli, S., Watson, N.V., 2014. Testosterone across successive competitions: evidence for a 'winner effect' in humans? Psychoneuroendocrinology 47, 1–9.

CHAPTER

The role of dopamine in the pathophysiology and treatment of apathy

17

T.T.-J. Chong[*,†,‡,1], **M. Husain**[§,¶]

[*]*Macquarie University, Sydney, NSW, Australia*

[†]*ARC Centre of Excellence in Cognition and its Disorders, Macquarie University, Sydney, NSW, Australia*

[‡]*Monash Institute of Cognitive and Clinical Neurosciences, Monash University, Clayton, VIC, Australia*

[§]*University of Oxford, Oxford, United Kingdom*

[¶]*John Radcliffe Hospital, Oxford, United Kingdom*

[1]*Corresponding author: Tel.: +61-2-9850-2980; Fax: +61-2-9850-6059, e-mail address: trevor.chong@mq.edu.au*

Abstract

Disorders of diminished motivation, such as apathy, are common and prevalent across a wide range of medical conditions, including Parkinson's disease, Alzheimer's dementia, stroke, depression, and schizophrenia. Such disorders have a significant impact on morbidity and quality of life, yet their management lacks consensus and remains unsatisfactory. Here, we review laboratory and clinical evidence for the use of dopaminergic therapies in the treatment of apathy. Dopamine is a key neurotransmitter that regulates motivated decision making in humans and other species. A large corpus of evidence suggests that it plays an important role in promoting approach behavior by attributing incentive salience to reward stimuli, and facilitating the overcoming of effort costs. Furthermore, dopaminergic neurons innervate several frontostriatal structures that mediate reward-guided behavior. Based on these findings, there are a priori reasons for considering dopamine in the treatment of disorders of diminished motivation. We highlight key studies that have attempted to use dopamine to manage patients with apathy, and that collectively offer cautious evidence in favor of its efficacy. However, many of these studies are small, unblinded, and uncontrolled, and utilize subjective, questionnaire-based measures of apathy. Given the development of novel paradigms which are able to objectively dissect motivational dysfunction, we are now well positioned to quantify the effect of specific classes of dopaminergic medication on reward- and effort-based decision making in apathy. We anticipate that such paradigms will lay the foundation for future studies to evaluate new and existing treatments for disorders of motivation, using sensitive measures of apathy as primary quantifiable end points.

Progress in Brain Research, Volume 229, ISSN 0079-6123, http://dx.doi.org/10.1016/bs.pbr.2016.05.007

Keywords

Motivation, Disorders of motivation, Apathy, Dopamine, Decision making, Effort, Reward

1 INTRODUCTION

Apathy is one of several disorders characterized by an impairment in motivation (Table 1). Some have proposed that such disorders lie on a continuum, from apathy on the milder end to akinetic mutism at its most severe (Marin and Wilkosz, 2005). Although the terminology for these disorders has been historically useful, many of these terms were defined on the basis of clinical observations over the last century. As such, they do not account for more contemporary discoveries in the biological sciences that have begun to distinguish different components of motivation. For example, "anhedonia" has been used to refer to multiple components of reward-based behavior, including the emotional experience of reward presentation; the anticipation of pleasurable outcomes; and the consumption of the desired good—however, extensive evidence now shows that these processes are dissociable (Berridge et al., 2009; Markou et al., 2013; Salamone et al., 2007; Smith et al., 2011; Treadway and Zald, 2011).

1.1 WHAT IS APATHY?

One of the earliest contemporary definitions of apathy was that it was characterized by "diminished goal-oriented behavior and cognition, and a diminished emotional connection to goal-directed behavior" (Marin, 1991); and, later, "the absence or lack of emotion, interest, concern, or motivation" (Marin, 1996). Over time, there have been several different conceptualizations of apathy, but a common feature is apathy as a disorder of motivation (Cummings et al., 1994; Levy and Dubois, 2006; Robert et al., 2002; Sockeel et al., 2006; Starkstein and Leentjens, 2008; Stuss et al., 2000). Indeed, one recent set of proposed diagnostic criteria requires a lack of motivation, resulting in significant clinical or functional impairment (Table 2; Mulin et al., 2011). This classification also preserves and elaborates on a feature of the earlier definition, which distinguishes several putative components of apathy, including behavioral, cognitive, and emotional elements. In schizophrenia, apathy is clustered alongside avolition and anergia in the Scale for Assessment of Negative Symptoms (SANS; Andreasen, 1984). For the purposes of this review, we consider diminished motivation to be a core feature of apathy, and we discuss studies which, unless otherwise indicated, have examined patients who have been diagnosed with apathy according to a standard set of diagnostic criteria (see also Chong et al., 2016 and Section 4.3.1 for a description of an apathetic patient involved in one of our tasks).

Table 1 Disorders of Diminished Motivation and Their Putative Definitions

Term	Definition(s)
Abulia	Impaired spontaneity in action and speech, with normal intellectual content, reduced range of movement, mental slowness, decreased attention in the presence of increased distractibility, and apathy (Bhatia and Marsden, 1994; Fisher, 1982).
Akinetic mutism	Lack of self-initiated motor or mental activity, and indifference even to biologically relevant stimuli (pain, hunger, thirst) in the presence of normal alertness (Cairns et al., 1941).
Anergia	Lack of perceived energy (Markou et al., 2013).
Anhedonia	The inability to experience pleasure (Barch and Dowd, 2010; Ribot, 1896).
Apathy	Disorder of motivation characterized by diminished voluntary and goal-directed behavior and cognition (Starkstein and Leentjens, 2008). The most recent working definition subdivides it into behavioral, cognitive, and emotional components (see Table 2). Specific deficits postulated to include intellectual curiosity, action initiation, self-awareness, emotion, and interest/enthusiasm (Cummings et al., 1994; Levy and Dubois, 2006; Marin et al., 1991; Robert et al., 2002; Sockeel et al., 2006; Starkstein and Leentjens, 2008; Stuss et al., 2000).
Autoactivation deficit (or athymhormia, psychic akinesia, reversible inertia)	Deficit in spontaneous activation of mental processing, observed in behavioral, cognitive, or affective domains, which can be totally reversed by external stimulation that activates normal patterns of response (Laplane and Dubois, 2001). Manifest as lack of self-initiated voluntary behavior (Levy and Dubois, 2006; van Reekum et al., 2005).
Avolition	Reduced ability to initiate and maintain goal-directed behavior (Foussias and Remington, 2010; Kraepelin, 1921).
Fatigue (central)	Lack of physical or mental energy related to abnormalities in motivational mechanisms (Chaudhuri and Behan, 2004).
Psychomotor retardation	Slowing of movement, or generally reduced tendency to engage in motor activity (Jones and Pansa, 1979; Sobin and Sackeim, 1997; Widlocher, 1983).

Table 2 Proposed Diagnostic Criteria for Apathy (Drijgers et al., 2010; Mulin et al., 2011; Robert et al., 2009)

For a diagnosis of Apathy the patient should fulfill criteria A, B, C, and D:

A. Loss of or diminished motivation in comparison to the patient's previous level of functioning and which is not consistent with his age or culture. These changes in motivation may be reported by the patient himself or by the observations of others.

B. Presence of at least one symptom in at least two of the three following domains for a period of at least 4 weeks and present most of the time.
Domain B1—Behavior: Loss of, or diminished, goal-directed behavior as evidenced by at least one of the following:
- *Initiation symptom*: loss of self-initiated behavior (for example: starting conversation, doing basic tasks of day-to-day living, seeking social activities, communicating choices).
- *Responsiveness symptom*: loss of environment-stimulated behavior (for example: responding to conversation, participating in social activities).

Domain B2—Cognition: Loss of, or diminished, goal-directed cognitive activity as evidenced by at least one of the following:
- *Initiation symptom*: loss of spontaneous ideas and curiosity for routine and new events (ie, challenging tasks, recent news, social opportunities, personal/family and social affairs).
- *Responsiveness symptom*: loss of environment-stimulated ideas and curiosity for routine and new events (ie, in the person's residence, neighborhood, or community).

Domain B3—Emotion: Loss of, or diminished, emotion as evidenced by at least one of the following:
- *Initiation symptom*: loss of spontaneous emotion, observed, or self-reported (for example, subjective feeling of weak or absent emotions, or observation by others of a blunted affect).
- *Responsiveness symptom*: loss of emotional responsiveness to positive or negative stimuli or events (for example, observer reports of unchanging affect, or of little emotional reaction to exciting events, personal loss, serious illness, emotional-laden news).

C. These symptoms (A–B) cause clinically significant impairment in personal, social, occupational, or other important areas of functioning.

D. The symptoms (A–B) are not exclusively explained or due to physical disabilities (eg, blindness and loss of hearing), to motor disabilities, to diminished level of consciousness, or to the direct physiological effects of a substance (eg, drug of abuse, a medication).

1.2 APATHY IS NOT DEPRESSION

Given the clinical manifestations of apathy, it is somewhat unsurprising that it is often conflated with depression. Indeed, they often overlap behaviorally, and apathy has been shown to be a harbinger of future depression (Starkstein et al., 2006). It is vital to appreciate, however, that apathy and depression are clinically and physiologically distinct entities (Kirsch-Darrow et al., 2006; Marin et al., 1993; Santangelo et al., 2013). Across multiple primary diseases, apathy has been found to be dissociable from other symptoms of depression, such as emotional distress, agitation, vegetative symptoms, suicidal ideation, hopelessness, and heightened sadness (Kirsch-Darrow et al., 2006; Levy et al., 1998; Starkstein et al., 2009). Thus, even though apathy and depression may share similar surface manifestations, they most likely arise from separate etiologies, which will be important in the development of future treatments tailored to both conditions.

1.3 APATHY IS INDEPENDENT OF COGNITIVE DYSFUNCTION

It is equally important to recognize that apathy does not merely reflect a generalized cognitive impairment. For example, in Parkinson's disease (PD), apathy is an isolated, independent, nonmotor symptom, even after adjusting for the severity of cognitive status and motor symptoms (Dujardin et al., 2014). General measures of attention and cognition (eg, MMSE or IQ) lack sensitivity for detecting apathy and typically show little change in apathetic individuals (Feil et al., 2003).

1.4 APATHY IS COMMON AND DEBILITATING

The exact prevalence of apathy is difficult to estimate, with cited rates being highly variable. For example, estimates of apathy have ranged from 17% to 70% in PD (Aarsland et al., 2009; Leentjens et al., 2008), 29% to 81% in Alzheimer's dementia (AD; Aarsland et al., 2001; Lyketsos et al., 2000; Marin et al., 1994; Migneco et al., 2001), and 10% to 71% in traumatic brain injury (Andersson et al., 1999; Kant et al., 1988). There are potentially several reasons for such variability. Because of variable diagnostic criteria, we lack an objective means to classify apathy, and it has traditionally been underrecognized (Landes et al., 2001). It may be confounded or confused with other phenomena: laziness, oppositional behavior, depression, or general emotional distress. In addition, although "apathy" as a symptom occurs in many diseases—including PD, dementia, schizophrenia, and depression—it tends to be overshadowed by the constellation of other symptoms which define the primary illness. Moreover, apathetic individuals have poor insight into their condition and are unable to advocate for themselves, and may therefore remain a silent population that is difficult to identify, study, and manage.

Nevertheless, the clinical impact of apathy is becoming increasingly well recognized. It is a significant contributor to poor outcome in neurologic and psychiatric populations, independent of depression. Apathy is associated with worsening social and functional impairment; decreased treatment responsiveness or compliance, poor awareness of behavioral and cognitive changes; poorer clinical outcome; and overall poorer quality of life (Boyle et al., 2003; Gerritsen et al., 2005; Mega et al., 1999; Pluck and Brown, 2002; Starkstein et al., 1993, 2001, 2006). Furthermore, it is associated with more rapid cognitive decline (Starkstein et al., 2006) and contributes above and beyond dementia severity in affecting basic activities of daily living (Zawacki et al., 2002). Apart from impacting upon the individual, it also contributes to caregivers' distress; dissatisfaction with caregiving; increased feelings of frustration; and disruptions to family life, which may compound patients' disability (Aarsland et al., 2007; Benoit et al., 1999; Campbell and Duffy, 1997; Kaufer et al., 1998; Lyketsos et al., 2002; van Reekum et al., 2005). Overall, therefore, apathy imposes high levels of economic, social, and physical burden and distress and frequently leads to earlier institutionalization than for similarly impaired patients without apathy (Moretti et al., 2002).

Despite its impact, only recently has apathy become an important subject of scientific enquiry. Treatment of the condition has not been the subject of many large-scale studies, and management strategies vary considerably. In addition to lifestyle and environmental interventions, a vast range of drugs have been used, depending on the patient and their primary disease. Of these treatments, there is a significant volume of preclinical literature supporting the involvement of dopamine in behavioral activation and motivation in nonhuman animals (Salamone and Correa, 2012). Here, therefore, we focus on the potential utility of dopamine as a treatment for apathy. In the following sections, we first consider the causal link between dopaminergic lesions and motivational deficits, before considering various attempts at using dopaminergic drugs to treat apathy in humans.

2 APATHY AS A DISORDER OF DOPAMINERGIC FUNCTION

2.1 DOPAMINERGIC DEFICITS IN NONHUMAN ANIMALS MODULATE REWARD AND EFFORT SENSITIVITY

A key feature of motivated behavior is that it requires one to decide whether to embark on a course of action for a given reward given the associated costs (Chong et al., 2016). Thus, motivation requires an animal to be sensitive to the rewards on offer for its actions ("reward sensitivity"), as well as the costs associated with it, such as the effort required to obtain it ("effort sensitivity"). This cost–benefit computation is thought to be underpinned by a distributed network of brain areas, including the ventral striatum, ventral pallidum, medial prefrontal and anterior cingulate cortices (ACC), and basolateral amygdala (Fig. 1; Farrar et al., 2008; Floresco and Ghods-Sharifi, 2007; Hauber and Sommer, 2009; Walton et al., 2003). The core of this network is composed of reciprocal connections between the basal ganglia and prefrontal cortex, particularly the dopaminergic mesocorticolimbic and nigrostriatal pathways (Levy, 2012).

The mesocorticolimbic dopamine system is considered to be central to the brain's reward and motivational circuitry (Robbins and Everitt, 2006; Salamone et al., 2006). It projects from the ventral tegmental area of the midbrain to a widespread area of cortical and subcortical regions, including the ventral striatum/nucleus accumbens (NAcc), medial prefrontal areas, the amygdala, and the hippocampus (Fig. 1). The NAcc is a subcortical structure comprising an inner core and outer shell, with the shell being one of the major projection areas of mesolimbic dopamine neurons, which also receives important connections from the hippocampus and amygdala (Sokoloff et al., 2006). Dopamine released in the NAcc is thought to play a central role in effort-based decisions, and some have proposed that the NAcc plays a critical role as a "limbic–motor interface" to translate motivation into action

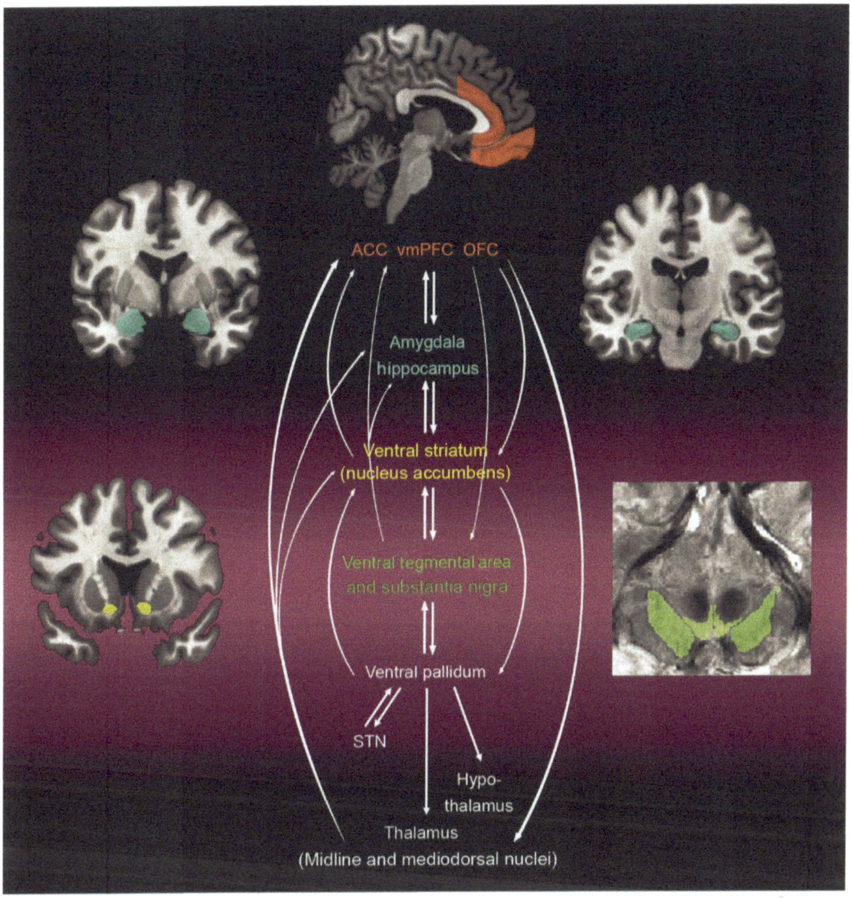

FIG. 1

Simplified schematic of the reward pathway in humans. The core of the mesocorticolimbic system is formed by basal ganglia nuclei (*shaded maroon*). Projections from the dopaminergic midbrain originate from the ventral tegmental area and substantia nigra and project to the ventral striatum (nucleus accumbens; *yellow*), prefrontal cortex (*red*), and limbic and other subcortical structures (amygdala and hippocampus, *blue*). The midsagittal section (*top*) illustrates the anterior cingulate cortex (ACC) superiorly and the ventromedial prefrontal cortex (vmPFC) inferiorly, with the orbitofrontal cortex (OFC) on the ventral surface of the brain. The coronal slices illustrate the amygdala nuclei (*top left, blue*), hippocampal formation (*top right, blue*), and ventral striatum (*bottom left, yellow*). The axial MRI of the midbrain illustrates the substantia nigra laterally and the ventral tegmental area medially (*bottom right, green*; as segmented in a recent 7T MRI study (Eapen et al., 2011)). STN, subthalamic nucleus.

(Mogenson et al., 1980). The prefrontal cortex, in particular ventromedial prefrontal areas and the ACC, is functionally interconnected with basal ganglia and limbic structures through different circuits (Mega and Cummings, 1994) and plays a critical role in reward processing, initiation, planning, and monitoring of goal-directed behavior (Fuster, 2008).

There is extensive evidence that a dopaminergic deficit, or selective lesions to the mesocorticolimbic system, results in less-motivated behavior, which resembles the behavior of patients with apathy. For example, typical studies in rodents require the animal to decide how much effort it is willing to invest for various rewards. Such paradigms may include operant conditioning tasks or dual alternative, effort discounting tasks (Chong et al., 2016). These tasks are able to quantify motivation in terms of the animal's sensitivity to available rewards—that is, how much reward is required to incentivize it to act. They are also able to quantify the animal's sensitivity to effort costs—that is, how much effort it is willing to exert to obtain those rewards. The animal's reward and effort sensitivities can then be compared before and after lesions to the mesocorticolimbic system. Typically, dopamine transmission is disrupted through systemic administration of low doses of dopamine antagonists, or selective dopamine depletion or antagonism (eg, with 6-hydroxy-dopamine, SCC 23390, ecopipam, haloperidol, flupenthixol; Salamone and Correa, 2012).

A vast volume of literature has been built on this approach, and the results from many of these tasks are strikingly similar. The overall pattern is that disrupting dopamine transmission reduces an animal's sensitivity to reward and increases its sensitivity to effort. Thus, it will require greater rewards to incentivize it to act, and it will be willing to invest less effort for given rewards. This is a consistent finding across a range of paradigms and can occur in the context of systemic dopaminergic depletion, disruption of dopaminergic input to the basal ganglia and/or frontal lobes, or from selective antagonism of basal ganglia and cortical dopamine receptors (Cousins and Salamone, 1994; Cummings, 1993; Denk et al., 2005; Farrar et al., 2010; Floresco et al., 2008; Hauber and Sommer, 2009; Mai et al., 2012; Nowend et al., 2001; Nunes et al., 2010, 2013; Pardo et al., 2012; Randall et al., 2012; Salamone and Correa, 2012; Salamone et al., 1991, 1994, 2003, 2007; Schweimer and Hauber, 2006; Sink et al., 2008; Walton et al., 2005). Together, this literature highlights the importance of the reward- and effort-related functions of dopaminergic systems.

2.2 DOPAMINERGIC DEFICITS IN HUMANS LEAD TO APATHY

In humans, apathy and lack of motivation have been reported following dopaminergic dysfunction and lesions to the mesocorticolimbic pathway. The most obvious example of this is patients with PD and Huntington's disease (Craufurd et al., 2001; Pederson et al., 2009), both of which are paradigmatic models of dopaminergic dysfunction. Although PD is characterized as a motor deficit stemming from nigrostriatal dysfunction, it is also associated with significant motivational deficits. For example, in patients with PD, striatal activity after monetary reward is reduced relative to healthy controls (Künig et al., 2000). Moreover, individuals with PD are willing to invest less effort than controls for low amounts of reward (Chong et al., 2015).

Apathy in PD has been linked to underactivity in the ventral striatum and disruption of basal ganglia circuitry due to midbrain neurodegeneration (Remy et al., 2005). In a study directly comparing PD patients scoring high on apathy vs those scoring low, apathy was associated with decreased responsivity to monetary gains in an extensive circuit involving the vmPFC, amygdala, striatum, and midbrain (Lawrence et al., 2011). This was thought to be caused by a reduction of dopaminergic afferents to the ventral striatum disrupting normal interactions among the frontal lobe, caudate, anterior cingulate circuits, and basal ganglia (Martínez-Horta et al., 2014). Dysfunction of this mesocorticolimbic dopaminergic pathway is therefore considered to be key to the pathophysiological basis of apathy in PD.

Other striatal lesions outside of PD have also been found to cause a profound apathetic state. For example, apathy occurs following strokes to the basal ganglia (Adam et al., 2012; Schmidt et al., 2008), while apathy, abulia, and akinetic mutism have all been reported following lesions to the globus pallidus, thalamus, and ACC (Oberndorfer et al., 2002; Tengvar et al., 2004).

Intriguingly, apathy in other patient populations also points to dopaminergic dysfunction. For example, although AD is not typically considered a disorder of dopamine, imaging studies have shown significantly decreased D2 receptor density and decreased dopamine reuptake. These findings are most pronounced in structures associated with the nigrostriatal and mesocorticolimbic tracts of AD patients, most notably the striatum (Mitchell et al., 2011). In addition, single-photon emission computed tomography (SPECT) studies in AD have found that apathy correlates with decreased activity in the ACC, and this relationship is independent of cognitive impairment (Craig et al., 1996; Migneco et al., 2001; Robert et al., 2006).

To summarize, data on human apathy are consistent with animal findings implicating central dopaminergic systems in the development of motivational deficits. Disrupting dopaminergic transmission within the mesocorticolimbic circuit is important in modulating reward- and effort-based decisions, which are important components in the pathogenesis of the amotivated, apathetic state (Bardgett et al., 2009; Chelonis et al., 2011; Krack et al., 2003; Ostlund et al., 2011; Salamone et al., 2007; Treadway et al., 2012).

3 DOPAMINE IN TREATING APATHETIC BEHAVIOR IN ANIMALS

The majority of studies implicating dopamine in animal models of motivation are based on dopamine antagonism, either through systemic administration of a dopamine antagonist or selective targeting of striatal or prefrontal structures. Surprisingly, however, relatively less work has been conducted on the effect of dopaminergic augmentation on motivation (Bardgett et al., 2009; Floresco et al.,

2008). Nevertheless, existing data suggest that stimulating the dopaminergic system, either nonselectively or with D1–D3 receptor agonists, can reverse experimentally induced deficits in reward and effort sensitivity.

3.1 NONSPECIFIC EFFECTS OF DOPAMINE

In rodent models of effort-based decision making, dopamine transmission is usually augmented by parenteral administration of D-*amphetamine*, an indirect dopamine agonist that increases synaptic dopaminergic levels. Using a T-maze procedure, animals in one study were required to choose between one arm offering a high reward for high amounts of effort, and another arm offering a low reward for less effort (Bardgett et al., 2009). Rodents that were rendered less motivated by the administration of D1 or D2 receptor antagonists shifted their preferences toward the low-effort/low-reward arm, but, importantly, D-amphetamine had the effect of restoring preferences for the higher effort offer (Bardgett et al., 2009). One limitation with D-amphetamine, however, is that it increases locomotor activity, and it is possible that restored preferences for the high effort arm could have been due to greater physical capacity (Salomon et al., 2006).

Apart from D-amphetamine, *bupropion* has been studied for its effects on inhibiting catecholamine and dopamine reuptake (Dwoskin et al., 2006). Bupropion is a drug commonly used in humans as an antidepressant. In rodents, the effect of bupropion has been tested with T-maze procedures and operant (fixed or progressive ratio) tasks. The administration of bupropion has been consistently found to increase preference for the high-effort/high-reward options, both in otherwise healthy rodents (Randall et al., 2015) and in those that develop effort-related impairments induced by tetrabenazine—a vesicular monoamine transporter (VMAT)-2 inhibitor that acts as a dopamine-depleting agent (Nunes et al., 2013; Randall et al., 2014; Yohn et al., 2015). However, D-amphetamine and bupropion are both relatively nonselective, and they increase levels of neurotransmitters other than dopamine (eg, serotonin and noradrenaline in the case of D-amphetamine).

3.2 RECEPTOR-SPECIFIC EFFECTS

Rather than nonspecifically raise plasma dopamine levels, and to control for the effects on other neurotransmitter systems, a more refined approach has been to target dopamine receptors with agonists that are selective to one or more dopamine receptor subtypes. Currently, there are five known dopamine receptors (D1–D5), which are classified as "D1-like" (D1, D5), or "D2-like" (D2–D4) according to their cellular transduction properties (Civelli, 1995). The distributions of these receptors differ considerably and are thought to reflect their differential roles in motor, cognitive, and limbic functions (Beaulieu and Gainetdinov, 2011; Bentivoglio and Morelli, 2005; Guillin et al., 2001; Weiner et al., 1991). D1 and D2 receptors are densely distributed within the frontotemporal cortices, limbic system, and striatum. The D3 receptor appears strategically distributed within the mesolimbic system, specifically the ventral striatum (in particular, the shell of the NAcc), midbrain, and pallidum. In contrast, the density of D4 and D5 receptors is much more limited, and their functions in the context of motivational processes remain less well defined (Beaulieu and Gainetdinov, 2011; Meador-Woodruff, 1994).

Given their distribution, D1–D3 receptors are thought to play important roles in regulating affective, reward-related, and motivational processes (Basso et al., 2005; de la Mora et al., 2010; Katz et al., 2006; Newman et al., 2012; Paolo and Galistu, 2012; Short et al., 2006; Sokoloff et al., 2006). For example, using an effort-based decision-making task, a recent study compared the efficacy of selective *D1 agonists* (SKF38393, SKF81297, and A77636) on reversing the effects of ecopipam, a selective D1/D5 receptor antagonist (Yohn et al., 2015a). Each of the D1 agonists administered significantly attenuated the effects of ecopipam, resulting in a shift in animals' preference toward exerting higher effort for higher reward vs exerting less effort for low reward.

Another approach to examine receptor specificity has been to overexpress dopamine D2 receptors, which has led to animals shifting their preference toward higher effort options in effort-based tasks (Trifilieff et al., 2013). In addition, *adenosine A2A antagonists* have been used to investigate motivation in animals based on their functional interaction with dopamine D2 receptors. Adenosine A2A receptors are primarily located in striatal areas, including the neostriatum and NAcc, and specifically reverse the effects of D2 antagonism. Although this interaction has traditionally been used to investigate motor functions related to parkinsonism, it has recently been discovered that adenosine A2A antagonists also affect motivated behavior. Specifically, they reverse the preference shift caused by D2 antagonism in rodents tested on both operant and T-maze choice procedures (Farrar et al., 2010; Mott et al., 2009; Nunes et al., 2010; Pardo et al., 2012; Salamone et al., 2009; Worden et al., 2009). These results implicate D2 receptors in the regulation of motivated behavior.

Following the discovery of D3 receptors, their relatively restricted distribution drew attention to their potential role in reward, particularly in the context of drug addiction (Newman et al., 2012). Indeed, the D3 receptor has been extensively investigated as a potential target to treat substance use disorders (the "D3 Receptor Hypothesis"). In addition to their important role in reward, more recent reports have uncovered their important contribution to effort-based decision making. For example, one study used a progressive ratio schedule to test the relative contributions of D1–D3 receptor stimulation, following dopaminergic cell loss in the substantia nigra pars compacta (SNc; Carnicella et al., 2014). The authors found that only the D3 agonist (PD-128907), but neither the D1 (SKF-38393) nor D2 (sumanirole) agonists, reversed the motivational deficits induced by the SNc dopaminergic lesions. Such effects are not universal (eg, Bardgett et al., 2009). Overall, however, D3 receptors seem to play an important role in the control of motivated behavior, and mediating the beneficial effects of dopamine agonists on the behavioral alterations induced by dopaminergic cell loss. This has led some to propose the D3 receptor as a specific therapeutic target for neuropsychiatric symptoms in several disorders (Sokoloff et al., 2006), including PD (Joyce, 2001; Leentjens et al., 2009).

Taken together, this body of data suggests that dopamine is capable of augmenting motivated behavior in animals, although the distinct role of specific receptor subtypes to this process remains to be further elaborated. Given the causal role that dopaminergic depletion appears to play in altering the animal's sensitivity to reward and effort, it seems intuitive that dopamine supplementation could be used to improve motivational impairments in humans. Next, we review the attempts that have been undertaken in humans to improve apathy by administering exogenous dopamine.

4 DOPAMINE IN THE TREATMENT OF HUMAN APATHY

The treatment of apathy currently lacks consensus, and the choice of pharmacotherapy is based principally on the primary disease. In this context, it is unsurprising that most studies examining the efficacy of dopamine for treatment of apathy come from patients with PD (Table 3; Leentjens et al., 2009). In contrast, dopaminergic drugs have rarely been trialed in conditions such as AD, in which anticholinesterase

Table 3 Dopamine Agonists Commonly Used for the Treatment of Parkinson's Disease

Generic Name	Trade Names	Dopamine Receptor Specificity[a]	Other Receptors
Ergoline derivatives			
Bromocriptine	Parlodel™, Cycloset™	**D2** > D3 (>D4 > D5 > D1)	5HT, α_1, α_2, β_1, β_2
Cabergoline	Caberlin™, Cabaser™	**D2** > D3 (>D5 > D4 > D1)	5HT, α_1, α_2
Pergolide	Permax™, Prascend™	**D2** > D1	5HT
Nonergoline derivatives			
Pramipexole	Sifrol™, Mirapex™, Mirapexin™	**D3** > D2 > D4	5-HT, α_2
Piribedil	Pronoran™, Trivastal™, Trastal™, Trivastan™, Clarium™	**D2, D3**	α_2
Ropinirole	Requip™, Repreve™, Ronirol™, Adartrel™	**D2**, D3, D4	Weak: 5-HT_2, α_2
Rotigotine	Neupro™	**D3** > **D4** > **D5** > **D2** > D1	5-HT, α_1, α_2, β_1, β_2, H_1
Other (antiviral)			
Amantadine	Symmetrel™	Poorly understood. Increases DA release; blocks DA reuptake	NMDA antagonist

[a]*Bold indicates greatest affinity.*

inhibitors are the mainstay of treatment (Berman et al., 2012). Similarly, in schizophrenia, antipsychotics are the primary class of drug used to treat apathy, even though the benefits of dopamine agonists on the negative symptoms of schizophrenia have long been recognized (Benkert et al., 1995; Bodkin et al., 2005; Jaskiw and Popli, 2004; Lindenmayer et al., 2013).

Although there are reports of dopamine being used for the treatment of apathy in disorders other than PD (such as stroke, traumatic brain injury, and depression), a significant gap in this literature is the lack of strong evidence in favor of this application (ie, Class I or II Evidence). The majority of reports involve small cohorts of individuals, are open label, and/or have not used apathy as a primary outcome measure. A likely reason for this is the underrecognition of apathy as a problem, and the difficulty in recruiting apathetic individuals for such studies. In addition, the vast majority of studies that attempt to monitor responses to treatment use one or more questionnaire-based tools, which lack the sensitivity to measure more objective metrics of motivation, such as break points or indifference points (Chong et al., 2016). As such, the effect of dopamine on specific components of apathy, such as reward or effort sensitivity, has remained poorly explored.

4.1 NONSELECTIVE DOPAMINE AUGMENTATION IN APATHY

The most direct, and least specific, method of augmenting the concentration of dopamine in humans is to administer *levodopa*—the precursor molecule of dopamine, and the mainstay of treatment for the motor symptoms of PD. One of the earliest studies to show an improvement in apathy on levodopa was conducted in 23 nondemented, nondepressed patients with PD (Czernecki et al., 2002). The main conclusion of this study was that patients were less apathetic when ON medication relative to OFF (mean daily dose 1115 mg), as measured using the Starkstein Apathy Scale.

Alternatively, presynaptic concentrations of dopamine can be increased by inhibiting its metabolism. *Monoamine oxidase-B (MAO-B) inhibitors*, such as selegiline and rasagiline, selectively target the predominant isoform of the MAO enzyme involved in the metabolic breakdown of dopamine in the brain (Fernandez and Chen, 2007). Although most often used in the treatment of PD, they have more recently been used in depression as well. A recent retrospective review of 181 patients with PD found that patients on selegiline or rasagiline were less likely to report apathy than those taking other antiparkinsonian agents (Zahodne et al., 2014). This complements other, much smaller, case series ($n < 5$), suggesting the utility of selegiline in stroke and traumatic brain injury, which came to similar conclusions (Marin et al., 1995; Newburn and Newburn, 2005).

Amantadine has been used to stimulate the release of dopamine and delay dopamine reuptake. However, its precise mechanism of action is not entirely clear, as it also has effects on glutamate, and is a potent NMDA receptor antagonist (Aoki and Sitar, 1988). Most reports of a beneficial effect of amantadine on apathy have involved small cohorts ($n < 6$) and mostly on patients with traumatic brain injury (Kraus and Maki, 1997; van Reekum et al., 1995).

The primary clinical application of levodopa, MAO-B inhibitors, and amantadine is in the treatment of the motor symptoms of PD, but their potent dopaminergic effects render them useful in off-label trials in managing apathetic symptoms. In addition to these drugs, other classes of medication have been trialed which also have potent dopaminergic effects, even though they are not principally utilized for these properties. For example, *methylphenidate* is a stimulant chemically related to amphetamine, which stimulates dopamine release (Seeman and Madras, 2002), and some studies have reported improvement of apathetic symptoms on this drug in AD (Herrmann et al., 2008; Padala et al., 2010). Similarly, *bupropion* (Wellbutrin™) is a catecholamine reuptake inhibitor most commonly prescribed as an antidepressant (Dwoskin et al., 2006). In animals, it significantly shifts preferences toward more effortful, more rewarding offers (see Section 3.1; Randall et al., 2015), and there is a suggestion in humans that it improved apathy in a small case series of patients with depression or organic brain disease, although it was unclear whether this was due to changes in depressive scores (Corcoran et al., 2004).

4.2 RECEPTOR-SPECIFIC DOPAMINE AGONISTS

The beneficial effects of levodopa on apathy may not be exclusively caused by the restoration of function to dopaminergic projections, as levodopa uptake and decarboxylation also occur, for example, in serotonergic neurons (Ng et al., 1971). Therefore, investigators have turned to more selective dopaminergic agonists to isolate the effect on postsynaptic dopamine receptors (Reichmann et al., 2006). One study examined the effect of a single dose of a highly selective D1 agonist (*dihydrexidine*, DAR-0100) on the negative symptoms of schizophrenia (George et al., 2007). This investigation failed to find any significant effects, but given the single dose and the absence of apathy as a primary end point, the utility of sustained D1 agonism specifically on apathy remains unknown. A more commonly encountered drug is *bromocriptine*, an ergot derivative dopamine agonist, and one of the earliest dopamine agonists to be used in the treatment of PD. It acts primarily on the D2 receptor, but is active at all receptor subtypes. Early studies on the use of bromocriptine in apathy were equivocal, and often involved patients concurrently taking other drugs, such as methylphenidate (Marin et al., 1995) or levodopa/benserazide (Debette et al., 2002).

More recently, nonergoline dopaminergic agonists have been developed which are in more common use as treatments for PD. Following the discovery of D3, and later D4 and D5 receptors, attention was drawn to the relatively restricted location of D3 receptors, seemingly related to dopaminergic functions associated with the mesolimbic system (see Section 3.2). Most modern nonergot dopamine agonists predominantly target the D2 and/or D3 receptors (Table 3). For example, *pramipexole* binds preferentially, and with high affinity, to the D3 receptor (Guttman and Jaskolka, 2001), although it also has agonist activity at pre- and postsynaptic receptors belonging to other receptors in the D2-like family (Piercey et al., 1996). *Piribedil* and *ropinirole* are both relatively selective D2/D3 agonists, which do not interfere with the serotonergic system. All of these agents have been reported to have some success in ameliorating apathy in PD (Czernecki et al., 2008; Oguro et al., 2014; Rektorova et al., 2008; Thobois et al., 2013) as well as in stroke (Kohno et al., 2010). In some of these studies, improvements in apathy might be difficult to disambiguate from accompanying improvement in mood, although they appear not to be correlated (Czernecki et al., 2008).

An informative study was recently conducted with the aim of performing a head-to-head comparison of the neuropsychiatric effects of levodopa, pramipexole, and ropinirole in PD (Pérez-Pérez et al., 2015). This was a large study of 515 nondemented patients, with apathy being one of several outcome measures assessed with the Neuropsychiatric Inventory. The overall conclusion was that both the frequency and severity of apathetic symptoms was less with pramipexole than either ropinirole or levodopa. This may be parsimonious evidence for the efficacy of D3 receptor agonists in the treatment of apathy.

4.3 DISSECTING THE EFFECT OF DOPAMINE ON OBJECTIVE METRICS OF MOTIVATION

In considering the preceding attempts at treating human apathy, an obvious feature of these studies is their heterogeneity, with several classes of dopaminergic drugs having been utilized across a range of disorders with varying efficacy. One of the limitations in understanding the role of dopamine in treating apathy is that apathy appears not to be a singular construct, but comprised of different elements, such as reward and effort sensitivity. However, current questionnaire-based methods are inherently limited in their ability to dissect the mechanisms of disordered motivation, and insufficiently sensitive to quantify or monitor any changes to effort- or reward-based decision making following treatment (Chong et al., 2016). Here, we review recent attempts at quantifying the effects of dopaminergic medication on different components of apathetic behavior.

4.3.1 Effects of dopamine on reward sensitivity in apathy

Based on animal data, one component of motivation appears to be impaired reward sensitivity. Consistent with this suggestion was a recent case study we recently reported on a patient (KD) who developed profound apathy following a rare, bilateral stroke affecting the globus pallidus, predominantly its internal components (GPi; Fig. 2A; Adam et al., 2012). Probabilistic diffusion tractography demonstrated that the region of the GPi that was particularly affected was strongly connected to the lateral orbitofrontal cortex and ventromedial prefrontal cortex—two areas which are significantly involved in reward sensitivity. Premorbidly, he was described as exuberant and outgoing, but, after his stroke, he became reticent and reserved. He became disinterested in others, had reduced spontaneity of thought and action, and lost his job. His scores on the Apathy Inventory were in the pathological range on the initiative and interest subscales (8/12, normal $\leq$4) (Robert et al., 2002). Importantly, however, he was not depressed, as reflected in his scores on several depression inventories, which were within the normal range (the Montgomery–Asberg

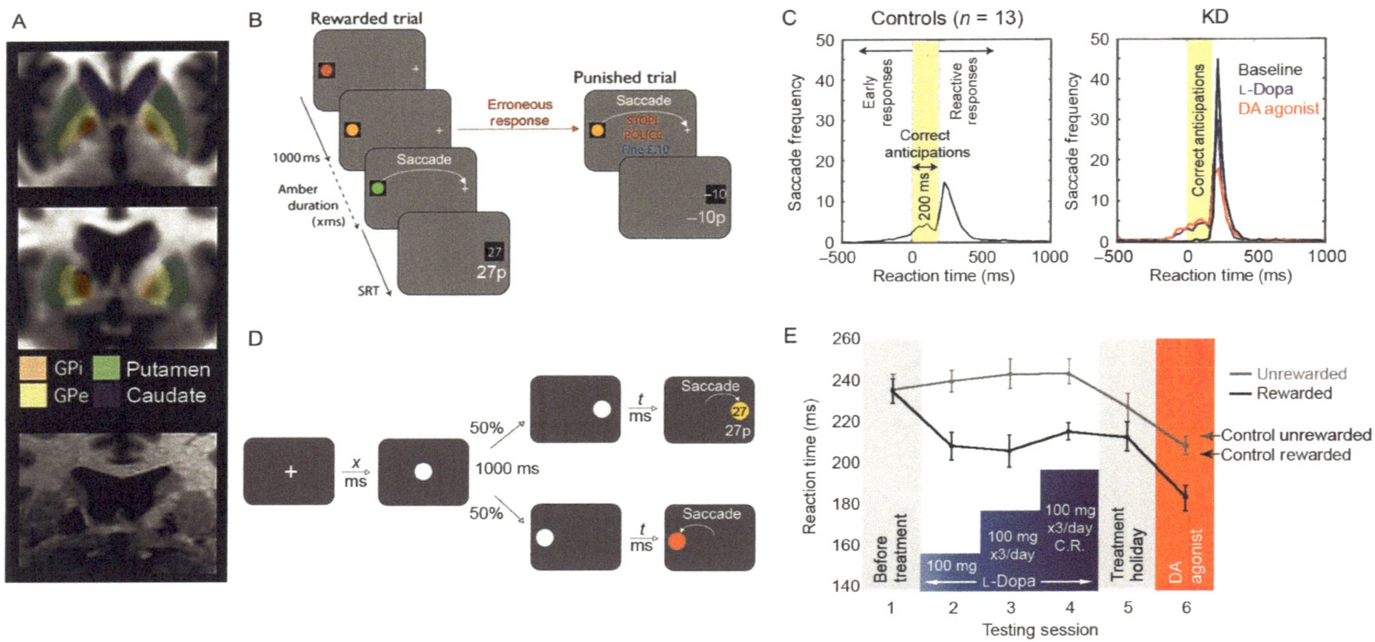

FIG. 2

See legend on opposite page.

FIG. 2

We examined the effects of dopamine on a patient (KD) with apathy caused by selective, bilateral lesions to the globus pallidus (Adam et al., 2012). (A) Sections demonstrating the extent of basal ganglia lesions. KD's GPi lesion was larger on the *left* than on the *right*. The lesions are projected onto boundaries of the GPi (*orange*), GPe (*yellow*), putamen (*green*), and caudate (*purple*). The *bottom left* coronal section is a close up at the level of the anterior commissure. (B) KD participated in two tasks examining reward sensitivity. In the traffic lights task (TLT), participants fixated a *circle* which successively turned *red*, *amber*, and *green*. They were required not to move their eyes until the onset of the green light; otherwise they receive a small, fixed penalty. To maximize reward, participants had to make a saccade to the contralateral target as quickly as possible after green light onset. Amber durations (x) were selected at random from a normal distribution. Reward was calculated with a hyperbolically decaying function with a maximum value of 150 pence (£1.50) at $t=0$. Thus to maximize reward subjects should program an eye movement to coincide with green light onset. However, amber durations were not constant and therefore they either had to take a risk (high reward or punishment) or wait for the green light before programming a saccade (low reward). (C) Traffic lights task (TLT): saccadic distributions. (A) Saccades for age-matched controls ($n=13$) performing the TLT showed two distinct distributions: an early, anticipatory distribution, and a later, reactive one made in response to green light onset. Early responses were divided into errors (saccades before the green light came on) and correct anticipations (saccades with <200 ms latency after the green light). Pretreatment, KD made mostly reactive saccades, and very few anticipatory saccades (*black*). After treatment with L-DOPA 100 mg (Madopar CR 125 mg) three times a day for 12 weeks, there was a dramatic increase in early responding in KD (*blue*). After 12 weeks treatment with a dopamine agonist (ropinirole XL, 4 mg once a day), KD's distribution of saccades looks most similar to that of control subjects (*red*). (D) In the directional saccadic reward task, participants attended a central fixation spot which was extinguished after 1000 ms. They then made a speeded saccade to a target to the *left* or *right* of fixation (equiprobable). One side was rewarded while the other received no reward. The rewarded side (RS) remained constant for an unpredictable number of trials before switching to the other side. (E) Results from the directional saccadic reward task. The control group ($n=12$, *arrows* to *side*) showed a preference for the rewarded target locations, with significantly shorter SRTs. KD showed no reward preference before treatment (Session 1). In Session 2, he was given a single dose (100 mg) of levodopa which led to a significant reward preference. This was maintained throughout chronic dopaminergic therapy (Sessions 3 Madopar 125 mg three times daily for 4 weeks, Session 4 Madopar Controlled Release 125 mg three times daily for 12 weeks). Following a treatment holiday (4 weeks), this reward preference was absent (Session 5). However, with subsequent treatment on the dopamine agonist ropinirole (1 mg three times a day), there was both a reestablishment of reward preference and significant decrease in latency to both rewarded and unrewarded targets. Error bars are ± 1 SEM (standard error of the mean).

Adapted from Adam, R., Leff, A., Sinha, N., Turner, C., Bays, P., Draganski, B., Husain, M., 2012. Dopamine reverses reward insensitivity in apathy following globus pallidus lesions. Cortex 49, 1292–1303.

Depression Rating Scale (Montgomery and Asberg, 1979), the Beck Depression Inventory (Beck et al., 1988), and the Hamilton rating scale for depression (Hamilton, 1960)).

KD's apathy was reflected in his performance on two oculomotor measures of motivation, which were specifically designed to probe reward sensitivity. In one task, the "Traffic Lights Task," KD fixated on a disc at the left or right of the screen, which successively turned red, amber, and green (Fig. 2B). The instant the disc turned green, he was required to make a speeded saccade to a target location on the opposite side of the screen. The faster the saccadic initiation time, the more he was rewarded up to a maximum of £1.50, according to an exponential falloff. Any preemptive saccades initiated prior to the onset of the green disc were penalized by a fixed, small amount (10p). In healthy participants, the distribution of reaction times is bimodal—although most responses are "reactive" and follow the onset of the green disc, a second peak of responses were due to anticipatory responses to the green disc. Up to 45% of responses in healthy controls were such "anticipatory" responses. In contrast, KD showed a unimodal response, with few attempts at initiating early saccades to maximize reward (<10%) (Fig. 2C).

The second task that KD performed was a directional reward-sensitive saccade task (Fig. 2D). This task required him to fixate a central cross and perform speeded saccades to targets to the left or right of fixation. The target locations were equiprobable, but only targets on one side were rewarded as a function of reaction time (with the equivalent exponentially decaying function as in the traffic lights task). The rewarded side was altered, without warning, every 10–14 trials. Reward sensitivity was measured as the difference in saccade reaction times to the rewarded and unrewarded sides. Controls showed a small, but significant, saccade reaction time advantage to the rewarded side. In contrast, however, KD showed no directional difference.

The decision was made to trial KD on dopamine supplementation with levodopa/benserazide (100/25 mg, Madopar™). He undertook both oculomotor tasks immediately prior to commencing his first dose and then 1 h after the administration of his first dose. Strikingly, after only one dose, he showed a significant improvement in his performance on both tasks. On the traffic lights task, he showed a restoration of the normal bimodal distribution seen in healthy controls (Fig. 2C). Similarly, on the directional reward-sensitivity task, KD showed a markedly significant preference for the rewarded side compared to the unrewarded side (211 vs 238 ms; Fig. 2E). Not only were these changes manifest only 1 h following his first dose, but his improvements in both tasks were sustained and continued over the following months while on medication—the proportion of early "anticipatory" responses in the traffic lights task reached a peak at 24 weeks (33.4%), and the advantage of the rewarded side increased in the directional task over the following 12 weeks.

The causal role of levodopa in ameliorating KD's reward sensitivity was demonstrated following a clinical decision to stop the levodopa, and switch him to the dopamine agonist, ropinirole. During the intervening "drug holiday" while KD was off medication, his performance on both oculomotor tasks returned back to pretreatment levels. The percentage of his anticipatory responses on the traffic lights task again declined back to baseline levels (<10%), and his preference for the rewarded side diminished back to pretreatment levels on the directional reward-sensitive task. However, after he was commenced on ropinirole (4 mg), his performance again improved on both tasks (Fig. 2C and E), to levels that appeared even greater relative to his performance on levodopa.

Importantly, the administration of levodopa/benserazide and ropinirole resulted not only in improved performance on the metrics of reward sensitivity but also in functional outcome. KD's clinical apathy improved when indexed against conventional apathy scales (the Apathy Inventory). He was also able to engage in more spontaneous conversation, had improved social interactions, was more interested in day-to-day events, and even managed to secure a job.

This case study demonstrates several points. First, it illustrates the utility of paradigms that can dissect a specific component of apathy—in this case, reward sensitivity—which can then be used as a proxy to measure motivation. Second, it is proof in principle for a strong causal relationship for dopamine in reversing reward insensitivity in a human model of apathy. Third, the restoration of KD's reward sensitivity correlated with clinical and functional improvements, as measured on traditional questionnaire-based measures, suggesting that reward sensitivity is an important component of apathy. Finally, it implies that selective dopamine agonist therapy (in this case with ropinirole) may be advantageous over less-selective dopamine supplementation (with levodopa), which suggests that future research should seek to clarify the differential role of dopamine receptors in the treatment of apathy.

4.3.2 Effects of dopamine on subclinical reward insensitivity

Although apathy represents the clinical manifestation of impaired motivational processes, it is unlikely to be an all-or-nothing phenomenon. Rather, early dysfunction in reward mechanisms may give rise to subtle impairments in motivation, which do not become clinically evident until they disrupt day-to-day function. Thus, another approach to determining the potential role of dopamine in treating apathy is to examine how it modulates subclinical changes in motivation. Indeed, given the evidence showing that lesions to dopaminergic pathways reduce reward sensitivity, one prediction is that all patients with PD should demonstrate varying levels of motivational deficits. However, subtle levels of motivational impairments might be ineffectively captured based on current self-report-based tools.

We have recently developed another oculomotor task that aims to probe subclinical changes in reward sensitivity in patients with PD and to examine the potential role of dopamine in ameliorating such deficits (Fig. 3; Manohar and Husain, 2015; Manohar et al., 2015). In this distractor-avoidance task, participants made a speeded saccade to a target location, while ignoring the presence of a distractor immediately preceding that target (Fig. 3A–C). Crucially, participants were provided with a monetary incentive for their performance. Prior to commencing each trial, an auditory precue was delivered to indicate the maximum reward that was available for an accurate saccade to that location (0, 10, 50p). As a measure of reward sensitivity, we measured autonomic arousal in the form of pupillary dilatation, which has the

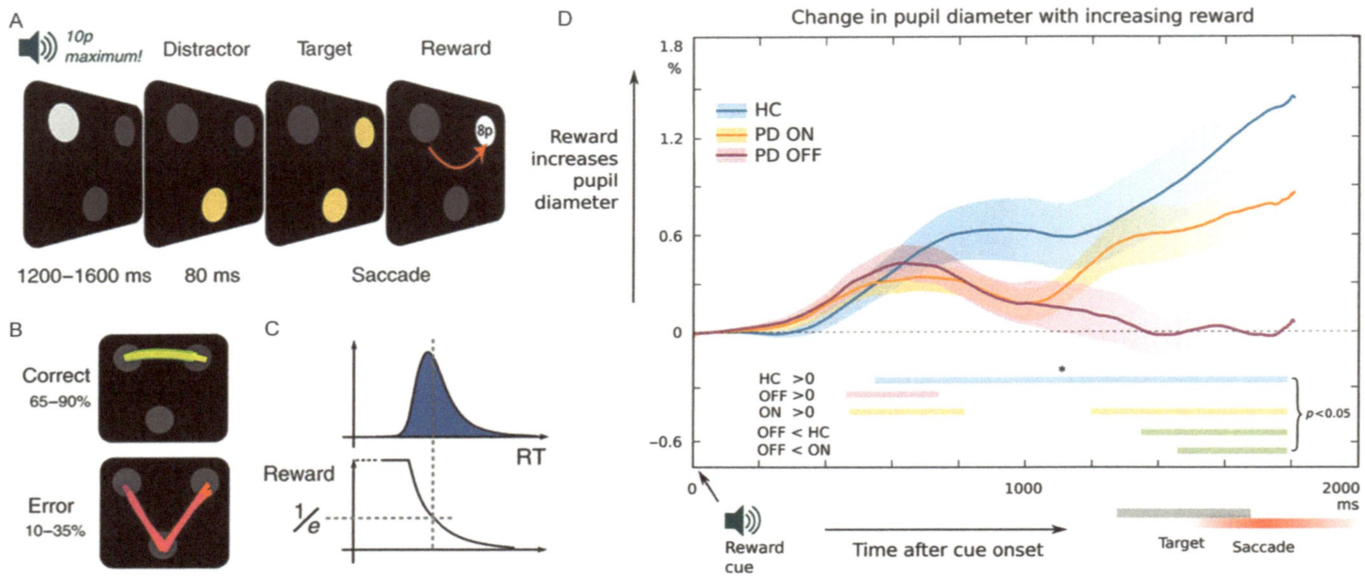

FIG. 3

Using a novel oculomotor reward sensitivity task, we measured autonomic responses to reward cues (Manohar and Husain, 2015). (A) Participants fixated an illuminated disc, and received an auditory cue indicating how much reward could be won by making a speeded eye movement (0, 10, 50p). After a variable delay, a saccade had to be made to the second of two other discs that illuminated, one slightly later than the other. (B) Correct saccades were those that went directly to the target, whereas on error trials an initial saccade was made to the distractor. Percentages indicate the range of proportion of correct and error trials over all participants. (C) Reward was numerically displayed at the target, based on speed, and scaled up by the amount on offer on that trial. The value fell off exponentially with increasing response time (measured from distractor onset until gaze arrived at the target), with adaptive time constants that maintained a constant average rate of reward. (D) The effects of reward on pupil size, given by linear regression at each time point. For each participant, the pupil traces were correlated with the incentive on the current trial. The correlation coefficient was plotted as a function of time. Positive values indicate that with higher incentives, the pupil was larger; conversely negative values indicate that reward made the pupil smaller. Comparisons of these coefficients with zero, and with each other, are shown. Reward increased pupil size in all three groups, but controls were significantly more reward sensitive than PD patients when OFF medication (unpaired comparison). Also, PD patients were more reward sensitive when ON compared with when OFF (paired comparison). All statistics are calculated for $p<0.05$ controlling for familywise error using permutation.

Adapted from Manohar, S.G., Husain, M., 2015. Reduced pupillary reward sensitivity in Parkinson's disease. NPJ Parkinson's Dis. 1, 15026.

advantage of being able to disambiguate the effect of dopamine on reward independent from its effects on motor function (Manohar and Husain, 2015).

We tested a group of nondemented, nonapathetic patients with PD over two counterbalanced sessions—ON and OFF their usual dopaminergic medication—and compared their performance to healthy, age-matched controls (Manohar and Husain, 2015). Patients were on either levodopa or a dopamine agonist. As predicted, controls demonstrated greater autonomic arousal in the form of pupillary dilatation to high vs low rewards. In contrast, PD patients OFF medication showed little differential response in their pupillary diameters to increasing reward. Crucially, however, these reward-sensitive pupillary responses were restored toward healthy levels when the identical patients were tested ON their usual medication (Fig. 3D). Together, these findings highlight that the autonomic responses to reward incentives in PD may be blunted, even in nonclinically apathetic patients, and that dopamine is effective in at least partially restoring these deficits.

4.3.3 Effects of dopamine on subclinical effort hypersensitivity

Based on the seminal work in animal studies of motivation, it is clear that motivation can be framed not only as reduced reward sensitivity but also as heightened sensitivity to effort (Chong et al., 2016). In a direct extension of this animal research, several human studies have now demonstrated that dopamine therapy increases the willingness of patients to exert effort for reward (Chong et al., 2015; Porat et al., 2014; Wardle et al., 2011). For example, we recently devised a novel paradigm to investigate the willingness of patients with PD to exert effort for reward, with effort being operationalized as the amount of force delivered to handheld dynamometers (Fig. 4; Chong et al., 2015). Notably none of these patients were clinically apathetic or depressed, as measured using standard clinical questionnaires (the Lille Apathy Rating Scale (Sockeel et al., 2006) and the Depression, Anxiety, and Stress Scales (Lovibond and Lovibond, 1995)).

The task was framed in the form of a game, the goal of which was to gather as many apples as possible from trees in an orchard (Fig. 4A and B). During the experiment, participants were presented with cartoons of apple trees and were instructed to accumulate as many apples as possible based on the combinations of stake and effort that were presented. Potential rewards were indicated by the number of apples on the tree (1, 3, 6, 9, 12, 15). Effort levels were individualized to each participant as a function of their maximum voluntary contraction (MVC) determined at the beginning of each experimental session. Effort requirements varied over six levels, from 60% to 110% MVC, in 10% increments. By referencing the effort levels in each session to each individual's maximum force, we were able to normalize the difficulty of each level across sessions and across individuals.

On each trial, participants had to decide whether they were willing to exert the specified level of effort for the specified stake. If they judged the particular combination of stake and effort to be "not worth it," they selected the "No" response, and the next trial would commence. If, however, they decided to engage in that trial, they selected the "Yes" option, and began delivering the required amount of force for the

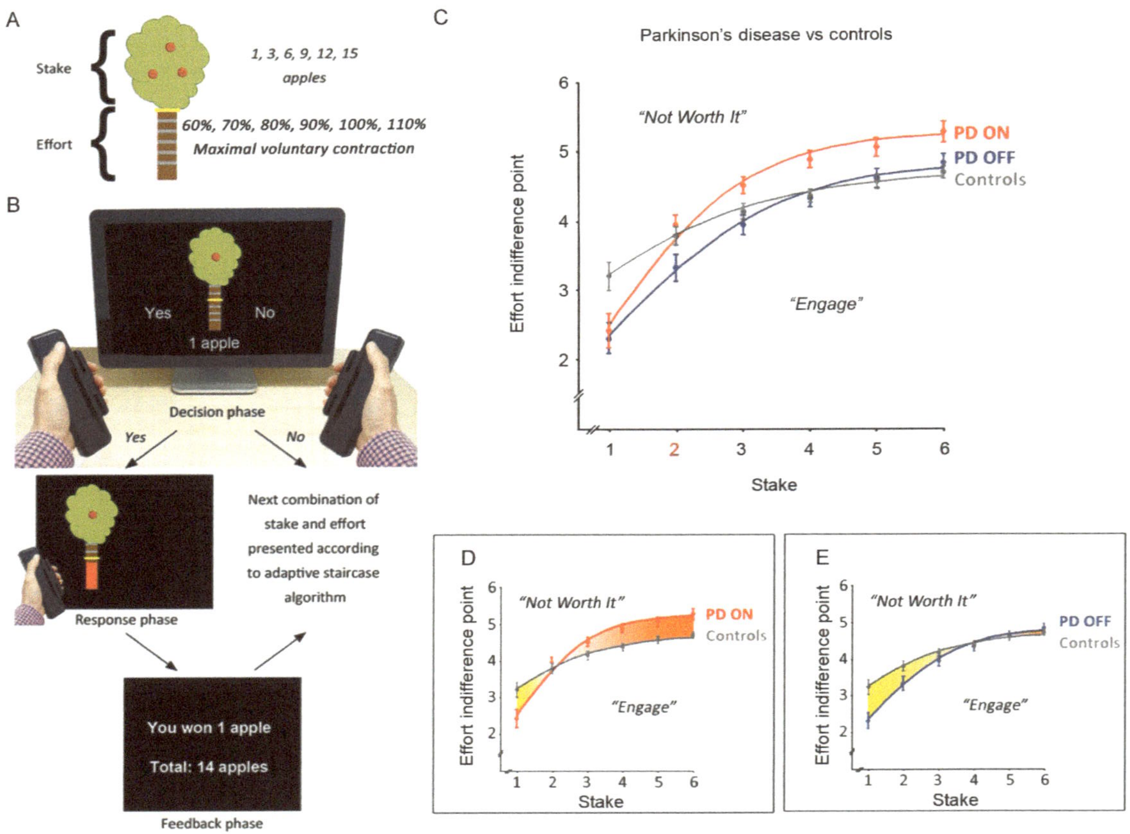

FIG. 4

See legend on opposite page.

apples on offer. By parametrically varying the combinations of effort and reward, and subsequently applying logistic regression techniques, we were able to determine, for each level of reward, the point at which participants accepted and rejected the offer on 50% of occasions—their effort "indifference points" (Bonnelle et al., 2015; Chong et al., 2015). These effort indifference points could then be used as a metric against which to benchmark each patient's motivation.

To determine the effect of dopaminergic medication on the willingness to exert effort for reward, we tested these patients ON and OFF their usual dopaminergic medication (which may have been either levodopa or a dopamine agonist). We found that, regardless of their medication status, patients with PD were willing to exert less effort when the stakes were low (Fig. 4C–E). This implied a degree of subclinical apathy, but only for the lowest rewards, that was not evident on standard clinical questionnaires. Furthermore, as predicted, patients OFF medication were willing to invest less effort for reward, but crucially this was ameliorated by dopamine, which had the effect of increasing the amount of effort that patients were willing to invest. Interestingly, relative to healthy controls, there was a reward-dependent effect, such that, at higher rewards, patients with PD ON dopaminergic medication

FIG. 4

The Apple Gathering Task (Chong et al., 2015). (A) In a typical trial, stakes were indicated by the number of apples on the tree, while the associated effort was indicated by the height of a *yellow bar* positioned at one of six levels on the tree trunk (as proportions of participants' MVCs. (B) On each trial, participants decided whether they were willing to exert the specified level of effort for the specified stake. If they judged the particular combination of stake and effort to be "not worth it," they selected the "No" response. If, however, they decided to engage in that trial, they selected the "Yes" response, and then had to squeeze a handheld dynamometer with a force sufficient to reach the target effort level. Participants received visual feedback of their performance, as indicated by the height of a *red force feedback bar*. To reduce the effect of fatigue, participants were only required to squeeze the dynamometers on 50% of accepted trials. At the conclusion of each trial, participants were provided with feedback on the number of apples gathered. (C) For each participant, we calculated their effort indifference points—the effort level at which the probability of engaging in a trial for a given stake was 50%. Regardless of medication status, patients had significantly lower effort indifference points than controls for the lowest reward. However, for high rewards, effort indifference points were significantly higher for patients when they were ON medication, relative not only to when they were OFF medication, but even compared to healthy controls. Inset: For clarity, PD data are replotted against control performance for patients (D) ON medication and (E) OFF medication. *Shading* denotes effort indifference points being greater for patients than controls (*orange*), or less for patients than controls (*yellow*). Error bars indicate ± 1 SEM.

Adapted from Chong, T.T.-J., Bonnelle, V., Manohar, S., Veromann, K.-R., Muhammed, K., Tofaris, G., Hu, M., Husain, M., 2015. Dopamine enhances willingness to exert effort for reward in Parkinson's disease. Cortex 69, 40–46.

were willing to invest even more effort than their age-matched counterparts. This echoes previous findings in animal studies showing that dopamine augmentation restored motivated behavior.

Other studies using different paradigms have documented similar effects in PD. For example, Porat and colleagues tested nonapathetic patients with PD on a Gain/Loss Effort Task, which is based on the progressive ratio tasks in animals (Chong et al., 2016; Porat et al., 2014). In this task, the authors separately measured the maximum amount of effort that participants are willing to expend to *either* increase monetary gain or avoid/minimize monetary loss. Effort in this paradigm was operationalized as the number of button presses on a keyboard, with the number of presses required to increase gain or avoid loss progressively increased in an exponential progressive ratio schedule.

Interestingly, the authors found a differential effect of dopamine as a function of patients' more affected side. Dopamine did indeed have the effect of increasing patients' willingness to exert effort. However, patients with a more affected right side were more willing to exert effort to maximize gain, whereas those with a more affected left side were more willing to exert effort to avoid loss. This asymmetry might reflect differential hemispheric involvement in PD—previous tracer studies have shown reduced uptake in the nigrostriatal system contralateral to the more affected side (Brooks, 2003; Djaldetti et al., 2006), which is most pronounced in the putamen, but also present in the caudate, ventral striatum, and frontal regions (Jokinen et al., 2009; Marie et al., 1995). These findings raise the suggestion that the effects of dopamine on motivation are sensitive to the nature of the reinforcer (positive or negative), and invite future studies of this distinction.

The effect of dopamine on incentivizing effort-based decisions has also been found in healthy, nonapathetic individuals. For example, the Effort Expenditure for Rewards Task (EEfRT) has been used to examine the effect of D-amphetamine on the willingness of healthy individuals to exert effort for reward (Wardle et al., 2011). This task, inspired by the T-maze tasks in rodents (Salamone et al., 2007), requires participants to choose between a high-effort/high-reward offer and a low-effort/low-reward option. The high-effort option requires 100 button presses in 21 s with the nondominant fifth digit, whereas the low-effort option requires 30 button presses with the dominant index finger in 7 s. For each successfully completed trial, the low-effort option was worth $1.00, whereas the value of the higher effort option was varied between $1.24 and $4.30. In the original version of the task, there was also a probabilistic component to the task, in which some trials were more likely to result in a payoff than others (12%, 50%, and 88%).

In this task, the proportion of trials in which participants chose the high-effort/high-reward offer was greater when they were on D-amphetamine vs placebo. There appeared to be a dose-dependent effect, such that it was only efficacious at a dose of 20 mg, but not 10 mg, relative to placebo. Further analyses were undertaken using a generalized regression technique (Generalized Estimating Equation modeling), which showed that D-amphetamine increased the willingness of volunteers to exert effort for monetary rewards particularly when reward probability was lower, suggesting a role for increased tolerance for probability costs. Amphetamine sped task performance, but its psychomotor effects did not significantly predict effects on decision making.

Together, although these studies were conducted on individuals without clinical apathy, they demonstrate the utility of dopamine in increasing sensitivity to reward and increasing the willingness of subjects to invest effort. These findings therefore represent proof in principle of the potential utility of dopamine in ameliorating key components of motivated decision making, and therefore apathetic behavior.

5 EXTENDING THIS WORK

5.1 EFFECTS OF DOPAMINE ON METRICS OF MOTIVATION

Salamone and colleagues have long argued that the primary effect of dopamine is to regulate effortful activity, allowing an animal to overcome response costs associated with pursuing valuable stimuli. The recent development of paradigms that are available to objectively quantify and track the progress of motivational disorders should act as an incentive for the development of new treatments and to determine the efficacy of existing drugs. Few of the drugs that have been trialed in the treatment of apathy have been systematically evaluated using sensitive measures of reward- or effort-based decisions, such as those described in the previous section. Based on these paradigms, future trials should focus on examining the effect of dopamine augmentation on modulating specific components of motivated decision making—such as reward or effort sensitivity—in order to relate them more closely to the clinical manifestations of apathy.

More broadly, although altered effort and/or reward sensitivity appears to be a core feature of the apathetic state, they need not be the only features that characterize it. Given the rich history of effort- and reward-based decision making in animals, applying the principles and paradigms developed in the animal literature would seem to be an obvious first step in defining the mechanisms underlying human apathy. However, it is entirely possible that there remain other deficits in decision making or executive function that characterize apathy, but which are yet to be defined. Defining the nature of such deficits would be a useful course for future research, as they may allow us to dissect different subtypes of apathetic behavior, and potentially clarify the distinction between the many terms that have been historically used to describe the phenotype of disordered motivation (Table 1). The dual goal of future research will therefore be to further clarify the role of effort- and reward-based decision making in human apathy, while defining other deficits which may be used to diagnose, monitor, and treat the apathetic state.

5.2 RECEPTOR SPECIFICITY

Clarifying the differential role of specific dopamine receptors to motivation will be integral to future work on developing targeted treatments for human apathy, which aim to maximize the effect on motivational deficits. For example, based on their distribution within the ventral striatum and other limbic regions of the brain, D3 receptors may be a particularly useful target for the treatment of apathy. Our single case study of ropinirole improving reward sensitivity in an individual with profound apathy would be a proof of principle that D2/D3 receptor agonism is a potentially effective treatment, and it would be useful to extend this to a larger cohort of apathetic individuals using similar paradigms.

5.3 TAILORING DOPAMINE TO SPECIFIC POPULATIONS

It remains unclear whether apathy in many psychiatric and neurological conditions in which it is encountered (eg, AD vs schizophrenia vs PD) represents the same phenotypic manifestation of the identical underlying pathology, or whether the motivational deficits in each of these conditions are subtly different. Such an issue would be important to clarify with more sensitive measures of motivation, as it would dictate the specific therapy that is used in treating these diseases, and in determining at what stage of each disease therapy should be initiated.

At present, treatment options defer to the primary illness. Thus, in PD, dopamine is a parsimonious treatment to manage both the apathetic symptoms as well as the motor manifestations. However, a much more complex management problem is posed by patients with schizophrenia, in which dopamine may potentially worsen psychotic symptoms (Lieberman et al., 1987). Some have proposed that increasing dopamine levels in schizophrenia in conjunction with concurrent dopamine D2 antagonism attenuates this risk, although such reports are largely anecdotal, and no controlled trials have been conducted to verify this (Angrist et al., 1982; Jaskiw and Popli, 2004; Levi-Minzi et al., 1991; Ohmori et al., 1993; Roesch-Ely et al., 2006; van Kammen and Boronow, 1988). In such a situation, clarifying the role of specific receptor subtypes to the pathology of apathy as well as in the pathology of the primary condition is imperative.

In addition to determining or developing the specific drugs to treat apathy, it is also important to clarify the dose-dependent relationship between dopamine and motivation in individual subjects. Several studies suggest that dopamine follows an inverted-U-shaped function, such that there exists an optimal point at which dopamine mediates particular cognitive functions (Cools and D'Esposito, 2011). Administering doses of dopaminergic medication in excess of the optimum may push apathetic individuals to the other end of the motivational spectrum, and potentially result in impulse control disorders commonly encountered in patients on dopamine agonists (Voon et al., 2009). The optimum dose of dopamine replacement in apathy is likely to vary across patients as a complex function of individual factors, such as genetically determined pharmacokinetic and pharmacodynamic effects. An important goal of future work will therefore be to develop methods capable of determining the dose of dopamine therapy that delivers the maximum therapeutic efficacy at the lowest tolerable doses for individual subjects. This is an important clinical issue, given that apathy is common in elderly patients (such as those with dementia), who in general are less tolerant of high doses of medication (Chong and D'Souza, 2013).

5.4 NONPHARMACOLOGICAL MEANS OF INCREASING DOPAMINE

In addition to pharmacological methods of increasing dopamine, several studies have suggested innovative ways to increase dopamine concentrations, involving noninvasive stimulation of the primary motor cortex. For example, transcranial direct current stimulation (tDCS) in rats has shown a 60% increase in dopamine concentration in the ipsilateral striatum (Tanaka et al., 2013), and a recent study in patients with PD has showed that bilateral tDCS over the primary motor cortex (with the cathode placed over the more affected side and anode over the less affected side) resulted in less subjective effort in a manual isometric force production task (Salimpour et al., 2015). In addition, transcranial magnetic stimulation (TMS) over primary motor cortex in PD has been shown to increase serum dopamine levels and result in improved motor performance (Khedr et al., 2007), and stimulation of the primary and supplementary motor areas in healthy individuals has reduced the subjective sense of physical effort (Chong, 2015; Takarada et al., 2014; Zénon et al., 2015). These suggest novel, nonpharmacological ways of increasing dopamine levels to reduce the subjective sense of effort, which may in turn aid those with clinical apathy.

6 CONCLUSION

The development of safe and effective therapies for apathy constitutes a pressing, unmet need. A rational approach to this goal is informed by the study of the components, circuitry and pharmacology of motivated behavior in human and nonhuman animals. Dopamine represents a useful and rational target for the treatment of apathetic symptoms across a wide range of psychiatric and neurological disorders.

The accelerating pace of basic and clinical neuroscience research promises to improve our understanding of apathy and its treatment with dopaminergic medication. Using the paradigms at our disposal, future research should focus on identifying the specific neural circuitry mediating the motivational effects of dopamine agonists and should employ tests of reward- and effort-based decision making to evaluate the utility of specific agonists for the treatment of apathy. Furthermore, by dissecting the phenomenon of motivation into its components (eg, reward vs effort sensitivity), it may be possible to refine targeted treatments tailored to individual populations, as a function of their major apathetic deficit.

Despite the promise of dopaminergic treatments of apathy, further large-scale, controlled clinical trials of potentially useful pharmacologic interventions are essential before any firm recommendations can be made. The growing body of empirical investigations on the neurobiology of apathy will likely prove helpful in providing a sound theoretical basis for the application of currently available treatments, as well as for the development of novel therapeutic interventions, that will ultimately allow us to determine which drugs to administer, and at what doses, for individual subjects to improve their objective deficits in motivated behavior.

ACKNOWLEDGMENTS

T.T.-J..C. is funded by the National Health and Medical Research Council (NH&MRC) of Australia (1053226). M.H. is funded by a grant from the Wellcome Trust (098282).

REFERENCES

Aarsland, D., Cummings, J.L., Larsen, J.P., 2001. Neuropsychiatric differences between Parkinson's disease with dementia and Alzheimer's disease. Int. J. Geriatr. Psychiatry 16, 184–191.

Aarsland, D., Brønnick, K., Ehrt, U., De Deyn, P.P., Tekin, S., Emre, M., Cummings, J.L., 2007. Neuropsychiatric symptoms in patients with Parkinson's disease and dementia: frequency, profile and associated care giver stress. J. Neurol. Neurosurg. Psychiatry 78, 36–42.

Aarsland, D., Marsh, L., Schrag, A., 2009. Neuropsychiatric symptoms in Parkinson's disease. Mov. Disord. 24, 2175–2186.

Adam, R., Leff, A., Sinha, N., Turner, C., Bays, P., Draganski, B., Husain, M., 2012. Dopamine reverses reward insensitivity in apathy following globus pallidus lesions. Cortex 49, 1292–1303.

Andersson, S., Krogstad, J., Finset, A., 1999. Apathy and depressed mood in acquired brain damage: relationship to lesion localization and psychophysiological reactivity. Psychol. Med. 29, 447–456.

Andreasen, N., 1984. Scale for the Assessment of Negative Symptoms (SANS). College of Medicine, University of Iowa, Iowa City.

Angrist, B., Peselow, E., Rubinstein, M., Corwin, J., Rotrosen, J., 1982. Partial improvement in negative schizophrenic symptoms after amphetamine. Psychopharmacology (Berl.) 78, 128–130.

Aoki, F.Y., Sitar, D.S., 1988. Clinical pharmacokinetics of amantadine hydrochloride. Clin. Pharmacokinet. 14, 35–51.

Barch, D.M., Dowd, E.C., 2010. Goal representations and motivational drive in schizophrenia: the role of prefrontal-striatal interactions. Schizophr. Bull. 36, 919–934.

Bardgett, M., Depenbrock, M., Downs, N., Points, M., Green, L., 2009. Dopamine modulates effort-based decision-making in rats. Behav. Neurosci. 123, 242.

Basso, A.M., Gallagher, K.B., Bratcher, N.A., Brioni, J.D., Moreland, R.B., Hsieh, G.C., Drescher, K., Fox, G.B., Decker, M.W., Rueter, L.E., 2005. Antidepressant-like effect of D2/3 receptor-, but not D4 receptor-activation in the rat forced swim test. Neuropsychopharmacology 30, 1257–1268.

Beaulieu, J., Gainetdinov, R., 2011. The physiology, signaling, and pharmacology of dopamine receptors. Pharmacol. Rev. 63, 182–217.

Beck, A., Steer, R., Garbin, M., 1988. Psychometric properties of the Beck Depression Inventory—25 years of evaluation. Clin. Psychol. Rev. 8, 77–100.

Benkert, O., Muller-Siecheneder, F., Wetzel, H., 1995. Dopamine agonists in schizophrenia: a review. Eur. Neuropsychopharmacol. 5 (Suppl.), 43–53.

Benoit, M., Dygai, I., Migneco, O., Robert, P.H., Bertogliati, C., Darcourt, J., Benoliel, J., Aubin-Brunet, V., Pringuey, D., 1999. Behavioral and psychological symptoms in Alzheimer's disease. Dement. Geriatr. Cogn. Disord. 10, 511–517.

Bentivoglio, M., Morelli, M., 2005. The organization and circuits of mesencephalic dopaminergic neurons and the distribution of dopamine receptors in the brain. Handbook of Chemical Neuroanatomy, vol. 21. Elsevier, Amsterdam, pp. 1–107.

Berman, K., Brodaty, H., Withall, A., Seeher, K., 2012. Pharmacologic treatment of apathy in dementia. Am. J. Geriatr. Psychiatry 20, 104–122.

Berridge, K.C., Robinson, T.E., Aldridge, J.W., 2009. Dissecting components of reward: 'liking', 'wanting' and learning. Curr. Opin. Pharmacol. 9, 65–73.

Bhatia, K.P., Marsden, C.D., 1994. The behavioural and motor consequences of focal lesions of the basal ganglia in man. Brain 117, 859–876.

Bodkin, J.A., Siris, S.G., Bermanzohn, P.C., Hennen, J., Cole, J.O., 2005. Double-blind, placebo-controlled, multicenter trial of selegiline augmentation of antipsychotic medication to treat negative symptoms in outpatients with schizophrenia. Am. J. Psychiatry 162, 388–390.

Bonnelle, V., Veromann, K.-R., Burnett Heyes, S., Sterzo, E., Manohar, S., Husain, M., 2015. Characterization of reward and effort mechanisms in apathy. J. Physiol. Paris 109, 16–26.

Boyle, P.A., Malloy, P.F., Salloway, S., Cahn-Weiner, D.A., Cohen, R., Cummings, J.L., 2003. Executive dysfunction and apathy predict functional impairment in Alzheimer disease. Am. J. Geriatr. Psychiatry 11, 214–221.

Brooks, D.J., 2003. Imaging end points for monitoring neuroprotection in Parkinson's disease. Ann. Neurol. 53, S110–S119.

Cairns, H., Oldfield, R.C., Pennybacker, J.B., Whitteridge, D., 1941. Akinetic mutism with an epidermoid cyst of the 3rd ventricle. Brain 64, 273–290.

Campbell, J.J., Duffy, J.D., 1997. Treatment strategies in amotivated patients. Psychiatr. Ann. 27, 44–49.

Carnicella, S., Drui, G., Boulet, S., Carcenac, C., Favier, M., Duran, T., Savasta, M., 2014. Implication of dopamine D3 receptor activation in the reversion of Parkinson's disease-related motivational deficits. Transl. Psychiatry 4, e401.

Chaudhuri, A., Behan, P.O., 2004. Fatigue in neurological disorders. Lancet 363, 978–988.

Chelonis, J.J., Johnson, T.A., Ferguson, S.A., Berry, K.J., Kubacak, B., Edwards, M.C., Paule, M.G., 2011. Effect of methylphenidate on motivation in children with attention-deficit/hyperactivity disorder. Exp. Clin. Psychopharmacol. 19, 145–153.

Chong, T.T.-J., 2015. Disrupting the perception of effort with continuous theta burst stimulation. J. Neurosci. 35, 13269–13271.

Chong, T.T.-J., D'Souza, W., 2013. Epilepsy in the elderly. In: Shorvon, S., Guerrini, R., Cook, M., Lhatoo, S. (Eds.), Oxford Textbook of Epilepsy and Epileptic Seizures. Oxford University Press, Oxford, pp. 201–210.

Chong, T.T.-J., Bonnelle, V., Manohar, S., Veromann, K.-R., Muhammed, K., Tofaris, G., Hu, M., Husain, M., 2015. Dopamine enhances willingness to exert effort for reward in Parkinson's disease. Cortex 69, 40–46.

Chong, T.T.-J., Bonnelle, V., Husain, M., 2016. Chapter 4—Quantifying motivation with effort-based decision-making paradigms in health and disease. In: Studer, B., Knecht, S. (Eds.), Progress in Brain Research, vol. 229. Elsevier, Amsterdam, pp. 71–100.

Civelli, O., 1995. Molecular biology of dopamine receptor subtypes. In: Bloom, F., Kupfer, D. (Eds.), Psychopharmacology: The Fourth Generation of Progress. Lippincott, Williams, & Wilkins, Philadelphia, pp. 155–161.

Cools, R., D'Esposito, M., 2011. Inverted-U-shaped dopamine actions on human working memory and cognitive control. Biol. Psychiatry 69, e113–e125.

Corcoran, C., Wong, M., O'Keane, V., 2004. Bupropion in the management of apathy. J. Psychopharmacol. 18, 133–135.

Cousins, M.S., Salamone, J.D., 1994. Nucleus accumbens dopamine depletions in rats affect relative response allocation in a novel cost/benefit procedure. Pharmacol. Biochem. Behav. 49, 85–91.

Craig, A.H., Cummings, J.L., Fairbanks, L., Itti, L., Miller, B.L., Li, J., Mena, I., 1996. Cerebral blood flow correlates of apathy in Alzheimer disease. Arch. Neurol. 53, 1116–1120.

Craufurd, D., Thompson, J.C., Snowden, J.S., 2001. Behavioral changes in Huntington disease. Cogn. Behav. Neurol. 14, 219–226.

Cummings, J., 1993. Frontal-subcortical circuits and human behavior. Arch. Neurol. 50, 873–880.

Cummings, J.L., Mega, M., Gray, K., Rosenberg-Thompson, S., Carusi, D.A., Gornbein, J., 1994. The Neuropsychiatric Inventory comprehensive assessment of psychopathology in dementia. Neurology 44, 2308–2314.

Czernecki, V., Pillon, B., Houeto, J.L., Pochon, J.B., Levy, R., Dubois, B., 2002. Motivation, reward, and Parkinson's disease: influence of dopatherapy. Neuropsychologia 40, 2257–2267.

Czernecki, V., Schüpbach, M., Yaici, S., Lévy, R., Bardinet, E., Yelnik, J., Dubois, B., Agid, Y., 2008. Apathy following subthalamic stimulation in Parkinson's disease: a dopamine responsive symptom. Mov. Disord. 23, 964–969.

de la Mora, M.P., Gallegos-Cari, A., Arizmendi-García, Y., Marcellino, D., Fuxe, K., 2010. Role of dopamine receptor mechanisms in the amygdaloid modulation of fear and anxiety: structural and functional analysis. Prog. Neurobiol. 90, 198–216.

Debette, S., Kozlowski, O., Steinling, M., Rousseaux, M., 2002. Levodopa and bromocriptine in hypoxic brain injury. J. Neurol. 249, 1678–1682.

Denk, F., Walton, M.E., Jennings, K.A., Sharp, T., Rushworth, M.F., Bannerman, D.M., 2005. Differential involvement of serotonin and dopamine systems in cost-benefit decisions about delay or effort. Psychopharmacology (Berl.) 179, 587–596.

Djaldetti, R., Ziv, I., Melamed, E., 2006. The mystery of motor asymmetry in Parkinson's disease. Lancet Neurol. 5, 796–802.

Drijgers, R.L., Dujardin, K., Reijnders, J.S.A.M., Defebvre, L., Leentjens, A.F.G., 2010. Validation of diagnostic criteria for apathy in Parkinson's disease. Parkinsonism Relat. Disord. 16, 656–660.

Dujardin, K., Langlois, C., Plomhause, L., Carette, A.S., Delliaux, M., Duhamel, A., Defebvre, L., 2014. Apathy in untreated early-stage Parkinson disease: relationship with other non-motor symptoms. Mov. Disord. 29, 1796–1801.

Dwoskin, L.P., Rauhut, A.S., King-Pospisil, K.A., Bardo, M.T., 2006. Review of the pharmacology and clinical profile of bupropion, an antidepressant and tobacco use cessation agent. CNS Drug Rev. 12, 178–207.

Eapen, M., Zald, D.H., Gatenby, J.C., Ding, Z., Gore, J.C., 2011. Using high-resolution MR imaging at 7T to evaluate the anatomy of the midbrain dopaminergic system. Am. J. Neuroradiol. 32, 688–694.

Farrar, A.M., Font, L., Pereira, M., Mingote, S., Bunce, J.G., Chrobak, J.J., Salamone, J.D., 2008. Forebrain circuitry involved in effort-related choice: injections of the GABA A agonist muscimol into ventral pallidum alter response allocation in food-seeking behavior. Neuroscience 152, 321–330.

Farrar, A.M., Segovia, K.N., Randall, P.A., Nunes, E.J., Collins, L.E., Stopper, C.M., Port, R.G., Hockemeyer, J., Muller, C.E., Correa, M., Salamone, J.D., 2010. Nucleus accumbens and effort-related functions: behavioral and neural markers of the interactions between adenosine A2A and dopamine D2 receptors. Neuroscience 166, 1056–1067.

Feil, D., Razani, J., Boone, K., Lesser, I., 2003. Apathy and cognitive performance in older adults with depression. Int. J. Geriatr. Psychiatry 18, 479–485.

Fernandez, H.H., Chen, J.J., 2007. Monoamine oxidase-B inhibition in the treatment of Parkinson's disease. Pharmacotherapy 27, 174S–185S.

Fisher, C.M., 1982. Honored guest presentation: abulia minor vs. agitated behavior. Clin. Neurosurg. 31, 9–31.

Floresco, S.B., Ghods-Sharifi, S., 2007. Amygdala-prefrontal cortical circuitry regulates effort-based decision making. Cereb. Cortex 17, 251–260.

Floresco, S.B., Tse, M.T.L., Ghods-Sharifi, S., 2008. Dopaminergic and glutamatergic regulation of effort- and delay-based decision making. Neuropsychopharmacology 33, 1966–1979.

Foussias, G., Remington, G., 2010. Negative symptoms in schizophrenia: avolition and Occam's razor. Schizophr. Bull. 36, 359–369.

Fuster, J.M., 2008. The Prefrontal Cortex: Anatomy, Physiology, and Neuropsychology of the Frontal Lobe. Academic Press, London.

George, M.S., Molnar, C.E., Grenesko, E.L., Anderson, B., Mu, Q., Johnson, K., Nahas, Z., Knable, M., Fernandes, P., Juncos, J., Huang, X., 2007. A single 20 mg dose of dihydrexidine (DAR-0100), a full dopamine D1 agonist, is safe and tolerated in patients with schizophrenia. Schizophr. Res. 93, 42–50.

Gerritsen, D., Jongenelis, K., Steverink, N., Ooms, M., Ribbe, M., 2005. Down and drowsy? Do apathetic nursing home residents experience low quality of life? Aging Ment. Health 9, 135–141.

Guillin, O., Diaz, J., Carroll, P., Griffon, N., Schwartz, J.C., Sokoloff, P., 2001. BDNF controls dopamine D3 receptor expression and triggers behavioural sensitization. Nature 411, 86–89.

Guttman, M., Jaskolka, J., 2001. The use of pramipexole in Parkinson's disease: are its actions D3 mediated? Parkinsonism Relat. Disord. 7, 231–234.

Hamilton, M., 1960. A rating scale for depression. J. Neurol. Neurosurg. Psychiatry 23, 56–62.

Hauber, W., Sommer, S., 2009. Prefrontostriatal circuitry regulates effort-related decision making. Cereb. Cortex 19, 2240–2247.

Herrmann, N., Rothenburg, L.S., Black, S.E., Ryan, M., Liu, B.A., Busto, U.E., Lanctôt, K.L., 2008. Methylphenidate for the treatment of apathy in Alzheimer disease: prediction of response using dextroamphetamine challenge. J. Clin. Psychopharmacol. 28 (3), 296–301.

Jaskiw, G.E., Popli, A.P., 2004. A meta-analysis of the response to chronic L-dopa in patients with schizophrenia: therapeutic and heuristic implications. Psychopharmacology (Berl.) 171, 365–374.

Jokinen, P., Helenius, H., Rauhula, E., Brück, A., Eskola, O., Rinne, J.O., 2009. Simple ratio analysis of 18F-fluorodopa uptake in striatal subregions separates patients with early Parkinson disease from healthy controls. J. Nucl. Med. 50, 893–899.

Jones, I.H., Pansa, M., 1979. Some nonverbal aspects of depression and schizophrenia occurring during the interview. J. Nerv. Ment. Dis. 167, 402–409.

Joyce, J.N., 2001. Dopamine D3 receptor as a therapeutic target for antipsychotic and antiparkinsonian drugs. Pharmacol. Ther. 90, 231–259.

Kant, R., Duffy, J., Pivovarnik, A., 1988. The prevalence of apathy following head injury. Brain Inj. 12, 87–92.

Katz, J.L., Kopajtic, T.A., Terry, P., 2006. Effects of dopamine D1-like receptor agonists on food-maintained operant behavior in rats. Behav. Pharmacol. 17, 303–309.

Kaufer, D.I., Cummings, J.L., Christine, D., Bray, T., Castellon, S., Masterman, D., MacMillan, A., Ketchel, P., DeKosky, S.T., 1998. Assessing the impact of neuropsychiatric symptoms in Alzheimer's disease: the Neuropsychiatric Inventory Caregiver Distress Scale. J. Am. Geriatr. Soc. 46, 210–215.

Khedr, E.M., Rothwell, J.C., Shawky, O.A., Ahmed, M.A., Foly, N., Hamdy, A., 2007. Dopamine levels after repetitive transcranial magnetic stimulation of motor cortex in patients with Parkinson's disease: preliminary results. Mov. Disord. 22, 1046–1050.

Kirsch-Darrow, L., Fernandez, H.F., Marsiske, M., Okun, M.S., Bowers, D., 2006. Dissociating apathy and depression in Parkinson disease. Neurology 67, 33–38.

Kohno, N., Abe, S., Toyoda, G., Oguro, H., Bokura, H., Yamaguchi, S., 2010. Successful treatment of post-stroke apathy by the dopamine receptor agonist ropinirole. J. Clin. Neurosci. 17, 804–806.

Krack, P., Batir, A., Van Blercom, N., Chabardes, S., Fraix, V., Ardouin, C., Koudsie, A., Limousin, P.D., Benazzouz, A., LeBas, J.F., Benabid, A.-L., Pollak, P., 2003. Five-year follow-up of bilateral stimulation of the subthalamic nucleus in advanced Parkinson's disease. N. Engl. J. Med. 349, 1925–1934.

Kraepelin, E., 1921. Dementia praecox and paraphrenia. J. Nerv. Ment. Dis. 54, 384.

Kraus, M.F., Maki, P.M., 1997. Effect of amantadine hydrochloride on symptoms of frontal lobe dysfunction in brain injury: case studies and review. J. Neuropsychiatry Clin. Neurosci. 9, 222–230.

Künig, G., Leenders, K.L., Martin-Sölch, C., Missimer, J., Magyar, S., Schultz, W., 2000. Reduced reward processing in the brains of Parkinsonian patients. Neuroreport 11, 3681–3687.

Landes, A.M., Sperry, S.D., Strauss, M.E., Geldmacher, D.S., 2001. Apathy in Alzheimer's disease. J. Am. Geriatr. Soc. 49, 1700–1707.

Laplane, D., Dubois, B., 2001. Auto-activation deficit: a basal ganglia related syndrome. Mov. Disord. 16, 810–814.

Lawrence, A.D., Goerendt, I.K., Brooks, D.J., 2011. Apathy blunts neural response to money in Parkinson's disease. Soc. Neurosci. 6, 653–662.

Leentjens, A., Dujardin, K., Marsh, L., Martinez-Martin, P., Richard, I., Starkstein, S., Weintraub, D., Sampaio, C., Poewe, W., Rascol, O., Stebbins, G., Goetz, C., 2008. Apathy and anhedonia rating scales in Parkinson's disease: critique and recommendations. Mov. Disord. 23, 2004–2014.

Leentjens, A., Koester, J., Fruh, B., Shephard, D., Barone, P., Houben, J., 2009. The effect of pramipexole on mood and motivational symptoms in Parkinson's disease: a meta-analysis of placebo controlled studies. Clin. Ther. 31, 89–98.

Levi-Minzi, S., Bermanzohn, P.C., Siris, S.G., 1991. Bromocriptine for "negative" schizophrenia. Compr. Psychiatry 32, 210–216.

Levy, R., 2012. Apathy: a pathology of goal-directed behaviour. A new concept of the clinic and pathophysiology of apathy. Rev. Neurol. (Paris) 168, 585–597.

Levy, R., Dubois, B., 2006. Apathy and the functional anatomy of the prefrontal cortex-basal ganglia circuits. Cereb. Cortex 16, 916–928.

Levy, M.L., Cummings, J.L., Fairbanks, L.A., Masterman, D., Miller, B.L., Craig, A.H., Paulsen, J.S., Litvan, I., 1998. Apathy is not depression. J. Neuropsychiatry Clin. Neurosci. 10, 314–319.

Lieberman, J.A., Kane, J.M., Alvir, J., 1987. Provocative tests with psychostimulant drugs in schizophrenia. Psychopharmacology (Berl.) 91, 415–433.

Lindenmayer, J.P., Nasrallah, H., Pucci, M., James, S., Citrome, L., 2013. A systematic review of psychostimulant treatment of negative symptoms of schizophrenia: challenges and therapeutic opportunities. Schizophr. Res. 147, 241–252.

Lovibond, S.H., Lovibond, P.F., 1995. Manual for the Depression Anxiety Stress Scales. Psychology Foundation, Sydney.

Lyketsos, C.G., Steinberg, M., Tschanz, J.T., Norton, M.C., Steffens, D.C., Breitner, J.C., 2000. Mental and behavioral disturbances in dementia: findings from the Cache County Study on memory in aging. Am. J. Psychiatry 157, 708–714.

Lyketsos, C.G., Lopez, O., Jones, B., Fitzpatrick, A.L., Breitner, J., DeKosky, S., 2002. Prevalence of neuropsychiatric symptoms in dementia and mild cognitive impairment: results from the cardiovascular health study. J. Am. Med. Assoc. 288, 1475–1483.

Mai, B., Sommer, S., Hauber, W., 2012. Motivational states influence effort-based decision making in rats: the role of dopamine in the nucleus accumbens. Cogn. Affect. Behav. Neurosci. 12, 74–84.

Manohar, S.G., Husain, M., 2015. Reduced pupillary reward sensitivity in Parkinson's disease. NPJ Parkinson's Dis. 1, 15026.

Manohar, S.G., Chong, T.T.-J., Apps, M., Batla, A., Stamelou, M., Jarman, P.R., Bhatia, K.P., Husain, M., 2015. Reward pays the cost of noise reduction in motor and cognitive control. Curr. Biol. 25, 1707–1716.

Marie, R., Barre, L., Rioux, P., Allain, P., Lechevalier, B., Baron, J., 1995. PET imaging of neocortical monoaminergic terminals in Parkinson's disease. J. Neural Transm. Park. Dis. Dement. Sect. 9, 55–71.

Marin, R.S., 1991. Apathy: a neuropsychiatric syndrome. J. Neuropsychiatry Clin. Neurosci. 3, 243–254.

Marin, R.S., 1996. Apathy: concept, syndrome, neural mechanisms, and treatment. Semin. Clin. Neuropsychiatry 1, 304–314.

Marin, R.S., Wilkosz, P.A., 2005. Disorders of diminished motivation. J. Head Trauma Rehabil. 20, 377–388.

Marin, R.S., Biedrzycki, R.C., Firinciogullari, S., 1991. Reliability and validity of the Apathy Evaluation Scale. Psychiatry Res. 38, 143–162.

Marin, R.S., Firinciogullari, S., Biedrzycki, R.C., 1993. The sources of convergence between measures of apathy and depression. J. Affect. Disord. 28, 117–124.

Marin, R.S., Firinciogullari, S., Biedrzycki, R.C., 1994. Group differences in the relationship between apathy and depression. J. Nerv. Ment. Dis. 182, 235–239.

Marin, R.S., Fogel, B.S., Hawkins, J., Duffy, J., Krupp, B., 1995. Apathy: a treatable syndrome. J. Neuropsychiatry Clin. Neurosci. 7, 23–30.

Markou, A., Salamone, J., Bussey, T., Mar, A., Brunner, D., Gilmour, G., Balsam, P., 2013. Measuring reinforcement learning and motivation constructs in experimental animals: relevance to the negative symptoms of schizophrenia. Neurosci. Biobehav. Rev. 37, 2149–2165.

Martínez-Horta, S., Riba, J., de Bobadilla, R.F., Pagonabarraga, J., Pascual-Sedano, B., Antonijoan, R.M., Romero, S., Mañanas, M.À., García-Sanchez, C., Kulisevsky, J., 2014. Apathy in Parkinson's disease: neurophysiological evidence of impaired incentive processing. J. Neurosci. 34, 5918–5926.

Meador-Woodruff, J.H., 1994. Update on dopamine receptors. Ann. Clin. Psychiatry 6, 79–90.

Mega, M., Cummings, J., 1994. Frontal-subcortical circuits and neuropsychiatric disorders. J. Neuropsychiatry Clin. Neurosci. 6, 358–370.

Mega, M.S., Masterman, D.M., O'Connor, S.M., Barclay, T.R., Cummings, J.L., 1999. The spectrum of behavioral responses to cholinesterase inhibitor therapy in Alzheimer disease. Arch. Neurol. 56, 1388–1393.

Migneco, O., Benoit, M., Koulibaly, P.M., Dygai, I., Bertogliati, C., Desvignes, P., Robert, P.H., Malandain, G., Bussiere, F., Darcourt, J., 2001. Perfusion brain SPECT and statistical parametric mapping analysis indicate that apathy is a cingulate syndrome: a study in Alzheimer's disease and nondemented patients. Neuroimage 13, 896–902.

Mitchell, R., Herrmann, N., Lanctôt, K., 2011. The role of dopamine in symptoms and treatment of apathy in Alzheimer's disease. CNS Neurosci. Ther. 17, 411–427.

Mogenson, G.J., Jones, D.L., Yim, C.Y., 1980. From motivation to action: functional interface between the limbic system and the motor system. Prog. Neurobiol. 14, 69–97.

Montgomery, S.A., Asberg, M., 1979. A new depression scale designed to be sensitive to change. Br. J. Psychiatry 134, 382–389.

Moretti, R., Torre, P., Antonello, R.M., Cazzato, G., Bava, A., 2002. Depression and Alzheimer's disease: symptom or comorbidity? Am. J. Alzheimers Dis. Other Demen. 17, 338–344.

Mott, A.M., Nunes, E.J., Collins, L.E., Port, R.G., Sink, K.S., Hockemeyer, J., Muller, C.E., Salamone, J.D., 2009. The adenosine A2A antagonist MSX-3 reverses the effects of the dopamine antagonist haloperidol on effort-related decision making in a T-maze cost/benefit procedure. Psychopharmacology (Berl.) 204, 103–112.

Mulin, E., Leone, E., Dujardin, K., Delliaux, M., Leentjens, A., Nobili, F., Dessi, B., Tible, O., Agüera-Ortiz, L., Osorio, R.S., Yessavage, J., 2011. Diagnostic criteria for apathy in clinical practice. Int. J. Geriatr. Psychiatry 26, 158–165.

Newburn, G., Newburn, D., 2005. Selegiline in the management of apathy following traumatic brain injury. Brain Inj. 19, 149–154.

Newman, A.H., Blaylock, B.L., Nader, M.A., Bergman, J., Sibley, D.R., Skolnick, P., 2012. Medication discovery for addiction: translating the dopamine D3 receptor hypothesis. Biochem. Pharmacol. 84, 882–890.

Ng, K., Chase, T., Colburn, R., Kopin, I., 1971. Dopamine: stimulation-induced release from central neurons. Science 172, 487–489.

Nowend, K.L., Arizzi, M., Carlson, B.B., Salamone, J.D., 2001. D1 or D2 antagonism in nucleus accumbens core or dorsomedial shell suppresses lever pressing for food but leads to compensatory increases in chow consumption. Pharmacol. Biochem. Behav. 69, 373–382.

Nunes, E.J., Randall, P.A., Santerre, J.L., Given, A.B., Sager, T.N., Correa, M., Salamone, J.D., 2010. Differential effects of selective adenosine antagonists on the effort-related impairments induced by dopamine D1 and D2 antagonism. Neuroscience 170, 268–280.

Nunes, E.J., Randall, P.A., Hart, E.E., Freeland, C., Yohn, S.E., Baqi, Y., Müller, C.E., López-Cruz, L., Correa, M., Salamone, J.D., 2013. Effort-related motivational effects of the VMAT-2 inhibitor tetrabenazine: implications for animal models of the motivational symptoms of depression. J. Neurosci. 33, 19120–19130.

Oberndorfer, S., Urbanits, S., Lahrmann, H., Kirschner, H., Kumpan, W., Grisold, W., 2002. Akinetic mutism caused by bilateral infiltration of the fornix in a patient with astrocytoma. Eur. J. Neurol. 9, 311–313.

Oguro, H., Kadota, K., Ishihara, M., Okada, K., Yamaguchi, S., 2014. Efficacy of pramipexole for treatment of apathy in Parkinson's disease. Int. J. Clin. Med. 5, 885–889.

Ohmori, T., Koyama, T., Inoue, T., Matsubara, S., Yamashita, I., 1993. B-HT 920, a dopamine D2 agonist, in the treatment of negative symptoms of chronic schizophrenia. Biol. Psychiatry 33, 687–693.

Ostlund, S.B., Wassum, K.M., Murphy, N.P., Balleine, B.W., Maidment, N.T., 2011. Extracellular dopamine levels in striatal subregions track shifts in motivation and response cost during instrumental conditioning. J. Neurosci. 31, 200–207.

Padala, P.R., Burke, W.J., Shostrom, V.K., Bhatia, S.C., Wengel, S.P., Potter, J.F., Petty, F., 2010. Methylphenidate for apathy and functional status in dementia of the Alzheimer type. Am. J. Geriatr. Psychiatr. 18 (4), 371–374.

Paolo, S.D., Galistu, A., 2012. Possible role of dopamine D1-like and D2-like receptors in behavioural activation and evaluation of response efficacy in the forced swimming test. Neuropharmacology 62, 1717–1729.

Pardo, M., Lopez-Cruz, L., Valverde, O., Ledent, C., Baqi, Y., Müller, C.E., Salamone, J.D., Correa, M., 2012. Adenosine A2A receptor antagonism and genetic deletion attenuate the effects of dopamine D2 antagonism on effort-related decision making in mice. Neuropharmacology 62, 2068–2077.

Pederson, K.F., Larsen, J.P., Alves, G., Aarsland, D., 2009. Prevalence and clinical correlates of apathy in Parkinson's disease: a community-based study. Parkinsonism Relat. Disord. 15, 295–299.

Pérez-Pérez, J., Pagonabarraga, J., Martínez-Horta, S., Fernández-Bobadilla, R., Sierra, S., Pascual-Sedano, B., Gironell, A., Kulisevsky, J., 2015. Head-to-head comparison of the neuropsychiatric effect of dopamine agonists in Parkinson's disease: a prospective, cross-sectional study in non-demented patients. Drugs Aging 32, 1–7.

Piercey, M.F., Hoffmann, W.E., Smith, M.W., Hyslop, D.K., 1996. Inhibition of dopamine neuron firing by pramipexole, a dopamine D3 receptor-preferring agonist: comparison to other dopamine receptor agonists. Eur. J. Pharmacol. 312, 35–44.

Pluck, G.C., Brown, R.G., 2002. Apathy in Parkinson's disease. J. Neurol. Neurosurg. Psychiatry 73, 636–642.

Porat, O., Hassin-Baer, S., Cohen, O.S., Markus, A., Tomer, R., 2014. Asymmetric dopamine loss differentially affects effort to maximize gain or minimize loss. Cortex 51, 82–91.

Randall, P.A., Pardo, M., Nunes, E.J., Lopez Cruz, L., Vemuri, V.K., Makriyannis, A., Baqi, Y., Muller, C.E., Correa, M., Salamone, J.D., 2012. Dopaminergic modulation of effort-related choice behavior as assessed by a progressive ratio chow feeding choice task: pharmacological studies and the role of individual differences. PLoS One 7, e47934.

Randall, P.A., Lee, C.A., Nunes, E.J., Yohn, S.E., Nowak, V., Khan, B., Shah, P., Pandit, S., Vemuri, V.K., Makriyannis, A., Baqi, Y., 2014. The VMAT-2 inhibitor tetrabenazine affects effort-related decision making in a progressive ratio/chow feeding choice task: reversal with antidepressant drugs. PLoS One 9, e99320.

Randall, P.A., Lee, C.A., Podurgiel, S.J., Hart, E., Yohn, S.E., Jones, M., Rowland, M., López-Cruz, L., Correa, M., Salamone, J.D., 2015. Bupropion increases selection of high effort activity in rats tested on a progressive ratio/chow feeding choice procedure: implications for treatment of effort-related motivational symptoms. Int. J. Neuropsychopharmacol, 18 (2), 1–11.

Reichmann, H., Bilsing, A., Ehret, R., Greulich, W., Schulz, J.B., Schwartz, A., 2006. Ergoline and non-ergoline derivatives in the treatment of Parkinson's disease. J. Neurol. 253, iv36–iv38.

Rektorova, I., Balaz, M., Svatova, J., Zarubova, K., Honig, I., Dostal, V., Sedlackova, S., Nestrasil, I., Mastik, J., Bares, M., Veliskova, J., Dusek, L., 2008. Effects of ropinirole on nonmotor symptoms of Parkinson disease: a prospective multicenter study. Clin. Neuropharmacol. 31, 261–266.

Remy, P., Doder, M., Lees, A., Turjanski, N., Brooks, D., 2005. Depression in Parkinson's disease: loss of dopamine and noradrenaline innervation in the limbic system. Brain 128, 1314–1322.

Ribot, T., 1896. La Psychologie des Sentiment. Felix Alcan, Paris.

Robbins, T.W., Everitt, B.J., 2006. A role for mesencephalic dopamine in activation: commentary on Berridge. Psychopharmacology (Berl.) 191, 433–437.

Robert, P.H., Clairet, S., Benoit, M., Koutaich, J., Bertogliati, C., Tible, O., Caci, H., Borg, M., Brocker, P., Bedoucha, P., 2002. The apathy inventory: assessment of apathy and awareness in Alzheimer's disease, Parkinson's disease and mild cognitive impairment. Int. J. Geriatr. Psychiatry 17, 1099–1105.

Robert, P.H., Darcourt, G., Koulibaly, M.P., Clairet, S., Benoit, M., Garcia, R., Dechaux, O., Darcourt, J., 2006. Lack of initiative and interest in Alzheimer's disease: a single photon emission computed tomography study. Eur. J. Neurol. 13, 729–735.

Robert, P., Onyike, C.U., Leentjens, A.F.G., Dujardin, K., Aalten, P., Starkstein, S., Verhey, F.R.J., Yessavage, J., Clément, J.P., Drapier, D., Bayle, F., 2009. Proposed diagnostic criteria for apathy in Alzheimer's disease and other neuropsychiatric disorders. Eur. Psychiatry 24, 98–104.

Roesch-Ely, D., Gohring, K., Gruschka, P., Kaiser, S., Pfuller, U., Burlon, M., Weisbrod, M., 2006. Pergolide as adjuvant therapy to amisulpride in the treatment of negative and depressive symptoms in schizophrenia. Pharmacopsychiatry 39, 115–116.

Salamone, J.D., Correa, M., 2012. The mysterious motivational functions of mesolimbic dopamine. Neuron 76, 470–485.

Salamone, J.D., Steinpreis, R.E., McCullough, L.D., Smith, P., Grebel, D., Mahan, K., 1991. Haloperidol and nucleus accumbens dopamine depletion suppress lever pressing for food but increase free food consumption in a novel food choice procedure. Psychopharmacology (Berl.) 104, 515–521.

Salamone, J.D., Cousins, M.S., Bucher, S., 1994. Anhedonia or anergia? Effects of haloperidol and nucleus accumbens dopamine depletion on instrumental response selection in a T-maze cost/benefit procedure. Behav. Brain Res. 65, 221–229.

Salamone, J.D., Correa, M., Mingote, S., Weber, S.M., 2003. Nucleus accumbens dopamine and the regulation of effort in food-seeking behavior: implications for studies of natural motivation, psychiatry, and drug abuse. J. Pharmacol. Exp. Ther. 305, 1–8.

Salamone, J.D., Correa, M., Mingote, S., Weber, S.M., Farrar, A.M., 2006. Nucleus accumbens dopamine and the forebrain circuitry involved in behavioral activation and effort related decision making: implications for understanding anergia and psychomotor slowing in depression. Curr. Psychiatr. Rev. 2, 267–280.

Salamone, J., Correa, M., Farrar, A., Mingote, S., 2007. Effort-related functions of nucleus accumbens dopamine and associated forebrain circuits. Psychopharmacology (Berl.) 191, 461–482.

Salamone, J., Correa, M., Farrar, A., Nunes, E., Pardo, M., 2009. Dopamine, behavioral economics, and effort. Front. Behav. Neurosci. 3, 13.

Salimpour, Y., Mari, Z.K.S., Shadmehr, R., 2015. Altering effort costs in Parkinson's disease with noninvasive cortical stimulation. J. Neurosci. 35, 12287–12302.

Salomon, L., Lanteri, C., Glowinski, J., Tassin, J.-P., 2006. Behavioral sensitization to amphetamine results from an uncoupling between noradrenergic and serotonergic neurons. Proc. Natl. Acad. Sci. U.S.A. 103, 7476–7481.

Santangelo, G., Trojano, L., Barone, P., Errico, D., Grossi, D., Vitale, C., 2013. Apathy in Parkinson's disease: diagnosis, neuropsychological correlates, pathophysiology and treatment. Behav. Neurol. 27, 501–513.

Schmidt, L., d'Arc, B.F., Lafargue, G., Galanaud, D., Czernecki, V., Grabli, D., Schüpbach, M., Hartmann, A., Lévy, R., Dubois, B., Pessiglione, M., 2008. Disconnecting force from money: effects of basal ganglia damage on incentive motivation. Brain 131, 1303–1310.

Schweimer, J., Hauber, W., 2006. Dopamine D1 receptors in the anterior cingulate cortex regulate effort-based decision making. Learn. Mem. 13, 777–782.

Seeman, P., Madras, B., 2002. Methylphenidate elevates resting dopamine which lowers the impulse-triggered release of dopamine: a hypothesis. Behav. Brain Res. 130 (1), 79–83.

Short, J.L., Ledent, C., Drago, J., Lawrence, A.J., 2006. Receptor crosstalk: characterization of mice deficient in dopamine D1 and adenosine A2A receptors. Neuropsychopharmacology 31, 525–534.

Sink, K.S., Vemuri, V.K., Olszewska, T., Makriyannis, A., Salamone, J.D., 2008. Cannabinoid CB1 antagonists and dopamine antagonists produce different effects on a task involving response allocation and effort-related choice in food-seeking behavior. Psychopharmacology (Berl.) 196, 565–574.

Smith, K.S., Berridge, K.C., Aldridge, J.W., 2011. Disentangling pleasure from incentive salience and learning signals in brain reward circuitry. Proc. Natl. Acad. Sci. U.S.A. 108, E255–E264.

Sobin, C., Sackeim, H.A., 1997. Psychomotor symptoms of depression. Am. J. Psychiatry 154, 4–17.

Sockeel, P., Dujardin, K., Devos, D., Deneve, C., Destee, A., Defebvre, L., 2006. The Lille apathy rating scale (LARS), a new instrument for detecting and quantifying apathy: validation in Parkinson's disease. J. Neurol. Neurosurg. Psychiatry 77, 579–584.

Sokoloff, P., Diaz, J., Le Foll, B., Guillin, O., Leriche, L., Bezard, E., Gross, C., 2006. The dopamine D3 receptor: a therapeutic target for the treatment of neuropsychiatric disorders. CNS Neurol. Disord. Drug Targets 5, 25–43.

Starkstein, S.E., Leentjens, A.F.G., 2008. The nosological position of apathy in clinical practice. J. Neurol. Neurosurg. Psychiatry 79, 1088–1092.

Starkstein, S.E., Fedoroff, J.P., Price, T.R., Leiguarda, R., Robinson, R.G., 1993. Apathy following cerebrovascular lesions. Stroke 24, 1625–1630.

Starkstein, S.E., Petracca, G., Chemerinski, E., Kremer, J., 2001. Syndromic validity of apathy in Alzheimer's disease. Am. J. Psychiatry 158, 872–877.

Starkstein, S.E., Jorge, R., Mizrahi, R., Robinson, R.G., 2006. A prospective longitudinal study of apathy in Alzheimer's disease. J. Neurol. Neurosurg. Psychiatry 77, 8–11.

Starkstein, S.E., Merello, M., Jorge, R., Brockman, S., Bruce, D., Power, B., 2009. The syndromal validity and nosological position of apathy in Parkinson's disease. Mov. Disord. 24, 1211–1216.

Stuss, D.T., Van Reekum, R.J.M.K., Murphy, K.J., 2000. Differentiation of states and causes of apathy. In: Borod, J.C. (Ed.), The Neuropsychology of Emotion. Oxford University Press, New York, pp. 340–363.

Takarada, Y., Mima, T., Abe, M., Nakatsuka, M., Taira, M., 2014. Inhibition of the primary motor cortex can alter one's "sense of effort": effects of low-frequency rTMS. Neurosci. Res. 89, 54–60.

Tanaka, T., Takano, Y., Tanaka, S., Hironaka, N., Kobayashi, K., Hanakawan, T., Watanabe, K., Honda, M., 2013. Transcranial direct-current stimulation increases extracellular dopamine levels in the rat striatum. Front. Syst. Neurosci. 7, 6.

Tengvar, C., Johansson, B., Sorensen, J., 2004. Frontal lobe and cingulate cortical metabolic-dysfunction in acquired akinetic mutism: a PET study of the interval form of carbonmonoxide poisoning. Brain Inj. 18, 615–625.

Thobois, S., Lhommée, E., Klinger, H., Ardouin, C., Schmitt, E., Bichon, A., Kistner, A., Castrioto, A., Xie, J., Fraix, V., Pellisier, P., Chabardes, S., Mertens, P., Quesada, J.-L., Bosson, J.-L., Pollak, P., Broussolle, E., Krack, P., 2013. Parkinsonian apathy responds to dopaminergic stimulation of D2/D3 receptors with piribedil. Brain 136, 1568–1577.

Treadway, M.T., Zald, D.H., 2011. Reconsidering anhedonia in depression: lessons from translational neuroscience. Neurosci. Biobehav. Rev. 35, 537–555.

Treadway, M.T., Buckholtz, J.W., Cowan, R.L., Woodward, N.D., Li, R., Ansari, M.S., Baldwin, R.M., Schwartzman, A.N., Kessler, R.M., Zald, D.H., 2012. Dopaminergic mechanisms of individual differences in human effort-based decision-making. J. Neurosci. 32, 6170–6176.

Trifilieff, P., Feng, B., Urizar, E., Winiger, V., Ward, R.D., Taylor, K.M., Martinez, D., Moore, H., Balsam, P.D., Simpson, E.H., Javitch, J.A., 2013. Increasing dopamine D2 receptor expression in the adult nucleus accumbens enhances motivation. Mol. Psychiatry 18, 1025–1033.

van Kammen, D.P., Boronow, J.J., 1988. Dextro-amphetamine diminishes negative symptoms in schizophrenia. Int. Clin. Psychopharmacol. 3, 111–121.

van Reekum, R., Bayley, M., Garner, S., Burke, I.M., Fawcett, S., Hart, A., Thompson, W., 1995. N of 1 study: amantadine for the amotivational syndrome in a patient with traumatic brain injury. Brain Inj. 9, 49–54.

van Reekum, R., Stuss, D., Ostrander, L., 2005. Apathy: why care? J. Neuropsychiatry Clin. Neurosci. 17, 7–19.

Voon, V., Fernagut, P.-O., Wickens, J., Baunez, C., Rodriguez, M., Pavon, N., Juncos, J.L., Obeso, J.A., Bezard, E., 2009. Chronic dopaminergic stimulation in Parkinson's disease: from dyskinesias to impulse control disorders. Lancet Neurol. 8, 1140–1149.

Walton, M.E., Bannerman, D.M., Alterescu, K., Rushworth, M.F., 2003. Functional specialization within medial frontal cortex of the anterior cingulate for evaluating effort-related decisions. J. Neurosci. 23, 6475–6479.

Walton, M.E., Croxson, P.L., Rushworth, M.F., Bannerman, D.M., 2005. The mesocortical dopamine projection to anterior cingulate cortex plays no role in guiding effort-related decisions. Behav. Neurosci. 119, 323–328.

Wardle, M.C., Treadway, M.T., Mayo, L.M., Zald, D.H., de Wit, H., 2011. Amping up effort: effects of D-amphetamine on human effort-based decision-making. J. Neurosci. 31, 16597–16602.

Weiner, D.M., Levey, A.I., Sunahara, R.K., Niznik, H.B., O'Dowd, B.F., Seeman, P., Brann, M.R., 1991. D1 and D2 dopamine receptor mRNA in rat brain. Proc. Natl. Acad. Sci. U.S.A. 88, 1859–1863.

Widlocher, D.J., 1983. Psychomotor retardation: clinical, theoretical, and psychometric aspects. Psychiatr. Clin. North Am. 6, 27–40.

Worden, L.T., Shahriari, M., Farrar, A.M., Sink, K.S., Hockemeyer, J., Müller, C.E., Salamone, J.D., 2009. The adenosine A2A antagonist MSX-3 reverses the effort-related effects of dopamine blockade: differential interaction with D1 and D2 family antagonists. Psychopharmacology 203, 489–499.

Yohn, S.E., Santerre, J.L., Nunes, E.J., Kozak, R., Podurgiel, S.J., Correa, M., Salamone, J.D., 2015a. The role of dopamine D1 receptor transmission in effort-related choice behavior: effects of D1 agonists. Pharmacol. Biochem. Behav. 135, 217–226.

Yohn, S.E., Thompson, C., Randall, P.A., Lee, C., Correa, M., Salamone, J.D., 2015b. The VMAT-2 inhibitor tetrabenazine alters effort-related decision making as measured by the T-maze barrier choice task: reversal with the adenosine A2A antagonist MSX-3 and the catecholamine uptake blocker bupropion. Psychopharmacology (Berl.) 232, 1313–1323.

Zahodne, L.B., Bernal-Pacheco, O., Bowers, D., Ward, H., Oyama, G., Limotai, N., Velez-Lago, F., Rodriguez, R.L., Malaty, I., McFarland, N.R., Okun, M.S., 2014. Are selective serotonin reuptake inhibitors associated with greater apathy in Parkinson's disease? J. Neuropsychiatry Clin. Neurosci. 24, 326–330.

Zawacki, T.M., Grace, J., Paul, R., Moser, D.J., Ott, B.R., Gordon, N., Cohen, R.A., 2002. Behavioral problems as predictors of functional abilities of vascular dementia patients. J. Neuropsychiatry Clin. Neurosci. 14, 296–302.

Zénon, A., Sidibé, M., Olivier, E., 2015. Disrupting the supplementary motor area makes physical effort appear less effortful. J. Neurosci. 35, 8737–8744.

CHAPTER

Changing health behavior motivation from I-must to I-want

18

S. Knecht*,†,[1], **P. Kenning**‡

**Institute of Clinical Neuroscience and Medical Psychology, Medical Faculty, Heinrich-Heine-University Düsseldorf, Düsseldorf, Germany*

†*Mauritius Hospital, Meerbusch, Germany*

‡*Heinrich-Heine-University Düsseldorf, Düsseldorf, Germany*

[1]*Corresponding author: Tel.: +49-2159-679-1537; Fax: +49-2159-679-1535, e-mail address: Stefan.Knecht@stmtk.de*

Abstract

In the past, medicine was dominated by acute diseases. Since treatments were unknown to patients they followed their medical doctors´ directives—at least for the duration of the disease. Behavior was thus largely motivated by avoiding expected costs associated with alternative behaviors (I-must).

The health challenges prevailing today are chronic conditions resulting from the way we chose to live. Traditional directive communication has not been successful in eliciting and maintaining appropriate lifestyle changes. An approach successful in other fields is to motivate behavior by increasing expected rewards (I-want). Drawing on neuroeconomic and marketing research, we outline strategies including simplification, repeated exposure, default framing, social comparisons, and consumer friendliness to foster sustained changes in preference. We further show how these measures could be integrated into the health care system.

Keywords

Neuroeconomics, Decision making, Behavioral change, Motivation, Public health, Life style, Drug adherence, Cardiovascular disease, Neurology

1 BACKGROUND

The spread of behaviorally mediated chronic vascular and metabolic disease presents a global crisis that so far we are ill equipped to deal with. In order to meet this crisis, we need to reconsider our model of how patients think, behave, and would be willing to change their behavior.

The traditional model of patient–physician interaction developed when the majority of disease was infectious or traumatic or for other reasons random, acute,

Progress in Brain Research, Volume 229, ISSN 0079-6123, http://dx.doi.org/10.1016/bs.pbr.2016.06.013

and transient. Patients were suffering and because of imminent risk to life either delegated decisions and measures to physicians or acted in strict accordance to their instructions. Motivation was based on "I must comply in order to avoid serious potential health costs." Patients thus seemed to behave in a more or less rational way, understand their stakes and optimize their utility given limited resources. This concept converges with the classic notion of the homo economicus as a rational and self-interested actor capable to make precise judgments toward defined ends (Kenning and Plassmann, 2005). This model of human decision making has not only been dominant in the economic, political, or legal field and the way physicians view their patients. This model also lies at the heart of our health technologies and institutions (Knecht, 2009). However, there is mounting evidence from behavioral science and in particular from behavioral economics that actual behavior deviates in systematic and predictable ways from the concept of rational actors (Mullainathan and Thaler, 2000; Strombach et al., 2016).

Deviations from rationality manifest in different fields: Probabilities are perceived in systematically distorted ways. People underweight outcomes that are merely probable relative to those that are certain (Kahneman and Tversky, 1979). Further, humans have great difficulties detecting changes in probabilities and tend to rely on past estimates (Achtziger et al., 2014). Their performance deteriorates even further when highly incentivized (Achtziger et al., 2015). Also, humans perceive risks in a highly distorted way. Despite similar or even lower hazards humans are more anxious about unknown relative to known risks (electric fields relative to auto exhaust) and risks outside relative to those within their control (skyscraper fire relative to smoking) (Slovic, 1987). Finally, people overrate their agency. They are generally overconfident that they can affect the state of a given environment even when they cannot or are highly unlikely to. Gamblers, for examples, tend to be certain that they can make the dice beat the odds (Johnson and Fowler, 2011).

In the medical world the deviation from rationality is reflected most prominently by the rise of behaviorally mediated chronic diseases. People who are physically inactive, eat too many calories but few fruits or vegetables, smoke, or drink excessively live 14 years less, and are cognitively more impaired than others (Khaw et al., 2008; Witte et al., 2009). Additionally, only 1 year after an acute event like a myocardial infarction half of patients will no longer take their medication as prescribed thereby multiplying their risk of death by four (Ho et al., 2006). Thus far physicians have failed to change patient behavior so as to effectively prevent or delay chronic disease. We assume that this failure is in large part related to a traditional model of patients as rational actors who focus on minimizing immediate costs. An alternative model would be that of biased actors who focus on sustained rewards. This is the model of customers. Their behavior deviates from rationality in predictable ways and can be addressed accordingly. Customers chose particular behaviors primarily not because they must but because they want to (Blyte, 2008).

2 CONCEPT

The gradual and probabilistic development of disease in the absence of manifest suffering increases choices and thus turns patients or patients-to-be into consumers and health into a commodity. From this perspective adopting health measures resembles buying a financial investment package. Readers may judge themselves if with such packages they feel like a homo economicus, who can make rational, ie, fully informed, competent, comprehensive, and consistent decisions providing with certainty the optimal utility. Knowledge and experience certainly help but are less than complete in the majority of people confronted with such choices (Linnet et al., 2010). Behavioral science has collected numerous examples of how limited and biased, ie, nonrational human decisions are and—for practical purposes—often need to be (for recent overviews see Kahneman, 2013). The extent of our deviations from rationality is reflected by the size of the marketing industry that caters to them. Rational actors would not need advertising in order to develop a specific preference for a product or service. However, according to the World Advertising Research Centers an estimated 500 billion US dollars per year are spent on advertising alone. Marketing strategies have been derived from theory, experimental evidence and—most importantly—from actual consumer behavior on the markets. Among other things these strategies have shifted from focusing on transactions, ie, individual sales to focusing on long-term relationships and retention of choices (Gordon, 1998). Rather than deploring the lack of rationality in patients or health consumers, we propose to review marketing strategies, examine how they address biased human decision making, and test which could be used for disease prevention. The advantage of marketing over traditional approaches is that a marketing perspective can help to redirect motivation from I-must to I-want or from costs to rewards which may be more effective at changing behavior long term by habit formation. Marketing is not limited to advertising but extends to shaping overall decision environments. More generally, marketing describes all activities at the interface between an organization and its customers. It comprises the process of planning and executing the conception, pricing, promotion and distribution of ideas, goods, and services (Blyte, 2008). As such it mostly builds on rewards while it does not exclude directing behavior by incurring costs. Table 1 juxtaposes concepts used in medical and marketing encounters.

3 IMPLEMENTATION

Medical doctors hesitate to adopt a marketing approach because they (i) do not perceive themselves as sellers but as authorities appealing to rationality, (ii) get paid for short- but not long-term outcomes, and (iii) are in a poor position for marketing because of their one-on-one working style and time-constraints. In interventional trials nonphysician personnel are often better at eliciting behavioral change (Cutrona et al., 2010). Therefore, one solution could be to establish independent organizations for

Table 1 Comparison of Conceptual Models

Medical Model of Patients	Marketing Model of Customers
Rationally responding to risk, ie, expected costs (I-must)	Biased and focusing on reward (I-want)
Capable of one-shot learning	Requiring multiple repetitions
– My doctor mentioned it in passing, so I will adopt this measure.	– Only repeated exposure makes customers habituated to and possibly like commodities (mere exposure-effect).
Amenable to facts	Amenable to social comparison (herding)
– Because atrial fibrillation can lead to blood clot formation I will benefit from anticoagulation.	– Things are good if many or important people do them. A best-selling novel is good simply because many people have bought it.
Comprehending probabilities	Biased perception of probabilities
– Patients have a clear idea what it means that a measure provides a 20% relative risk reduction over 5 years.	– Customers are very poor at probabilities and therefore buy lottery tickets or damage waivers, although they are certain to lose money.
Nondiscounting	Discounting
– Patients will accept costs now in proportion to distant and uncertain benefits.	– The more distant and uncertain benefits are, the less costs consumers will accept now.
Focused	Confused
– Patients follow their physician without being affected by what other "experts," celebrities or their spouses claim.	– The more options or notions are offered the less likely a decision will be taken. Therefore simplify!
Certain	Uncertain
– Doctors are always right and never exaggerating nor misinformed.	– Only continuous exposure can build and maintain some extent of trust.
Insensitive to price in time, work, or money	Sensitive to price
– Patients never miss a renewal of their prescription even when seeing their doctor and getting the drug is extremely cumbersome for them.	– If customers cannot perceive a difference between A and B while A is more comfortable or cheaper, they will go for A—particularly if A is a lot cheaper.
Consistent preferences	Inconsistent preferences
– Once patients have realized they need to change they will maintain lifestyle changes for the rest of their life.	– Customers are fickle and therefore can be tricked into offers like "buy-now-pay-later" although these are much more expensive than immediate payments.
Insensitive to situational frames	Susceptible to choice architecture (nudges)
– If advised accordingly patients will opt-out of perceived norms and quit smoking although everybody around them still smokes.	– Customers prefer default options that require little active decisions making.

health facilitation that can be contracted in cases of behaviorally mediated risk or disease as determined in out-patient settings, hospitals, or rehabilitation facilities. Physicians could characterize the risk profile, advise patients to enroll in a facilitation program, and communicate with the provider.

Given that marketing is paid by sales and that sales are needed to feedback on marketing strategies, how should health facilitation providers be financed? One possibility would be to use follow-up on the initial risk profile over time, preferably using objective measures like body mass index, levels of cotinine, vitamin C, and carbohydrate-deficient transferrin, long-term blood pressure measurement, and actimetry. The physician who engaged the provider could thus supply a measure of success that can be weighted by healthcare cost effectiveness criteria and remunerated accordingly.

How could a health facilitation provider operate? The philosophy—as in most marketing approaches—would be to shift from patients needing to PULL to being PUSHED (or at least "nudged") help with adjusting their lifestyle (Sunstein, 2014). Ultimately the goal would be to make people want to do what they need to do. Neurobiological research indicates that such retraining is quite possible even at the level of brain reward system activation (Deckersbach et al., 2014).

Health facilitation is not such a far fetched goal considering that socioeconomic elites already evolved healthier lifestyles and also conveyed relevant values and techniques to their children (Frederick et al., 2014). Many individuals actually are willing to adopt healthier behaviors but often simply lack skills and support (Borland et al., 2012). Potentially helpful marketing techniques comprise assessment of consumer behavior, advertising, framing, taxing, financial incentives, social comparison, and consumer friendliness.

4 STRATEGIES

Analysis of marketing data is a starting point for evidence-based interventions into markets. Epidemiological patterns of cigarette smoking identify targets groups in greatest need for support (Amerson et al., 2014). Information about medication adherence and nonadherence can indicate potential for intervention such as direct-to-consumer advertising as being used in the United States and New Zealand. This has been shown to increase medication use (for review as well as concerns see Wang and Kesselheim, 2013).

Health campaigns are classical tools in primary prevention to inform and motivate large audiences to change their health behavior. They use organized communication activities and feature repeated, varied, and prominently placed messages in multiple channels similar to commercial advertising campaigns. These channels often comprise television and radio commercials complemented by print materials such as posters, booklets, and brochures. Campaign strategies frequently pay homage to motivational theories like communication-persuasion matrix or more specifically agenda setting, diffusion of innovation, health belief model, self-efficacy, social cognitive theory, or the transtheoretical model (for review see Atktin and Rice, 2013). Generally health campaigns affect cognitive outcomes, but less so attitudes, and even less actual behavior (Atktin and Rice, 2013). One reason may be that for financial reasons health campaigns are usually limited to short time periods. As a consequence they are limited in affecting behavior by learning.

Permanent health messages often take the form of fear appeals such as graphic health warning on cigarette packages. There is, however, little convincing evidence for broad effectiveness of these messages. Rather, when acknowledging the threat, but feeling helpless what to do, people may engage in defensive action and continuation of the health risk behavior (Ruiter et al., 2014). Indeed, eliciting shame, anger, or distress seem more effective in reducing smoking than fear and disgust (Bogliacino et al., 2015).

Framing of health messages in various types of communication has been studied by research on biases in decision making. Classic work has shown that people are more likely to agree to measures when medical problems are framed in terms of probability of living rather than in terms of probability of dying (McNeil et al., 1982). Higher consent rates are achieved when effects are expressed as number of people needed to treat to prevent one case of disease rather than percentages or equivalent postponements of disease (Halvorsen et al., 2007). Most health messages can be classified as loss- or gain framed. Overall, loss-framed messages seem to appeal best to involved and informed individuals like health professionals, ie, those who frame most health messages. Conversely, gain-framed messages work best with people who are less involved, little informed or risk-averse (Wansink and Pope, 2015). For them gain-framed information provides actionable messages. Further, particularly in elderly, decisions for medication to prevent potential future disease are highly sensitive to how potential but immediate side effects are presented (Fried et al., 2011). Such information could be easily integrated into marketing information strategies for health facilitation.

Taxation is a long-established tool to regulate markets on a governmental level. It modulates behavior by adding weight to the cost side of the cost–benefit evaluation during decision making. A 50% increase in price through taxation cuts tobacco consumption by 20% with the largest impacts in the poor (Jha and Peto, 2014). People—unlike industry pressure groups—generally accept disincentivizing market interventions like bans or taxes on harmful products as drugs, alcohol, or even sugar sweetened soda in Mexico (Boseley, 2014).

Financial incentive marketing, conversely, adds weight to the benefit side of the cost–benefit evaluation and has been piloted in developing countries that pay people to attend health programs (Scott et al., 2011). Thus the Mexican conditional cash transfer program, Progresa elicited a 4% decline in municipality-level mortality (Barham and Rowberry, 2013). Financial incentives have also been used in developed countries and found successful in weight loss (John et al., 2011) and in smoking cessation (Halpern et al., 2015; Volpp, 2006). However, financial incentives combined with peer networks did not succeed in making older people walk more (Kullgren et al., 2014). Generally financial incentives were found to be most effective in poor people (Mantzari et al., 2015). Because people in developed economies are more financially saturated here conditional cash payment programs, which have only limited budgets, often employ lottery schemes. Using such a lottery-based approach Kimmel and colleagues managed to improve medication adherence in patients at risk for poor adherence (Kimmel et al., 2012). This suggests that such extrinsic rewards could have particular benefits in individuals with weak autonomous motivation for behavior (Ryan and Deci, 2000). However, it has been argued that financial incentives could induce crowding out effects, ie, that they can diminish intrinsic motivation (Strombach et al., 2015). Moreover, there is occasional resistance to financial incentive schemes based on the notion that they reward the feckless rather than the responsible. This is remarkable since no such discussion arises when cost for medication is involved (Marteau and Mantzari, 2015).

Implementation strategies include precommitment and goal shielding. They can be viewed as an intermediary between I-want and I-must because they involve wanting to having to. Forming an implementation intention promotes the attainment of different types of goals by establishing an "if-then plan" (Achtziger et al., 2008). For example, to precommit myself to riding bicycle with a helmet, I could attach it to my bicycle lock so when I unlock the bicycle I will have the helmet in my hand and nowhere else to put but on my head. Goal shielding in this example could involve keeping a raincoat ready so that when rain tempts me to defect my plan of bicycle riding the goal can be maintained. Another variant would be self-control commitment. For example, Schwartz and colleagues had grocery shoppers, who were already enrolled in an incentive program that discounts the price of eligible groceries, put their discount on the line and only retrieve it once they had increased their purchases of healthy food by 5 percentage points above their household baseline (Schwartz et al., 2014). They obtained a 3.5% increase in healthy grocery items purchased in the intervention group.

Social comparison is a mainstay of marketing and can also be exploited for health facilitation since health behaviors also spread in social networks (Christakis and Fowler, 2007, 2008). Social contagion of health behaviors can be triggered by celebrities, peer groups, as well as friends and families (Cram et al., 2003). Success has been achieved by programs using church-based social relationships (for review see Peterson et al., 2002). There is conceptual evidence that self-definitions as an individual or a group can increase self-regulation. Thus people are likely to better resist a temptation to smoke, if they define themselves as "quitters" as part of their identity (Berkman et al., 2015). Social support interventions were successful in improving glycemic control in diabetes (Piette et al., 2013). Social ties are also an effective part of transfer packages used in poststroke care to maintain and enhance use of paretic limbs. Patients sign behavioral contracts with therapists, practice problem solving to overcome perceived barriers and use weekly telephone calls to exchange on progress (Taub et al., 2013).

Consumer-friendliness marketing strategies build on making healthy behaviors as convenient as possible—if possible more convenient than unhealthy behaviors—by reducing physical, cognitive, and emotional costs (King et al., 2014). For example, to render medication adherence easy, Lee and colleagues sent out medications in time-specific blister packs to community-based patients taking at least four chronic medications. They achieved a 95% adherence rate compared to 65% in the standard care group at 6 months (Lee et al., 2006). Already some medical centers have set up interdisciplinary medication adherence programs involving motivational interviewing combined with medication adherence electronic monitors and back-reporting to patients, physicians, and nurses (Lelubre et al., 2015).

5 TRANSLATION INTO PRACTICE

Save for regulatory governmental measures like taxation most other of the earlier mentioned facilitation strategies have been experimental and isolated. Comprehensive health facilitation providers do not yet fit into the established set of medical institutions and may arouse opposition on grounds of patient confidentiality, medical responsibility, or other concerns. Therefore their range of action should be limited to low-controversy and high-impact behaviors like smoking cessation, physical activity, calorie-limited and varied diet, drinking moderation, and adherence to a core set of medication like anticoagulants or antihypertensives. Because of larger return on investment institutionalized comprehensive health facilitation may be best piloted in high-risk populations like patients after stroke or heart attack who are at greatest risk for new vascular events unless they effectively change their health behavior.

Because marketing builds on principles, creativity, exploration, and feedback we cannot tell what successful marketing strategies for healthier behaviors will ultimately look like. However, based on the general principles we would not expect successful programs to place much weight on explaining medical problems in depth, use threats, or attempt to relate percentages of risks reduction. We would expect them to use a limited number of simplified but pithy messages that will be repeated in a varied fashion over time. We also would expect a successful program to relate and rehearse behavioral shortcuts. We further expect social comparison and involvement of family and friends to play a role. Additionally, loyalty programs would probably be used to support sustained behaviors. We would expect providers to address customers via multiple passive as well as interactive channels and in an individualized, toolbox-based way, such that, for example, varied strategies will be used for different groups of still-smoking customers. Finally, since marketing extends to logistics, health facilitation providers would be likely to also push efficiency of services like coordination of near-home health activities, appointment reminding, and simplification of medication including use of fixed-combination pills and home delivery of customized blister-packaged medication.

6 PERSPECTIVE

Could a marketing approach do more harm than good? Doctor–patient relations require trust which also involves respect for incompliance (Eyal, 2014). A marketing approach could corrupt such a relation. Therefore, it would be important to instantiate health facilitation as an option provided by a third party.

Table 2 Bullet Points

- Behaviorally mediated vascular and metabolic diseases have become a dominant challenge across the globe.
- Since the traditional concept of patients as rational actors may have contributed to the failure to prevent unhealthy lifestyles, alternative working models need to be considered. Here we propose to address patients as long-time customers and use marketing approaches for behavior change.
- Promising marketing tools to address unhealthy behaviors include simplification, repeated exposure, default framing, social comparisons, and consumer-friendliness.
- Marketing approaches could be integrated and remunerated in systematic health facilitation programs as an adjunct to present efforts at improving health behaviors.

Further, marketing has been and still is critical in spreading many unhealthy behaviors like smoking or overeating (Knecht et al., 2008). When health-focused strategies threaten vested interests corporations tend to develop countermeasures (Saloojee and Dagli, 2000). For example, tobacco media campaigns now use opaque semiotics deprecating health claims to undermine health warnings (University of Bath, Tobacco Control Research Group, 2012). Marketing for harmful consumption has rightly made people suspicious of marketing in general and could offset health-focused advertising. However, there is no historical evidence for an advertising campaign having an inverse effect. Moreover, as we pointed out, marketing comprises many more strategies than advertising (Table 2).

Using marketing or market mechanisms to improve health is no fail-proof remedy for behaviorally mediated disease. However, marketing can address human vignettes that medicine seems to have missed so far. To this end we should start to change our views on behavioral change as well as our technologies and eventually even our institutions.

REFERENCES

Achtziger, A., Gollwitzer, P.M., Sheeran, P., 2008. Implementation intentions and shielding goal striving from unwanted thoughts and feelings. Personal. Soc. Psychol. Bull. 34 (3), 381–393.

Achtziger, A., Alós-Ferrer, C., Hügelschäfer, S., Steinhauser, M., 2014. The neural basis of belief updating and rational decision making. Soc. Cogn. Affect. Neurosci. 9 (1), 55–62.

Achtziger, A., Alós-Ferrer, C., Hügelschäfer, S., Steinhauser, M., 2015. Higher incentives can impair performance: neural evidence on reinforcement and rationality. Soc. Cogn. Affect. Neurosci. 10 (11), 1477–1483.

Amerson, N.L., Arbise, B.S., Kelly, N.K., Traore, E., 2014. Use of market research data by state chronic disease programs, Illinois, 2012–2014. Prev. Chronic Dis. 11, E165, 1–8.

Atktin, K.C., Rice, R.E., 2013. Theory and principles of public communication campaigns. In: Rice, R.E., Atktin, K.C. (Eds.), Public Communication Campaigns. Sage, Thousand Oaks, CA, pp. 3–19.

Barham, T., Rowberry, J., 2013. Living longer: the effect of the Mexican conditional cash transfer program on elderly mortality. J. Dev. Econ. 105, 226–236.

Berkman, E., Livingston, J.L., Kahn, L.E., 2015. Finding the 'self' in self-regulation: the identity-value model. Available at SSRN: http://ssrn.com/abstract=2621251 or http://dx.doi.org/10.2139/ssrn.2621251.

Blyte, J., 2008. Essentials of Marketing, fourth ed. Pearson Education, London.

Bogliacino, F., Codagnone, C., Veltri, G.A., Chakravarti, A., Ortoleva, P., Gaskell, G., Ivchenko, A., Lupianez-Villanueva, F., Mureddu, F., Rudisill, C., 2015. Pathos & ethos: emotions and willingness to pay for tobacco products. PLoS One 10 (10), 1–25.

Borland, R., Partos, T.R., Yong, H.-H., Cummings, K.M., Hyland, A., 2012. How much unsuccessful quitting activity is going on among adult smokers? Data from the International Tobacco Control Four Country cohort survey: prevalence of quitting activity among smokers. Addiction 107 (3), 673–682.

Boseley, S., 2014. Mexico enacts soda tax in effort to combat world's highest obesity rate Health officials in the United States look to Mexico's new law as an experiment in curbing sugar consumption. The Guardian. https://www.theguardian.com/world/2014/jan/16/mexico-soda-tax-sugar-obesity-health.

Christakis, N.A., Fowler, J.H., 2007. The spread of obesity in a large social network over 32 years. N. Engl. J. Med. 357 (4), 370–379.

Christakis, N.A., Fowler, J.H., 2008. The collective dynamics of smoking in a large social network. N. Engl. J. Med. 358 (21), 2249–2258.

Cram, P., Fendrick, A.M., Inadomi, J., Cowen, M.E., Carpenter, D., Vijan, S., 2003. The impact of a celebrity promotional campaign on the use of colon cancer screening: the Katie Couric effect. Arch. Intern. Med. 163 (13), 1601.

Cutrona, S.L., Choudhry, N.K., Stedman, M., Servi, A., Liberman, J.N., Brennan, T., Fischer, M.A., Brookhart, M.A., Shrank, W.H., 2010. Physician effectiveness in interventions to improve cardiovascular medication adherence: a systematic review. J. Gen. Intern. Med. 25 (10), 1090–1096.

Deckersbach, T., Das, S.K., Urban, L.E., Salinardi, T., Batra, P., Rodman, A.M., Arulpragasam, A.R., Dougherty, D.D., Roberts, S.B., 2014. Pilot randomized trial demonstrating reversal of obesity-related abnormalities in reward system responsivity to food cues with a behavioral intervention. Nutr. Diabetes 4, e129.

Eyal, N., 2014. Using informed consent to save trust. J. Med. Ethics 40 (7), 437–444.

Frederick, C.B., Snellman, K., Putnam, R.D., 2014. Increasing socioeconomic disparities in adolescent obesity. Proc. Natl. Acad. Sci. 111 (4), 1338–1342.

Fried, T.R., Tinetti, M.E., Towle, V., O'Leary, J.R., Iannone, L., 2011. Effects of benefits and harms on older persons' willingness to take medication for primary cardiovascular prevention. Arch. Intern. Med. 171 (10), 923–928.

Gordon, I., 1998. Relationship Marketing: New Strategies, Techniques, and Technologies to Win the Customers You Want and Keep Them Forever. John Wiley & Sons, Canada.

Halpern, S.D., French, B., Small, D.S., Saulsgiver, K., Harhay, M.O., Audrain-McGovern, J., Loewenstein, G., Brennan, T.A., Asch, D.A., Volpp, K.G., 2015. Randomized trial of four financial-incentive programs for smoking cessation. N. Engl. J. Med. 372 (22), 2108–2117.

Halvorsen, P.A., Selmer, R., Kristiansen, I.S., 2007. Different ways to describe the benefits of risk-reducing treatments: a randomized trial. Ann. Intern. Med. 146 (12), 848–856.

Ho, P.M., Spertus, J.A., Masoudi, F.A., Reid, K.J., Peterson, E.D., Magid, D.J., Krumholz, H.M., Rumsfeld, J.S., 2006. Impact of medication therapy discontinuation on mortality after myocardial infarction. Arch. Intern. Med. 166 (17), 1842–1847.

Jha, P., Peto, R., 2014. Global effects of smoking, of quitting, and of taxing tobacco. N. Engl. J. Med. 370 (1), 60–68.

John, L.K., Loewenstein, G., Troxel, A.B., Norton, L., Fassbender, J.E., Volpp, K.G., 2011. Financial incentives for extended weight loss: a randomized, controlled trial. J. Gen. Intern. Med. 26 (6), 621–626.

Johnson, D.D.P., Fowler, J.H., 2011. The evolution of overconfidence. Nature 477 (7364), 317–320.

Kahneman, D., 2013. Thinking, Fast and Slow. Farrar, Straus and Giroux, New York. 1st pbk. e.

Kahneman, D., Tversky, A., 1979. Prospect theory: an analysis of decision under risk. Econometrica 47 (2), 263.

Kenning, P., Plassmann, H., 2005. Neuroeconomics: an overview from an economic perspective. Brain Res. Bull. 67 (5), 343–354.

Khaw, K.-T., Wareham, N., Bingham, S., Welch, A., Luben, R., Day, N., 2008. Combined impact of health behaviours and mortality in men and women: the EPIC-Norfolk prospective population study. PLoS Med. 5 (1), e12.

Kimmel, S.E., Troxel, A.B., Loewenstein, G., Brensinger, C.M., Jaskowiak, J., Doshi, J.A., Laskin, M., Volpp, K., 2012. Randomized trial of lottery-based incentives to improve warfarin adherence. Am. Heart J. 164 (2), 268–274.

King, D., Thompson, P., Darzi, A., 2014. Enhancing health and wellbeing through 'behavioural design'. J. R. Soc. Med. 107 (9), 336–337.

Knecht, S., 2009. Overcoming systemic roadblocks to sustainable health. Proc. Natl. Acad. Sci. U. S. A. 106 (28), E80.

Knecht, S., Ellger, T., Levine, J.A., 2008. Obesity in neurobiology. Prog. Neurobiol. 84 (1), 85–103.

Kullgren, J.T., Harkins, K.A., Bellamy, S.L., Gonzales, A., Tao, Y., Zhu, J., Volpp, K.G., Asch, D.A., Heisler, M., Karlawish, J., 2014. A mixed-methods randomized controlled trial of financial incentives and peer networks to promote walking among older adults. Health Educ. Behav. 41 (1 Suppl.), 43S–50S.

Lee, J.K., Grace, K.A., Taylor, A.J., 2006. Effect of a pharmacy care program on medication adherence and persistence, blood pressure, and low-density lipoprotein cholesterol. JAMA 296 (21), 2563.

Lelubre, M., Kamal, S., Genre, N., Celio, J., Gorgerat, S., Hugentobler Hampai, D., Bourdin, A., Berger, J., Bugnon, O., Schneider, M., 2015. Interdisciplinary medication adherence program: the example of a university community pharmacy in Switzerland. BioMed Res. Int. 2015, 1–10.

Linnet, J., Gebauer, L., Shaffer, H., Mouridsen, K., Møller, A., 2010. Experienced poker players differ from inexperienced poker players in estimation bias and decision bias. J. Gambl. Issues 24 (24), 86–100.

Mantzari, E., Vogt, F., Shemilt, I., Wei, Y., Higgins, J.P.T., Marteau, T.M., 2015. Personal financial incentives for changing habitual health-related behaviors: a systematic review and meta-analysis. Prev. Med. 75, 75–85.

Marteau, T.M., Mantzari, E., 2015. Public health: the case for pay to quit. Nature 523 (7558), 40–41.

McNeil, B.J., Pauker, S.G., Sox, H.C., Tversky, A., 1982. On the elicitation of preferences for alternative therapies. N. Engl. J. Med. 306 (21), 1259–1262.

Mullainathan, S., Thaler, R.H., 2000. Behavioral economics. National Bureau of Economic Research, Inc. NBER Working Paper 7948. http://www.nber.org/papers/w7948.

Peterson, J., Atwood, J.R., Yates, B., 2002. Key elements for church-based health promotion programs: outcome-based literature review. Public Health Nurs. 19 (6), 401–411.
Piette, J.D., Resnicow, K., Choi, H., Heisler, M., 2013. A diabetes peer support intervention that improved glycemic control: mediators and moderators of intervention effectiveness. Chronic Illn. 9 (4), 258–267.
Ruiter, R.A.C., Kessels, L.T.E., Peters, G.J.Y., Kok, G., 2014. Sixty years of fear appeal research: current state of the evidence. Int. J. Psychol. 49 (2), 63–70.
Ryan, R.M., Deci, E.L., 2000. Self-determination theory and the facilitation of intrinsic motivation, social development, and well-being. Am. Psychol. 55 (1), 68–78.
Saloojee, Y., Dagli, E., 2000. Tobacco industry tactics for resisting public policy on health. Bull. World Health Organ. 78 (7), 902–910.
Schwartz, J., Mochon, D., Wyper, L., Maroba, J., Patel, D., Ariely, D., 2014. Healthier by precommitment. Psychol. Sci. 25 (2), 538–546.
Scott, A., Sivey, P., Ait Ouakrim, D., Willenberg, L., Naccarella, L., Furler, J., Young, D., 2011. The effect of financial incentives on the quality of health care provided by primary care physicians. In: The Cochrane Collaboration (Ed.), Cochrane Database of Systematic Reviews. John Wiley & Sons, Ltd., Chichester, UK.
Slovic, P., 1987. Perception of risk. Science (New York, N.Y.) 236 (4799), 280–285.
Strombach, T., Hubert, M., Kenning, P., 2015. The neural underpinnings of performance-based incentives. J. Econ. Psychol. 50, 1–12.
Strombach, T., Strang, S., Park, S.Q., Kenning, P., 2016. Chapter 1—Common and distinctive approaches to motivation in different disciplines. In: Studer, B., Knecht, S (Eds.), Progress in Brain Research, vol. 229. Elsevier, Amsterdam, pp. 3–23.
Sunstein, C.R., 2014. Nudging: a very short guide. In: International Geoscience and Remote Sensing Symposium, vol. 1, pp. 1–5.
Taub, E., Uswatte, G., Mark, V.W., Morris, D.M., Barman, J., Bowman, M.H., Bryson, C., Delgado, A., Bishop-McKay, S., 2013. Method for enhancing real-world use of a more affected arm in chronic stroke: transfer package of constraint-induced movement therapy. Stroke 44 (5), 1383–1388.
University of Bath, Tobacco Control Research Group, 2012. Be Marlboro: targeting the world's biggest brand at youth. Tobacco Tactics. http://www.tobaccotactics.org/index.php/Be_Marlboro:_Targeting_the_World%27s_Biggest_Brand_at_Youth.
Volpp, K.G., 2006. A randomized controlled trial of financial incentives for smoking cessation. Cancer Epidemiol. Biomarkers Prev. 15 (1), 12–18.
Wang, B., Kesselheim, A.S., 2013. The role of direct-to-consumer pharmaceutical advertising in patient consumerism. Virtual Mentor 15 (11), 960–965.
Wansink, B., Pope, L., 2015. When do gain-framed health messages work better than fear appeals? Nutr. Rev. 73 (1), 4–11.
Witte, V., Fobker, M., Gellner, R., Knecht, S., Floel, A., 2009. Caloric restriction improves memory in elderly humans. Proc. Natl. Acad. Sci. U. S. A. 106 (4), 1255–1260.

SECTION

Conclusions

5

CHAPTER

Motivation: What have we learned and what is still missing?

19

B. Studer[1], S. Knecht

Mauritius Hospital, Meerbusch, Germany

Institute of Clinical Neuroscience and Medical Psychology, Medical Faculty, Heinrich-Heine-University Düsseldorf, Düsseldorf, Germany

[1]Corresponding author: Tel.: +49-2159-679-5114; Fax: +49-2159-679-1535, e-mail address: bettina.studer@stmtk.de

Abstract

This final chapter deliberates three overarching topics and conclusions of the research presented in this volume: the endurance of the concept of extrinsic vs intrinsic motivation, the importance of considering subjective costs of activities when aiming to understand and enhance motivation, and current knowledge of the neurobiological underpinnings of motivation. Furthermore, three topics for future motivation research are outlined, namely the assessment and determinants of intrinsic benefits, the reconciliation of activity-specific motivation models with generalized motivation impairments in clinical populations, and the motivational dynamics of groups.

Keywords

Motivation, Conclusions, Open questions, Extrinsic, Intrinsic, Brain, Costs, Groups

The chapters of this volume provided an up-to-date account of motivation theory, demonstrated novel directions in the assessment of motivation and its determinants in healthy and clinical populations, presented recent findings on the neurobiological correlates of motivation and goal-directed behavior, and discussed and tested novel strategies on enhancing motivation in a variety of application domains. In this final chapter, we briefly reflect on some overarching topics and insights emerging from the research presented in the chapters of this volume and outline three topics for future motivation research.

Progress in Brain Research, Volume 229, ISSN 0079-6123, http://dx.doi.org/10.1016/bs.pbr.2016.07.001

1 THREE OVERARCHING INSIGHTS GAINED THROUGH THIS VOLUME

1.1 ENDURANCE OF THE CONCEPT OF EXTRINSIC VS INTRINSIC MOTIVATION

The distinction between intrinsic and extrinsic (sources of) motivation (eg, Deci, 1971; Deci and Ryan, 1985; White, 1959) has a long tradition in psychology (for reviews, see Strombach et al., 2016, this volume; Oudeyer et al., 2016, this volume) and intuitive appeal. The persistence of this proposition in the field of motivation research is detectable in several chapters of this volume. For instance, the benefit–cost framework of motivation for specific activity that we propose in Studer and Knecht (2016, this volume) distinguishes intrinsic and extrinsic benefits (and costs), with intrinsic benefits defined as positive feelings experienced during performance of the activity itself (eg, enjoyment, pleasure, feeling of accomplishment), and extrinsic benefits defined as instrumental gains, rewards, or goals achieved through performance of the activity. Nafcha et al. (2016, this volume) review previous evidence that (experience of) personal control over ones environment is intrinsically rewarding and can drive behaviors with no or only small extrinsic benefits. They postulate that this intrinsic benefit of control is what motivates habitual behavior, and describe recent empirical findings supporting this hypothesis. Losecaat Vermeer et al. (2016, this volume) and we (Studer et al., 2016, this volume) argue that competing against an opponent—or oneself—might carry both an intrinsic benefit (ie, challenge is enjoyable) and an extrinsic benefit (ie, winning against someone, or improving once own performance, is rewarding). The functional magnetic resonance imaging (fMRI) study by Widmer et al. (2016, this volume) investigated and compared the impact of an extrinsic monetary reward and performance feedback as an "intrinsic reward" upon motor skill learning and neural activations. They found that monetary reward and randomly presented performance feedback, but not performance feedback provided systematically for good training trials only, were associated with stronger activation of the ventral striatum and better overnight skill consolidation. The interplay between extrinsic, externally set incentives, intrinsic motivation, and behavioral output, was discussed in depth by Strang et al. (2016, this volume). This topic was controversially debated in the extant theoretical and application-focused literature: on one hand, decades of behavioral research has shown that extrinsic incentives (for instance, food rewards) are effective in modulating behavior in animals and humans (eg, Morales et al., 2016, this volume; Skinner, 1963; Toppen, 1965), and performance-dependent compensation and other monetary incentives schemes are used widely in our society (eg, Gerhart and Fang, 2015). On the other hand, it has been argued that extrinsic, externally set incentives diminish intrinsic motivation for a target activity and could therefore negatively impact performance of that activity (eg, Deci, 1971; Deci et al., 1999; Lepper et al., 1973). Through a careful review of previous evidence, Strang et al. clarify under which conditions extrinsic, externally set incentives positively affect behavior and overall motivation. They concluded that extrinsic incentives are most effective in situations where intrinsic motivation (or in other words, the anticipated intrinsic benefit) is low, the target activity is easy and when extrinsic benefits are of a social nature (eg, positive verbal feedback). Meanwhile, extrinsic incentives are less effective, or may even reduce performance of a target activity, when intrinsic motivation is high and when the target activity is prosocial behavior. With regard to potential application, it is also noteworthy that removal of extrinsic incentives often leads to a drop in performance. If used for motivation enhancement in long-term interventions (eg, health facilitation), externally set incentives might thus have to be sustained over long periods.

1.2 THE IMPORTANCE OF COSTS

A second overarching conclusion that can be drawn from the work presented in this volume is that motivation theories, particularly those focusing on motivation for specific activities, should consider not only the subjective benefits (classically referred to as motivators) but also the (intrinsic and extrinsic) subjective costs of an activity. In some psychological theories of motivation (eg, in self-determination theory; Deci, 1980; Deci and Ryan, 2000) as well as in many of the questionnaire-based measures used to investigate and quantify motivation in healthy and clinical human populations, the cost dimension has received little attention. Meanwhile, the trade-off between benefits and costs is a core topic in (neuro-)economic models of human motivation and behavior (eg, Strang et al., this volume) and motivation research in animals (Hull, 1943; Morales et al., 2016, this volume). Building upon this tradition, recent neuroscience research has often used behavioral arrays of motivation, which test the willingness of a human or animal to exert physical effort (as an intrinsic cost) for extrinsic rewards of different magnitudes (see, for instance, Bernacer et al., 2016, this volume; Morales et al., 2016, this volume; Salamone and Correa, 2012; Salamone et al., 2007). As reviewed by Chong et al. (2016, this volume), these behavioral measures have proven very useful and informative in identifying neurobiological underpinnings of motivation, are sensitive to individual differences, and hold great promise as more sensitive tools for the assessment and dissociation of motivation impairments in different clinical conditions. Convergent, a novel study by Bernacer et al. (2016, this volume) demonstrated that subjective effort costs of treadmill running differentiates between individuals with an active lifestyle and individuals with a sedentary lifestyle. Relatedly, Kroemer et al. (2016, this volume) discuss how a high body mass index makes movement metabolically more costly, which is likely to increase perceived effort and thus decrease motivation for physical activity in obese individuals. As a final example, Morales et al. (2016, this volume) successfully used a behavior paradigm with an effort component to investigate the role of opioid signals in different aspects of food motivation. Together these chapters demonstrate how considering subjective expected costs of a target activity, alongside of subjective expected benefits, is extremely useful for understanding an individual's motivation for performance of that activity and how to enhance it.

1.3 MOTIVATION IN THE BRAIN: DOPAMINE AND BEYOND

A third overarching topic of this volume is the neurobiological underpinnings of motivation and goal-directed behavior. Through careful reviews of extant neuroscience findings and novel original investigations, the chapters included in Section 3 (and elsewhere) of this volume provided an up-to-date account of the brain structures and neurotransmitter systems implicated in motivation and goal-directed behavior. Let us briefly recount some core insights. First, with regard to brain structures, extant neuroscience research revealed that a distributed network of brain areas, including the midbrain, ventral and dorsal striatum, pallidum, prefrontal cortex, anterior cingulate cortex, and basolateral amygdala, is implicated in the computation of the subjective expected benefits and costs of an actions and benefits–costs comparison. As reviewed in Chong et al. (2016, this volume) and Kroemer et al. (2016, this volume), lesions of structures of this network, in particular of the nucleus accumbens, medial prefrontal cortex, and amygdala, reduce willingness to work for a reward and lead to a preference of actions with low effort costs. Lesions to areas of this network also affect sensitivity to the (extrinsic) benefits or rewards outside of an effort context (eg, Clarke et al., 2008; Coutureau et al., 2009; Manohar and Husain, 2016; Studer et al., 2015). These lesion results are corroborated by a large body of functional neuroimaging studies on decision making. For instance, as Bernacer et al. (2016, this volume) point out, fMRI studies using effort-based decision-making paradigms find that activity patterns in the striatum, anterior cingulate, and ventromedial prefrontal cortex, as well as the motor cortex and supplementary motor areas, reflect the level of required physical effort associated with choice options (eg, Croxson et al., 2009; Prevost et al., 2010). In their own novel study, Bernacer and colleagues also find sensitivity for required effort in the ventrolateral prefrontal cortex. Furthermore, as we review in Studer and Knecht (2016, this volume), activity patterns in this network were found to covary with factors determining the expectancy and value of extrinsic rewards and their integrated overall benefit. Finally, Kroemer et al. (2016, this volume) presented an intriguing neurocomputational model in which fluctuations in motivation and behavior over time are caused by signal fluctuations in key nodes of this network.

Second, several chapters examined the role of the neurotransmitter dopamine in motivation. As pointed out by Chong and Husain (2016, this volume), dopaminergic neocortical and nigrostriatal pathways are at the core of the network described earlier. And, a large body of research in nonhuman animals shows that (systemic or region-targeted) disruption of dopaminergic transmission leads to a decreased motivation to work for a given reward, or in other words, a shift toward a reduced benefit and increased costs evaluation (see Chong and Husain, 2016, this volume; Kroemer et al., 2016, this volume for reviews). Chong and Husain (2016, this volume) reasoned that this implied central role of dopamine in signaling (extrinsic) incentives to act and in facilitating the overcoming of effort costs is consistent with the finding that apathy is common in neurological disorders affecting the dopaminergic system (such as Parkinson disease). Kroemer and colleagues (2016, this volume) highlighted the convergence between the aforementioned results from animal studies and recent human data on dopaminergic transmission during an effort-based decision-making paradigm by Treadway et al. (2012). But, Kroemer et al. also point out that while the importance of dopaminergic signaling for effort-related aspects of motivation is now well established, much less is known about the neurobiological substrates of other motivational dimensions. Relevant to this point is the novel study by Morales et al. (2016, this volume) on the functions of opioid signaling in food motivation, which provides a highly interesting addition to previous knowledge. Morales and colleagues build on an argument originally made by Salamone and colleagues (Salamone and Correa, 2012) that dopamine is primarily implicated in activational and directional aspects of food motivation (or in other words, instrumental behavior), rather than in hedonic aspects of food consumption. In two well-designed experiments, Morales et al. tested the effects of the opioid receptor antagonist naloxone upon instrumental and consummatory behavior in rats. They find that disruption of opioid signaling reduces rats' liking of a preferred food and decrease their willingness to lever press (ie, exert effort) for that preferred food. This suggests that the opioid system might be involved in computing anticipated pleasure of food consumption, or more generally speaking, expected hedonic value (ie, intrinsic benefit) of an activity.

Finally, this volume also demonstrated how knowledge of the neurobiological underpinnings may be used for clinical and nonclinical application. For instance, in their fMRI study, Widmer et al. (2016, this volume) found that providing concurrent performance feedback and monetary reward during a motor learning task raised ventral striatum activation, and that stronger responsiveness in the striatum to these incentives was associated with better overnight skill consolidation. These results suggest that increasing ventral striatal activity during motor training through verbal or monetary reward could help improve consolidation of the motor skill. Further, Chong and Husain (2016, this volume) argue that the aforementioned findings from animal research on dopaminergic functions in effort-related motivation imply dopamine agonists as a primary candidate for pharmacological treatment of apathy. Reviewing extant research using this intervention, they conclude that the studies conducted to date offer some evidence for (selective) dopamine agonist therapy being effective in ameliorating apathy in human patients, but also highlight that more well-controlled clinical studies are required.

2 OUTLOOK: THREE TOPICS FOR FUTURE RESEARCH

2.1 UNDERSTANDING AND ENHANCING THE INTRINSIC BENEFIT OF AN ACTIVITY

Intrinsic motivation—or, in the terminology of the benefit–cost framework proposed in Studer and Knecht (2016, this volume), the intrinsic benefit of an activity—is a popular topic in the psychological, educational, rehabilitation, and healthcare literature. And, what determines intrinsic motivation and how intrinsic motivation can be enhanced remain two hot topics in current motivation research (see for instance Nafcha et al., 2016, this volume; Oudeyer et al., 2016, this volume). Yet, the vast majority of extant studies investigating the neural correlates of motivation have used extrinsic incentives. Therefore, intrinsic benefits and their neurobiological underpinnings remain less well understood. Furthermore, although several behavioral (eg, for how long an activity is performed during a free-choice period) and questionnaire assays of intrinsic motivation have been established through laboratory and field studies (eg, Deci et al., 1999; Pelletier et al., 1995; Vallerand et al., 1992), intrinsic benefits or motives such as enjoyment, curiosity, control over the environment, novelty (perceived), competence, and interest are—in our opinion—more difficult to identify, quantify, and understand in real-life scenarios than extrinsic benefits, due to their more abstract nature. We hope that future research will continue to shed light on the determinants of intrinsic benefits and how intrinsic motives can be integrated into current neurobiological and (neuro-)economic models of human motivation.

2.2 ACTIVITY-SPECIFIC MODELS VS GENERALIZED MOTIVATION IMPAIRMENTS IN CLINICAL POPULATIONS

Most chapters of this volume discussed motivation in specific contexts and for specific activities or tasks and assumed that the degree of motivation in such situation is determined by (subjective evaluation of) activity-specific aspects (ie, possible rewards, effort requirements, perceived autonomy and control, etc.). One open question is how such activity-specific models of motivation, and the findings obtained through research utilizing these models, can be reconciled with the more generalized (ie, situation-unspecific) and enduring motivation deficits observed in clinical contexts (including apathy and the motivation-affecting syndromes depression and fatigue). One current proposition is that such disorders are characterized by a systematic shift in the subjective evaluation of benefits and/or of (effort) costs, and/or a change in the benefit–cost comparison, and that such alterations could be captured and dissociated through use of effort-based decision-making paradigms (see Chong et al., 2016, this volume; Chong and Hussain, 2016, this volume). As reviewed by Chong and colleagues, results from first studies using this methodology in patient samples are promising; however, how suitable and insightful such effort-based decision-making paradigms are as diagnostic tools in real-life clinical application has yet to be established. Future research should also further investigate how different neurobiological correlates of these clinical conditions, for instance hyperarousal in depression, hypoarousal in fatigue (Hegerl and Ulke, 2016, this volume), and dopaminergic dysfunctions in Parkinson disease and other neurological conditions with high prevalence of apathy (Chong and Husain, 2016, this volume), relate to dissociable or shared alterations in motivation.

2.3 SOCIAL INFLUENCES ON MOTIVATION AND GROUP BEHAVIOR

Psychological theories and contemporary economic models recognize the importance of social factors in determining motivation and guiding behavior. Social approval, fairness, reciprocity, social status, and reputation have all been postulated to be strong motivators of human behavior (see Losecaat Vermeer et al., 2016, this volume; Strombach et al., 2016, this volume, for reviews of the relevant literature). Furthermore, social comparison is known to affect the subjective valuation of extrinsic benefits and costs (eg, Báez-Mendoza and Schultz, 2013; Bault et al., 2011; Grygolec et al., 2012) and perception of one's own skill and performance (Corcoran et al., 2011; Vostroknutov et al., 2012). Previous research has also shown that humans do not only take their own benefits and costs into account, but also what benefits and costs their actions bring for others (eg, Crockett et al., 2014). Note, however, that these lines of research focus on how social factors affect motivation and behavior of an individual. A topic that received less attention in this volume, and in neuroscience research on motivation in general, is what drives motivation and behavior of groups. Of course, a group can simply be seen as multiple individuals, each with their own motivation. But in every-day life, we also observe situations where the motivation and behavior of group members appear to arise, or at least be enhanced, though an interactive dynamic. Think for instance of eruptions of violence in crowds of soccer supporters. Such group dynamics are deliberated in the social psychology literature. For instance, it has been shown that group interaction enhances individuals' propensity to take risks ("risky-shift phenomenon"; Kogan and Wallach, 1967) and polarize individuals' attitudes (eg, Keating et al., 2016; Myers and Lamm, 1976). Future research might examine how such group dynamics fit into current psychological and economic models of motivation, and how they might be utilized for motivation enhancement. Further, it would be interesting for future studies to assess the neurobiological correlates of such interactional social influences upon motivation and behavior.

REFERENCES

Báez-Mendoza, R., Schultz, W., 2013. The role of the striatum in social behaviour. Front. Neurosci. 7, 1–14.

Bault, N., Joffily, M., Rustichini, A., Coricelli, G., 2011. Medial prefrontal cortex and striatum mediate the influence of social comparison on the decision process. Proc. Natl. Acad. Sci. U.S.A. 108, 16044–16049.

Bernacer, J., Martinez-Valbuena, I., Martinez, M., Pujol, N., Luis, E., Ramirez-Castillo, D., Pastor, M.A., 2016. Chapter 5—Brain correlates of the intrinsic subjective cost of effort in sedentary volunteers. In: Studer, B., Knecht, S. (Eds.), Progress in Brain Research, vol. 229. Elsevier, Amsterdam, pp. 103–123.

Chong, T.T.-J., Bonnelle, V., Husain, M., 2016. Chapter 4—Quantifying motivation with effort-based decision-making paradigms in health and disease. In: Studer, B., Knecht, S. (Eds.), Progress in Brain Research, vol. 229. Elsevier, Amsterdam, pp. 71–100.

Chong, T.T.-J., Husain, M., 2016. Chapter 17—The role of dopamine in the pathophysiology and treatment of apathy. In: Studer, B., Knecht, S. (Eds.), Progress in Brain Research, vol. 229. Elsevier, Amsterdam, pp. 389–426.

Clarke, H.F., Robbins, T.W., Roberts, A.C., 2008. Lesions of the medial striatum in monkeys produce perseverative impairments during reversal learning similar to those produced by lesions of the orbitofrontal cortex. J. Neurosci. 28, 10972–10982.

Corcoran, K., Crusius, J., Mussweiler, T., 2011. Social comparison: motives, standards, and mechanisms. In: Chadee, D. (Ed.), Theories in Social Psychology. Wiley-Blackwell, Oxford, UK.

Coutureau, E., Marchand, A.R., Di Scala, G., 2009. Goal-directed responding is sensitive to lesions to the prelimbic cortex or basolateral nucleus of the amygdala but not to their disconnection. Behav. Neurosci. 123, 443–448.

Crockett, M.J., Kurth-Nelson, Z., Siegel, J.Z., Dayan, P., Dolan, R.J., 2014. Harm to others outweighs harm to self in moral decision making. Proc. Natl. Acad. Sci. U.S.A. 111, 17320–17325.

Croxson, P.L., Walton, M.E., O'reilly, J.X., Behrens, T.E.J., Rushworth, M.F.S., 2009. Effort-based cost–benefit valuation and the human brain. J. Neurosci. 29, 4531–4541.

Deci, E.L., 1971. Effects of externally mediated rewards on intrinsic motivation. J. Pers. Soc. Psychol. 18, 105.

Deci, E.L., 1980. The Psychology of Self-Determination. Heath, Lexington, MA.

Deci, E.L., Ryan, R.M., 1985. Intrinsic Motivation and Self-Determination in Human Behavior. Springer Science & Business Media, New York, NY.

Deci, E.L., Ryan, R.M., 2000. The "what" and "why" of goal pursuits: human needs and the self-determination of behavior. Psychol. Inq. 11, 227–268.

Deci, E.L., Koestner, R., Ryan, R.M., 1999. A meta-analytic review of experiments examining the effects of extrinsic rewards on intrinsic motivation. Psychol. Bull. 125, 627–668. discussion 692–700.

Gerhart, B., Fang, M., 2015. Pay, intrinsic motivation, extrinsic motivation, performance, and creativity in the workplace: revisiting long-held beliefs. Ann. Rev. Org. Psychol. Org. Behav. 2, 489–521.

Grygolec, J., Coricelli, G., Rustichini, A., 2012. Positive interaction of social comparison and personal responsibility for outcomes. Front. Psychol. 3, 1–13.

Hegerl, U., Ulke, C., 2016. Chapter 10—Fatigue with up- vs downregulated brain arousal should not be confused. In: Studer, B., Knecht, S. (Eds.), Progress in Brain Research, vol. 229. Elsevier, Amsterdam, pp. 239–254.

Hull, C., 1943. Principles of Behavior: An Introduction to Behavior Theory. Appleton-Century, New York, NY.

Keating, J., Van Boven, L., Judd, C.M., 2016. Partisan underestimation of the polarizing influence of group discussion. J. Exp. Soc. Psychol. 65, 52–58.

Kogan, N., Wallach, M.A., 1967. Risky-shift phenomenon in small decision-making groups: a test of the information-exchange hypothesis. J. Exp. Soc. Psychol. 3, 75–84.

Kroemer, N.B., Burrasch, C., Hellrung, L., 2016. Chapter 6—To work or not to work: Neural representation of cost and benefit of instrumental action. In: Studer, B., Knecht, S. (Eds.), Progress in Brain Research, vol. 229. Elsevier, Amsterdam, pp. 125–157.

Lepper, M.R., Greene, D., Nisbett, R.E., 1973. Undermining children's intrinsic interest with extrinsic reward: a test of the "overjustification" hypothesis. J. Pers. Soc. Psychol. 28, 129.

Losecaat Vermeer, A.B., Riečanský, I., Eisenegger, C., 2016. Chapter 9—Competition, testosterone, and adult neurobehavioral plasticity. In: Studer, B., Knecht, S. (Eds.), Progress in Brain Research, vol. 229. Elsevier, Amsterdam, pp. 213–238.

Manohar, S.G., Husain, M., 2016. Human ventromedial prefrontal lesions alter incentivisation by reward. Cortex 76, 104–120.

Morales, I., Font, L., Currie, P.J., Pastor, R., 2016. Chapter 7—Involvement of opioid signaling in food preference and motivation: studies in laboratory animals. In: Studer, B., Knecht, S. (Eds.), Progress in Brain Research, vol. 229. Elsevier, Amsterdam, pp. 159–187.

Myers, D.G., Lamm, H., 1976. The group polarization phenomenon. Psychol. Bull. 83, 602–627.

Nafcha, O., Higgins, E.T., Eitam, B., 2016. Chapter 3—Control feedback as the motivational force behind habitual behavior. In: Studer, B., Knecht, S. (Eds.), Progress in Brain Research, vol. 229. Elsevier, Amsterdam, pp. 49–68.

Oudeyer, P.-Y., Gottlieb, J., Lopes, M., 2016. Chapter 11—Intrinsic motivation, curiosity and learning: theory and applications in educational technologies. In: Studer, B., Knecht, S. (Eds.), Progress in Brain Research, vol. 229. Elsevier, Amsterdam, pp. 257–284.

Pelletier, L.G., Fortier, M.S., Vallerand, R.J., Tuson, K.M., Briere, N.M., Blais, M.R., 1995. Toward a new measure of intrinsic motivation, extrinsic motivation, and amotivation in sports: the sport motivation scale (sms). J. Sport Exerc. Psychol. 17, 35–53.

Prevost, C., Pessiglione, M., Metereau, E., Clery-Melin, M.-L., Dreher, J.-C., 2010. Separate valuation subsystems for delay and effort decision costs. J. Neurosci. 30, 14080–14090.

Salamone, J.D., Correa, M., 2012. The mysterious motivational functions of mesolimbic dopamine. Neuron 76, 470–485.

Salamone, J., Correa, M., Farrar, A., Mingote, S., 2007. Effort-related functions of nucleus accumbens dopamine and associated forebrain circuits. Psychopharmacology (Berl.) 191, 461–482.

Skinner, B.F., 1963. Operant behavior. Am. Psychol. 18, 503.

Strang, S., Park, S., Strombach, T., Kenning, P., 2016. Chapter 12—Applied economics: the use of monetary incentives to modulate behavior. In: Studer, B., Knecht, S. (Eds.), Progress in Brain Research, vol. 229. Elsevier, Amsterdam, pp. 285–301.

Strombach, T., Strang, S., Park, S.Q., Kenning, P., 2016. Chapter 1—Common and distinctive approaches to motivation in different disciplines. In: Studer, B., Knecht, S. (Eds.), Progress in Brain Research, vol. 229. Elsevier, Amsterdam, pp. 3–23.

Studer, B., Knecht, S., 2016. Chapter 2—A benefit–cost framework of motivation for a specific activity. In: Studer, B., Knecht, S. (Eds.), Progress in Brain Research, vol. 229. Elsevier, Amsterdam, pp. 25–47.

Studer, B., Manes, F., Humphreys, G., Robbins, T.W., Clark, L., 2015. Risk-sensitive decision-making in patients with posterior parietal and ventromedial prefrontal cortex injury. Cereb. Cortex 25, 1–9.

Studer, B., Van Dijk, H., Handermann, R., Knecht, S., 2016. Chapter 16—Increasing self-directed training in neurorehabilitation patients through competition. In: Studer, B., Knecht, S. (Eds.), Progress in Brain Research, vol. 229. Elsevier, Amsterdam, pp. 367–388.

Toppen, J.T., 1965. Effect of size and frequency of money reinforcement on human operant (work)behavior. Percept. Mot. Skills 20, 259–269.

Treadway, M.T., Buckholtz, J.W., Cowan, R.L., Woodward, N.D., Li, R., Ansari, M.S., Baldwin, R.M., Schwartzman, A.N., Kessler, R.M., Zald, D.H., 2012. Dopaminergic mechanisms of individual differences in human effort-based decision-making. J. Neurosci. 32, 6170–6176.

Vallerand, R.J., Pelletier, L.G., Blais, M.R., Briere, N.M., Senecal, C., Vallieres, E.F., 1992. The academic motivation scale: a measure of intrinsic, extrinsic, and amotivation in education. Educ. Psychol. Meas. 52, 1003–1017.

Vostroknutov, A., Tobler, P.N., Rustichini, A., 2012. Causes of social reward differences encoded in human brain. J. Neurophysiol. 107, 1403–1412.

White, R.W., 1959. Motivation reconsidered: the concept of competence. Psychol. Rev. 66, 297–333.

Widmer, M., Ziegler, N., Held, J., Luft, A., Lutz, K., 2016. Chapter 13—Rewarding feedback promotes motor skill consolidation via striatal activity. In: Studer, B., Knecht, S. (Eds.), Progress in Brain Research, vol. 229. Elseiver, Amsterdam, pp. 303–323.

Index

Note: Page numbers followed by "*f*" indicate figures, "*t*" indicate tables, "*b*" indicate boxes, and "*np*" indicate footnotes.

N

O

P

R

S

T

V

W

Y

Z

Kolibri 360 GmbH
Ferdinand-Jühlke-Straße 7
99095 Erfurt, Deutschland
produktsicherheit@kolibri360.de